The Stability of Belief

The Stability of Belief

How Rational Belief Coheres with Probability

Hannes Leitgeb

OXFORD
UNIVERSITY PRESS

OXFORD
UNIVERSITY PRESS

Great Clarendon Street, Oxford, OX2 6DP,
United Kingdom

Oxford University Press is a department of the University of Oxford.
It furthers the University's objective of excellence in research, scholarship,
and education by publishing worldwide. Oxford is a registered trade mark of
Oxford University Press in the UK and in certain other countries

First Edition published in 2017

Published in the United States of America by Oxford University Press
198 Madison Avenue, New York, NY 10016, United States of America

British Library Cataloguing in Publication Data
Data available

Library of Congress Control Number: 2016957782

ISBN 978–0–19–873263–1

Preface

With a little luck there are three things for you to take away from this book.

The first one, I hope, is a promising theory of rational (all-or-nothing) belief and rational (numerical) degrees of belief. Its key ingredient is a stability conception of rational belief that does not originate with this essay: in fact, the goal of epistemic stability has been emphasized throughout the history of epistemology, such as in work done by René Descartes, David Hume, Charles Sanders Peirce, and others, not mentioning recent work.[1] The thought that is new to this essay, I believe, is that stability can be turned into a precisely formulated bridge principle that relates rational all-or-nothing belief with rational degrees of belief. While belief is subject to logical norms, degrees of belief are subject to probabilistic norms, and stability norms hold the two of them together. Moreover, the logical norms and the stability norms are not independent of each other: if belief is stable enough, it follows to be closed under logical consequence; and given the right background assumptions, also vice versa. Depending on one's starting point, the logic of belief reflects, or yields, the stability of belief.

Which leads me to my second take-home message: a proposal of how to do logic and probability theory simultaneously. This is not exactly a new ambition either: for example, George Boole, and later Ernest Adams and Dorothy Edgington, are examples of scholars who developed logic and probability theory jointly.[2] In recent years, however, one may also observe the contrary trend: various areas of philosophy, such as epistemology, philosophy of science, and the philosophy of language, seem to bifurcate into two different directions so far as the more technical work in these areas is concerned. One of these directions is the 'Bayesian' one, which aims at probabilistic accounts of rationality, science, and meaning; the other is the more traditional logical direction. This bifurcation is accompanied by different sorts of tensions: logical tensions, as in the Lottery-like Paradoxes that will keep us busy in this book, but also social tensions between two 'styles' of reasoning and their corresponding communities. It is a defining feature of this book that I will commit myself to both directions at the same time: a rational agent's degrees of belief will be assumed to satisfy the axioms of probability, and the same agent's all-or-nothing beliefs will be assumed to be consistent and closed under logic. I do not think that

[1] More details on this history can be found in Loeb (2010). As far as very recent work is concerned, at the final stage of the publication process I came across Krista Lawlor's (2014) excellent article on the stability of belief, which—unfortunately—I did not have time and place any more to discuss in this book. But I can recommend her article very much, and she develops various similar points about belief on independent grounds.

[2] For more on the interaction between the fields of logic and probability theory, see Leitgeb (2016).

the logical tensions between the two kinds of beliefs can be eliminated completely, but they can at least be alleviated by shifting them to a place where they do not hurt as much any more. That place will be: context. Rational all-or-nothing belief will turn out to be relative to context, which will be the compromise required for bridging the divide. On the social side, I hope that this book will contribute to a peace project of mutual engagement between logic and probability theory that will benefit both parties to the debate. Along the way, the book will also introduce many of the essentials of the two streams of formal epistemology and hence may take over, in parts, the role of a textbook, too.

Thirdly, whatever ultimately the merits or shortcomings of the theory that I am going to develop may be, this essay should also serve a purpose that goes beyond them: to constitute, hopefully, an illuminating case study of mathematical (or formal) philosophy—philosophy done with the help of mathematical methods. If the theory is successful at all, then this success will be brought out fully by its formal precision and deductive structure, and the same will be true of any of the flaws that it might have. There is a great tradition of doing philosophy with this kind of methodology:[3] Rudolf Carnap's work is a paradigm case. And so far as the joint study of rational belief and probability is concerned in particular, Isaac Levi's work serves as a model. This book sees itself in the tradition of both of them.

Let me conclude my little reflections here with a note on how to read this book. I still like to think that philosophers may at times take the effort of reading a monograph as it is, and that they would do so in all of its details. But the chapters of this book have also been designed so that they can be read, more or less, independently. I trust that anyone reading the book as a whole will excuse the occasional redundancies that have resulted from this design strategy. All of the chapters invoke mathematical symbols and mathematical patterns of reasoning, but the mathematics required is normally light and self-explanatory. The exception is Chapter 4, which is logically and mathematically a bit more demanding; the chapter may easily be skipped at first reading, though I recommend a quick scan of just its first three subsections. Similarly, the appendices to Chapters 1, 2, and 4, and the appendix to the book (Appendix D) may be bypassed without fear of losing the overall thread of reasoning (but of course some valuable additional insights might get lost). A detailed summary of the contents of all of the chapters can be found in section 1.7 at the end of the introductory Chapter 1. Specialists on the topic of this essay might well start reading from there and skip all of the sections of Chapter 1 before section 1.7. On the other hand, I hope Chapter 1 will serve as a useful general introduction to the debate on belief vs degrees of belief for anyone who is not familiar with it as yet.

There are way too many people to thank without whom this book would not exist. Let me at least try: David Makinson is the person whose extensive comments over the years have had the greatest impact on this book—thanks so much, David. I am

[3] For more on this, see e.g. Leitgeb (2013c).

also extremely grateful to Albert Anglberger, Horacio Arló-Costa, Brendan Balcerak Jackson, Magdalena Balcerak Jackson, Alexandru Baltag, Hanoch Ben-Yami, Luc Bovens, Richard Bradley, Seamus Bradley, Johannes Brandl, Peter Brössel, Edward Buckner, Catrin Campbell-Moore, Fabrizio Cariani, Jennifer Carr, David Chalmers, Jake Chandler, John Collins, Eleonora Cresto, Vincenzo Crupi, Erik Curiel, Hans Czermak, Georg Dorn, Kenny Easwaran, Philip Ebert, Anna-Maria Eder, Lee Joseph Elkin, David Etlin, Christian Feldbacher, Branden Fitelson, Haim Gaifman, Chris Gauker, Leon Geerdink, Nina Gierasimczuk, Norbert Gratzl, Alan Hájek, Volker Halbach, Stephan Hartmann, Frederik Herzberg, Alexander Hieke, Markus Hierl, Ole Hjortland, Wes Holliday, Leon Horsten, Franz Huber, Laurenz Hudetz, Humphrey, Simon Huttegger, James Joyce, Kevin Kelly, Aviv Keren, Cornelia Kroiss, Martin Krombholz, Maria Lasonen-Aarnio, Isaac Levi, Hanti Lin, Fenrong Liu, Yang Liu, Louis Loeb, Sebastian Lutz, Aidan Lyon, Tim Lyon, Alexandru Marcoci, Rosella Marrano, Michael Bennett McNulty, Edgar Morscher, Julien Murzi, Ronald Ortner, Oskar, Fabio Paglieri, Rohit Parikh, Arthur Paul Pedersen, Richard Pettigrew, Lavinia Picollo, Jonas Raab, Wlodek Rabinowicz, Martin Rechenauer, Jan-Willem Romeijn, Tobias Rosefeldt, Hans Rott, Olivier Roy, Gil Sagi, Gerhard Schurz, Teddy Seidenfeld, Sonja Smets, Martin Smith, Stanislav Speranski, Wolfgang Spohn, Jan Sprenger, Julia Staffel, Florian Steinberger, Johannes Stern, Corina Strößner, Scott Sturgeon, Patrick Suppes, Paul Thorn, Johan van Benthem, Barbara Vetter, Kevin Walkner, Christian Wallmann, Paul Weingartner, Jonathan Weisberg, Philip Welch, Charlotte Werndl, Greg Wheeler, Robert Williams, Jon Williamson, Timothy Williamson, Reinhard Wolf, Lena Zuchowski, various anonymous referees for my journal articles on the topic, two anonymous readers of my book draft, and quite simply all of my colleagues at the Munich Center for Mathematical Philosophy, which constituted the perfect academic environment for a work like this. As always, I want to thank especially the members of the Luxemburger Zirkel for their support and friendship over the years. I am particularly grateful to the organizers and participants of three reading groups on this book in 2014/2015: one at the London School of Economics (organized by David Makinson and Alex Marcoci), one at the University of Salzburg (organized by Charlotte Werndl), and one at LMU Munich. Since 2010 I have presented parts of this book to audiences at Tilburg, Glasgow, St Andrews, Hejnice, Carnegie Mellon, Konstanz, Paris, Amsterdam, Bayreuth, Salzburg, Vienna, Düsseldorf, Jerusalem, Barcelona, Nancy, Stockholm, Bochum, Cologne, Stanford, Tutzing, Buenos Aires, Berlin, Frankfurt, Stirling, Bristol, Leipzig, Cambridge, Ghent, Groningen, Rutgers, Helsinki, Hangzhou, Beijing, Bern, Venice, Hamburg, Columbia University, LSE, Rome, Fraueninsel (Chiemsee), Aarhus, Ann Arbor, Bonn, and Warwick: many thanks to all of the organizers and participants. In particular, I would like to thank Sven Ove Hansson for inviting me to give the Theoria Lecture at Stockholm on 14 June 2013, Jan Sprenger and Dominik Klein for inviting me to give the Descartes Lectures at Tilburg on 20–22 October 2014, and the Aristotelian Society and the Mind Association for inviting me to speak at their Joint Session at Warwick on 12 July 2015. On some occasions, I had the pleasure to be helped by invaluable oral or written commentaries on my talks given by Kevin Kelly, Hanti Lin, Branden Fitelson, Julia Staffel, Patrick Suppes, Hanoch Ben-Yami, Alexandru Baltag, Sonja Smets, Richard Pettigrew, Jan-Willem Romeijn, Nina Gierasimczuk, and Gerhard Schurz: I would like to thank them in particular. This work has been supported generously by the Alexander von Humboldt Foundation through an Alexander von Humboldt Professorship. I am very grateful to my editor Peter Momtchiloff and his colleagues at OUP for making the publication process run so smoothly. Last, but certainly not least, I would like to thank my wonderful family for their unconditional love, especially Conny, Basti, and Vicky. I dedicate this book to my parents Margit and Helmuth Leitgeb who supported my mathematical and philosophical inclinations without knowing what to make of them.

Finally, I should address previous work on which parts of this book are based.

Chapter 1 is new, except for its appendix which is a modified version of 'The Review Paradox: A Note on the Diachronic Costs of Not Closing Rational Belief Under Conjunction', *Nous* 48(4) (2014), 781–93.

Chapter 2 is a revision and extension of my earlier article 'The Humean Thesis on Belief', *Proceedings of the Aristotelian Society* 89(1) (2015), 143–85. Its appendix is new.

Chapter 3 is a revision and extension of my earlier article 'The Stability Theory of Belief', *Philosophical Review* 123(2) (2014), 131–71.

Chapter 4 is a thoroughly revised, restructured, and extended version of 'Reducing Belief *Simpliciter* to Degrees of Belief', *Annals of Pure and Applied Logic* 164(12) (2013), 1338–89. One of the biggest changes is that in the journal article I still aimed to defend a reductive account according to which rational all-or-nothing belief would reduce to rational degrees of belief. I do not do so any longer, which is one reason (but not the only one) why Chapter 4 needed to differ from the journal article. The appendix to Chapter 4 is new.

Chapter 5 is new.

Chapter 6 is also new except for its section 6.4, which is based on 'A Way Out of the Preface Paradox?', *Analysis* 74(1) (2014), 11–15 but which also contains new parts.

The appendix to the book is an adaptation of 'A Lottery Paradox for Counterfactuals without Agglomeration', *Philosophy and Phenomenological Research* 89(3), 605–36.

I thank the editors concerned (and Cornell University and Duke University in the case of the *Philosophical Review*) for permission to use this material.

Hannes Leitgeb

Munich
January 2017

Contents

List of Figures

List of Tables

1

Introduction

This is a normative study of rational belief. Its central question will be:

> What do a perfectly rational agent's beliefs and degrees of belief have to be like in order for them to cohere with each other?

Ultimately, my answer to that question will be:

> A perfectly rational agent believes a proposition if and only if she assigns a stably high degree of belief to it.

I will make that answer precise, I will show what implications it has, and I will determine, in turn, from which of its implications it can be derived.

If I had to summarize the main findings of the book in a couple of sentences, then I would say: there is a stability account of rational belief according to which all-or-nothing belief is pretty much what you might have thought it is, numerical degrees of belief are pretty much what you might have thought they are, and the two of them relate to each other pretty much in ways in which you might have thought they would. At least this will be so as long as the space of possibilities on which both belief and degrees of belief are defined is reasonably 'coarse-grained'. If the space of possibilities is very fine-grained, then degrees of belief will not be affected, but the theory predicts that a rational agent's beliefs will have to be more cautious in that case than one might have thought they are.

But before I can turn to perfect rationality and other normative matters, I will need to prepare the ground by some metaphysical and conceptual considerations: on the *nature* of belief, and on *concepts* of belief and what they denote. In other words: I need to clarify first what we are talking about and how we are going to talk about it. No normative epistemology of belief without the metaphysics and semantics of belief! Afterwards, I will turn to the specific normative presumptions and aims of this study, and I will present the traditional norms that are supposed to govern all-or-nothing belief and degrees of belief jointly. Finally, I will survey the overall structure of this book and its various chapters.

I should warn you that most of the assumptions, arguments, and conclusions in this chapter will remain rather vague. This should be fine just as a starting point, and things will get more precise later, beginning with the appendix to this chapter.

1.1 The Nature of Belief

What is belief?[4] Synonymously: what is holding something to be true? It is folklore in philosophy, and in (most of) the cognitive sciences more generally, to regard belief as a certain kind of mental state:

> Assumption 1: Belief is a *propositional* attitude of cognitive agents: an agent's belief *that X* is a mental state that has as its content the *proposition that X*.

For instance, if I believe that my wife is on the second floor, then I am in a certain mental state that has as its content the proposition that my wife is on the second floor.[5]

In principle, the 'cognitive agents' in Assumption 1 might well be animal or artificial agents, but ultimately I will be interested mostly in agents who have at least human capacities. Some of my assumptions below will concern only or predominantly such agents.[6]

It is also part of folk psychology that what distinguishes belief from other propositional attitudes, such as an agent's desire that *X* or an agent's supposition that *X*, is the special function of belief: the functional or causal roles that beliefs play in an agent's mental life.[7] These roles are often spelled out in teleological or normative or evaluative terms—what belief aims at, which rational commitments it involves, what ideals it is subject to, and for what acts it ought to be a necessary condition. Such characterizations of belief can be understood as yielding a description of belief as *the propositional attitude the function of which is to reach the goal so-and-so and to satisfy the norms so-and-so and to realize the valuable state so-and-so*, or to achieve all of that at least to a great extent and in normal circumstances.[8]

[4] I will remain neutral as to whether 'belief' in this section refers to all-or-nothing belief or an assignment of numerical degrees of belief or to belief in some other sense. The examples that I will give for purposes of illustration will, however, be examples of categorical or all-or-nothing belief. I will take up this topic in detail in section 1.2.

[5] Throughout this book I will not pay proper attention to the issue of belief contents having *indexical* components, such as those expressed by '*my* wife' and 'is (*now*) on the second floor (of *our* house)'. From Chapter 2 onwards, propositions will be considered as sets of possible worlds, where I am going to suppress the option of understanding the worlds in question as 'centered' on an agent and a point of time (as suggested by Lewis 1979a). Similarly, I will not have anything to say about the contribution that proper names make to (linguistic expressions of) belief, as famously discussed by Kripke (1979).

[6] I will not deal at all with 'group agents' and with what might be called their 'social (collective) beliefs'. But see Cariani (2016) for an application of the theory that will be developed from Chapter 2 onwards to judgement aggregation.

[7] See Armstrong (1968) for a classic source on functionalism about mental states in general. The functions of belief could also be analysed in evolutionary or quasi-evolutionary terms (see e.g. Millikan 1984): as those for which belief got selected. But this would take me too far off topic.

[8] I am going to avoid any discussion on how great 'great extent' is, and what the relevant notion of normality is meant to be: whether statistical normality or prototypical normality or some kind of normative notion again. It will certainly involve a modal or counterfactual component, but fortunately these issues won't be particularly important for the rest of the monograph. For a survey and systematization of different notions of normality, see Strößner (2015).

Methodologically, such a description can be viewed as defining a special theoretical term—the term for belief—along the lines of Lewis's (1970) proposal in 'How to Define Theoretical Terms':[9] the theory *Th*[*T*] that is meant to define implicitly the meaning of a theoretical term *T* is translated into the base clause of a definite description of the form ιR *Th*[*R*] by which the term *T* can then be defined explicitly: $T = \iota R$ *Th*[*R*].[10]

In the case of belief, it is the folk-psychological theory of belief and its defining functions that is turned in this way into an explicit definition of belief. If the resulting definition succeeds, that is, if its defining definite description refers, belief will indeed reach the goal so-and-so and satisfy the norms so-and-so and realize the valuable state so-and-so at least to a great extent and in normal circumstances. So if the definition succeeds, belief cannot be too far off track normatively. That much the present view shares with Davidson on belief (as in Davidson 1975).[11] The definition of belief itself, however, is not a normative statement: the normative force of all of the normative expressions within the body clause of the defining definite description is cancelled by terms such as 'reach' or 'satisfy' or 'realize' in the context of which the normative terms occur.

Later I will return to the question of whether the corresponding functionalist definition of belief does succeed in that sense. But first I will need to fill in some details on the functions that belief is meant to serve by its very definition.

One of the central such functional roles of belief is *epistemic* in so far as it concerns how belief relates to truth:

> Assumption 2: Belief is an agent's *representation* of what the world is like; it aims at the *truth*.[12]

Equivalently, in the helpful terminology that was triggered by Anscombe's (1957) work, one may say that beliefs have a direction of fit towards the world:[13] if I believe that my wife is presently on the second floor, but actually she is not, then it is my belief that is false and that ought to be revised, rather than that the world would have to be revised by somehow making my wife be on the second floor. I will turn to this in more detail

[9] See also Lewis (1966) on this topic, which deals especially with the functional definition of terms for mental states.

[10] ι is the definite description operator. If the definite description in question got analysed away, according to Russell's famous proposal, in terms of a combination of quantified statements, then the result of applying that analysis to the definition in question would not be an explicit definition any more. It is only when the definition description operator is taken as a logical primitive that a definition of the form $T = \iota R$ *Th*[*R*] is indeed an explicit one.

[11] In other respects, however, my account of belief will not be committed at all to Davidson's interpretivist view of belief.

[12] See e.g. Wedgwood (2002) for a detailed analysis of this feature of belief. Belief's aiming-at-the-truth will be the topic of section 5.1 in Chapter 5. Apart from truth there might be further aims of belief: e.g. in section 5.2 of Chapter 5, I will suggest that all-or-nothing belief might also aim at degrees of belief in a way that I am going to clarify in that section.

[13] In the case of beliefs about mental states, the 'world' would have to include mental states. But I will disregard the case of beliefs about mental states, such as the case of introspective beliefs, throughout this book.

in Chapter 5, but essentially this means that there is a valid norm or value statement that relates belief to truth, where the exact logical form of this statement is a matter of debate. One version would be to regard the following as correct: *to believe a proposition if and only if the proposition is true.*[14] Another one would be: holding the content or informativeness of one's beliefs fixed, it is *epistemically better to believe truths than falsehoods*. In any case, I assume that it is constitutive of belief to be causally related to other mental states and acts so that such epistemic norms or value statements are realized to a great extent and in normal circumstances. In the first version from before, belief would have to be such that *a proposition is believed if and only if it is true* at least in normal circumstances and for many propositions. In the other version, belief would have to be such that it happens to be above some appropriate threshold on the 'epistemic betterness scale', at least normally and to a great extent.

For example, belief seems to be related in such ways to perception and to receiving information through communication: if I perceive my wife to be on the ground floor, or if a source that I regard as trustworthy tells me that this is so, then receiving that kind of evidence will indeed normally make me adjust my beliefs accordingly. That is just one aspect of the functional role of belief. And indeed perception—as understood in the epistemological tradition in which '*a* perceives that *X*' is meant to entail 'it is true that *X*'—is veridical by definition (just as this is the case for knowledge), and testimony by sources that I regard as trustworthy is at least likely to convey the truth in normal cases (or so I hope). If a mental state does not occupy such causal pathways that are, mostly and normally, truth-conducive, then the state is not one of belief, as it belongs to the function of belief to produce true representations (even when this goes wrong occasionally).[15] For the same reason, normally, my beliefs are not completely under my control either:[16] I cannot simply decide to believe *X* without any sensitivity towards the truth of *X*, or the resulting mental state would not count as belief.

I regard belief's aiming-at-the-truth norm as grounding also typical evidentialist norms, such as that belief ought to be justified in light of an agent's total evidence.[17] That is also the reason why I will not complement the truth-focused Assumption 2 with any special evidentialist assumptions on belief: by realizing the aiming-at-the-truth norm, belief also realizes evidence-related norms.

At the same time, and of equal importance, there is also a *pragmatic* side to belief, which concerns how belief relates to what we do:

14 This would mean to give correctness a wide-scope reading: cf. Chan (2013, introduction, pp. 4–5).

15 Compare Velleman (2000). Horwich (2013) argues that our commitment to the truth norm for belief is manifested in our practice of gathering (ever more) evidence.

16 See e.g. Williams (1973) for more on this thesis of doxastic involuntarism. Not everyone agrees: e.g. van Fraassen (1984) defends voluntarism about belief.

17 Not everyone agrees with this: but see Wedgwood (2012) and Joyce (n.d.) for corresponding arguments for the priority of truth-related norms for belief over evidence-related norms for belief.

> Assumption 3: If combined appropriately with an agent's desires (and subject to a ceteris paribus clause), belief should commit an agent to rational *action*.[18]

I assume that it is a defining feature of belief to satisfy this norm at least normally and to a great extent. For example, if I believe my wife to be on the second floor, and I wish to talk to her, then normally this will make me climb the stairs to the second floor (unless I have strong conflicting reasons to do otherwise), and indeed this will be the rational choice. Desire's direction of fit is precisely the reverse of that of belief: if the world is presently not so that I can speak to my wife, then from my present point of view it is the world (or my bodily position therein) that ought to change. And it is constitutive of belief to assist the realization of such desires by informing the agent *how* they can be realized best (e.g. how my bodily position needs to change). This is sometimes expressed as: if one believes that X, then one will normally act *as if* X were the case. I will turn to this in much more detail in section 2.3.3 of Chapter 2 and section 6.1 of Chapter 6.

Other than such input and output functions, beliefs also have a major causal role to play with respect to *each other*: one that has to do with their overall *coherence*. Let me quote Bratman on this:

> Assumption 4: 'An agent's beliefs are subject to an ideal of integration. Other things equal one should be able to agglomerate one's various beliefs into a larger, overall view; and this larger view should satisfy demands for consistency and coherence' (Bratman 1999, p. 17).

Stated in less obviously normative terms, it belongs to the nature of belief that an agent's various beliefs in different propositions are part of a system: a coherent whole of pieces of belief that fit together in some sense or which are manipulated by certain cognitive mechanisms in order to make them fit together. For instance, my belief that my wife is on the second floor might have been derived from my belief that she is getting a book from one of the shelves on the second floor. In turn, on the basis of this derived belief, I might drop my belief that she will hear me when I call her (since I am on the ground floor).

Belief is such that joint patterns of presences and absences of beliefs may cause one to add beliefs, or to abandon beliefs or, in any case, to modify one's overall belief system.[19] These mental acts of coherence 'maintenance' either reflect the existing coherence amongst one's beliefs or aim at restoring it, at least partially. I will return to the normative question of what precisely this ideal of coherence might consist in later in this chapter. I suggest not to read too much as yet into Bratman's terms 'consistency'

[18] For instance, Davidson (1963) presents a version of such a belief–desire model of action.

[19] I emphasize 'absences' of beliefs as well, since there are cases in which the presence of a belief *taken together with the absence of another belief* makes one draw a conclusion, and indeed to do so justifiably. This nonmonotonicity feature of belief will become prominent in Chapter 4.

or 'coherence': I am yet to clarify what notions of consistency and coherence they might express.

All of the constitutive functional properties of belief so far have been quite indifferent as to whether the agent in question is human or not.[20] In contrast, the next, and final, item on my list of functional properties of belief is (more or less, and as far as we know) distinctive of human agents:

> Assumption 5: If an agent is capable of linguistic discourse, then what is expressed by the agent's sincere *assertions* should be her beliefs.[21] An agent ought: *to assert* sincerely that *X* only if she believes that *X*.[22]

For example, if my daughter drops by and asks me where her mother (my wife) is, then normally I will assert—in line with my corresponding belief, and co-determined by it—that she is on the second floor. And in normal circumstances my daughter will infer from this, and indeed correctly so, that I have the corresponding belief. Grounding such speech acts is just a special case of the output function of belief, which was covered already by the 'Action' Assumption 3. But since assertion is such a salient expression of belief, and one that highlights the crucial *social* role that belief can play, it is worth putting on record separately. I will deal with this in detail in section 6.2.

Let us leave the functional specification of belief at that. I am not claiming that the list is complete, only that the entries on it are correct.[23] Indeed, Chapter 2 will start with an argument that one fundamental feature of belief is missing from this list of constitutive properties: the *stability* of belief. But the list so far will be good enough for my present purposes.

Nor should the functional properties be considered independent of each other: since belief is meant to be such that it satisfies all of these constraints simultaneously, each of the constraints needs to be satisfied in a way that is compatible with all the other ones. For instance, the coherence or 'fitting together' aspect of beliefs needs to be compatible with their 'aiming at the truth' aspect: that is, mechanisms of inference or belief revision must aim at being truth-conducive, and at least normally and to a great extent they must succeed in reaching that aim if the intended definition of belief is to succeed at all. Or take assertion and aiming at truth: when I assert that my wife is on the second floor, my daughter may normally justifiably take me to aim at speaking the truth, because belief aims at the truth and assertion ought to express belief. And so on.

[20] In an earlier publication (Leitgeb 2004), I argued that even agents with very simple neural net-like cognitive architectures are capable of having justified beliefs and of drawing justified inferences.

[21] The descriptive sentence that gets uttered expresses a proposition, that proposition gets asserted, and the act of asserting that proposition expresses the speaker's belief in that proposition.

[22] Searle (1969) defends a version of this. By Assumption 5 I do not want to rule out that there are even stronger constraints on assertion, such as: an agent ought *to assert that X only if she* knows *that X* (where knowledge is normally taken to entail belief). Williamson (2000, ch. 11) defends this knowledge norm on assertion and discusses its relation with other norms on assertion. I will return to this in section 6.2.

[23] For other such lists of 'platitudes' on belief, see Bratman (1999, ch. 2, s. II), Engel (2000, ch. 1), Fantl and McGrath (2009, p. 141), and Ross and Schroeder (2014, s. 2).

As we are going to see later, due to various kinds of theoretical pressure, there will be a constant temptation to drop some of these features as being constitutive of belief. For example, in section 6.3 of Chapter 6 I will discuss *acceptance*, a mental state closely related to belief that satisfies all of the assumptions above except for the second one: acceptance does not necessarily aim at the truth.[24] In my view this just means that acceptance is not belief (not every instance of acceptance is an instance of belief). I regard it as crucial for beliefs to be mental states that occupy a central place at the intersection of theoretical *and* practical cognition: they play an epistemic role with respect to the inputs to a cognitive system, a pragmatic role with respect to the outputs of such a system, and a 'maintenance' role with respect to each other and to further internal states and acts of the system. None of this is negotiable, as far as I am concerned, or the mental states in question will not be *belief* states.

Similarly, I understand e.g. Kaplan's (1981, 1996) and Maher's (1993) 'assertion views' of belief as reducing the constitutive properties of belief to the one expressed by the 'Assertion' Assumption 5: consequently, these views will not suffice to count as views of *belief* in my terms. And the like. So I shall resist the temptation of giving up on any of the constitutive properties of belief above.

However, this resistance will also be subject to some qualifications. In particular, here is yet another assumption about belief that one can find in parts of the literature but which will only be satisfied *partially* by the theory that I will develop. Let me quote Bratman again:

> Assumption* [NOT GENERALLY SATISFIED by my theory]: 'Reasonable belief is, in an important way, context independent: at any one time a reasonable agent normally either believes something (to degree n) or does not believe it (to that degree). She does not at the same time believe that p relative to one context but not relative to another' (Bratman 1999, p. 18).

I will have much more to say about contexts in section 3.3 of Chapter 3, but roughly: there are two notions of context in the relevant literature. One is semantic: a context in that sense might determine the content of the belief term, just as contexts are taken to determine the content of an indexical expression. That is not the understanding that I am going to favour, and it is not Bratman's understanding in this quotation above. The other notion of context is an epistemic one: the context in which an agent is reasoning will involve the sum of the same agent's practical interests, her focus of attention, what is salient to her, and the like, at a point in time. According to the theory that I am going to develop from Chapter 2 onwards, an agent's *numerical degree-of-belief* function

24 In parts of formal epistemology and general philosophy of science, the term 'acceptance' is used differently from my current usage: it is used there as a technical term that is meant to express something 'like' or in the same ballpark as belief, without any special additional connotation. I am going to avoid using the term in that sense. In this book (especially in section 6.3), 'acceptance' will be used as it is understood in other parts of epistemology, the philosophy of mind, and in general philosophy of science: as expressing a mental state that is like belief in all pragmatic aspects but not necessarily in all epistemic respects.

may or may not be independent of the context in which the agent is reasoning: the theory will be silent about this. But I will normally assume that an agent's degrees of belief are context-*in*sensitive. However, the theory *will* entail that an agent's rational *categorical* or *all-or-nothing* beliefs do depend on her context of reasoning in that sense of the term. This will be one of the worries about the theory that I will deal with later, especially in section 3.3 of Chapter 3.[25]

There will be further occasions in this book to touch upon metaphysical matters—on reduction and supervenience, dispositional vs occurrent beliefs, and the like—though fortunately I will not have to deal with these topics in much metaphysical detail. But for now let me turn to more conceptual issues.

1.2 Concepts of Belief

Belief states can be ascribed to agents by means of different concepts.

On the one hand, there are some syntactic issues that we need to get out of the way: belief concepts may be expressed either by a sentential operator, as in '*it is believed by agent so-and-so [to degree x] at time t that X*', or by a predicate, as in '*X is believed by agent so-and-so [to degree x] at time t*' (or with the help of a function symbol, which is much like the predicate case). In the former case, '*X*' is a placeholder for a sentence, and the belief operator is syntactically of the same type as the necessity operator 'it is necessary that' in modal logic. In the latter case, '*X*' is a placeholder for a singular term, that is, the name of a proposition, and once all the free parameters in 'is believed by agent so-and-so [to degree *x*] at time *t*' have been filled in, the resulting phrase determines a property of propositions *X*. Fortunately, these syntactic distinctions between operators and predicates will not matter much in anything that follows, since I am not going to talk about belief in terms of a fully formalized language anyway. Instead I will use natural language for that purpose, augmented by portions of the language of mathematics.[26] Accordingly, I will take the liberty of switching between operator and predicate ways of talking about belief interchangeably.

On the other hand, there is also a distinction between concepts that is much more crucial as far as the goals of this monograph are concerned: some concepts of belief occupy *different scales of measurement*,[27] or at least they appear to do so according to the surface structure of natural language. Moreover, these different concepts of belief on different scales also belong to different intellectual traditions. In particular:

[25] Not everyone shares Bratman's context-independence view about belief, though: for instance, Thomason (1986) and Nozick (1994, pp. 96–100) argue for the context-*dependence* of rational belief.

[26] So I will follow the tradition of e.g. belief revision theory (as in Gärdenfors 1988) or standard probability theory, which do not rely on a fully formalized language either. This is in contrast with e.g. dynamic epistemic/doxastic logic (see van Ditmarsch et al. 2008, Segerberg 2001, Leitgeb and Segerberg 2007), probability logic (see Leitgeb 2016 for a survey), or formal systems for probabilistic dynamic update (see Baltag and Smets 2008, van Benthem et al. 2009), all of which study doxastic update in fully formalized logical languages.

[27] More on scales of measurement in general can be found in Krantz et al. (1971).

> Assumption 6: There are different belief concepts, including a *categorical* or *classificatory* concept of belief and a *numerical* or *degree of* belief concept.

The categorical concept of belief occupies a categorical or nominal scale of measurement. It is also expressed by means of terms such as 'qualitative belief', 'all-or-nothing belief',[28] 'belief *simpliciter*', 'flat-out belief', 'plain belief', 'binary belief', 'holding true', 'regarding true', 'taking to be the case', or simply 'belief' (of a given agent, at a given point of time). Sometimes, 'categorical' is taken as the opposite of 'conditional', which is *not* what I have in mind here: indeed, *conditional all-or-nothing* belief will be studied in detail in Chapter 4. (But I will mostly ignore conditional all-or-nothing belief in the present context.) Focusing just on one proposition X, when one ascribes belief to an agent in categorical terms, one says that (i) the agent *believes that* X, or (ii) the *agent believes that* $\neg X$ (that is, *not* X),[29] or (iii) the agent *neither believes that* X *nor* $\neg X$, that is, she suspends judgement on X (she is 'agnostic' about X). If the agent in question is perfectly rational, then, presumably, (i) and (ii) cannot obtain simultaneously, in which case for each proposition X precisely one of the three cases must obtain.

This categorical concept of belief constitutes the standard in traditional epistemology, philosophy of mind, classical cognitive psychology, classical artificial intelligence, and philosophical logic. It is also applied often in natural language discourse ('I believe that . . . ', 'I don't believe that . . . ').

I will not enter any debate on what exactly the cognitive implementation of all-or-nothing belief in human agents might be like: whether belief in X might consist in some mental representation being held in one's mental 'belief box', such that this representation expresses X,[30] or whether belief in X might consist in the disposition to generate certain neural patterns of activation in certain regions in the brain, or whatever else. That is to be settled by psychology and neuroscience.

It is much more important for my purposes that there is also an alternative way of understanding the constitutive assumptions about belief from the previous section: in terms of a numerical concept of *degree of belief*, which is usually assumed to occupy a so-called absolute numerical scale—the only manner of transforming degrees of belief without changing their meaning is to apply the identity map. The concept can also be expressed by means of terms such as 'quantitative belief', '(numerically) graded belief', 'partial belief', '(numerical) credence', and 'degree of confidence' (in each case of a given agent, at a given point of time).[31] Some of these terms can also be used to ascribe

[28] The qualification 'all-or-nothing' does not rule out suspension of judgement—it only means that for each proposition X, either one believes that X or one does not.

[29] Alternatively, I will say in such a case: 'the agent *dis*believes that X'. So I take 'believes that $\neg X$' and 'disbelieves that X' to be synonymous. Not everyone accepts this synonymy: see e.g. Russell (1993).

[30] Fodor (1975) is a classical case of such a representationalist account of belief.

[31] In view of the existence of the term 'partial belief' for numerical belief, I am going to avoid referring to categorical belief by means of 'full belief', since that might be taken to suggest that categorical belief would have to coincide with maximum partial belief. But that would be a mistake. I will return to this in section 1.6 when I discuss the 'Certainty Proposal'.

beliefs to an agent on a different level of measurement, but when one uses them in order to ascribe belief in the sense that I have in mind right now, then one intends them to express that the agent *believes X to a degree of x* or with $x \cdot 100$ per cent, where 'x' denotes a real number in the unit interval $[0, 1]$. That number is supposed to measure the strength of the agent's belief in the proposition X. Typically, 'X is believed to degree 1' means that the agent is certain that X is true, 'X is believed to degree 0' corresponds to the agent being certain that $\neg X$ is true (and thus X is false), and any degree of belief in between these two values represents the agent's strength of belief in X lying in between the two extreme cases.

The degree-of-belief concept is the one that dominates talk of belief in subjective probability theory, decision theory, economics, certain parts of artificial intelligence, and all areas denoted by a 'Bayesian' term: Bayesian epistemology, Bayesian philosophy of science, Bayesian psychology, Bayesian neuroscience, and the like.

For some philosophers the most immediate way of understanding Assumptions 1–5 will be in categorical terms. But our assumptions may just as well be read numerically. (1) A degree-of-belief function assigns numerical strengths of belief to propositions. E.g. I might assign a high degree of belief to the proposition that my wife is on the second floor. (2) The degree of belief in X ought to be as close as possible to the truth value of X.[32] And indeed, e.g. if I perceive my wife to be on the ground floor, then my degree of belief in her being on the second floor will normally be pushed towards 0 or even be set to 0 itself. (3) In combination with an agent's utility measure, a degree-of-belief function commits the agent to rational action.[33] E.g. given my high degree of belief in the proposition that my wife is on the second floor, the expected utility of walking upstairs in order to meet her will be high, which will make me commit (ceteris paribus) to the corresponding course of action. (4) An agent's degrees of belief in different propositions need to cohere with each other. E.g. if my degree of belief in my wife's being on the ground floor increases, my degree of belief in her being on the second floor ought to decrease, and indeed normally it does. Finally, (5) an agent ought to assert that X only if her degree of belief in X satisfies an appropriate constraint, such as being high enough, or the like.

I will not go into any further details on the cognitive implementation of human degrees of belief either: for example, whether the degrees of belief of an agent might be constituted somehow by her preferences over actions (as suggested perhaps by certain decision-theoretic representation theorems), or whether an agent might have degrees of belief in some other, perhaps more robust, sense.[34]

In the following, in any context in which the distinction between all-or-nothing belief and graded belief is salient, the unqualified term 'belief' shall always mean the

[32] As maintained by epistemic decision theory: see e.g. Joyce (1998). I will return to this in section 5.1.

[33] This is one of the basic tenets of traditional (pragmatic) decision theory; see e.g. Joyce (1999).

[34] More about the metaphysics of rational degrees of belief, and on the metaphysical interpretation of decision-theoretic representation theorems in particular, can be found in Christensen (2004).

former. But in any more general context, such as that of our discussion of Assumptions 1–5, 'belief' shall either stand for all-or-nothing belief, or for degree of belief, or for both of them simultaneously. Each of the two belief concepts is defined as *the propositional attitude the function of which is*... It is just that the '...' part needs to be filled in respectively: either by the categorical or by the numerical version of Assumptions 1–5. And, once again, if one of the two concepts were not defined in that way, then the concept in question should not be considered a concept of *belief* (but maybe of some other mental state).[35]

If the two definitions of these two concepts of belief—the categorical and the graded one—are so similar to each other in structure and content, do these concepts therefore denote the same propositional attitude or at least aspects of the same such attitude? Or does the all-or-nothing way of filling '...' within '*the propositional attitude the function of which is*...' determine an entity different from the one that is determined by filling '...' by means of quantitative terms? I will deal with these questions in the next section.[36]

1.3 Elimination, Reduction, Irreducibility

Actually, things are a bit more complicated, for there are really three broad options here: (i) at least one of the two—the categorical concept of belief or the degree-of-belief concept—is empty in the sense of not referring to anything at all. (ii) Or both refer, and they refer to the same phenomenon. (iii) Or both refer, but they do not refer to the same phenomenon.

The way in which I will understand reference here is a bit loose: I will say that 'agent a believes X at time t' refers to a mental state—or, more briefly: '(all-or-nothing) belief' refers—when there is a unique natural[37] mental phenomenon type that satisfies categorical versions of Assumptions 1–5. That phenomenon type an agent a must exemplify in order for her to believe X at t.

Accordingly, I will say that 'agent a believes X to degree x at time t' refers to a mental state—or, more briefly: 'degree of belief' refers—when there is a unique natural mental phenomenon type that satisfies numerical versions of Assumptions 1–5. That phenomenon type an agent a must exemplify in order for her to believe X to such a degree x at t.

[35] Of course, even a concept with a totally different definition could still *refer* to the same phenomenon as one of our belief concepts. But in my terminology this would not make that concept a *belief* concept: it would refer to belief, but *intensionally* it would differ from a concept of belief.

[36] I should emphasize that there are various further concepts of belief over and above the categorical and numerical ones, some of which we will also encounter in later chapters: conditional belief or belief on an ordinal scale (see Chapter 4), Spohnian ranking functions (cf. Spohn 2012) or belief on a ratio scale, qualitative probability or degree-of-belief on an ordinal scale (see section 5.2), and more. But for present purposes it will be best just to focus on plain categorical belief vs numerical degree of belief.

[37] I will say more about the qualification 'natural' in the next section, when I turn to Lewis on natural properties.

Instead of using the term 'refers', I might just as well have used terms such as 'represents', 'is about', or 'describes'. In any case, one should not necessarily think of reference here as restricted to the semantic relationship between a proper name and what it refers to.[38]

Failure of reference in either case would mean that there is no phenomenon as required, respectively.

Identity of reference would mean that the natural mental phenomena in question are equal to each other or that they can at least be understood as aspects of a joint, and more comprehensive, natural phenomenon. In terms of an analogy (one that is going to recur): I would also say that 'physical body y is warm at t' (as understood relative to a certain context and experiencing person) and 'physical body y has numerical temperature z (in degrees centigrade) at t' refer to the same phenomenon—something like average kinetic energy of motions of microscopic particles of z at t. That is so although 'y is warm at t' only manages to express a particular coarse-grained aspect of what that phenomenon is like, while 'y has numerical temperature z at t' describes the phenomenon in much more specific terms.

Finally, *difference* of reference would correspond to 'a believes X at time t' and 'a believes X to degree x at time t' speaking about different natural phenomena which, for standard explanatory purposes, one would not even want to understand as distinct aspects of one joint underlying natural phenomenon. In the same sense, 'warm' differs clearly in reference from e.g. 'has a weight of so-and-so many grams'.

Each of the three broad options (i)–(iii) will allow for different specifications, some of which will be familiar from other areas in which questions of elimination, reduction, and irreducibility are discussed: the paradigm case being typical debates in the philosophy of mind about the conceptual relationships between discourse in mentalistic and physicalistic terms, and about the metaphysical relationships between mental states and physical states.[39] When I turn now to possible specifications of these options (i)–(iii) in the following sections, it should be understood that my list of such specifications will remain incomplete. I will restrict myself only to those cases that will be particularly salient as far as my own purposes are concerned.

Here is what my main conclusions will be: as things stand, there are no good reasons to regard the categorical concept of belief as empty, and the same holds for the numerical concept of degree of belief. Both of them manage to refer to something, and hence neither of the concepts ought to be eliminated for lack of reference. If they happen ultimately to refer to (aspects of) the same type of mental state, then there is

[38] If one wanted to make 'reference' of belief concepts perfectly precise, one would have to return to syntactic questions first. How should one talk about belief or degrees of belief, once all of the usual ambiguities of natural language would have been cleaned up? By means of a definite description that defines (i) a predicate, or (ii) a sentential operator, or (iii) a function symbol? 'Reference' would mean something else in each case.

[39] Spohn (2012, s. 10.1) makes a similar comparison of positions concerning belief and degrees of belief to positions concerning the mind–body problem.

a lot of pressure on the all-or-nothing concept to be reduced to the numerical one: it is plausible that the concept of all-or-nothing belief will then be definable on the basis of the concept of degree of belief. And while there will be no need to eliminate the categorical concept, for most purposes speaking in terms of the degree-of-belief concept will be the superior choice. On the other hand, if the binary and the numerical concept of belief refer to different types of mental states (even in my broad sense of 'reference' just sketched), then one ought to take the possibility very seriously that the mental phenomena of all-or-nothing belief and graded belief are ontologically independent of each other: they are not just distinct, but one can be instantiated without the other one being instantiated, too.

Let me turn to option (i) first.

1.3.1 The Elimination (without Reduction) Option (i): At Least One of the Two Concepts of Belief is Empty

Since each of the two concepts of belief is given by a definite description of the form *the propositional attitude the function of which is . . .*, this option (i) might hold for one of two possible reasons: (i.i) Either because there is nothing that satisfies the body clause '. . . ' of one of these definite descriptions, and hence the description's existence condition fails. (i.ii) Or there is more than one phenomenon that satisfies the '. . .' part of one of the definite descriptions, and hence the description's uniqueness condition fails. In either case, depending on one's favourite analysis of definite descriptions, atomic statements about categorical or graded belief would be either false or truth-valueless, which would strongly support the elimination of the respective concept of belief from scientific and philosophical discourse. Call this the *Elimination (Without Reduction) Option* (i) concerning either all-or-nothing belief or graded belief or both.

Following Lewis (1970) and Papineau (1996),[40] the second worry (i.ii) about lack of uniqueness may be obviated by assuming that the variable '*R*' that is contained in '*the propositional attitude R the function of which is . . .* ' is ranging over *natural* entities only, in this case, natural mental states (certain natural relations between agents, times, and propositions): entities that 'carve nature at its joints'. If there were too many of the relevant entities around, then the definite description would have trouble picking out one of them. But once all the 'gerrymandered' states are excluded from the start, then it might well be that only one *R* is left that will do the job. Indeed, Lewis argues that in all realistic cases of theoretical terms in science and philosophy, the corresponding move makes it unlikely that there is more than one realizer of the defining clauses of these terms. In the following, I will take for granted that he is right about this as far as the definitions of our two concepts of belief are concerned, and that both of our definite

[40] See Papineau (1996, n. 4).

descriptions for belief do come equipped with the required quantifier restrictions to natural mental states.[41]

At least prima facie, the other worry (i.i) about existence seems unlikely to apply: given what appears to be one instance after another of successful applications of the two belief concepts in everyday life, philosophy, and science, could it really be that one of them does not manage to pick out any natural phenomenon at all? If my daughter asks me whether I believe her mother to be on the second floor, and I nod in approval, do we fail to refer to anything at all? (Analogously, if she asks me whether I am more than 80 per cent confident that her mother is on the second floor.) And even if that were so for applications of these concepts in such everyday situations, what about the roles that (more or less) the same folk-psychological concepts play in theories in the cognitive sciences? Is it really the case that all of these theories, or substantial parts thereof, are empty?

Clearly, this is but a defeasible way of arguing against (i.i): to quote a well-worn example, the concept of phlogiston enjoyed some initial success in eighteenth-century chemistry, too, and yet today's chemists do not believe any more that it refers to some natural substance. But then again it is not clear by any means that the psychological theory that constitutes the theoretical concept of belief—and to which Assumptions 1–5 belong—has been discredited empirically in any way that would resemble the case of phlogiston theory.

Much closer to the subject matter, a part of the connectionist literature on cognition by neural networks maintains that belief in the sense of folk psychology simply does not exist, and that therefore the corresponding concept (or concepts) of belief ought to be eliminated from scientific and philosophical discourse. Churchland (1981) is a classic source on this eliminativist view; he regards the ontology of folk psychology to be illusory as a whole. But so far most of cognitive science does not seem to have followed that eliminativist proposal: instead, artificial neural networks are embraced as additional tools or models in the study of cognitive states and processes, including belief, desire, intention, perception, inference, memory, and the like.

A different 'existence worry' concerns the question of whether one of the assumptions on belief in section 1.1 might actually be in tension with another one. For instance: cognitive dissonance theory[42] is a psychological theory that holds that we constantly

[41] The alternative would be to replace a Lewisian definition of belief based on a definite description either by a Ramsification of the belief term (as discussed by Lewis 1970 himself) or by a Carnapian definition of belief on the basis of a so-called epsilon term (see Carnap 1961): both of these alternatives would only demand the existence of a mental state type that satisfies our assumptions from above, but they would not demand uniqueness. I will ignore these options here, since my own arguments starting with Chapter 2 do not seem to leave much space for the existence of various alternative kinds of categorical belief, with the exception of the mental state of acceptance that will be discussed in Chapter 6. (Belief will, however, be diagnosed to depend on the context.) But see Pettigrew (2015) for arguments for the contrary thesis that there might actually be several (natural) categorical types of belief each of which satisfies at least some of the assumptions on belief from section 1.1.

[42] Cf. Festinger (1957).

seek consistency in our beliefs (and further attitudes). So far this is perfectly in line with the 'Integration' Assumption 4 from section 1.1. But the theory also suggests that sometimes, in order to restore or maintain consistency, we behave irrationally: e.g. we might change some of our beliefs without good reason and, therefore, without any concern for truth.[43] So the worry would be that by satisfying Assumption 4 we might fail to satisfy our 'Truth-Aiming' Assumption 2 from before. Whether we do so is ultimately an empirical question. As long as the assumptions are co-satisfied at least in normal circumstances and to a great extent, my intended definition of belief in terms of a definite description that incorporates all of these assumptions simultaneously will still do its job.

Finally, there are some philosophical arguments that attack the existence of especially *all-or-nothing* belief as a natural mental state: e.g. Christensen (2004) argues that the categorical belief concept does not refer to anything that 'cuts nature at its joints' (see Christensen 2004, p. 97); it does not pick out any epistemically important (truth-aiming) property and, especially if rational belief is meant to be closed under logic, it does not pick out any pragmatically important property (in decision-making) either. Where the categorical concept of belief is useful at all, it would therefore have to be so for other reasons. Christensen's arguments for these claims are based on a couple of examples, some of which derive from Preface-Paradox-like situations.[44] This is ultimately a matter of weighting pros and cons, of course, but at this point these examples do not seem sufficient to me to defeat the prima facie plausible inference from the apparent usefulness of the belief concept—as it is understood in everyday contexts, science, and philosophy—to the existence of belief as a natural state of mind. As I will try to show later in this book (sections 3.4, 3.5, 6.4), the insights that we can get from studying the Lottery Paradox and the Preface Paradox do not require us to re-evaluate the prima facie plausibility of that inference. Therefore, I think it is fair to say that the burden of proof still lies on the side of the advocates of the existence failure option (i.i), and a heavy burden it is.

In any case, I will state this as yet another (at least prima facie plausible) assumption:

> Assumption 7: Both the categorical concept of belief and the degree-of-belief concept refer and they do so uniquely (whether to the same phenomenon or not).

So I take both categorical belief and the degree-of-belief assignment (of an agent at a time) to be real. And both of them are kinds of belief: both of them need to satisfy

[43] I am grateful to Lena Zuchowski and Michael Bennett McNulty for bringing up that example.

[44] See Christensen (2004, s. 4.4) for these examples and his arguments. While Christensen attacks the naturalness of all-or-nothing belief, he also thinks that categorical belief-talk might still be useful for some purposes, although it is not quite clear to me for what purposes exactly. For instance, presumably, he cannot have in mind any epistemically important purposes, since he does not consider the all-or-nothing concept of belief to pick out an epistemically important property. At those places at which Christensen acknowledges the amenities of binary belief-attributions, his position might actually be closer to the Reduction (without Elimination) Option that I will discuss under (ii) in the next section.

versions of our Assumptions 1–5. Categorical beliefs and assignments of degrees of belief are in the same 'business', as it were.[45]

Which leads us back to the two remaining options from before. (ii) Both the concept of categorical belief and the concept of degree of belief refer, and they refer to the same phenomenon. (iii) Or both refer, but they do not refer to the same phenomenon. Which is it? Let me turn to option (ii) now.

1.3.2 The Reduction Option (ii): Both Concepts of Belief Refer, and they Refer to the Same Phenomenon

In case (ii), there is just one natural mental state out there about which—or about certain aspects of which—one may talk either by means of the categorical belief concept or by means of the numerical one. Even if that is so, it does not by itself mean that we should be able to reduce one of the two concepts to the other one (for example, by explicit definition), or that one of the two concepts ought to be eliminated in favour of the other. Perhaps one of them might in principle be reducible to the other, but the reduction may still not be practically feasible: it might be too complicated, or it might have to involve linguistic expressions that are simply not available to us, or the like. And perhaps one of them could be eliminated in principle, but it is still handy to have it around; or really both of them should be eliminated in favour of a third one (which shares reference with both of them). And so forth.[46]

That being said, if case (ii) obtains, then I do think this will at least exert prima facie pressure towards either the reduction or the elimination of one of the two concepts—and, presumably, that concept will be the *categorical* one.

Compare what Carnap has to say about such situations in which, ultimately, categorical or classificatory concepts happen to compete against numerical or quantitative ones:

> Among the kinds of concept used in science, three are of special importance. We call them classificatory, comparative, and quantitative concepts . . . In prescientific thinking classificatory concepts are used most frequently. In the course of the development of science they are replaced in scientific formulations more and more by concepts of the other two kinds, although they remain always useful for the formulation of observational results. (Carnap 1950a, pp. 8–9)

> Classificatory concepts are the simplest and least effective kind of concept. Comparative concepts are more powerful, and quantitative concepts still more; that is to say, they enable us to give a more precise description of a concrete situation and, more important, to formulate more comprehensive general laws. (Carnap 1950a, p. 12)

[45] Not everyone agrees: e.g. Buchak (2014) argues that belief and degrees of belief are not, or at least not quite, in the same business, because they are responsive to different aspects of evidence.

[46] I should emphasize from the start that my discussion of reduction and elimination in this book will not go into any detail—doing otherwise would take me too far off topic. E.g. I will not try to explain what kind of constraint *reduces-to* amounts to across metaphysically possible worlds, or the like. I will have just a tiny bit more to say about reduction in Appendix C, but that will be it. More about some of the metaphysical issues concerning belief and degrees of belief can be found in Sturgeon (2015).

He adds that sometimes a numerical and a categorical concept come as a pair:

> In many cases a quantitative concept corresponds to a classificatory concept. Thus temperature corresponds to the property Warm; and the concept of a distance of less than five miles corresponds to the relation of proximity. (Carnap 1950a, p. 9)

When Carnap says that the categorical concept of being warm (with the context and subject of experience being fixed) and the numerical concept of temperature 'correspond' to each other, this is just another way of saying that they refer to (aspects of) the same natural phenomenon. And while he thinks the categorical concept might still be of limited use, such as in the description of observation results, he regards the numerical concept to be theoretically superior in allowing for greater precision and facilitating the formulation of general laws. Therefore, Carnap claims, the numerical concept is ultimately bound to surpass its categorical sibling in the development of science. At best, one may hope to reconstruct the classificatory concept on the basis of the quantitative one, such as in Carnap's characterization of *proximity* in terms of *distance of less than five miles* (where 'five' may be replaced by any other contextually determined numeral). In this way, the classificatory concept would remain part of the language of science, but it would also become reducible to its quantitative partner concept by means of a stipulative definition: whatever the original meaning of proximity might have been, *from now on, we might just as well understand proximity as distance of less than five miles*. Accordingly, on the ontological level, one might start viewing proximity to be *nothing but* distance of less than five miles, and warmth to be *nothing over and above* high enough temperature: the members of each pair refer to the same phenomenon. At the same time it is to be acknowledged in each case that the quantitative concept captures that phenomenon in the more fine-grained, sophisticated, and 'scientific' manner. For instance: claiming that x is close to y and y is close to z does manage to express something about their respective distances. But the numerical concept of distance allows for a more detailed description of the 'extents' of closeness of x to y, of y to z, and of how they relate to each other (where the exact meaning of 'more detailed' is determined by the respective scales).

Of course, Carnap's claim of the stepwise replacement of classificatory concepts in science by concepts on more complex scales is an empirical claim that would need to be supported empirically (by data from the history of science). But the claim does seem plausible enough. Assuming that such long-term scientific developments do in fact indicate scientific progress, it would follow that numerical concepts are indeed more useful scientifically in the long run than their classificatory counterparts. In the case of belief, the upshot of these considerations would be: if both the categorical concept of belief and the numerical concept of degree of belief refer to the same entity—if they 'correspond' to each other—then at least prima facie the pressure will be on the former rather than the latter. The concept of categorical belief might still prove its worth in terms of its simplicity ('Classificatory concepts are the simplest . . . kind of concept') and its continuity with commonsensical ascriptions of belief wherever appropriate

('In prescientific thinking classificatory concepts are used most frequently... they remain always useful for the formulation of observational results'). But ultimately the categorical concept of belief will be, at best, reducible to the concept of quantitative belief by stipulative definition: whatever the original definition of categorical belief in terms of *the (all-or-nothing) propositional attitude the function of which is...* might have been, *from now on, we might just as well understand belief in X in terms of degrees of belief in the way that...*

In the simplest case, this reduction might proceed analogously to the case of proximity: *belief in X is to be understood as one's degree of belief in X exceeding some (contextually determined) threshold.* Or in more ontological terms: all-or-nothing belief may easily end up being viewed as nothing but high enough degree of belief.[47] Which is one possible interpretation of the descriptive version of a normative principle about belief and degrees of belief that we are going to encounter later in this chapter and that will remain a topic throughout the whole monograph: the so-called Lockean thesis on belief.[48]

Or things are more complicated: it might still be that all-or-nothing belief is nothing but degrees of belief being so-and-so, but the 'being so-and-so' does not coincide straightforwardly with credence exceeding a threshold. After all, belief is a phenomenon that has certainly a much more complex internal structure than warmth or proximity, which might well become manifest in terms of a more complicated pattern of reduction to partial belief. E.g. Frankish (2004, 2009) defends the view that flat-out belief in X is an intentional disposition that is indeed 'realized in' and causally active 'in virtue of' partial belief and desire (see Frankish 2004, p. 65): so flat-out belief is nothing over and above partial belief and desire. However, the manner in which categorical belief in X is realized in degrees of belief is much more complicated than the degree of belief in X being sufficiently high: it rather corresponds to a sufficiently high confidence in one's having adopted a so-called 'premising policy' with respect to X.[49] In at least one reading of Frankish's theory, this is a reductive account in which all-or-nothing belief reduces to partial belief and desire.[50]

47 I am extrapolating from Carnap's proposal here, since he does not actually discuss the case of belief vs degree of belief in the quoted passages. Elsewhere (Carnap 1963, pp. 972–3), he does discuss the closely related issue of all-or-nothing acceptance vs degree of rational credibility. The conclusions that he draws there might actually be closer to Jeffrey's eliminativist position, which I am going to discuss next. While Carnap still does not deny there that 'sometimes rules of acceptance may be useful', he also thinks that rules of acceptance are 'inadequate' as 'ultimate rules for inductive reasoning', by giving us 'in some respect too much, in another respect too little' in the field of practical reasoning.

48 See Foley (1993) for more on this thesis. I will return to it in section 1.6.

49 So Frankish (2004, 2009) considers the belief in X to be something like a context-insensitive disposition to take X as a premise for theoretical and practical reasoning, combined with the desire to adhere to that premise, and combined also with some kind of epistemic, that is, truth-related interest in that premise. In my terms, this will not quite count as belief, nor as (plain) acceptance, but it will rather be close to what I will call *accepted belief*—a case of one's believed proposition being accepted. See section 6.3 in Chapter 6 for more on this.

50 Perhaps Frankish only wants to claim that all-or-nothing belief reduces to partial belief and desire on the level of tokens, but not on the level of types: belief tokens are degree-of-belief tokens, but belief types are

In any case, we may subsume all of these specifications of option (ii) under what might be called the *Reduction (without Elimination) Option* concerning all-or-nothing belief. All-or-nothing belief can be reduced to degrees of belief, but the concept of all-or-nothing belief is not eliminated.

Richard Jeffrey, Carnap's student, took these considerations one step further by recommending simply the elimination of the all-or-nothing concept of belief:

> By 'belief' I mean the thing that goes along with valuation in decision-making: degree-of-belief, or subjective probability, or personal probability, or grade of credence. I do not care what you call it because I can tell you what it is, and how to measure it, within limits . . . Nor am I disturbed by the fact that our ordinary notion of belief is only vestigially present in the notion of degree of belief. I am inclined to think Ramsey sucked the marrow out of the ordinary notion, and used it to nourish a more adequate view. (Jeffrey 1970, pp. 171–2)

Jeffrey's idea does not seem to be that the ordinary categorical concept of belief lacks reference and *therefore* ought to be eliminated. The categorical and the numerical concept of belief do share reference (at least in parts: that's the 'marrow'). However, while the phenomenon in question can only be described in opaque or incomplete or even somewhat confused terms by means of the categorical concept, the numerical—indeed, probabilistic—concept of belief does not suffer from the same shortcomings. And *that* is why one ought to drop the categorical concept in favour of the numerical one:

> The notions of belief and disbelief are familiar enough but, I find, unclear. In contrast, I find the notion of subjective probability, for all its (decreasing) unfamiliarity, to be a model of clarity . . . I continue to avoid talk about knowledge and acceptance of hypotheses, trying to make do with graded belief. (Jeffrey 1970, p. 183)

So in contrast with the reductionist view from before, Jeffrey does not even aim at reconstructing belief on the basis of degrees of belief: he simply goes for the latter and eliminates the former altogether. Call this the *Elimination by Reduction Option* concerning categorical belief. Jeffrey thinks that what is natural and important about belief can be reduced to degrees of belief. Because this is so, one might just as well eliminate the concept of all-or-nothing belief altogether.[51]

Clearly, this eliminativist (but still referential) view of belief is a much more radical response to option (ii) than the merely reductionist one. In terms of burden of proof again, the eliminativist's burden is much greater: first, he would need to show that everything that had been useful about more than two thousand years of talk in terms of all-or-nothing belief can be 'sucked out' and expressed in terms of the numerical belief concept. The 'talk' in question could be everyday talk, but more crucially it would

not degree-of-belief types. If so, then, in this respect, his view would actually be closer to the Davidsonian version of the Irreducibility Option (iii) that I am going to discuss in the next section.

[51] One might want to add: the more confused the categorical concept of belief is in Jeffrey's eyes, the closer Jeffrey's position will be to an eliminativist one of the same breed as Churchland's in option (i).

include a great lot of scientific and philosophical talk of belief. And, secondly, he would have to show that expressing the same insights using the original categorical concept of belief would be too misleading to be maintained. All of that is definitely conceivable, but it will need a lot of work to be shown: work that has not been done, at least not yet. At least as things stand, elimination seems too costly to be true. Which leaves us with the non-eliminative reductionist account from before.

As I have mentioned already, none of these arguments comes with more than just prima facie or defeasible support: prima facie, the assumption that both concepts of belief refer to the same phenomenon seems to support the reducibility of categorical belief to numerical belief. *As things stand*, eliminating the categorical concept of belief is too costly.

What would it take to undermine such defeasible arguments while still assuming option (ii)? Other than independent support for Jeffrey's eliminativist proposal, or arguing that there is a third coreferential concept of belief to which both categorical and numerical belief give way, there would be yet another way of turning the tables: showing that the reduction must actually proceed in the opposite direction. Perhaps it only *seemed* to be the case that the numerical concept of belief was the more fine-grained, sophisticated, and 'scientific' one. What if graded belief could itself be understood in terms of all-or-nothing belief[52] or maybe in terms of a combination of all-or-nothing belief and some 'worldly' numerical concept (e.g. flat-out belief in the objective chance of *X* being so-and-so)?[53] Call this the *Elimination by Reduction Option* concerning *graded* belief. Obviously, this would relieve the pressure on all-or-nothing belief as being the theoretically inferior concept, and instead it would now be quantitative belief that would have to be regarded as deriving from its categorical counterpart. I will not discuss this any further; the prospects of any such proposal would have to be judged by its exact details.[54]

Instead, let me turn to option (iii) now.

1.3.3 The Irreducibility Option (iii): Both Concepts of Belief Refer, But Not to the Same Phenomenon

According to this final case (iii), both the all-or-nothing concept of belief and the concept of degree of belief refer to natural phenomena, but to numerically distinct ones. So there must be (at least) two types of belief states: one is the categorical belief type, the other one is the degree of belief type. Neither of the two concepts of belief is (conceptually) reducible to the other, and neither of the two states of belief is (ontologically) reducible to the other. Call this the *Irreducibility Option* concerning

[52] See Harman (1986, ch. 3) for a proposal of that kind.

[53] See Holton (2014) for a corresponding proposal.

[54] But see Frankish (2009, s. 2) for some objections to the viability of any such reduction of partial belief to flat-out belief.

categorical belief and numerical belief. Just as with the previous options, this option also allows for several, mutually exclusive, specifications.

Here is one, which might be called the *Anomalous Monism Option* about belief and degrees of belief, in analogy with Donald Davidson's famous anomalous monism about the mental and the physical realm.[55] According to this option, the term 'phenomenon' in the title of this section is actually ambiguous between 'Both concepts of belief refer, but not to the same phenomenon *type*' and 'Both concepts of belief refer, but not to the same phenomenon *tokens*': the Anomalous Monism Option would hold the former but not the latter. The thought would be: while belief talk and degree of belief talk give rise to fundamentally different ways of classifying natural occurrences, and while categorical belief states and graded belief states are distinct qua types of states, they would still coincide on the level of tokens. For instance, in the simplest case: if a person believes X at time t, and if she also assigns a certain degree of belief to X at that time t, then there might be *one* mental state *token* that instantiates both the person's belief in X and her assignment of a degree of belief to X at that point of time—hence the 'monism'. However, because of the severe differences on the levels of concepts and types, there would not be any laws by which categorical belief and numerical belief would relate to each other—which is the 'anomalous' aspect of this position. All of this might be compatible even with some form of supervenience or dependency of one kind of belief on the other, for instance, of all-or-nothing belief on numerical belief, just as Davidson claims mental state types supervene on physical state types: there could not be any difference concerning all-or-nothing belief without a difference concerning numerical belief.[56]

Obviously, just as in the corresponding discussion of Davidson's original anomalous monism, a lot of this would be open to debate: the exact meaning of supervenience, the question whether there can be supervenience without reduction, the question whether the 'x is nothing but (nothing over and above) y' phrase that I have used before would be compatible with supervenience without reduction, and so forth. I will not enter the discussion of any of these topics here.

For my own purposes it is more important to point out that there is also a strong *dis*analogy between the 'belief vs degrees of belief' case and the 'mental vs physical' case, which casts some doubts upon 'anomalous belief monism' that go beyond those concerning Davidson's original anomalous monism: it is quite simply far from clear that the qualitative concept of belief and the quantitative one belong to 'fundamentally different' classification systems (as Davidson's argument would require). In fact, unlike

[55] Davidson (1970) is the first presentation of the theory, which later got extended and modified in various respects.

[56] On the normative side, there are indeed theories of rational belief (or acceptance) and subjective probability according to which an agent's complete belief state is a function of—and in that sense *supervenes on*—degrees of belief: Lin and Kelly (2012a, 2012b) are an example. I will return to this in Appendix C, where I will give two arguments for the thesis that an agent's rational all-or-nothing belief set is *not* a function of the same agent's rational degree-of-belief function (or at least not of the agent's degree-of-belief function alone).

the mental vs physical concept pair, it seems the two belief concepts do have a lot in common: both seem to refer to intentional states (belief *in proposition* X, degree of belief *in proposition* X), both seem to be defined by definite descriptions that involve normative expressions (as explained before), and both seem to refer to mental states that are in the same 'business' (as also explained before). Which makes this Anomalous Monism Option, if anything, look less plausible than Davidson's.

So here is then an alternative manner of filling in the details of the present Irreducibility Option (iii): assume all-or-nothing belief and degree-of-belief assignments to be ontologically independent *both on the level of types and on the level of tokens.* One may be instantiated without the other one being instantiated as well. And yet both types of belief would be such that they aim at truth, influence action, and so on. Call this the *Independence Option* concerning all-or-nothing belief and graded belief.

Wouldn't any such 'double bill' account of belief be unfounded or excessive? Not necessarily. It is well-known from dual-process theories of the mind[57] that similar or even identical mental phenomena may well result through distinct mental processes along distinct mental paths. Similarly, what if the human mind had *two* belief systems: one for categorical belief, the other one for degrees of belief?

The corresponding hypothetical story about this option might, for instance, be told along the following lines. (See Figure 1.1 for an illustration.) Say, within one and the same cognitive system, there are two distinct belief systems or two 'belief boxes', as some cognitive psychologists would say. Say, in principle, one could even surgically

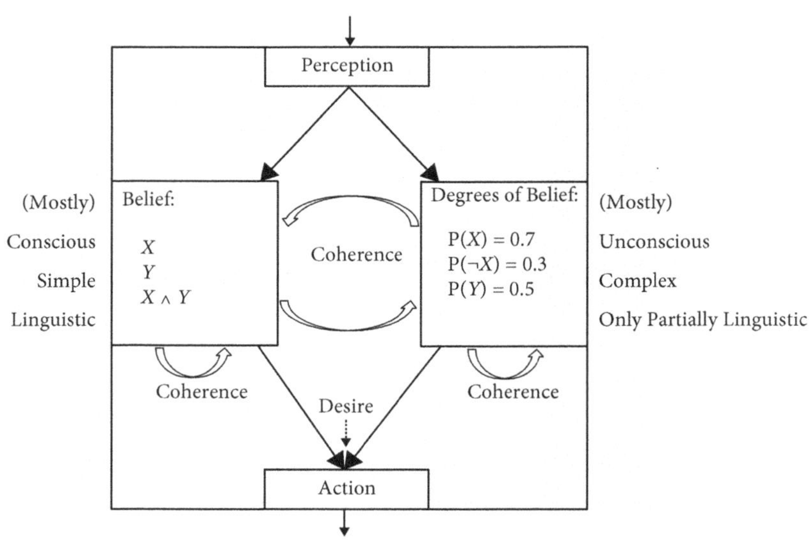

Figure 1.1. The Independence Option: an example

[57] See Evans (2008) and Frankish (2004) for overviews of such theories.

remove one of them without destroying the other. (Even though the remaining belief system might lose some efficiency as a result.) Each of the two systems would be fed by perception and other sources of evidence. Each of them would be able to commit the agent to action by collaborating with the agent's desires. Each of them would have its own coherence 'maintenance' mechanisms. Each of them would be capable of being expressed by assertion. And yet in other respects the two belief systems would differ.

For instance: the states produced by the categorical belief system could be structurally simpler than those of its degree-of-belief companion, much in the sense in which the categorical scale of measurement is simpler than the numerical scale. For the same reason, the categorical system would be easier to access consciously: introspection might work well for simple categorical beliefs while it might be hard, if not impossible, in the case of the much more demanding numerical beliefs. In the words of Foley (2009, p. 45), we would find ourselves 'overwhelmed' and 'inundated' if we had to deal consciously with the more finely qualified degrees of confidence. Which would also explain why it seems much easier to answer a 'Do you believe that *X*?' question than its 'What is your degree of belief in *X*?' counterpart. (I set aside all issues of the reliability of introspection here.) The categorical belief system might also be more intimately involved with language processing and reasoning in language, due to the discrete all-or-nothing structure of its states. On the other hand, the greater complexity of the degree-of-belief system would make it the superior decision-maker in complex situations (such as numerical betting scenarios). Accordingly, the two systems might have developed subject to different evolutionary pressures: while the flat-out belief system might have been selected for its simplicity, which was a prerequisite for conscious reasoning and the affinity with language, the degree-of-belief system might have been selected for its ability to act automatically and yet (instrumentally) rationally in complex environments. And so on and so forth; there might be further *differentiae specificae* by which the one belief system might differ from the other.

But both of them would still be systems that generate *beliefs*. Natural language, the language of science, and the language of philosophy would offer two distinct concepts of belief by which one may talk about beliefs generated by either of the two systems. Beliefs of both types would be 'explanatorily salient psychological states' (in the terms of Frankish 2009), which is also why it would be foolish to eliminate either of the two concepts by which we can refer to them. For any agent who possesses this kind of cognitive architecture, categorical belief and numerical belief will be ontologically independent—the agent can have one type of belief without having the other (even in the same proposition). What is more, to the extent that one of the two belief systems is capable in principle of functioning successfully without input from the other one, the system in question will also be *systemically independent* of the other one: it does not just exist independently of the other, it is also able to do its work successfully without the other. This would be one conceivable instance of the present Independence variant of the Irreducibility Option (iii).

All of that is quite close to how Frankish (2004) contrasts what he calls 'strand 2' (flat-out) beliefs with 'strand 1' (partial) belief, or belief in the 'supermind' with belief in the 'basic mind'. However, there is also an essential difference: while Frankish regards strand 2 beliefs to exist *in virtue of* strand 1 beliefs—flat-out beliefs to be 'realized in' partial beliefs, which is an instance either of our previous Reduction Option (ii) or of the Anomalous Monism version of the Irreducibility Option (iii)—the present Independence version of the Irreducibility Option (iii) considers the two 'strands' of belief to be ontologically independent even on the level of mental state tokens.

Ross and Schroeder (2014) also defend a non-reductive account that belongs to the current broad Irreducibility Option (iii).[58] They regard all-or-nothing belief in X as a defeasible disposition to treat X as true in reasoning, where 'reasoning' includes practical reasoning based on (probabilistic) degrees of belief along the lines of standard decision theory. So their view is close to Levi (1980), Frankish (2004), Weatherson (2005), and Fantl and McGrath (2009) in thinking that believing or accepting a proposition involves taking the proposition as a premise for (certain types of) reasoning, but without committing themselves in any way to Frankish's, Weatherson's, and Fantl and McGrath's inclinations for reducing categorical belief to degrees of belief and additional practical features. (Levi does not hold a reductionist view either.) Ross and Schroeder's view also seems to be consistent with the dual-belief Independence variant of option (iii) that I explained before in some detail, but of course they do not commit themselves to anything like it.[59]

Let me return one last time to that Independence variant of option (iii) again. Let us assume just for the sake of the argument that we are dealing with agents whose beliefs are indeed distributed over two independent belief systems of the type described. Then, in spite of their independence, the two belief systems would normally have to *harmonize* with each other: after all they would still be serving one and the same cognitive agent, and if they differed regularly in their recommendations to that agent, the agent would be bound to face serious normative dilemmas. One system would tell the agent to do A, whereas the other system would recommend doing not A: how should that conflict be resolved? That is where coherence enters the picture. The Humean thesis on belief that I will introduce in Chapter 2 will constitute a coherence norm that may be applied in such a case. If a dual-belief system satisfies that Humean thesis, then the two belief systems will be in a special kind of stability or equilibrium state to which each of the two systems may contribute individually (where the extent of their individual contributions will depend on the context), and where their individual

[58] Weisberg (n.d.) defends another non-reductive view of belief that belongs to the present category.

[59] Ross and Schroeder's view also differs from the account of belief that I am going to develop starting in Chapter 2. In section 6.3 of Chapter 6 I will deal with acceptance of X as a mental process in which X is taken as a premise for reasoning, including practical reasoning based on degrees of belief. But I will regard the resulting state of acceptance (and also what I will call 'accepted belief') as distinct from all-or-nothing belief. I understand Ross and Schroeder to give an account of acceptance or accepted belief rather than belief itself.

contributions 'fit together' (sufficiently). For instance, as we will see in section 2.3.3, the Humean thesis will imply that the set of actions that are rationally permissible in the sense of decision theory based on all-or-nothing beliefs always has non-empty intersection with the set of actions that are rationally permissible in the sense of Bayesian decision theory. Indeed, the former will always be a superset of the latter. So while Bayesian decision may be able to make more specific recommendations, it will never happen that the degree-of-belief system permits an action that is not also permitted by the all-or-nothing belief system. In this sense, the two systems will be practically coherent. I am going to study many further such aspects of coherence between all-or-nothing beliefs and degrees of belief in this essay. And, as mentioned before, how exactly beliefs and degrees of belief are going to cohere will be seen to depend also on the context in which e.g. the decision-making takes place: what the agent is interested in, how cautious she wants to be in her decisions, and the like.

This essay will be an extended defence of the Humean 'stability thesis' of Chapter 2 being the right coherence norm for belief and degrees of belief. But of course other coherence norms and corresponding states of coherence are conceivable, too: as always, it is a matter of argumentation. In any case, if an agent exemplifies a dual-belief architecture as described before, something must be in place that makes sure doxastic 'Buridan's ass' situations are avoided, at least normally and to sufficient extent.

Clearly, this Independence variant of the Irreducibility Option (iii) is purely hypothetical. But it is at least conceivable and, at best, a serious candidate hypothesis about what the cognitive architecture of belief and degree of belief might look like, were option (iii) the case.[60]

I will not be able to make any informed proposal on which of the two broad options (ii) or (iii) is the *more* plausible one. And I should not, as ultimately this is a scientific question: it is a matter of empirical investigation by cognitive psychologists, neuroscientists, and more, whether *the (all-or-nothing) propositional attitude the function of which is . . .* is identical to *the (graded) propositional attitude the function of which is . . .* To the best of my knowledge, science has not determined an answer as yet.[61]

[60] In order to work out this kind of Independence proposal in full detail, lots of additional questions would have to be answered. For instance: is there a corresponding all-or-nothing desire box and a corresponding graded desire box? My initial guess would be: yes, and they are close companions to the respective belief boxes on the same scale. Or: is there another system for belief on the comparative or ordinal scale? My preliminary answer would be: both the all-or-nothing belief box and the degree-of-belief box come equipped with mechanisms for comparative beliefs already. On the categorical side, the relevant notion is conditional belief, which I will deal with in Chapter 4. As will become clear from that chapter, if there is an all-or-nothing belief system at all, then it should be viewed as a system for *conditional* all-or-nothing belief from the start. On the numerical side, the notion in question is qualitative probability or probability orderings of propositions, as explained e.g. in section 5.2.1. Once again these two kinds of comparative belief need to cohere with each other, and what coherence for them might amount to will be explicated in Chapter 5 (from section 5.2.1 onwards).

[61] As far as philosophers voicing their opinions on this matter are concerned, e.g. Weatherson (2005, p. 420) is pessimistic about the Independence variant of option (iii): 'There is no evidence to believe that

That said, Appendix C will make an attempt at getting a bit closer to an answer on normative grounds—so far as the belief and degrees of belief of *perfectly rational* agents are concerned. I will give two reasons there to believe that rational categorical belief does not supervene on rational degrees of belief alone, which, presumably, would then also rule out that rational categorical belief reduces to rational degrees of belief being so-and-so. For that reason, it does seem to me that the Independence Option from (iii) might fit the normative account of belief that will be developed in this book better than the Reduction Option from (ii) does.[62] One might think that this could not be so because of the usual logical barriers to is–ought inferences: one cannot derive logically a non-trivial normative statement about belief from a descriptive premise about belief, or so it seems; contrapositively, one cannot derive the denial of a descriptive premise about belief (lack of supervenience for all agents) from the denial of a normative statement about belief (lack of supervenience for perfectly rational agents). But if a descriptive premise is so general that it concerns belief by all agents whatsoever, then it also concerns beliefs by *perfectly rational* agents.[63] This is like: if a statement holds in all metaphysically possible worlds, then it also holds in all normatively perfect worlds (assuming they are all metaphysically possible). Accordingly, a normative statement about belief, if phrased as a statement about perfectly rational agents, may well contradict a general statement about all types of agents whatsoever and their beliefs. In particular: if *perfectly rational* agents' beliefs can be argued not to be reducible to such agents' degrees of belief, then this does yield an argument against the *general* reducibility of belief to degrees of belief.[64]

If I had to put my cards on the table, I would regard the Independence Option from (iii) to be the most plausible one for actual human agents. But it is quite clear that a normative theory of rational belief and degrees of belief such as the one to be developed in this book will not be able to settle all of the conceptual and metaphysical issues to do with belief. Various different conceptual and metaphysical proposals about belief and degrees of belief will ultimately be compatible with the normative theory in this book, and I will not be able to drive a wedge between any of them.

the mind contains two representational systems, one to represent things as being probable or improbable and the other to represent things as being true or false.' But there is no evidence for a contrary thesis either. At this point, any empirical verdict about this matter is premature.

62 However, my Appendix C will leave open whether rational all-or-nothing belief supervenes on, and perhaps reduces to, rational graded belief taken together with certain other aspects of a perfectly rational agent's mental state, such as their attention, interests, and the like.

63 I am assuming here that perfectly rational agents belong to the universe of discourse over which we are quantifying. For instance, I might get this by assuming the universe of discourse to be the same at every possible world, and hence for perfectly rational agents to exist (though presumably as non-concrete entities) even at the actual world.

64 There are also weaker forms of prima facie pressure that can proceed from is to ought: if a normative theory of belief were such that most real-world human agents would turn out to be highly and systematically irrational in most circumstances, then this should put at least some weak prima facie pressure on any such normative theory.

It is time to sum up. If both the categorical belief concept and the degree-of-belief concept refer, which I assume to be the case, then either they refer to the same mental phenomenon, or the two concepts refer to distinct phenomena. In the former case, it is plausible that categorical belief is reducible to, and indeed nothing but, numerical belief or an aspect thereof. Yet the concept of all-or-nothing belief should not be eliminated, as it allows for a simpler and more commonsensical way of speaking about degrees of belief, which might still be useful for certain purposes. In the other case, it is quite plausible that the two belief phenomena are indeed ontologically independent, even when they are normally co-located within one and the same human cognitive system. The greater simplicity of categorical beliefs might then come in handy for conscious access and language-related purposes, while the greater complexity of degree-of-belief assignments might prove more useful for complex decision-making.

So much for the metaphysical and the conceptual side of belief. Now let me approach the actual topic of this book: the normative side of belief.

1.4 Norms for Belief: How Should Beliefs Cohere?

Both epistemology and decision theory are concerned in parts with norms involving belief. The relevant norms in epistemology are epistemic in the sense of the 'Truth-Aiming' Assumption 2 from before: they are meant to guide belief to truth or they can be used to evaluate belief with regard to truth. In contrast, the respective norms from decision theory are pragmatic as required by the 'Action' Assumption 3 (and the 'Assertion' Assumption 5): they are supposed to guide belief and desire to rational action (including linguistic discourse) or they are applied to evaluate them in that respect. Finally, the coherence among beliefs, as considered by the 'Integration' Assumption 4, ought to be subject to both epistemic and pragmatic constraints at the same time: beliefs ought to relate to each other so that they aim at the truth and facilitate rational action.

In previous sections, I characterized belief as the natural mental state that obeys all of these norms at least to a great extent and in normal circumstances. From the next chapter onwards, the norms themselves will take centre stage, and in particular those concerning the coherence of belief. Such coherence norms (in their strict, non-defeasible versions) tell us what one ought to believe, what one is permitted to believe, and what one is forbidden to believe, *given certain belief circumstances*. Or, as I will often prefer to say, but which I regard as equivalent: what a *perfectly rational* agent must believe (what is necessary for such an agent to believe), what a perfectly rational agent can believe (what is possible for such an agent to believe), and what a perfectly rational agent cannot believe (what is impossible for such an agent to believe), given certain belief circumstances. This is premised on the normative operators in question being sufficiently idealized and on perfectly rational agents being subjects that ordinary agents ought to approximate, as it were, in the ideal limit.[65]

Perfectly rational agents are meant to relate to actual human agents in an analogous way as morally perfect worlds relate to the actual world in the semantics of deontic logic: ultimately, one is interested in normative constraints on the actual case, but expressing these constraints by means of properties of ideal cases is often a helpful simplification. Or to use yet another analogy: it is sometimes relatively easy to describe the limit value that a real-valued function or sequence approximates, while it would be very hard to describe *how* exactly the function or sequence does so. By talking about perfectly rational agents without dealing much with *how* real-world agents ought to approximate them, I am sweeping lots of interesting, important, and complex normative issues under the carpet.[66] Methodologically I see this as the instantiation of a divide-and-conquer strategy. Determining how a *perfectly rational agent's* beliefs and degrees of belief cohere with each other will prove to be tricky enough, and coming up with a (hopefully) good proposal should certainly constitute some kind of progress. This does not mean that it would not be important to complement such findings later with a story of how the beliefs and degrees of belief of *actual* human agents ought to relate to those of these 'ideal limit' agents. In fact, this is an extremely important topic, but very different questions arise from it—ones that I will, fortunately, be able to bracket in what follows. Questions, such as: what if an ordinary agent *cannot* perfectly approximate an ideal agent due to its cognitive limitations and hence *should* not perfectly approximate an ideal agent? (Assuming a less idealized sense of 'should', and given the validity of a corresponding Ought–Can principle.) What partial approximation will then be the normatively right one? In particular: which shortcuts is a boundedly rational agent allowed to take for the sake of 'satisficing'?[67] All of this needs to be dealt with, ultimately, but not in the present monograph in which I will deal almost exclusively with the ideal limit case directly.[68]

That said, perfectly rational agents in my sense also share various properties with real human agents: in particular, as far as their beliefs are concerned, they are meant

[65] Williamson (2000, ch. 10) gives reasons that undermine the closely related equivalence between *evidential probabilities* on the one hand and *subjective degrees of belief of a perfectly rational being* on the other: but these reasons are to do with evidential probabilities about subjective probabilities—the second-order evidential probability of propositions about the first-order subjective degrees of belief of certain agents to be so-and-so. Since I will focus solely on first-order beliefs about the 'world' in this monograph, maintaining the analogous equivalence that is required in my case should be rather unproblematic.

[66] So far as rationality constraints on belief are concerned, some of them are discussed in Harman (1986), MacFarlane (2004), and Steinberger (n.d.).

[67] Simon (1956) is the original source of much of the corresponding literature on bounded rationality.

[68] I interpret the arguments given by Harman (1986, ch. 2) as supporting the thesis that it is very hard to state general logical (or probabilistic) norms on the beliefs of real-world human agents. This is due to the great number and variety of circumstances that might inflict ceteris paribus clauses on any such general proposal. (However, see Field 2009 for an attempt at such a proposal.) But that does not rule out that it might still be relatively easy to say how logic and probability theory relate to the rationality of *perfectly rational* agents. And it does not mean either that it would not be important to settle that question.

to satisfy all of the assumptions on belief as well as all of the assumptions that are yet to come. The same applies to metaphysical considerations: for example, there is an Independence specification of option (iii) from above that applies to perfectly rational agents in the same manner in which it applies to real human agents; if it holds, then the all-or-nothing beliefs of perfectly rational agents will be ontologically independent of their degrees of belief. And so forth. I should also stress that my perfectly rational agents will not have to be ideal in each and every respect: for instance, they are not assumed to be omniscient or omnipotent. Their beliefs do not have to be instances of knowledge. Their degrees of belief do not have to track objective 'worldly' chances. And so on. Instead, by 'perfectly rational' I only mean that these agents perfectly satisfy the *coherence ideal* in the 'Integration' Assumption 4 from section 1.1. One might also say: the agents that I am going to deal with are only assumed to be perfectly rational *inferentially* in satisfying various logical closure conditions on categorical belief, various probabilistic closure conditions on degrees of belief, and some bridge postulates for how their categorical beliefs and degrees of belief relate to each other. More about this in sections 1.5 and 1.6.

In other respects, my perfectly rational agent may differ substantially from the more ordinary ones. For instance: assume the Independence variant of the Irreducibility Option (iii) again to apply both to human agents and to the perfectly rational agents that they ought to approximate (in some sense). Now the following might be the case: if one were to surgically remove the all-or-nothing belief system of a *human* agent, then her overall performance would suffer drastically. This might be so because the human degree-of-belief system would regularly be incomplete (certain propositions not having a degree of belief) or because it would be affected by other shortcomings, and the human all-or-nothing belief system might play a major role in filling in these gaps and in helping to sort out some of these shortcomings. However, if one were to remove the all-or-nothing belief system of a *perfectly rational* agent, then her overall performance might not be hampered at all. Because of the perfect state of such an agent's degree-of-belief system, every task that might have been carried out by the agent's categorical belief system could always be taken over, at least in principle, by the numerical system itself. I will leave open whether this is actually so, but it is certainly conceivable and maybe even likely, since the degree-of-belief system is likely to be more complex than its categorical counterpart. And if things were like that, then human agents and the ideal agents after whom they ought to strive would indeed differ at least in terms of some of their *modal* properties.

But note that even if this were so, it would not show that the belief system would be redundant for *human* agents. Nor would it show that perfectly rational agents could not have had a categorical belief system in the first place: after all, redundancy does not entail non-existence (nor irrationality). Nor would it show that when we describe in a normative theory the type of perfectly rational agent that human agents ought to approximate, these perfect agents would necessarily lack all-or-nothing belief systems: since the all-or-nothing belief systems of human agents might play a crucial role in the course of their approximating ideal agents, it might be much more useful to think of these ideal agents as having an all-or-nothing belief system, too, which could then be approximated by their human counterparts. *Approximate an ideal agent without a categorical belief system* is not particularly helpful or informative if one has a categorical belief system and needs to use it. *Approximate an ideal agent whose categorical and numerical belief systems relate to each other in such-and-such a way* is much more to the point. It would only be once a human agent has actually reached the state of perfect rationality (if they ever do), that their system of categorical beliefs would have become superfluous. Another analogy might help here: clearly the existence of referees is vital for football matches by ordinary football players. But in terms of fairness, ordinary football matches should be such that in principle referees would be superfluous. Were we to describe the normatively ideal football game, it might be still be useful to include a description of the referees, if only because the existence of referees will still be crucial in order for actual football games to approximate ideal ones.

Let me now turn to the coherence ideal for rational belief in more detail. If we combine the 'Integration' Assumption 4 from section 1.1 with the 'Concepts of Belief' Assumption 6 from section 1.2, then, depending on how the term 'belief' is interpreted, we are led to three different ways of making Assumption 4 more precise.

The first one is concerned solely with the coherence of all-or-nothing beliefs:[69]

> Assumption 4′: An agent's *all-or-nothing* beliefs are subject to an ideal of integration. Other things equal one should be able to agglomerate one's various *all-or-nothing* beliefs into a larger, overall view; and this larger view should satisfy demands for consistency and coherence.

The second one is about the coherence of degrees of belief:

> Assumption 4″: An agent's *degrees* of belief are subject to an ideal of integration. Other things equal one should be able to agglomerate one's various *degrees* of belief into a larger, overall view; and this larger view should satisfy demands for consistency and coherence.

And the third one pertains to the coherence *between* an agent's all-or-nothing beliefs *and* her degrees of belief—assuming the agent has both of them, as seems to be the case with human agents:

> Assumption 4‴: An agent's *beliefs* and *degrees of belief* are subject to an ideal of integration. Other things equal one should be able to agglomerate one's various *all-or-nothing* beliefs and *degrees* of belief into a larger, overall view; and this larger view should satisfy demands for consistency and coherence (between *all-or-nothing* beliefs and *degrees* of belief).

[69] In each case, I will paraphrase the original quotation from Bratman (1999, p. 17).

Given the 'Reference' Assumption 7 from before, none of these assumptions is empty. The 'Belief Integration' Assumption 4′ states one of the defining features of all-or-nothing belief which we assume to be a real phenomenon. The 'Degree of Belief Integration' Assumption 4″ does the same for degrees of belief. And the idea will now be that the 'Belief vs Degree of Belief Integration' Assumption 4‴ ought to be counted as constitutive of belief, too, but this time of both kinds of belief simultaneously.

If the Reduction Option (ii) from the last section obtains, then Assumption 4‴ should hold for trivial reasons, or if one prefers, for *conceptual* or *metaphysical* reasons: in that case, the two concepts of belief refer to the same phenomenon or aspects of the same phenomenon, which is why there is no question about the coherence of the phenomena themselves (or rather, of *the* phenomenon). Of course, 'they' need to cohere! Similarly, since neither of the two belief concepts is to be eliminated, as we argued in the last section, talking about that phenomenon in all-or-nothing belief terms ought to cohere with talking about it in degree-of-belief terms at least to a great extent and in normal circumstances. What we have to determine then, as epistemologists or decision-theorists, is a consistent, plausible, and unified normative manner of speaking about the one belief phenomenon in *categorical* and *numerical terms* simultaneously.

On the other hand, if the Independence Option in (iii) from the last section is the right one, then the 'Belief vs Degree of Belief Integration' Assumption 4‴ applies in a more substantial sense: in that case, the two concepts of belief refer to distinct phenomena, but as long as these two phenomena coexist within one and the same agent, they better cohere with each other at least to a great extent and in normal circumstances. For by their very nature, they are in the same 'business': both aiming at the truth, both committing the agent to action, and so on. As independent as the two kinds of belief states may be ontologically, their underlying belief systems serve one and the same agent when fulfilling these functional roles, and if they do not cooperate with each other while fulfilling these roles, the agent whom they are meant to serve might turn out to be incoherent overall. For any such agent, there must be a consistent, plausible, and unified system of norms for her *categorical beliefs* and *numerical beliefs* simultaneously.

This leads us, finally, to the central question of this book:

> What do a perfectly rational agent's beliefs and degrees of belief have to be like in order for them to cohere with each other?

Finding an answer to this question will tell us more about the concept of belief and the nature of belief: this is so because belief was defined as the propositional attitude the function of which is to reach the goal so-and-so and to satisfy the norms so-and-so and to realize the valuable state so-and-so (in the sense of our Assumptions), or to achieve all of that at least to a great extent and in normal circumstances. In other words: belief is the attitude that approximates, to a great extent and in normal circumstances, belief by a perfectly rational agent. And answering the question above will tell us more about beliefs held by perfectly rational agents.

Even more importantly, finding an answer will help us formulate epistemological and decision-theoretic norms that will jointly apply to our beliefs and degrees of belief such that ultimately both of them will aim at the truth, support rational action, and cohere with each other. So there are normative, conceptual, and metaphysical reasons to be interested in the question. The goal of this book is to develop and defend an answer to the question in terms of what I am going to call the *Humean thesis on belief* (and its equivalents): a stability conception of rational belief.

1.5 The Route to an Answer

Of course, there are different possible ways of approaching any such comprehensive account of coherence for belief and degrees of belief. For instance: one might first determine coherence as understood in the 'Degree of Belief Integration' Assumption 4″ from section 1.4 (e.g. by assuming the axioms of subjective probability and more). Then one would somehow determine coherence as understood in the 'Belief vs Degree of Belief Integration' Assumption 4‴. That is: what coherence between all-or-nothing beliefs and degrees of belief would have to be like. And finally one would somehow try to derive from these two assumptions a notion of coherence as understood in the 'Belief Integration' Assumption 4′—coherence among all-or-nothing beliefs. The obvious problem with this strategy is that we do not really know as yet what coherence between all-or-nothing beliefs and degrees of belief is meant to be like. So one of the premises would need to be supplemented first. Indeed, that is precisely what I am going to do in Chapter 2: amongst other things, I will derive coherence postulates for rational all-or-nothing belief—e.g. the logical closure of rational belief—essentially from well-known coherence assumptions for rational degrees of belief together with a new proposal concerning the coherence between the two kinds of belief: the Humean thesis on belief.

In the other chapters of this book my strategy will be a different one: first I will determine coherence as understood in the 'Belief Integration' Assumption 4′—coherence among categorical beliefs. Then I will determine coherence as understood in the 'Degree of Belief Integration' Assumption 4″—coherence among degrees of belief. And from this, in conjunction with some auxiliary hypotheses, I will aim to derive what coherence as understood in Assumption 4‴ must be—coherence between categorical beliefs and degrees of belief. In a sense, this way of proceeding will be more straightforward, as I will be able to build on already existing and sufficiently detailed accounts of coherence for categorical beliefs and for degrees of belief taken separately. With respect to these existing accounts I will simply pick the standard default options that are available in the relevant literature.

Let me put these default proposals on record for now as my final two assumptions (the exact details of which will be filled in later by other chapters). The 'Belief Integration' Assumption 4′ is usually specified as follows:

> Assumption 8: The coherence norms on all-or-nothing belief are precisely what the canonical literature on the logic of belief takes them to be: (a) synchronically, the set of beliefs of a perfectly rational agent is consistent and closed under logic (in the sense of doxastic or epistemic logic; cf. Hintikka 1962).[70] (b) Diachronically, belief change of a perfectly rational agent is governed by the axioms of belief revision (in the sense of so-called AGM belief revision: cf. Alchourrón et al. 1985 and Gärdenfors 1988).

I will neglect the diachronic part of Assumption 8 for the time being (which will be the topic of Chapter 4). But it is easy enough to sketch at least the synchronic aspect of Assumption 8: the set of propositions believed by a perfectly rational agent at a time is meant to include all logical laws; it is supposed not to include any logical contradictions; for every believed proposition X it is taken to include every proposition Y that follows from X logically; and finally the agent's belief set is assumed to be closed under conjunction. Closure under conjunction of a perfectly rational agent's belief set at a time means: if propositions X and Y are believed by the agent at the time, then also their conjunction $X \wedge Y$ is believed by the agent at the same time (and the same must apply then also to the conjunction of any *finite* number of believed propositions). All of these assumptions taken together constitute, or rather are equivalent to, the combined consistency and logical closure of such an agent's set of beliefs.

In this essay I will be following e.g. Stalnaker (1984) in taking propositions to be sets of possible worlds. This is mainly for simplicity, but also because some of the formal theories on which I will build make the same assumption.[71] The set of all worlds, which I will always denote by 'W', should be thought of as the set of *logically possible* worlds, but where possibilities may be individuated in a rather coarse-grained manner; more about this below. I will always denote propositions by capital letters, such as 'X', 'Y', 'Z', or 'A', 'B', 'C'. A proposition is then a contradiction if and only if it is the empty set of worlds; logical entailment between propositions is the subset relation between sets of worlds; and the conjunction of propositions is given by their intersection. If we use slightly more formal terms, e.g. the closure of rational belief under conjunction can, then, be stated as: if $Bel(X)$ and $Bel(Y)$, then $Bel(X \cap Y)$. Occasionally, I will also write '$X \wedge Y$' instead of '$X \cap Y$' (as I did in the previous paragraph)—and the like. Belief in a proposition, such as the set X or the set Y of worlds, corresponds to the belief that the

70 The standard literature on doxastic and epistemic logic, such as Hintikka (1962), is very much concerned also with introspective belief and knowledge: logical principles of positive or negative introspection. In this essay I will leave out considerations on introspective rational belief. Therefore, for my purposes, none of the typical principles on introspection will have to be included in Assumption 8.

71 So I will not deal with more demanding accounts of propositions as structured entities or the like, except for a brief remark on propositions as hyperintensions in section 3.3 of Chapter 3.

actual world is amongst the members of that set. While real human agents normally need to draw inferences, that is, perform certain mental processes in order to close parts of their belief sets under conjunction—before the inference they did not believe $X \cap Y$ as yet, but they do so after the inference—a perfectly rational agent's belief is already closed under conjunction from the start, and that will be so at any point in time. In this sense, closure under conjunction is a *synchronic* property of a perfectly rational agent's belief set, as is logical closure more generally.[72]

One might think that assuming logical closure in this sense would be overly demanding even for perfectly rational agents, as the set of propositions believed by a perfectly rational agent might be thought to be necessarily infinite in that case. But, first of all, demanding even an infinite set of beliefs might well be forgivable in the case of a properly *perfectly rational* agent. Secondly, the consistency and logical closure of a belief set do *not* just by themselves entail the existence of infinitely many believed propositions: for example, an agent who only believes the proposition that is given by the set of all worlds already satisfies consistency and logical closure—for the agent does not believe the empty set, and the only proposition entailed by the believed set of all worlds is the set of all worlds again. Thirdly, and most importantly, one might well take quantifiers such as 'all logical laws' or 'for every ... proposition' in the two paragraphs before to be restricted contextually to all laws or propositions *that can be expressed or apprehended given some finite coarse-grained space of possibilities*. Indeed, from the next chapter, and in most detail in Chapter 3, I will argue that normally, in everyday contexts of reasoning, even an inferentially perfectly rational agent may only attend to a coarse-grained partition of possible cases or distinctions. For instance, the agent may only be interested in which of three logically independent propositions A, B, C is the case, and not more, which would correspond to a *finite* space of $8 = 2^3$ coarse-grained 'worlds' that range from the $A \wedge B \wedge C$ case to the $\neg A \wedge \neg B \wedge \neg C$ case. This set W of eight worlds may be regarded as the set of all logically possible worlds for the language of propositional logic with precisely three propositional letters (which correspond to 'A', 'B', 'C'); it is just that the language itself will not play any role, only the worlds will do so. The agent thereby suppresses any *distinctions* between e.g. different $A \wedge B \wedge C$-ways the world might be, but she does not ignore any *way* the world might be: every such way belongs to one (and only one) of the coarse-grained worlds. The space of coarse-grained logical possible worlds corresponds to a partition of the space of maximally fine-grained logical possible worlds (if such exist at all).[73] If restricted in such manner, the synchronic part of the 'Logic' Assumption 8 will not

[72] With the exception of Chapter 4, I will not deal with infinite operations on propositions, since the underlying set of worlds will be assumed to be finite. However, in Chapter 4 I will assume that a perfectly rational agent's set of believed propositions is even closed under infinite conjunction or intersection.

[73] Here is an alternative understanding of the 'given' set W of possible worlds that I will *not* embrace for most of this essay: taking W to be the set of all logically possible worlds in a context in which the agent *takes certain propositions to be given already*. The agent has accepted them as premises, and in that context her reasoning will proceed under these premises. For instance, the agent might have accepted that $\neg A$, in which case all A-worlds will be excluded from W. I will deal with acceptance in that sense in section 6.3 of

be particularly demanding any more: there will be the unique logically true set of all (eight coarse-grained) worlds; the agent will not believe the uniquely determined, logically false, empty set of worlds; with every believed set of (coarse-grained) worlds the agent will believe each of its supersets (consisting at most of eight coarse-grained worlds); and if X and Y are sets of (coarse-grained) worlds believed by the agent, then so is their intersection. For instance, if our rational agent believes the proposition A, that is, the set $\{A \wedge B \wedge C, A \wedge B \wedge \neg C, A \wedge \neg B \wedge C, A \wedge \neg B \wedge \neg C\}$ that consists of four worlds, and if she also believes the proposition B, that is, the set $\{A \wedge B \wedge C, A \wedge B \wedge \neg C, \neg A \wedge B \wedge C, \neg A \wedge B \wedge \neg C\}$, which also consists of four worlds, then she also believes their intersection: the set $\{A \wedge B \wedge C, A \wedge B \wedge \neg C\}$. At least in such coarse-grained contexts, the usual worries concerning epistemic logic's assumption of 'logical omniscience'[74] lose much of their bite: there is just one logical truth to be rationally believed, and checking for logical implication amounts to a mere test for subsethood, neither of which is particularly delicate given a reasonably small number of coarse-grained possibilities.

There is a strong case to be made for the view that the consistency and logical closure of rational belief belong to the *default* assumptions about rational all-or-nothing belief. Standard epistemic and doxastic logic certainly assume them: starting with Hintikka (1962), through all of the treatments of rational belief operators based on normal modal axiomatic systems with their standard possible worlds semantics, to the (static subsystems of the) more recent accounts of dynamic epistemic or dynamic doxastic logic.[75] Other than philosophical logic, a great part of the tradition in epistemology and philosophy of science has emphasized the roles of consistency and logical closure: knowledge, or at least being in the position to know, seems to satisfy these constraints, cf. Williamson (2000). Hempel (1962) and Levi (1967) are early sources in the epistemology of belief or acceptance in which consistency and logical closure ('deductive cogency') are assumed. Scientific theories, which are presumably what scientists hold to be true, are reconstructed as consistent and deductively closed sets of sentences according to the logical empiricists' (such as Carnap's or Hempel's) syntactic view on theories.[76] Scientists themselves seem to logically draw conclusions from law-like hypotheses, auxiliary assumptions, observation results, and the like, and

Chapter 6, but I will also distinguish acceptance there from belief. So this is not my intended understanding of W as long as only belief is concerned.

[74] See e.g. Fagin et al. (1995, ch. 9) and Christensen (2004, ch. 6) for more on this.

[75] See e.g. van Ditmarsch et al. (2008) and Leitgeb and Segerberg (2007).

[76] One might worry here that every scientific theory might actually fail outside some bounded domain of applicability, and that scientists would actually know that some of the logical consequences they draw from even our best theories are false, though the theories themselves cannot tell us explicitly what the boundaries of their own regimes of applicability are. (I owe this worry to Erik Curiel.) I will be able to make sense of worries like that as follows. First of all, I will argue (as developed in detail in Chapter 3) that rational belief is sensitive to contexts, rational belief is always closed logically within a context, but one cannot always draw logical inferences from rational belief as given in one context to rational belief as given in another. Contexts in this sense might correspond, roughly, to domains of applicability of scientific theories. Similarly, acceptance—including the acceptance of a scientific theory—is context-sensitive and closed under logical

they seem to reason and act upon them. The same seems to be the case outside of the academic context: in the courtroom, judges or jury members are supposed to logically draw conclusions from police findings, witness reports, expert verdicts, and relevant background information. In everyday contexts of reasoning or argumentation, we seem to logically draw inferences based on perception, supposition, or communication, upon which we might end up acting. And in all of these cases something would be diagnosed to have gone wrong if a contradictory proposition were to be derived: a rule of rationality or a rationality commitment would have been broken.[77]

In a nutshell: the synchronic part of the 'Logic' Assumption 8 does come with significant prima facie support. Theories arguing *against* this synchronic part of Assumption 8 had better include an error theory of why at least in many of the cases mentioned before people are not irrational in doing what they do, or why it might at least appear as though they were not irrational in doing what they do. So much for now on the specification of the 'Belief Integration' Assumption 4′ in terms of Assumption 8.

Next I specify the 'Degree of Belief Integration' Assumption 4″:

> Assumption 9: The coherence norms on degrees of belief are precisely what the canonical Bayesian literature takes them to be (see e.g. Howson and Urbach 1989, Earman 1992, Bovens and Hartmann 2003): (a) synchronically, the degree-of-belief assignment of a perfectly rational agent satisfies the axioms of probability. (b) Diachronically, degree-of-belief change of a perfectly rational agent is given by conditionalization, that is, by taking probabilities conditional on the evidence (or by something that is reasonably close to conditionalization, such as Jeffrey conditionalization; cf. Jeffrey 2004).

While subjective probability theory is of course not the only normative theory of numerical degrees of belief,[78] it is clearly the default option again. As far as the literature concerning degrees of belief in epistemology and philosophy of science is concerned, the probabilistic view of rational degrees of belief has been the dominating paradigm at least since Carnap's (1950a) work on inductive logic, if not before (with work done by Frank P. Ramsey or Bruno de Finetti). There are famous pragmatic arguments for the thesis that rational degrees of belief must be governed by the axioms of probability, such as the classical Dutch book arguments or arguments based on decision-theoretic representation theorems.[79] But there are also epistemic

consequence only within a context: which will be worked out in section 6.3, where I am going to distinguish between belief and acceptance.

77 Indeed, Levi (2002) regards rationality postulates such as the consistency and logical closure of belief as expressing a doxastic *commitment* that a real-world agent carries around without necessarily being able to live up to it all the time.

78 The Dempster–Shafer theory (see e.g. Yager and Liu 2008) or Spohn's (2012) ranking theory are alternative accounts of degrees of belief; see various of the papers collected in Huber and Schmidt-Petri (2009) for more on this.

79 See Howson and Urbach (1989) and Earman (1992) for more on this.

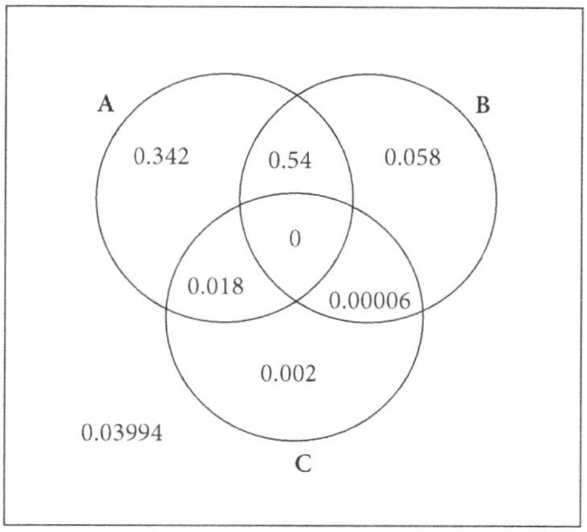

Figure 1.2. A simple probability measure

arguments for the same thesis that derive from theorems of the form: one's degrees of belief satisfy the axioms of probability if, and only if, they minimize inaccuracy, that is, they approximate truth to the greatest possible extent (in a sense that can be made formally precise).[80]

Without going into any details, and with a focus on simplicity of presentation, let me explain briefly the gist of subjective probability theory. Let W be our set of eight coarse-grained logically possible worlds from before. Then, for instance, Figure 1.2 depicts a probability measure P on this set W. (This example measure will reappear later in Chapters 3 and 4.) The eight elementary (largest undivided) regions in this Euler-Venn diagram represent the eight coarse-grained possible worlds that correspond to the logical combinations ranging from $A \wedge B \wedge C$ to $\neg A \wedge \neg B \wedge \neg C$ again. Throughout the book, when I discuss rational degrees of belief and all-or-nothing belief simultaneously, I will always assume that P and *Bel* 'inhabit' the same logical space: rational all-or-nothing belief and rational degrees of belief are given relative to the same set of worlds. Each of the eight coarse-grained worlds in our example is assigned a real number between 0 and 1 now (or a percentage between 0 and 100),[81] so that all of these numbers taken together sum up to 1 (or the percentages to 100). The degree of belief that is assigned to a proposition or set of worlds is given by the sum of numbers that are associated with the elementary regions in that set. For example, the degree of belief in B as being given by P in Figure 1.2 is $0.54 + 0.058 + 0.00006 + 0$ which is 0.59806 (or about 60 per cent). Accordingly, the degree of belief in $\neg B$, that is, the negation of B—the set of (coarse-grained) worlds that are not included in B—is 1 minus that number: 0.40194 (or about $100 - 60 = 40$ per cent). Clearly,

[80] More on such epistemic arguments and on what is meant by 'minimizing inaccuracy' can be found in Joyce (1998) and Leitgeb and Pettigrew (2010b). I will return to this literature in section 5.1 of Chapter 5. For a general survey on arguments for probabilism, see Hájek (2009).

[81] More precisely, one should say: each singleton set of any such world is assigned such a number.

if P is an agent's degree-of-belief function, then the corresponding agent regards it as more likely than not that B is the case, that is, as more likely than not that the actual world belongs to B. On the other hand, for instance, the agent's degree of belief in $A \wedge B \wedge C$ in this example is 0, which means that the agent rules out that possibility with complete certainty. The subjective probability of a logical truth or the set of all worlds is 1 (since all the numbers must sum up to 1), whereas the probability of a logical contradiction or the empty set is always 0 (as nothing gets summed up). If Y is entailed by, that is, is a superset of X, then the probability of Y must be greater than or equal to that of X. (If anything, more numbers are summed up in the case of Y than in the case of X.) Finally, if X and Y do not overlap at all—if they have empty intersection—then the probability of their disjunction or union $X \cup Y$ is the sum of the probability of X and the probability of Y. Or in slightly more formal terms: it holds that $P(X \cup Y) = P(X) + P(Y)$, which is called the finite additivity principle for probability measures. It is such principles that constitute the synchronic part of the 'Probability' Assumption 9. The degrees of belief that a perfectly rational agent distributes over propositions at a time are assumed to satisfy collectively the axioms of probability, such as the finite additivity principle. I should emphasize again that these axioms are not meant to describe how *actual* human agents distribute their degrees of belief, but only how perfectly rational agents do so. While e.g. an actual human agent might forget or ignore or simply fail to understand that A is tautological and thus not assign a degree of 1 to it, a perfectly rational agent will not be affected by any such epistemic impairments. This said, in recent years Bayesian psychology has also become one of the leading approaches to the psychology of actual human agents (cf. Oaksford and Chater 2007).

On the diachronic side, one feature that all the so-called (subjective) Bayesian or (subjective) probabilistic accounts of rational belief have in common is that they highlight the role of conditionalization. Suppose that an agent whose degree-of-belief function at a time is given by P in Figure 1.2 receives a piece of evidence: proposition C. How should her degree-of-belief assignment change given that piece of information? Here is the idea: first of all, all degrees of belief corresponding to $\neg C$-worlds are to be wiped out, that is, set to 0. By receiving evidence C the agent has learned to rule out $\neg C$ completely.[82] Since the original or prior probability of C was less than 1 in this case, the remaining numbers after ruling out $\neg C$ do not sum up to 1 any more. This can be corrected by multiplying each of the remaining non-zero degrees of belief of elementary regions with the same constant number, such that overall the resulting numbers after multiplication sum up to 1 again. It is easy to see that the constant factor that does the trick is determined uniquely: it is nothing but $\frac{1}{P(C)}$. (The prior

[82] It is this step that can be treated more 'softly' by means of Richard Jeffrey's refinement of conditionalization: Jeffrey's update allows for *partially* ruling out $\neg C$ by assigning merely a high degree of belief to C without that degree necessarily being equal to 1 or 100%. I will return to Jeffrey update in Appendices A and B.

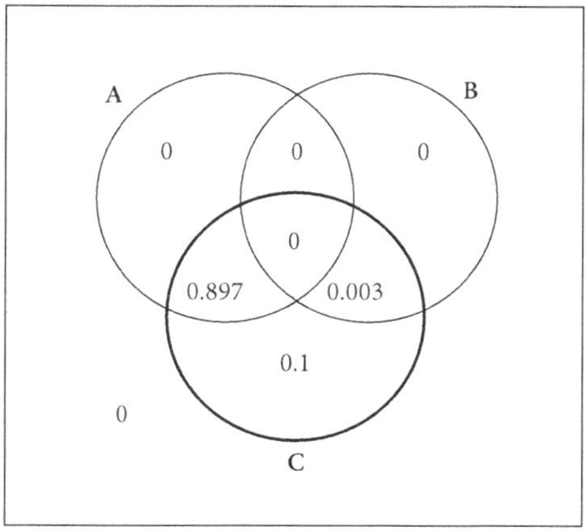

Figure 1.3. The same measure conditionalized on C

probability of C is greater than 0 in our example, so this fraction is well-defined.) The new or posterior probability measure that emerges from this update process is denoted by '$P(\cdot|C)$' and is called: the result of conditionalizing P on C. In our example, the corresponding probability measure is depicted by Figure 1.3. For instance, the posterior probability $P(B|C)$ of B after conditionalizing P on C is $0.003 + 0 + 0 + 0$, which is 0.003 or 0.3%. In this case, learning C has disconfirmed B significantly from the viewpoint of our agent, since her probability in B decreased from 0.59806 to 0.003 in the course of the update. One can also show easily that instead of presenting conditionalization as such a two-step process—setting the values of worlds that are ruled out to 0, and then 'renormalizing' values so that they sum up to 1 again—one can determine conditional probabilities more directly, and equivalently, by use of what is called the ratio formula: the conditional probability $P(B|C)$ of B given C is nothing else than $\frac{P(B\cap C)}{P(C)}$, and that is how conditional probabilities are actually defined officially in standard probability theory (when the denominator, $P(C)$, is greater than 0, or otherwise the ratio would be undefined). This should suffice as a first and preliminary explanation of the manner in which the 'Probability' Assumption 9 specifies the 'Degree of Belief' Assumption 4″ from before. More details and the exact statement of the axioms of probability can be found in any textbook on Bayesianism as well as in Chapters 3 and 4.

By using terms such as 'standard', 'default', and 'canonical' in Assumptions 8 and 9, I do not want to say that any of these norms are *sacrosanct* or beyond doubt: not at all. All of them have been attacked on various different grounds, and for all of them one finds multiple variations and alternatives in the literature (some of which I will mention in due course).[83] But then again, if the 'experts' on either side were asked

[83] For a recent survey of such attacks and defences, see Christensen (2004). Christensen himself ultimately argues against the 'Logic' Assumption 8 but in favour of the 'Probability' Assumption 9.

'What are the coherence norms on belief?', then Assumptions 8 and 9 would be their *first* answers. That should be reason enough to take them as the starting point in the search for coherence between the two types of belief. As far as the 'Logic' Assumption 8 is concerned, that is what I am going to do in the chapters after Chapter 2, and as far as the 'Probability' Assumption 9 is concerned, that is what I will do throughout the essay.

The strategy in Chapter 2 of determining coherence norms for *belief* from coherence norms on *degrees of belief* and from a coherence bridge norm for *beliefs and degrees of belief* might in principle have led to revisionary consequences—perhaps the norms on categorical belief might *not* have ended up including consistency and logical closure (although in fact they do in Chapter 2). On the other hand, my approach in Chapter 3 and in subsequent chapters is more conservative. There the rule will be: don't mess with the norms of either type of belief taken in isolation! These will be the rules of my game, and the game is defined by what seem to be reasonable and well-motivated assumptions. The game will be lost if, on these grounds, coherence cannot be determined jointly for flat-out belief and partial belief in a consistent, plausible, and unified manner. One of the challenges, therefore, will be to determine ways around some of the notorious paradoxes that seem to affect the joint coherence of beliefs and degrees of belief once Assumptions 8 and 9 from above are in place: most notably, the Lottery Paradox (Kyburg 1961) and the Preface Paradox (Makinson 1965), to which I will turn in some detail in Chapter 3 and section 6.4 of Chapter 6. (I will already state the Lottery Paradox, if only briefly, in the next section.) But I hope to show that ways around these paradoxes can be found, and the game will not be lost. Moreover, the win will be robust in the sense that even if only parts of Assumptions 8 and 9 are combined with each other, it is always more or less the same joint theory of rational belief and degrees of belief that will emerge. This will follow from the results in Chapters 3–6, not all of which will be based on precisely the same assumptions.

Before I turn to some proposals in the literature on how to specify the 'Belief vs Degree of Belief Integration' Assumption $4'''$, let me close with a few remarks on the 'Logic' Assumption 8 that are aimed at those who are familiar with the logic of belief but who are critical of it. Once again I will focus solely on its synchronic aspects here, that is: the assumption that perfectly rational belief is consistent and closed under logical consequence.

It is not a huge surprise that logicians who like possible worlds accounts of belief are generally fond of the logical closure and consistency of belief: essentially, the former is just a semantic way of expressing the latter.[84] And determining rational belief by partitioning logically possible worlds into those that are (for all that one believes) *live* possibilities—the so-called doxastically accessible worlds, or the viable candidates for being the actual world—and those which are not is certainly a highly appealing picture. It's a picture that conforms to the combination of two views: information is given by

[84] This will be made more precise in Chapters 2–4.

ruling out possibilities, and belief is given by having information (and being ready to act upon it).

But there are also other reasons for logical closure and consistency. Let me just mention one here: *simplicity*. As argued in section 1.3, whether the Reduction Option (ii) or the Irreducibility Option (iii) on belief is the correct one, one of the distinguishing features of categorical belief compared to numerical belief is its greater simplicity, which may give categorical belief an advantage for some purposes and a disadvantage for others. Considerations of simplicity are not 'merely' pragmatic here, as they pertain to one of the distinguishing features of all-or-nothing belief and hence to the nature of all-or-nothing belief. Now, simplicity is precisely what the standard logic of belief or the standard possible worlds semantics picture delivers: take propositions to be subsets of the set of all possible worlds again, and assume the set of all worlds to be finite (thus including sufficiently coarse-grained worlds). Then for each proposition—each set of worlds—our perfectly rational agent will either believe it or not believe it (that is, disbelieve it or suspend judgement on it). In principle, the resulting overall *set of believed sets of worlds* could become quite complicated. But given logical closure, that belief set can always be determined uniquely from *one non-empty set of worlds* consisting precisely of the live possibilities (as is easy to show). That is a simplification by one level in the cumulative hierarchy of sets: from a set of sets of worlds to a set of worlds. And that simplification is achieved by the logic of belief. I will return to this point in later chapters.[85]

The appendix to this chapter will add a diachronic consideration in favour of the most controversial feature of the logical closure of rational belief: closure of belief under conjunction. And the reason for logical closure that will be most relevant for my own purposes will be determined in Chapter 2: once rational belief is postulated to be sufficiently stable, the logical closure of belief will simply follow from this. Therefore, as mentioned before, the logical closure aspect of Assumption 8 will not so much be a premise in Chapter 2 but really a conclusion.

1.6 Bridge Principles for Rational Belief and Rational Degrees of Belief

The goal of this book is to specify the remaining 'Belief vs Degree of Belief Integration' Assumption $4'''$ from section 1.4: to determine the exact sense in which an agent's beliefs and degrees of belief are subject to an ideal of integration. In Chapter 2 I will defend a bridge principle about rational belief and rational degrees of belief to precisely this effect: the Humean thesis on belief. It says:

[85] Incidentally, an analogous point can be made about subjective probability. In principle, an *assignment of degrees of belief to arbitrary sets of worlds* could be quite complicated. But given the axioms of probability (and a finite total set of worlds), any such assignment is determined uniquely by an *assignment of degrees of belief to worlds* (rather than to sets of worlds).

The *Humean Thesis*: It is rational to believe a proposition just in case it is rational to assign a stably high degree of belief to it.

It is a bridge principle for qualitative and quantitative belief in the sense that it is a principle which involves the concept of all-or-nothing belief and the concept of degree of belief simultaneously, and hence builds a bridge between the two of them. Since it invokes the normative term 'rational', it is a normative bridge principle. The expression 'stably high degrees of belief' will be understood in terms of degrees of belief that remain high under certain salient conditionalizations, but the exact details of this proposal are to be explained in Chapter 2. In the same chapter I will also make clear why the Humean thesis does not have to be understood as a reductive claim of any sort.

Let me instead turn now to the proposals for bridge principles on categorical and numerical belief that are already available in the relevant literature.[86] Most of the more traditional proposals belong to one of the following categories or are at least very close to a proposal in one of those categories.

1.6.1 The Certainty *or* Probability 1 Proposal

According to this norm,[87] a proposition X is believed by a perfectly rational agent just in case the agent assigns the maximal possible degree of belief to X, where rational degrees of belief are assumed to satisfy the axioms of subjective probability (in line with Assumption 9). Or, as I will say more briefly:

$$Bel(X) \text{ iff } P(X) = 1.$$

[86] Hilpinen (1968) gives a very nice summary of the traditional bridge principles for rational belief and rational degrees of belief. Swain (1970) collects many important primary sources including early versions of the theories by Levi, Kyburg, and Jeffrey. Christensen (2004) and Huber and Schmidt-Petri (2009) do the same, respectively, for the more recent theories. Spohn (2012, s. 3.1) gives yet another survey of such bridge principles.

[87] Roorda (1995) calls this 'the received view', and Gärdenfors (1986a) is a representative of this view. But actually it is not so easy to find proponents of it in the literature on this topic. If anything, often only its left-to-right direction is being adopted, and even that is usually subject to certain qualifications: e.g. Levi (1980) accepts the left-to-right direction for his so-called credal probability measures (though expressed in terms of 'knowledge' rather than 'belief'). But that is in a context in which a credal state may involve more than just one credal probability measure, and where propositions of credal probability 1 are not meant to be incorrigible (as they may cease to have probability 1 in the future in view of possible future revisions of one's corpus of knowledge). Van Fraassen (1995), Arló-Costa (2001), and Arló-Costa and Parikh (2005) also take the left-to-right direction of the Probability 1 Proposal for granted. However, they do not just presuppose standard probability measures but so-called primitive conditional probability measures (or Popper functions): probability measures that allow for the conditionalization on sets of probability 0; see Makinson (2011) for a recent overview. As they show, one can then always find so-called belief cores which are propositions with particularly nice logical properties; and by taking supersets of those one can define elegant notions of qualitative belief in different variants and strengths. Since all such belief cores have absolute probability 1, they end up with the left-to-right direction of the proposal above. Clarke (2013) does regard belief as entailing a credence of 1, but only once the agent's 'global' credence function has been conditionalized on some propositions that are determined contextually. Finally, in a spirit similar to that of the Probability 1 Proposal, Williamson (1998) suggests determining so-called epistemic probabilities by conditionalization on the conjunction of everything one knows, which has the consequence that if A is known then A has epistemic probability 1. But of course knowledge is not belief and epistemic probabilities are not meant to be degrees of belief either.

Here, '$Bel(X)$' is short for 'X is believed', and '$P(X) = 1$' is short for 'X is assigned (the maximal) degree of belief 1'. It is to be understood that both sides of the equivalence are meant to apply to one and the same agent at one and the same time: it is the same agent who has such a belief and assigns such a degree of belief at the same instant. The whole equivalence is tacitly universally quantified ('for all X: . . .'). I am going to use abbreviations like that throughout this book. Finally, the overall statement is a norm, because 'perfectly rational' is a normative (or evaluative) expression.

If the Probability 1 Proposal is granted, then from it—in conjunction with the axioms of probability—one can derive the basic logical axioms for belief. For the set of propositions with probability 1 is closed under logical consequence (from finite sets of premises). Additionally, a contradiction cannot be believed, since by the laws of probability a contradiction is assigned probability 0. In this way, at least the synchronic part of the 'Logic' Assumption 8 can actually be derived from this kind of bridge principle.

This said, in spite of this logical attraction, the proposal still seems wrong, at least if taken as a principle that is supposed to be generally valid. One problem concerns all-or-nothing belief and rational betting behaviour (as discussed by Roorda 1995): for example, I honestly believe that I will be in my office tomorrow. But I would refrain from accepting a bet on this if I were offered €1 if I were right, and if I were to lose €1,000,000 if not. However, by the Certainty Proposal my degree of belief in me being in my office tomorrow would have to be 1 (because of my all-or-nothing belief), and by the standard Bayesian understanding of probabilities in terms of betting quotients, this would mean that I would in fact have to accept the bet on X that will give me €1 if I am right and that will cost me €1,000,000 if I am wrong. And yet I feel perfectly rational in having my belief and refraining from accepting the bet. Combining the Certainty Proposal with what seem to be standard cases of all-or-nothing belief, belief would seem to commit an agent to all sorts of irrational action—such as accepting weird bets—and problems like that might emerge for virtually every proposition X that one believes to be true. In other words: the Certainty Proposal seems to be in conflict with Assumption 3 on how (all-or-nothing) belief should dispose one to act rationally.[88] Roorda (1995) also gives further arguments against the proposal: for instance, I might believe the contingent proposition that Millard Fillmore was a President of the United States, and also the logically true proposition that Millard Fillmore either was or was not a President of the United States (a proposition of the form $A \vee \neg A$). But intuitively I would not want to invest the same strength of belief in the two propositions, and again there does not seem to be anything irrational about that. Or here is another argument against the Certainty Proposal from a more conceptual point of view: it is one thing to rationally believe something to be the

[88] Alternatively, one could abandon the standard interpretation of subjective probabilities in terms of betting quotients in this case, but breaking in this way with the mainstream Bayesian tradition would come with a huge price of its own.

case—to plausibly expect something to be true—but another to be certain of it. If certainty is captured by assigning the maximal degree of belief of 1 or 100 per cent, then rational belief does not necessarily coincide with certainty (although it might do so in special cases). Or another argument: by the proposal, all believed propositions would need to have probability 1. But once a proposition is assigned probability 1, its probability cannot be decreased any more by update by conditionalization (on propositions with positive probability), which leads to further worries that are well-known, e.g. from the debate on the so-called Old Evidence Problem.[89] Finally: if the Certainty Proposal were right, then beliefs would simply not seem to be robust enough to survive the presence of even minor uncertainties that almost inevitably occur in the real world and that will lead to probabilities ever so slightly below 1—which should be worrisome in itself.

Summing up: the Probability 1 Proposal seems too restrictive. It should be possible for a perfectly rational agent to believe X even in a case when she is not assigning the maximal possible degree of belief to X. This does not exclude the *possibility* of cases where X is believed and X is also assigned probability 1—it is just that this should not be so by necessity.

1.6.2 The Lockean Thesis

The obvious suggestion of how to avoid the problem that seemed to affect the Certainty Proposal is to weaken its right-hand side.[90] That is: to maintain for every perfectly rational agent that having the belief that X is equivalent to assigning to X not necessarily the maximum possible subjective probability but merely one that is above some threshold s (that is less than 1):

$$Bel(X) \text{ iff } P(X) > s.$$

If one holds P fixed, this Lockean threshold s might be said to measure either the belief-agent's or the belief-ascriber's cautiousness with respect to qualitative belief: the greater s, the more cautious the agent's beliefs will be, for more is then demanded of believed propositions with respect to their subjective probability. Vice versa, the lower s, the braver the agent's beliefs will be, for less is demanded of believed propositions with respect to their probability.

89 See Glymour (1980).

90 The most famous defender of the Lockean thesis is Kyburg; see e.g. Kyburg (1961), the last chapter of Kyburg (1970b) which contains a comparison between Kyburg's account and some rival ones, including Levi's, and the much more recent Kyburg and Teng (2001). I should add that for Kyburg the probability of a proposition is actually an *interval* of real numbers, but not much hangs on this as far as present purposes are concerned. More recently, Foley (1993) has argued in favour of the Lockean thesis, and Hawthorne and Bovens (1999) and Hawthorne and Makinson (2007) have studied logical systems of absolute belief and conditional belief, respectively, that result from taking versions of the Lockean thesis for granted. Also Sturgeon (2008) defends the Lockean thesis but combines it with an understanding of categorical belief as 'thick confidence', where 'thick' contrasts with 'point-valued' (subjective probability).

This proposal was termed the 'Lockean thesis' by Richard Foley (1993) who traces it back to John Locke's *Essay Concerning Human Understanding*. If the corresponding threshold s is greater than or equal to $\frac{1}{2}$—which is the standard assumption—then belief is equivalent to high enough subjective probability (where the exact meaning of 'high enough' depends on the context). This does look right, at least at first glance.

On the other hand, the thesis also leads to a worry about logical coherence: at least so long as the Lockean threshold can be chosen freely, the probability of $X \cap Y$ might well drop below a chosen threshold even in a case in which the probabilities of X and Y do not. This is illustrated by the famous Lottery Paradox:[91] consider a fair lottery with, say, 100 tickets, which is certain to take place; the agent is aware of all of that. Set the threshold value s to 0.98. Then for each ticket i the proposition that i will not win will have to be believed by the agent, by the uniformity or 'fairness' of the agent's subjective probability measure taken together with the Lockean thesis for that threshold s. From the closure of rational belief under conjunction it will then follow that the agent will have to believe that ticket 1 will not win and ticket 2 will not win and . . . and ticket 100 will not win. But that conjunctive proposition has probability 0 and hence is not to be believed, by the Lockean thesis again. So we have a contradiction. The Lockean thesis seems to be in conflict with the 'Logic' Assumption 8 and the 'Probability' Assumption 9 from section 1.5, or, if Assumption 9 is taken to be beyond doubt, it seems to be in conflict with Assumption 8.

Kyburg's own reaction to his paradox was to sacrifice one of the standard logical closure properties of qualitative rational belief, that is, the closure of belief under conjunction, while keeping the Lockean thesis intact. In view of my own rules of the game, as I presented them before, this will not be the option that I am going to take. As it happens, one can show that the Lockean thesis *can* be combined with the logical closure of rational belief *if only the choice of the threshold value s is assumed to depend on the probability measure P in question*. It is just that all-or-nothing belief will end up context-sensitive in this way (and in other ways). I will discuss all of this in detail in Chapter 3. So I will propose to maintain *both* the Lockean thesis and the logical closure of rational belief but to contextualize rational belief instead. Indeed, the Lockean thesis and the logical closure of rational belief will both be seen to follow from the Humean thesis in Chapter 2.

1.6.3 Decision-Theoretic Accounts

The Probability 1 Proposal and the Lockean thesis are pretty much the simplest possible bridge principles for rational qualitative and quantitative belief that one can think of. If both of them are problematic, at least without adding some further qualifications, then one natural way out would be to look for a more complex set of joint principles

[91] The paradox goes back to Kyburg (1961). See also Wheeler (2007) and Douven and Williamson (2006) for more discussion of it.

for the two kinds of rational belief. One way of realizing this ambition is by using a decision-theoretic account.[92]

The underlying idea is this: consider *rationally believing* X as some kind of action (the action $bel\,X$). Which actions should you take? As Bayesian decision theory has it, only those that maximize expected utility, or those for which the expected utility is greater than that of some relevant alternative actions, or the like. So given an agent's subjective probability measure P, and given also some utility measure u that assigns some sort of numerical utilities to the outcomes of belief acts at worlds (maybe 'epistemic' or 'theoretical' utilities rather than 'practical' ones), it should be the case that

$$Bel(X) \text{ iff } \sum_{w \in W} [P(\{w\}) \cdot u(bel\,X, w)] \text{ has property so-and-so.}$$

On the right-hand side of this equivalence, '$\sum_{w \in W} [P(\{w\}) \cdot u(bel\,X, w)]$' refers to the expected utility of the act of believing X, which is defined by summing up over all possible worlds w the utility $u(bel\,X, w)$ of that act in that world, where the act's utility at w is weighted by the subjective probability $P(\{w\})$ of that world w being the actual world. 'so-and-so' needs to be replaced appropriately, so that the expected utility of the act of believing X is high enough when compared to the expected utilities of alternative actions in some relevant class. That class could be e.g. $\{bel\,X, bel\,\neg X, suspend\ on\ X\}$, or maybe something else. As usually the case in decision theory, one does not have to read the right-hand side of the equivalence as something that a rational agent would have to compute consciously and in precisely these numerical terms; the proposal would not necessarily demand beliefs to be the outcomes of conscious explicit decisions at all. Instead, describing a rational agent as maximizing the expected utility of some doxastic act could rather be interpreted as an ascription of as-if rationality: the perfectly rational agent's mental state is such *as if* she had gone through the required computation and comparison of expected utilities.[93]

What beliefs a perfectly rational agent will have according to this proposal will not just depend on the agent's degrees of belief but also, crucially, on what her utility measure is like (and also on the set of relevant alternative acts). Depending on the properties of u, such a decision-theoretic account might well collapse into one of the previous proposals: for instance, Hempel's (1962) classical decision-theoretic account of belief or acceptance turns out to be equivalent to the Lockean thesis from above for a Lockean threshold $s = \frac{1}{2}$.[94] Or such a decision-theoretic account might differ completely from any of the previous suggestions. Indeed, for certain utility measures u, a proposition X might end up being 'rationally' believed by an agent while the agent's degree of belief in X would be less than $\frac{1}{2}$. But that would seem to be in conflict with

92 Such accounts can be found, e.g. and in different forms, in Hempel (1962), Levi (1967), Kaplan (1981, 1996), Lehrer (1990), Maher (1993), and Frankish (2004).

93 See Christensen (2004, ch. 5) for a discussion of such matters.

94 Easwaran (2016), Fitelson (n.d.), Dorst (n.d.), and Leitgeb (n.d.) have refined and extended Hempel's result in recent work. I will return to this in section 5.1.

Assumption 2 from section 1.1: in so far as an agent's degrees of belief aim at the truth, $P(X)$ being less than $\frac{1}{2}$ would mean that the agent takes $\neg X$ to be closer to the truth than X, for it would be the case then that $P(\neg X) > P(X)$. But how can the agent then believe X? If she does, it seems she does not aim at the truth any more on the all-or-nothing side of belief, which would run against one of the constitutive features of belief.[95]

Furthermore, there is nothing in the decision-theoretic picture just by itself that would guarantee that any of the standard logical closure properties of rational belief would follow from it or even be consistent with it: e.g. while believing X might maximize expected utility, and while believing Y might do so, too, it might be the case that believing $X \cap Y$ does not. If the logic of belief is taken as a given (as in our case), then one way of accommodating this in such a decision-theoretic context would be to compute not the expected utilities of single belief-acts but either the act of choosing a unique *logically strongest believed proposition* (like Levi's 1967 'strongest accepted sentence'), which is taken to entail all other believed propositions, or the act of choosing a full belief *system* or *theory* that is required to be closed logically *by fiat* (as is the case e.g. in Maher 1993). But this generates new worries: it is not clear by itself why the expected utility of believing a proposition Y would be high just because Y is *entailed* by the logically strongest believed proposition (or by a theory) X whose expected utility is indeed maximal or salient in some other sense. So decision-theoretic accounts may also come in conflict with the 'Logic' Assumption 8 from section 1.5. Whether they do will depend on the exact properties of u and the underlying decision-theoretic framework again.[96]

1.6.4 *The* Nihilistic Proposal

Finally, if no proposal seems to work, one might draw the conclusion that there are no general and informative bridge principles at all relating rational belief and rational degree-of-belief assignments. Spohn (2012, section 10.1) suggests a view like that. Roorda (1995) seems to be close to such a position, too (although he adds to this the view that graded belief possesses some kind of priority over belief, which e.g. Spohn does not): 'The depressing conclusion . . . is that no explication of belief is possible within the confines of the probability model.'[97] One metaphysical view of belief that might fit this normative diagnosis would be the Anomalous Monism

[95] Maher (1993, s. 6.2.4) argues that scientific theories can be accepted rationally without assigning to them a probability greater than $\frac{1}{2}$. I agree: but only because I understand 'acceptance' so that it does not entail belief and it does not necessarily aim at the truth, as I will explain in section 6.3 of Chapter 6.

[96] In Chapter 5 I will present a decision-theoretic account of belief that will turn out to be equivalent to the stability account of belief that is developed in the rest of this book.

[97] Roorda himself goes on to suggest an explication of rational belief that is relative to a *set* of subjective probability measures—which he calls the 'extended probability model'—rather than just one probability measure as standard Bayesianism has it. Sturgeon (2008) makes a similar move, but without reducing all-or-nothing belief to sets of probability measures. In contrast, I am going to bite the bullet and stick to just one subjective probability measure on the degree-of-belief side.

variant of the Irreducibility Option (iii) from section 1.3.3. Another one would be what Spohn (2012, p. 202) calls 'separatism', which I take to be the position that categorical belief differs from graded belief both on the level of tokens and on that of types, and there are no bridge principles relating the two; yet, in Spohn's view, the two of them may still exemplify some kind of pre-established harmony.

Clearly, the main worry about this (lack of a) proposal is: if binary belief and graded belief are both real and kinds of *belief*, as we had assumed before, can it really be the case that there is no general coherence norm that would constrain the two types of belief in an informative and transparent manner? This is more or less the same worry reinstantiated in normative terms that applied already to the Anomalous Monism Option in our discussion of option (iii). Indeed, Spohn (2012, p. 202) admits to 'sense the absurdity of this position'.

In what follows, I will aim to fuel this worry by developing such a general joint principle on rational qualitative and quantitative belief. However, this will be achieved by means of a stability account of rational belief that does not coincide with any of the proposals that have been mentioned so far.

This list of bridge norms on belief and degrees of belief is *far* from being exhaustive. Most notably, recently, Lin and Kelly (2012a, 2012b) have developed a beautiful theory of rational categorical belief (or acceptance) and numerical belief that does not belong to any of the four categories either. A preliminary discussion of how their theory differs from the one in this monograph can be found in Lin and Kelly (2013) and Leitgeb (2013b).[98]

1.7 What is to Come

The overall structure of this book is as follows.

Appendix A (the appendix to the present chapter) gives an argument to the effect that if rational belief is not closed under conjunction, then revising it cannot quite proceed in the way in which one would expect. Indeed, it seems that without closure under conjunction, belief revision for rational all-or-nothing belief cannot be accomplished without taking one's rational degrees of belief into account. This would mean that although rational all-or-nothing belief might still be ontologically independent of rational graded belief—one might be instantiated without the other—it could not be systemically independent in the sense discussed under option (iii) from section 1.3: the belief system will be required to be 'told' by the degree-of-belief system how to revise all-or-nothing beliefs, and hence it could not work successfully without the degree-of-belief system being around and functioning. Or the other way around: if one

[98] Further excellent work on the topic of rational belief vs degrees of belief is on its way or has appeared recently: Fitelson (n.d.), Pettigrew (2016), unpublished work by Alexandru Baltag and Sonja Smets on this topic, and more. In parts of this unpublished work, Alexandru Baltag shows that the stability theory of belief that is to be developed in this book can also be derived from joint assumptions on rational belief, *knowledge*, and their interaction. So there is yet another starting point from which the theory of rational belief of this book can be derived.

regards it as plausible that the belief system of a rational agent is capable of revising all-or-nothing beliefs independently of the agent's degree-of-belief system, then the appendix to Chapter 1 will amount to an argument in favour of the closure of rational belief under conjunction (which is probably the most controversial part of the 'Logic' Assumption 8 from section 1.5).

Chapter 2 on 'The Humean Thesis on Belief' will determine an answer to our central question. That answer will be: the categorical beliefs of a perfectly rational agent cohere with her degrees of belief just in case *belief is equivalent to stably high subjective probability*. I will call this equivalence thesis the 'Humean thesis on belief', since it will be motivated by some considerations on belief regarding Hume's *Treatise of Human Nature* (following Louis Loeb's interpretation of Hume). Stably high subjective probability will be explicated as subjective probability that remains high enough under salient instances of conditionalization (which will relate to Brian Skyrms's work on probabilistic resiliency). I will show that the Humean thesis—taken together with the axioms of probability for degrees of belief, and assuming that the contradictory proposition is not believed—entails three plausible consequences on rational belief. First, the logical closure of belief: the logic of belief may thus be viewed as a manifestation of the stability of belief. Second, the so-called Lockean thesis on belief: that is, it is rational to believe a proposition just in case it is rational to assign a high enough degree of belief to it. Third: the Humean thesis also entails the compatibility between decisions based on all-or-nothing beliefs and those made in line with Bayesian decision theory. A brief appendix (B) to the chapter will explain where the required stability of belief might emerge from: if not from the representation of causal relationships (as in an example from section 2.3.1) or from a priori judgements (e.g. concerning the simplicity of inductive hypotheses as in Example 18 from Chapter 4), it might be the iterated impact of evidence itself that leads to stability.

The remaining chapters of the book derive (more or less) the same joint theory of rational belief and degrees of belief as Chapter 2, but they do so from alternative starting points. These starting points will (hopefully) be plausible in themselves. One of these assumptions that I am going to make throughout all of the chapters is that the axioms of subjective probability govern a rational agent's distribution of degrees of belief over propositions (that is, the 'Probability' Assumption 9 from section 1.5). In Chapter 3 the basic additional assumptions will be: one direction of the Lockean thesis and the logical closure of belief. In Chapter 4 I will start from the other direction of the (conditional variant of the) Lockean thesis combined with the standard AGM theory of belief revision. Chapter 5 will proceed from the logical closure of belief together with some postulates of epistemic decision theory that will formalize the idea that all-or-nothing belief aims at either truth or subjective probability or both. (What that means exactly will be explained there in proper detail.) Section 6.1 of Chapter 6 takes the logical closure of belief and the coherence between decisions based on all-or-nothing beliefs and decisions in line with Bayesian decision theory as its starting points. And section 6.2 starts from joint assumptions on belief and assertability. Each of these different sets of assumptions will be proven equivalent to the Humean thesis on belief (modulo some details concerning the choice of thresholds and the like). The equivalences in question will follow from so-called representation theorems. These theorems will state that a pair $\langle Bel, P\rangle$ of a rational agent's belief set *Bel* and her degree-of-belief function *P* at a given point in time satisfies one of our sets of plausible assumptions if, and only if, the pair meets a certain formal condition that is easy to handle and to determine. And that formal condition will, in turn, be equivalent to the Humean thesis on belief. The first instances of such formal conditions (most importantly, the so-called '*P*-stability of an agent's logically strongest believed proposition') will be stated in Appendix B. All of these equivalence results taken together will support the robustness of the stability account of belief that I will defend in Chapter 2: given various different sets of assumptions on rational belief and degrees of belief that are each plausible independently, the stability of belief in the sense of Chapter 2 is simply unavoidable.

Since some of the starting points of the chapters after Chapter 2 are actually consequences of the Humean thesis in Chapter 2, it is also possible to view the purpose of Chapters 3–6 as to yield 'recovery arguments' for the Humean thesis in the sense of Koellner (2013). In Koellner's case, his recovery arguments are supposed to support certain set-theoretic axioms (of determinacy): first these axioms are shown to have some plausible consequences; then one proves that the axioms themselves can be recovered, that is, derived, from some of their consequences taken together as bundles. As Koellner points out, this type of argumentative strategy is not normally available in the empirical sciences, but if he is right, it is available in the a priori domain of the philosophy of set theory. One way of interpreting the results of this book is that the same manner of argumentation is also available in the a priori domain of normative epistemology.

Now let me comment in more detail on the specific contents of Chapters 3–6.

As mentioned before, Chapter 3 on 'Logical Closure and the Lockean Thesis' does not start from the stability of belief but from three alternative assumptions: the consistency and logical closure of rational belief; the axioms of probability for rational degrees of belief; and (the left-to-right direction of) the Lockean thesis. I will show that this combination of principles is satisfiable (and indeed non-trivially so) and that the principles are jointly satisfied if and only if rational belief is equivalent to stably high rational degree of belief. So given the axioms of subjective probability and the consistency of rational belief, (a version of) the Humean thesis on belief, which had been a premise in Chapter 2, can be recovered from two of its own consequences: logical closure and the Lockean thesis. Thus it turns out that the stability of belief may also be seen as a manifestation of these other principles. The logical closure of belief and the Lockean thesis are attractive features of this theory of belief, and these features will be exemplified, amongst other examples, in an application of the theory to the Lottery Paradox in section 3.4 (and, to a first approximation, also to the Preface Paradox in section 3.5). On the other hand, the chapter will also point to what is probably the main worry about the emerging stability account of belief: a strong context-sensitivity of rational belief. The underlying notion of context can be understood in two ways: semantically, as a context of belief ascription, and non-semantically, as the belief-subject's own context of reasoning. Both interpretations will be compatible with the theory, but I will focus on the second non-semantic one. I will argue that we should be able to live with rational belief being context-sensitive in either of the two senses (or both).

Chapter 4 on 'Conditional Belief and Belief Dynamics' turns to a conditional notion of all-or-nothing belief—belief in *Y given X*—which may be viewed as (entailing) a disposition for belief change. I will prove that the following combination of assumptions is equivalent to an extension of the Humean thesis for unconditional belief to a stability conception of conditional belief: the axioms of subjective probability, the axioms of AGM belief revision for conditional belief, and the right-to-left direction of the Lockean thesis for conditional belief (that is, conditional belief entails high enough conditional probability). In particular, the so-called Preservation Axiom for belief revision will be found to express the stability of conditional belief, much as logical closure was seen before to reflect the stability of unconditional belief. The other purpose that this chapter will serve is to develop in full formal detail all of the technical machinery that is required to support the main formal claims made in this book. Furthermore, while I will restrict myself to the case of finitely many possible worlds in all other chapters, I will also deal with the case of infinitely many worlds (and propositions) in Chapter 4.

In Appendix C I will determine which additional assumptions it would take to make categorical rational belief supervene on rational degrees of belief. We will find that these additional assumptions do not look very plausible. Moreover, I will use a result by Lin and Kelly (2012b) to point out one of the consequences that adding the supervenience of rational belief on rational degrees of belief to the postulates of my stability theory would have: rational categorical belief change could not proceed in line with AGM belief revision theory. This would thus contradict the diachronic part of my 'Logic' Assumption 8 from section 1.5. Lin and Kelly (2012b) use this as an argument against AGM, but I will argue instead that rational all-or-nothing belief does not supervene on rational degrees of belief. This will in turn entail that rational all-or-nothing belief does not reduce to rational degrees of belief either. However, I will leave open whether or not rational all-or-nothing belief supervenes on (and perhaps reduces to) rational degrees of belief *and some practical features given by context* taken together; that is a question that I will not be able to settle.

Chapter 5 on 'Stability and Epistemic Decision Theory' begins with an accuracy argument in favour of the Humean thesis on belief: assuming the axioms of subjective probability and the logical closure and consistency of rational belief, the thesis that the so-called expected epistemic utility of rational belief is stably positive turns out to be equivalent to the Humean thesis on belief again. The corresponding notion of expected epistemic utility is motivated and explained beforehand. Then I consider a second way of measuring the accuracy of belief: by determining how well belief approximates degrees of belief. In that second case, I turn to conditional belief for perfectly rational agents again, and I assume that conditional belief is given by a doxastic preference ordering over possible worlds, as belief revision theory or nonmonotonic reasoning has it (and as presupposed in Chapter 4). Given that, I am going to answer the question: what could it mean to say that a doxastic preference ordering over worlds approximates a subjective probability measure more accurately than another? As I am going to show, the answer to this question will ultimately determine a theory of belief that is equivalent to the Humean thesis on belief again. All of this will be done in a way that is similar to arguments for Bayesianism in so-called epistemic decision theory.

Chapter 6 on 'Action, Assertability, Acceptance' deals with three topics of practical rationality. In section 2.3.3 of Chapter 2 it was shown that the axioms of subjective probability taken together with the Humean thesis on belief entail the compatibility between decisions based on all-or-nothing beliefs and those in line with Bayesian decision theory. In section 6.1 I will prove that given the axioms of subjective probability, the Humean thesis on belief is actually equivalent to the logical closure and consistency of belief taken together with this kind of decision-theoretic compatibility. So this is yet another recovery result for the Humean thesis. In section 6.2, I will deal with the assertability of propositions and of indicative conditionals, including the famous Ramsey test for conditionals. I will relate assertability to both graded and categorical belief, and I will show that a major part of the principles for conditional belief from Chapter 4 turns out to be equivalent in this way to plausible principles for the assertability of conditionals. In rough-and-ready terms (and suppressing details), the categorical assertability of an indicative conditional will follow to be equivalent to its corresponding conditional probability being stably high. The section will also demonstrate that this stability theory of assertability and belief is able to recover some of Frank Jackson's independent views on assertability and robustness. Section 6.3 of the chapter will be devoted to acceptance as a mental act that is closely related to belief and yet distinct from it. However, acceptance will be argued to be stable in a way that is similar to the stability of belief. As a by-product, I will show that the Preservation Axiom for belief revision can be derived from a plausible joint assumption about acceptance and belief. The final section 6.4 of this chapter suggests a way out of the Preface Paradox that is consistent with the emerging stability theory of belief.[99]

By their very nature, the representation theorems in this book will be formal: a pair $\langle Bel, P \rangle$ that satisfies certain formal constraints is proven to satisfy yet another set of formal constraints and vice versa. The intended interpretation and application of such theorems is of course in terms of a perfectly rational agent's belief set *Bel* and her degree-of-belief function *P* at a point of time. But as long as *Bel* and *P* are *anything* that jointly meet the one set of constraints, they will also meet the other set of constraints, by mathematical proof. This feature allows for alternative applications of the formal

[99] So I will deal with the Preface Paradox in two parts: first in section 3.5 and then again in section 6.4.

findings in this book. For instance: assume *conditional all-or-nothing belief*—belief in Y given X—to be replaced (or reinterpreted) by *truth of the counterfactual conditional* $X \mathrel{\Box\!\!\rightarrow} Y$. Accordingly, assume *conditional degree of belief*—$P(Y|X)$—to be replaced (or reinterpreted) by *conditional objective chance of Y given X*. And suppose that *truth of counterfactuals* and *conditional chance* were to satisfy analogous formal constraints as those imposed on *Bel* and *P* in Chapter 4 of this book: then perhaps some interesting conclusions could be derived on counterfactuals and conditional chance based on the theorems in the previous chapters. That is essentially what I will be doing in Appendix D. The same strategy may give rise to further alternative applications of the findings in this book, but I will restrict myself to counterfactuals and chance in the appendix.

In more detail: Appendix D will develop a stability account of counterfactual truth in which truth of a counterfactual relates to objective chance in an analogous way as belief relates to subjective probability in the main part of this book. I will start the discussion with a new lottery-style paradox on counterfactuals and chance. What seem to be plausible premises concerning the truth values of ordinary counterfactuals, the conditional chances of possible but non-actual events, a bridge principle relating them, and a fragment of the logic of counterfactuals lead to contradiction. Unlike the usual lottery-style paradoxes, logical closure under conjunction—that is, in this case, the rule of Agglomeration of (consequents of) counterfactuals—will not play a role in the derivation and will not be entailed by these premises either. I will sketch four obvious but problematic ways out of the dilemma, and I will end up with a new resolution strategy that is non-obvious but (I hope) less problematic: contextualism about what counts as a proposition. This proposal will save us from the paradox, it will save each premise in at least some context, and it will be motivated by independent considerations from measure theory and probability theory in which it is a standard move not to count each and every set of possibilities as a measurable event. If the argument in Appendix D is sound, then whether a counterfactual expresses a proposition will be just as context-dependent as all-or-nothing belief is found to be in the main part of this essay.

Appendix A

The Review Argument

On the Diachronic Costs of Not Closing Rational Belief under Conjunction

In this appendix I argue that giving up on the closure of rational belief under conjunction would come with a substantial price. Either rational belief is closed under conjunction, or else the epistemology of belief has a serious diachronic deficit over and above the synchronic failures of conjunctive closure. The argument for this, which can be viewed as a sequel to the Preface Paradox, is called the 'Review Argument';[100] it is presented in four distinct but closely related versions.

In order to get just a quick impression of what is going on here, it would be perfectly sufficient to take a look only at the first argument in section A.2 and the first argument in section A.3—the second argument of each section is but a generalization of the first one in the same section.

A.1 Closing Rational Belief under Conjunction

Is rational (all-or-nothing) belief—the set of propositions believed by a perfectly rational agent—bound to be closed under conjunction? There are quite a few philosophers who think the answer is *no* (such as, famously, Henry Kyburg). They do so in spite of a great tradition in doxastic/epistemic logic according to which the closure of belief under conjunction counts as a fundamental rationality postulate (cf. Hintikka 1962, Levi 1967). In the eyes of these philosophers, the logical tradition suffers from 'conjunctivitis' (Kyburg 1970a).

Some of the opponents of conjunctive closure are impressed by the Lockean thesis (cf. Foley 1993) which says that it is rational to believe a proposition X if and only if the subjective probability of X is greater than some threshold s, where the threshold in question may be vague and depend on the context. Formally: there is an s, such that for all X,

$$Bel(X) \text{ iff } P(X) > s.$$

100 In Leitgeb (2014d) the argument was called the 'Review Paradox'. But it is perhaps safer to call it merely an 'argument'.

And then they point to the obvious existence of cases in which $P(A) > s$, $P(B) > s$, whilst $P(A \cap B) \not> s$, so that, by the Lockean thesis, it must hold that $Bel(A)$, $Bel(B)$, but *not* $Bel(A \cap B)$—the propositions A and B are to be believed, though their conjunction $A \cap B$ is not.

Others might argue against the closure of belief under conjunction on grounds of paradoxes such as the Preface Paradox (cf. Makinson 1965): it does not seem irrational for an author to claim in the preface of her book that she will have made some mistakes in the subsequent chapters, and at the same time to claim, and thus presumably believe, each of the statements $A_1, \ldots, A_n$ that are being made in these chapters. Closure under conjunction would seem to leave the author with a belief in the contradictory statement $\neg(A_1 \cap \ldots \cap A_n) \cap (A_1 \cap \ldots \cap A_n)$, which would certainly not be rational. Hence, closure seems to be wrong.[101]

In the following, I am going to argue that giving up on closing rational belief under conjunction would come with a serious *diachronic* price. The argument for this will be presented in four closely related versions: the second one will generalize the first one, and the fourth one will generalize the third one.

My main target in this appendix will be philosophers who are non-radical Bayesians: they assume that a rational agent's degree-of-belief function is a probability measure, they think that a rational agent updates by conditionalization (or, more generally, by Jeffrey conditionalization), but they are not in the business of eliminating the categorical concept of belief. Typically, they are happy to sacrifice the closure of rational all-or-nothing belief under conjunction, if they regard this as a sacrifice at all. Given some additional auxiliary assumptions (such as P1 in section A.2), I will show that such philosophers will also be committed to deviating drastically from standard *diachronic* norms on all-or-nothing belief change. Roughly: rational categorical belief update could not proceed without help from the degree-of-belief side. My target philosophers might not worry about this—in fact, they might cheerfully embrace this consequence—but they should be aware that they are on their way to sacrificing the epistemology of all-or-nothing belief as something that might, in principle, have a life of its own.[102]

A.2 The Argument

Let us presuppose that we intend to describe an agent's doxastic states both qualitatively, in terms of categorical belief ascriptions, and quantitatively, by means of ascribing numerical degrees of belief; abandoning either of the two kinds of ascriptions is not an option. Therefore, when the agent receives some piece of evidence X,

[101] Later in this book, in sections 3.5 and 6.4, I will argue that the Preface Paradox does *not* rule out closure of rational belief under conjunction.

[102] So these assumptions will effectively rule out a rational agent's having a dual-belief architecture such as described under the Independence variant of the Irreducibility Option (iii) in section 1.3.3.

we should be able to express what is going on doxastically in qualitative and in quantitative terms simultaneously or in parallel: if she learns (or updates on) X, then something is the case that will be expressed qualitatively, and at the same time something is the case that will be expressed quantitatively.

I consider some (inferentially) perfectly rational agent. Let t be an arbitrary point of time, let Bel_t be the set of propositions believed by the agent at t, and let P_t be the same agent's degree-of-belief function at t; analogously for some point of time t' after t and the corresponding $Bel_{t'}$, $P_{t'}$. I will assume, without further justification, that the degree-of-belief function of a perfectly rational agent must always be a subjective probability measure (in line with Assumption 9 from section 1.5 of Chapter 1).

Each version of our argument will proceed from three premises. In the first version, P1 is a bridge principle that tells us something about how the agent's degrees of belief and her beliefs relate to each other. P2 expresses a qualitative feature of update by evidence. P3 states how the agent updates in quantitative terms. These are the premises in more detail:

P1 *If the degrees of belief that the agent assigns to two propositions are identical, then either the agent believes both of them or neither of them.* That is:

For all X, Y: if $P_t(X) = P_t(Y)$ then

$$Bel_t(X) \text{ iff } Bel_t(Y).$$

P2 *If the agent already believes X, then so far as the effects are concerned that updating has on all-or-nothing beliefs, updating on the piece of evidence X (learning X) does not change the agent's system of all-or-nothing beliefs at all.* That is:

For all X: if the evidence that the agent obtains between t and $t' > t$ is the proposition X, but it holds already that $Bel_t(X)$, then for all Y:

$$Bel_{t'}(Y) \text{ iff } Bel_t(Y).$$

P3 *When the agent updates (learns), then so far as the effects are concerned that updating has on degrees of belief, updating on X is captured probabilistically by conditionalization on X.* That is:

For all X (with $P_t(X) > 0$): if the evidence that the agent obtains between t and $t' > t$ is the proposition X, then for all Y:

$$P_{t'}(Y) = P_t(Y \mid X).$$

P1 expresses that if two propositions X and Y are assigned the same degrees of belief by the same perfectly rational agent at the same point in time, then the agent must treat X and Y equally with regard to belief at that point in time. For instance, every supporter of the Lockean thesis must accept this: for if $P(X)$ is identical to $P(Y)$, then either both of them will exceed the threshold s in the Lockean thesis or neither of them will.[103] More generally, every theory of belief and degrees of belief according to which belief in X supervenes, or functionally depends, on the probability of X will deliver P1 as a consequence. But P1 holds on yet more general grounds: supervenience would mean that there could not be a difference in the belief status of a proposition without a difference in the probability of that proposition, which is a matter of comparing different degree-of-belief functions, and different belief sets, with each other. But P1, which only concerns one degree-of-belief function and belief set at the time, is strictly weaker than that: if $P(X) = P(Y)$, then P1 leaves open whether X and Y are believed or not: it demands only that the two propositions have the same belief status, that is, both of them are to be believed or neither of them. Indeed, *'I believe X but I do not believe Y, even though the two of them are equally likely for me.'* sounds odd independently of the fate of the Lockean thesis or of some other principle of supervenience that might relate P and *Bel*. Or once again in other terms, from the point of view of the central epistemological goal of truth approximation (and disregarding other more pragmatic goals that an agent might have): if the probabilities of X and Y are the same, then this means that the agent's estimates of the truth values of X and Y are the same; since rational belief aims at the truth (cf. Wedgwood 2002), how could a perfectly rational agent not assign the same belief status to the two propositions? P1 will also follow from the theory of belief that I will develop later in this book (as will follow immediately from considerations concerning the Lockean thesis in Chapters 2 and 3).[104]

P2 should be quite convincing as well: it states that if a perfectly rational agent already believes X, and if she then updates on X as a piece of evidence—and there are no other simultaneous 'non-learning' changes[105]—her set of believed propositions will remain the same. *I: 'I believe X to be the case.' You: 'X is the case.' I: 'Oh my goodness, now I need to change some of my beliefs.'* does sound odd again. Accordingly, in the standard

[103] I am assuming here, in line with the standard interpretation of the Lockean thesis: although the value of the Lockean threshold might depend on the context, within one context one and the same threshold is to be used for all propositions. If this were not so, then it would be questionable whether even pretty uncontroversial aspects of the 'Coherence' Assumption 4 from section 1.1 would be satisfied: whether beliefs in different propositions would cohere with each other. For instance, assume that X and Y are propositions, $X \subseteq Y$, that is, X entails Y, and $P(X) = 0.85$ and $P(Y) = 0.9$: if X came with 'its own' Lockean threshold, say, 0.8, and Y had 'its own' Lockean threshold of 0.95, then X would be believed while Y would not be, even though X logically entails Y.

[104] There is one respect in which things are actually more complex than I make them sound here. As I am going to argue from Chapter 2, and in most detail in Chapter 3, rational belief is context-dependent. If explained in terms of the Lockean thesis: in one context the Lockean threshold might be set cautiously, say, to 0.98, while in a different context the Lockean threshold might be set more bravely, say, to 0.8. The difference between the two contexts might be due to different stakes or to whatever else suggests to the agent to be cautious in the one context but brave in the other. If the degree of belief of a proposition X is, say, 0.9 independently of contexts, then X would be believed in the latter context but not in the former one—contrary to P1. So P1 should really be claimed to hold only *within one context*. If restricted to just one context, the argument in this appendix will go through again. But I will leave this to one side here.

[105] I am grateful to Branden Fitelson here who rightly urged me to add this qualification. In the terminology of the later chapters: P2 will hold in my theory as long as the context of reasoning does not change.

purely qualitative theory of belief revision (cf. Gärdenfors 1988), if X is a member of the agent's present (and consistent) set K of believed propositions, then the revision $K * X$ of K by evidence X is demanded to be K again. Since X had already been believed, receiving it as a piece of evidence should not change anything as far as all-or-nothing beliefs are concerned; the agent simply ought to retain her current belief set.[106] The same is assumed by the less idealized theory of so-called belief base revision (see Hansson 1999) in which, unlike standard belief revision theory, the closure of belief sets under conjunction is not presupposed. P2 will also follow from the theory to be developed later in this book (as explained in Chapter 4).

P3 is the standard Bayesian postulate on probabilistic update. There are some justifications for it in the Bayesian literature, but I will not go into them here. P3 is contained in the theory of belief in this essay (compare Assumption 9 in section 1.5).[107]

Clearly, not everyone will buy each of these premises. In particular, some Bayesians might dismiss all-or-nothing belief revision (and its postulate P2) from the start. Some people in the more logically inclined all-or-nothing belief camp might dismiss Bayesian update (and its postulate P3) from the start. But *if* one presupposes that we intend to describe an agent's doxastic states both qualitatively and quantitatively, as I do, and if one buys into the standard assumptions on belief revision on either side, then P2 and P3 look fine at least at first glance. Also the bridge principle P1 that relates degrees of belief and belief seems plausible. This yields P1–P3. From them the closure of belief under conjunction will follow. Ultimately, the upshot of this will be the following. If we take P1 and P3 for granted, then there are just two possibilities: either one sticks to the logical tradition concerning rational categorical belief and retains the synchronic norm of perfectly rational agents' belief sets being closed under conjunction. Or one rejects one of the standard diachronic norms of rational categorical belief change to the effect that rational all-or-nothing belief revision cannot proceed independently of rational degree-of-belief update. Since, in contrast, rational degrees of belief can be updated independently of rational all-or-nothing beliefs (by P3), the epistemology of degrees of belief would follow to be prior to the epistemology of all-or-nothing beliefs in that case.

Here is then, in a nutshell and stated at first only informally, the argument: assume that a perfectly rational agent believes A and B but does not believe $A \cap B$. Let the

106 In belief revision terms, the 'update' in this case is an especially unproblematic case of belief *expansion*. Belief revision *proper*, in which the agent learns a piece of evidence that contradicts some of the agent's previous beliefs, is not at issue here at all. See Chapter 4 for more on this.

107 A premise weaker than P3 would actually be sufficient for running my first argument in this appendix: the weaker premise that after updating on X it holds for all Y that the probability of $X \cap Y$ after the update is equal to the probability of Y after the update. Or equivalently: the probability of X is 1 after the update. Of course, the standard Bayesian manner of bringing this about is by conditionalizing the agent's degree-of-belief function on X: what this adds to the probability of X being 1 after the update is that also the ratios of probabilities of propositions within X stay the same in the course of update. Since I am committing myself to the standard Bayesian norms on degrees of belief anyway (recall Assumption 9 in section 1.5), I have formulated P3 in terms of conditionalization right from the start. Additionally, the other three arguments in this appendix do depend on the details of the update procedure. I am grateful to Chris Gauker for alerting me to the possibility of using the weaker premise in the first argument.

agent's initial degree of belief in A lie strictly between 0 and 1.[108] Suppose the agent then receives A as a piece of evidence: when the agent updates on A, by P3, her subjective probability in B will become identical to her probability in $A \cap B$. By P1, the agent must thus have the same doxastic all-or-nothing attitude towards B and $A \cap B$ after the update. But by P2 her doxastic all-or-nothing attitude towards each of B and $A \cap B$ must be the same after updating on A as it had been before. Initially, by assumption, the agent believed B but did not believe $A \cap B$. Contradiction. Hence, given P1–P3, a failure of closure of belief under the conjunction of A and B leads to the absurd conclusion that the agent cannot update on A: something that should be perfectly unproblematic. Therefore, either closure of belief under conjunction must hold, or one of P1–P3 needs to be given up.[109]

One may illustrate what is going on here in terms of a sequel to the Preface Paradox: assume with the paradox that the author believes each of $A_1, \ldots, A_n$ without believing $A_1 \cap \ldots \cap A_n$. Let m be the maximal number less than n, such that the author believes $A_1 \cap \ldots \cap A_m$ without believing $A_1 \cap \ldots \cap A_{m+1}$; clearly, there must be such a number m in the Preface Paradox situation. Finally, suppose that someone writes a review of the author's book in which the reviewer strengthens the author's case for $A_1 \cap \ldots \cap A_m$, without saying anything at all about A_{m+1} or any other of the author's theses (maybe the reviewer is simply not interested in them): '*What I can say about this book is that* $A_1 \cap \ldots \cap A_m$ *definitely is the case.*' Assume that the author is rationally absorbing this report—updating on the proposition $A_1 \cap \ldots \cap A_m$ qualitatively, and, if stated in quantitative terms, updating on $A_1 \cap \ldots \cap A_m$ by conditionalization: then given the additional assumption that P1–P3 are the case, one encounters a contradiction, as

[108] The existence of such a proposition A should be unproblematic: for instance, I rationally believe that I will be in my office tomorrow, even though I would not accept a bet on this proposition by which I would win one euro if I were to be in my office tomorrow, while I would lose a million euro if not. By the standard Bayesian interpretation of degrees of belief in terms of betting quotients, this shows that it is rationally possible for me to believe a proposition without assigning to that proposition the maximal degree of belief of 1. Note also that the 'extreme version of the Lockean thesis'—$Bel(X)$ iff $P(X) = 1$—would in fact guarantee the closure of rational belief under conjunction from the start; there would be nothing left to argue for in this appendix. For more on this, see the discussion of what I called the Certainty Proposal in section 1.6.

[109] In their effort to criticize the Lockean thesis, Lin and Kelly (2012b, pp. 958–61) also present a puzzle in which an agent's probability measure is updated by a proposition that is already believed. But there are several differences: they consider a particular example measure that proves it possible to run into a problem, where I am interested in an argument with general premises and an absurd conclusion that shows that one will always run into a problem given the general premises and an arbitrary failure of the closure of belief under conjunction. They apply the Lockean thesis, which I do not. Their preservation principle of 'hypothetico-deductive monotonicity', which they show to be invalidated in their example, is a bridge principle for probability and belief that differs from my purely qualitative preservation principle P2 which is just the rather trivial 'if A is in K (and K is consistent), then $K * A = K$' (in belief revision terms, where K is the belief set and $*$ is the belief revision operator). Unlike them, I do not presuppose that belief is functionally determined by a probability measure. Finally, closure under conjunction is not their concern, while it is the central topic of this appendix. In contrast to the additional versions of our paradox that will be stated further later, Lin and Kelly restrict themselves to update by conditionalization, and they do not deal with the potential vagueness of thresholds in bridge principles for belief and probability. This said, their case is very similar to ours in addressing *static* postulates on belief and probability (such as the Lockean thesis) from a *dynamic* point of view.

follows from the considerations above with A being $A_1 \cap \ldots \cap A_m$, and B being A_{m+1}. It seems that the author cannot rationally take in a perfectly positive review of her book. Call this the *Review Argument*.

Before I make the underlying reasoning formally precise, I introduce a second version of the argument in which some of the concepts used in P1 and P3 will be relaxed a bit while the contents of P1 and P3 will be strengthened. Learning evidence with certainty, as covered by P3, is rarely the case in the real world, whereas learning evidence with some probability α just a little short of 1 is much more plausible. Our new P3* will take care of this. Accordingly, P1* will extend P1 to cases in which the degrees of belief of two propositions are sufficiently close to each other without being strictly identical, where 'sufficiently close' will be treated as a vague term. By these changes I will be able to avoid replies to the argument above of the form: sure, the agent cannot rationally update by conditionalization in the story from before, but conditionalization is artificial anyway.

This second version of our argument will proceed from three premises again, amongst which the second premise P2* will simply coincide with P2 from above (which is why I will not state P2* again):

P1* *For almost all numbers s', if the degrees of belief that the agent assigns to two propositions are sufficiently similar to s', then either the agent believes both of them or neither of them.* That is:

For almost all $0 \leq s' \leq 1$, for all X, Y: if both $P_t(X)$ and $P_t(Y)$ are sufficiently close to s', then

$$Bel_t(X) \text{ iff } Bel_t(Y).$$

P3* *When the agent learns, this is captured probabilistically by Jeffrey conditionalization* (see Chapter 11 of Jeffrey 1965, or section 3.2 of Jeffrey 2004). That is:

For all X (with $P_t(X) > 0$): if between t and $t' > t$ the evidence that the agent obtains leads her to impose the probabilistic constraint

$$P_{t'}(X) = \alpha,$$

then for all Y:

$$P_{t'}(Y) = \alpha \cdot P_t(Y \mid X) + (1 - \alpha) \cdot P_t(Y \mid \neg X).$$

P1* is a strengthening of P1 that allows for cases in which two propositions X and Y are assigned only sufficiently similar degrees of belief by a perfectly rational agent, and yet the agent must still treat X and Y equally with regard to belief. Once again, every supporter of the Lockean thesis must accept this: as long as s' is not equal to the threshold s itself, it holds that if both $P(X)$ and $P(Y)$ are sufficiently close to s', then either both of them will exceed s (when $s' > s$) or neither of them will (when $s' < s$). Therefore, P1* holds, where in this case 'almost all' means: *all except for one* (that is, s).

In order to be able to derive P1* from the Lockean thesis, it would not be possible to omit this qualification of 'almost all', for if $P(X)$ is very close to s but less than s, whereas $P(Y)$ is very close to s but greater than s, then X is not to be believed according to the Lockean thesis whereas Y is. However, just as it was the case for P1, also P1* may be expected to hold on far more general grounds than just the Lockean thesis.

The terms 'almost all' and 'sufficiently close' in P1* are meant to express vague concepts, but that should not bother us too much: the potential vagueness of the threshold in the Lockean thesis is generally not perceived to be a problem either.[110] In fact, there is an even stronger correspondence to the literature on vagueness: in the terminology of that literature, P1* says more or less (ignoring possible complications from the 'almost all' quantifier) that belief is *tolerant* with respect to degrees of belief (see Shapiro 2006, p. 8); but I leave this to one side now. I will not presuppose any particular semantic method of making the vague expressions 'almost all' and 'sufficiently close' more precise (whether in terms of a supervaluation semantics, or a measure-theoretical 'Lebesgue measure understanding' of 'almost all', or whatever else). For the argument below, amongst other possibilities, the following manner of making P1* 'crisp' would do: for all degree-of-belief functions P_t and belief sets Bel_t of a perfectly rational agent, for all numbers $0 \leq s' \leq 1$ except for one, there is some ('small') number $1 - \epsilon$, such that for all propositions X, Y: if both $|P_t(X) - s'| < 1 - \epsilon$ and $|P_t(Y) - s'| < 1 - \epsilon$, then it holds that $Bel_t(X)$ if and only if $Bel_t(Y)$.[111]

P3* is one of the usual diachronic Bayesian postulates. In the extreme case in which $\alpha = 1$, Jeffrey conditionalization simply turns into standard conditionalization on the evidence. In this sense, the original postulate P3 is actually but a special case of P3*.

110 Compare Foley (1993, p. 142): 'in itself this [the vagueness of the threshold in the Lockean thesis] isn't a serious objection to the Lockean thesis. It only illustrates what should have been obvious from the start—namely, the vagueness of belief-talk.' It would be interesting to investigate this vagueness aspect of the Lockean thesis in more detail, but for the sake of simplicity I will simply ignore it in this book except for the present appendix. Instead I will always take the threshold numeral in the Lockean thesis to be 'crisp' in all other parts of this book.

111 One might worry that P1* would be susceptible to a Sorites-type of reasoning that would lead to absurdity: start with a belief in a proposition X_1 that has probability x. Then find another proposition X_2 whose probability is $x-\delta$, where δ is sufficiently small as to make no difference to whether something counts as a belief or not (by the lights of P1*). Then find another proposition X_3 the probability of which is $x - 2\delta$. And so on. One might believe that eventually one would find a proposition X_k whose probability is small enough not to count as believed. If so, somewhere along the way there would have to be a pair of 'adjacent' propositions, differing in probability by only ϵ, with the first believed but the second disbelieved, contra P1*. Fortunately this is not actually the case: first of all, the tolerance principle that is enshrined in P1* only holds for almost all numbers, not for all of them, which is why there would be no guarantee for this sequence of reasoning steps to go through for each of x, $x - \delta$, $x - 2\delta$, and so on. Secondly, and more importantly, there is no guarantee either that at each of the steps a modification of the probability in question by one and the same amount of δ would count as 'sufficiently small'. P1* demands only for almost all x the existence of some such δ, but not necessarily the same such δ for different x. For instance, consider the Lockean thesis with a threshold of 0.9: subtracting a δ of 0.05 from an initial probability x of 0.96 would work precisely one time without changing belief into disbelief, but then for the resulting second probability $x - \delta$, that is, 0.91, subtracting by 0.05 would not be licensed any more by P1*, though subtracting e.g. by a δ' of 0.005 would be. No Sorites problem emerges from this. (I am grateful to an anonymous referee of Leitgeb 2014d for bringing this to my attention.)

As in the previous argument, assuming that a perfectly rational agent's beliefs in A and B are not closed under conjunction will again entail an absurd conclusion: the agent cannot update on A in such a way that the probability of A becomes close to 1. In the Review Argument situation, the author cannot update on a friendly review of the form: '*What I can say about this book is that I can very much confirm* $A_1 \cap \ldots \cap A_m$.'

I will now spell out the argument in full formal detail, where I will deal with both variants of the argument at the same time. Given either P1–P3 or P1*–P3*, suppose some perfectly rational agent's beliefs at time t_0 are such that

$$Bel_{t_0}(A), Bel_{t_0}(B), \text{ but } \textit{not } Bel_{t_0}(A \cap B).$$

I also presuppose that $0 < P_{t_0}(A) < 1$.

Assume that the agent receives precisely evidence A between t_0 and t_1: in qualitative terms, this means that the evidence that the agent obtains between t_0 and $t_1 > t_0$ is the proposition A; in quantitative terms, it means that the evidence that the agent obtains between t_0 and t_1 leads her to impose the probabilistic constraint

$$P_{t_1}(A) = \alpha$$

for some α that is either identical to 1, in the first version of the argument, or at least close to 1, in the second version. I presuppose the qualitative and the quantitative ways of describing the agent's evidence to be applicable simultaneously.

Leaving the exact value of 'α' open for the moment, consider next the following thought experiment: think of α gradually tending towards 1. Then, with increasing α, it must be the case that $P_{t_1}(B)$ and $P_{t_1}(A \cap B)$ will get ever closer to $P_{t_0}(B \mid A)$. For, by P3*, learning proceeds by Jeffrey conditionalization, and hence

$$P_{t_1}(B) = \alpha \cdot P_{t_0}(B \mid A) + (1 - \alpha) \cdot P_{t_0}(B \mid \neg A),$$

which tends towards $P_{t_0}(B \mid A)$ when α tends towards 1. And by the definition of conditional probability, the same holds for:

$$P_{t_1}(A \cap B) = \alpha \cdot P_{t_0}(A \cap B \mid A) + (1 - \alpha) \cdot P_{t_0}(A \cap B \mid \neg A) = \alpha \cdot P_{t_0}(B \mid A).$$

Therefore, when α tends towards 1, both $P_{t_1}(B)$ and $P_{t_1}(A \cap B)$ tend towards the same number $P_{t_0}(B \mid A)$. In the extreme case $\alpha = 1$ (as covered by P3), it simply holds that $P_{t_1}(B) = P_{t_1}(A \cap B) = P_{t_0}(B \mid A)$. Either way, there must be an α so close to 1 that the degrees of belief that the agent assigns to B and $A \cap B$ at t_1 are sufficiently similar to the number $s' = P_{t_0}(B \mid A)$. In the second version of the argument, I am simply going to suppose that this number s' is amongst the 'almost all numbers' over which P1* quantifies; so this is really another modest constraint on what A, B, and P_{t_0} are like. For instance, if 'almost all' means 'all except for s', then the additional assumption will be that A, B, and P_{t_0} are so that $P_{t_0}(B \mid A) \neq s$, and I add this assumption to the presumed failure of the closure of *Bel* under the conjunction of A and B. The additional constraint is modest then in the sense that $P_{t_0}(B \mid A)$ can still be 'almost' any number: any number with just one exception.

Now assume that the agent's evidence imposes on her the probabilistic constraint from above for such an α. From $P_{t_1}(B)$ and $P_{t_1}(A \cap B)$ being equal or at least sufficiently close to $s' = P_{t_0}(B \mid A)$, it follows with $t = t_1$ and P1/P1* that

$$\text{(i.i) } Bel_{t_1}(B) \text{ iff } Bel_{t_1}(A \cap B),$$

and the agent's updating on A entails with $t = t_0$, $t' = t_1$, $Bel_{t_0}(A)$ (by assumption), and P2/P2* that both

$$\text{(ii.i) } Bel_{t_1}(B) \text{ iff } Bel_{t_0}(B)$$

and

$$\text{(ii.ii) } Bel_{t_1}(A \cap B) \text{ iff } Bel_{t_0}(A \cap B)$$

must be the case.

By assumption again, it holds that $Bel_{t_0}(B)$, which implies with (ii.i) and (i.i) that

$$Bel_{t_1}(A \cap B),$$

but then again, by assumption, $Bel_{t_0}(A \cap B)$ does not hold, which entails with (ii.ii) that

$$\text{not } Bel_{t_1}(A \cap B).$$

So we end up with a contradiction.

With P1–P3 or P1*–P3* and some failure of closing belief under conjunction being in place (as well as a minor additional assumption in the second version of the argument as mentioned before), it cannot happen that our perfectly rational agent adapts to evidence as described: she cannot update, in qualitative terms, on the proposition A, and at the same time, as far as the probabilistic side is concerned, update by conditionalizing on A or by Jeffrey conditionalizing on A with an α sufficiently close to 1.

Before I turn to the conclusions that one ought to draw from this, I will briefly discuss two further variants of the argument in which premises P1/P1* are modified.[112]

A.3 A Variation

Let us replace P1 from the last section by this principle:

Q1 *If the degree of belief that the agent assigns to a proposition is identical to* 1, *then the agent believes the proposition.* That is:

For all X: if $P_t(X) = 1$, then $Bel_t(X)$.

[112] This variation of the argument was suggested to me by David Makinson in personal communication.

Q1 is entailed by some theories of belief or acceptance. Indeed, '*I assign maximal degree of belief to X (I am certain that X is true), but I do not believe X*' sounds strange again.

Accordingly, replace P1* from the last section by

Q1* *If the degree of belief that the agent assigns to a proposition is sufficiently close to 1, then the agent believes the proposition.* That is:

For all X: if $P_t(X)$ is sufficiently close to 1, then $Bel_t(X)$.

Here, 'sufficiently close' is a vague term again. For the argument that follows, for instance, it would be sufficient to make Q1* crisp by: for all degree-of-belief functions P_t and belief sets Bel_t of a perfectly rational agent, there is some number α (that is 'close' to 1), such that for all propositions X: if $P_t(X) > \alpha$, then it holds that $Bel_t(X)$. This will then amount to an instance of the right-to-left direction of the Lockean thesis.

Q1 follows from our original P1 and the additional assumption that there exists at least one proposition of probability 1 (e.g. a tautology) that is believed by the agent. Q1* follows from P1* given the same additional assumption together with the premise that 1 is amongst the 'almost all' numbers s' over which P1* quantifies.

Other than Q1/Q1*, I will only presuppose P2 (= P2*) and P3/P3* as used before; so Q2 = P2, Q3 = P3, Q2* = P2 = P2*, Q3* = P3*.

Now we reason as follows: assuming Q1/Q1*, Q2 (= Q2*), and Q3/Q3*, suppose there exist two propositions A, B of positive probability that are probabilistically independent of each other if measured relative to a perfectly rational agent's degree-of-belief function at t_0. That is:

$$P_{t_0}(A \cap B) = P_{t_0}(A) \cdot P_{t_0}(B),$$

or equivalently

$$P_{t_0}(B) = P_{t_0}(B \mid A) = P_{t_0}(B \mid \neg A) \text{ and } P_{t_0}(A) = P_{t_0}(A \mid B) = P_{t_0}(A \mid \neg B).$$

And let us suppose again that the agent believes each of A and B at t_0 but does not believe their conjunction:

$$Bel_{t_0}(A), Bel_{t_0}(B), \text{ not } Bel_{t_0}(A \cap B).$$

Finally, assume that the agent's stream of evidence makes her update first on A (between t_0 and t_1), and then on B (between t_1 and t_2), taking each of their probabilities either to 1—in the Q1–Q3 version—or very close to 1—in the Q1*–Q3* version—where we exploit the independence of A and B and apply Q3/Q3* first for a suitable α and then for a suitable α'. It follows from the properties of conditionalization (Q3) and Jeffrey conditionalization (Q3*) that the independence of A and B will not be affected by this sequence of updates.

Formally: whatever the value of α is like, updating first on A leaves the probability of B the same, by B being independent of A relative to P_{t_0}:

$$P_{t_1}(B) = \alpha \cdot P_{t_0}(B \mid A) + (1-\alpha) \cdot P_{t_0}(B \mid \neg A) = \alpha \cdot P_{t_0}(B) + (1-\alpha) \cdot P_{t_0}(B) = P_{t_0}(B).$$

At the same time, the probability of A becomes α, of course:

$$P_{t_1}(A) = \alpha.$$

Furthermore, A is still independent of B relative to P_{t_1} by the definition of conditional probability and B being independent of A at P_{t_0}, as follows from

$$P_{t_1}(A \cap B) = \alpha \cdot P_{t_0}(A \cap B \mid A) + (1-\alpha) \cdot P_{t_0}(A \cap B \mid \neg A) = \alpha \cdot P_{t_0}(A \cap B \mid A) = \alpha \cdot P_{t_0}(B)$$

and $\alpha \cdot P_{t_0}(B)$ being equal to $P_{t_1}(A) \cdot P_{t_1}(B)$, by what was shown above.

For similar reasons as before, updating P_{t_1} on B now leaves the probability of A the same while the probability of B becomes α':

$$P_{t_2}(A) = P_{t_1}(A) = \alpha \text{ and } P_{t_2}(B) = \alpha'.$$

In the case of the argument from Q1–Q3, of course, both α and α' are 1, and then the two updates are nothing but instances of conditionalization on A and B, respectively.

In any case, by Q2/Q2* and assuming α and α' to be equal to, or at least sufficiently close to, 1, the agent must continue to believe each of A, B, while still not believing their conjunction $A \cap B$. But if α and α' are 1 or sufficiently close to 1, then also the probability of $A \cap B$ must be 1 or sufficiently close to 1; for it follows from the axioms of probability that $P(A \cap B) \geq P(A) + P(B) - 1$. (If the original α and α' are not close enough to 1, let them be even closer!) Thus, $A \cap B$ must in fact be believed by the agent in view of Q1/Q1* from above. Contradiction. Therefore, given either Q1–Q3 or Q1*–Q3* and a failure of closing belief under the conjunction of two probabilistically independent propositions, the agent could not update on these propositions one after the other, which is again absurd.

In the Review Argument situation, this would correspond to the reviewer stating (in the Q1–Q3 case) '*I can say that* $A_1 \cap \ldots \cap A_m$ *is definitely the case. The same holds for* A_{m+1}' or (in the Q1*–Q3* case) '*I can very much confirm* $A_1 \cap \ldots \cap A_m$. *I can also very much confirm* A_{m+1}', where the author's claims $A_1 \cap \ldots \cap A_m$ and A_{m+1} happen to be independent of each other as measured by the author's degree-of-belief function. Another pair of reviews that our poor perfectly rational author is not able to enjoy.

A.4 Conclusions

What I have shown in section A.2 on the basis of P1–P3 and P1*–P3* was: if $Bel_{t_0}(A)$, $Bel_{t_0}(B)$, and not $Bel_{t_0}(A \cap B)$ (and $0 < P_{t_0}(A) < 1$), then *our perfectly rational agent can never simultaneously update her beliefs by A and also update her degree-of-belief function by assigning the maximal or at least some sufficiently high probability to A*. (In the 'sufficiently high' case, this was subject to a weak additional constraint on $P_{t_0}(B \mid A)$ that I will simply suppress in what follows.) Similarly, in the last section, I showed that if one relies on Q1–Q3 or Q1*–Q3*, *a perfectly rational agent can never* *update in the respective manner first on A, and then on B, where the two propositions are probabilistically independent.*

Obviously, this is absurd: why couldn't a perfectly rational agent update on evidence in these ways? How else should e.g. the author in the Review Argument react to the positive reviews of his book as described in sections A.2 and A.3? Either the relevant premises cannot all be true, or

$$Bel_{t_0}(A), Bel_{t_0}(B), \text{ and not } Bel_{t_0}(A \cap B)$$

cannot hold if the agent in question is perfectly rational.

Which one should be given up? As always, different philosophers might give different diagnoses: a radical Bayesian, such as Richard Jeffrey, might take the whole misery to be yet another indication that the concept of all-or-nothing belief itself ought to be abandoned; they might say that not even dropping the closure of belief under conjunction can save the epistemologist of belief, and all qualitative talk of 'learning (or updating on) a proposition' needs to be given up accordingly. I will not argue against this way out of the Review Argument here, but following it would certainly be against the basic assumptions of this appendix and of this book more generally. And it would come with a very high price, as explained in section 1.3.2.

Or all-or-nothing *belief* in and all-or-nothing *learning* of (or *update* on) a proposition are to be kept as concepts, but one of the premises from before is rejected or conjunctive closure is retained. Which one is it?

Perhaps P1/P1*/Q1/Q1* should get rejected, which would mean that belief and degrees of belief would not line up as nicely as e.g. the defenders of the Lockean thesis might have thought. It would not be good enough to know then that a perfectly rational agent believes two propositions to the same, or pretty much the same, degree, in order to infer that she would not believe one of these propositions without believing the other; nor would it be sufficient to know that such an agent assigns the maximal or at least a super-high degree of belief to a proposition in order to conclude that the agent believes that proposition to be true. In the case of the argument from section A.2, in spite of the fact that $P_{t_1}(B)$ is identical or very close to $P_{t_1}(A \cap B)$ after updating on A, it would not be ruled out any more that B is believed by the agent while $A \cap B$ is not; accordingly, mutatis mutandis, for the arguments from the last section.

Or P2 (= P2* = Q2 = Q2*) is being attacked, in which case one should be prepared to accept changes of belief that are grounded in evidence (the propositional contents of) which had been believed from the start. This would go against standard presumptions on qualitative belief revision. But not just that: effectively it would mean that if there is a special system of all-or-nothing beliefs that is part of the agent's overall cognitive system (as in the Independence version of option (iii) from section 1.3.3), then that system would not be able to register the occurrence of certain pieces of evidence—because there would not be any changes of belief about them—even when these pieces of evidence might rationally trigger changes of belief in some other propositions. And such changes would be triggered in cases such as the Review Argument situation if we still grant the combined consequences of probabilistic update (given P3/P3*/Q3/Q3*) and the assumption of P1/P1*/Q1/Q1*. In order to carry out the required revisions of belief, the system of beliefs would need some help and guidance from the agent's system of degrees of belief. In other words: the belief system could not be systemically independent of the degree-of-belief system. For the same reason, the epistemology of belief would not be able to distinguish between cases in which some believed proposition comes along as evidence and nothing ought to be done about this by the agent, and the same believed proposition comes along as evidence and some of the agent's beliefs ought to be revised. For instance, if the evidence has the form that is described by P3* with an α that hardly exceeds the agent's present degree of belief in X, then, presumably, the agent's system of beliefs should not be affected. But if α is really close to 1, then the agent's belief system might be affected, even though in both cases X would have been believed by the agent even before the probabilistic update. So far as the argument from section A.2 is concerned, with P2/P2* being dropped, one would no longer be able to conclude that the agent's belief in B and her disbelief in $A \cap B$ are being preserved when the agent receives the believed proposition A as input; analogously for the arguments from the last section.

One way of putting some pressure on P2 might be to question its validity as far as it applies to *doxastic* (or *modal*) belief contents: for instance, at first one might believe both X and also that it might be the case that not X, but after receiving X as a piece of the evidence one might end up believing X without believing that it might be the case that not X. Or first one believes X and also that *there is a chance that not* X, while when the evidence comes along, one believes X but no longer that there is a chance that not X. If so, then in either of these cases receiving X as evidence would in fact trigger some change of belief, and hence P2 would be false.[113] However, even if this were the case, it would not be clear at all whether this would solve the problem: for the only instances of P2 that were required in order to get the arguments going were about belief contents of the form $A_1 \cap \ldots \cap A_m$ or A_{m+1} which might well be non-doxastic (and non-modal) propositions about, say, the Dead Sea or celestial bodies or natural numbers, depending on what the book in the (sequel to the) Preface Paradox is about. Accordingly, if P2 were restricted just to propositions of that sort, would not the same problematic reasoning go through as before? Furthermore, in the Jeffrey conditionalization versions of the arguments, the evidence did not actually have to push the probabilities of the propositions in question to a degree of 1: hence believing that it might be the case that not X as well as believing that there is a chance that not X might both be rational before and after receiving the evidence, which means that in these cases there are not any obvious changes of beliefs with respect to doxastic (or modal) belief contents either.

[113] I am grateful to an anonymous referee of Leitgeb (2014b) for raising this concern.

In any case, giving up on P2 (= P2* = Q2 = Q2*) would certainly be bad news for those who subscribe to the traditional laws of the rational dynamics of all-or-nothing belief, if they also aim to play by my rules of the game and hence do not reject simultaneous and interlocking descriptions of belief dynamics in qualitative *and* quantitative terms.

Or P3/P3*/Q3/Q3* is denied, which would go against the Bayesian mainstream and my previous Assumption 9 in section 1.5.

Or one returns to the principle of closure of rational belief under conjunction, which, just as dropping any of the previous premises, would have the virtue of saving a perfectly rational agent's beliefs from dynamic incoherence as exemplified by the considerations already given. That is, given the previous premises: saving the agent from the embarrassment of challenging her belief in B or her lack of belief in $A \cap B$ when the evidence strengthens her degree of belief in a proposition A which she already believes to be true. By closure, our perfectly rational agent would simply never find herself in a position at time t_0 in which she believes A and B without also believing $A \cap B$.[114]

Amongst these options, restoring closure of belief under conjunction and/or abandoning P2 (= P2* = Q2 = Q2*) seem to be the most promising emergency exits, and not just because failure of conjunctive closure and P2 (= P2* = Q2 = Q2*) have been the only assumptions that remained invariant throughout all four versions of the argument. Hence, retaining both the concepts of belief and degree of belief in our epistemology, and taking the other premises for granted, the short story is: *if rational belief has the synchronic property of not necessarily being closed under conjunction, then also the rational dynamics of all-or-nothing belief must be quite different from what it is usually taken to be.* Even when a qualitative theory of belief abandons the requirement of closure under conjunction, maybe in order to be closer to a probabilistic theory of belief, differences between the two still emerge when we pass to belief change. Either the traditional epistemology of rational belief preserves closure under conjunction, or it has a more serious problem than it is normally thought to have. Either one takes one step back to the tradition or one moves even further away from it, with not much space left in between. One man's Modus Ponens about this will be another man's Modus Tollens.

In this book I will go with the tradition and preserve closure of rational belief under conjunction (as included in my Assumption 8 from section 1.5).

114 One can prove all of P1–P3, P1*–P3*, Q1–Q3, Q1*–Q3* to be consistent with closure of belief under conjunction. All of these principles can be shown to follow from the joint theory of belief and degrees of belief to be defended in this book: the details are worked out in Chapters 2–4. That theory will be found to have a great variety of models, including also a great variety of models in which some proposition is being believed in spite of its probability being less than 1.

2

The Humean Thesis on Belief

How does rational all-or-nothing belief relate to rational degrees of belief? How do they cohere with each other? In this chapter I will defend an answer to these questions in terms of what I am going to call the *Humean thesis on belief*: it is rational to believe a proposition just in case it is rational to have a stably high degree of belief in it. Or, more precisely: *a perfectly rational agent believes a proposition just in case she has a stably high degree of belief in it.*

In section 2.1, I will motivate the thesis by some considerations on the stability of belief that derive from Hume's *Treatise*, even though ultimately my goals are systematic, not historical. In section 2.2, I will make the Humean thesis formally precise, which is going to lead me to an explication of the mutual coherence between rational belief and rational degrees of belief. Section 2.3 is devoted to the justification of that explication in terms of three of its consequences (given also some background assumptions): the logic of belief (logical closure of rational belief), the Lockean thesis on belief, and the coherence between decision-making based on rational all-or-nothing beliefs and Bayesian decision-making. The upshot will be that there is a stability account of belief that builds a plausible bridge between the formal epistemology of all-or-nothing belief and the formal epistemology of degrees of belief.

The formulation of the stability theory of belief from this chapter will be complemented by various other sets of postulates on belief in subsequent chapters which will all turn out to be (more or less) equivalent to the theory that I am going to develop now.

2.1 Introduction

I am, by far, not the first one to highlight the role that stability plays for belief states. In particular, if Louis Loeb (2001, 2002, 2010) is right, then David Hume's theory of belief in *A Treatise of Human Nature* (Hume 1978) amounts to a stability account of belief.[115]

[115] In the more recent literature on belief, Holton (2014), Wedgwood (2012), Ross and Schroeder (2014), and Leitgeb (2013a, 2014a)—corresponding to my Chapters 3 and 4, respectively—also defend stability views. But these papers differ from the present chapter in various respects. Holton treats both beliefs and intentions as stable 'coordination points' that are not readily being given up. But he also suggests reducing talk about subjective probabilities to talk about all-or-nothing beliefs, which is not what I am going to do. Wedgwood (2012) regards outright belief in a proposition as a stable disposition to assign a practical credence of 1 to that proposition, where practical credences are distinguished from theoretical credences; in contrast, I will not split degrees of belief into practical ones and theoretical ones. (Though maybe

This is how Hume himself presents what is generally regarded as the Humean conception of belief:

an opinion or belief is nothing but an idea, that is different from a fiction, not in the nature or the order of its parts, but in the *manner* of its being conceived . . . An idea assented to *feels* different from a fictitious idea, that the fancy alone presents to us: And this different feeling I endeavour to explain by calling it a superior *force*, or *vivacity* . . .

In other words: an idea assented to, that is, a belief, is characterized by its special liveliness ('force', 'vivacity'), and that is also how Hume's view on belief usually gets summarized in the relevant literature: beliefs are lively ideas.

But, actually, the manner in which he continues the quotation differs from what one might have expected of Hume orthodoxy of this kind:

. . . or *solidity*, or *firmness*, or *steadiness*. (*Treatise*, section VII, part III, book I)

Hume qualifies belief here by means of terms that belong to a different category than that of liveliness: the category of *stability* ('solidity', 'firmness', 'steadiness') or resilience. The same is the case in the following quotation:

its true and proper name is *belief*, which is a term that every one sufficiently understands in common life. [. . .] It gives them [the ideas of the judgement] more force and influence; makes them appear of greater importance; infixes them in the mind; and renders them the governing principles of all our actions. (*Treatise*, section VII, part III, book I)

Having more force and influence, and greater importance, corresponds to the liveliness aspect of belief, whereas 'infixedness' corresponds to stability again. Although Hume does not say so explicitly, one may speculate that liveliness makes belief powerful enough to govern an agent's actions, whereas its stability ensures that the required degree of liveliness is being maintained for a sufficient period of time e.g. until an action is fully executed, and under a sufficiently great variety of conditions. Both

Wedgwood's practical credences can be understood as degrees of acceptance as developed in section 6.3.) Ross and Schroeder's (2014, s. 2.3) stability claim is this: 'A fully rational agent does not change her beliefs purely in virtue of an evidentially irrelevant change in her credences or preferences.' This is not formalized in any way, but, depending on their understanding of 'evidentially irrelevant change in her credences', their thesis might well correspond to some instance of the Humean thesis scheme $HT^r_{\mathcal{Y}}$ in section 2.2. On the other hand, the special Humean thesis HT^r_{Poss} for which I am going to argue below will have the consequence that rational belief is context-sensitive: for instance, if one's willingness to take risks changes from one context to the next, then this may also change one's beliefs. If changes like that count as 'evidentially irrelevant changes', then my favourite Humean thesis will differ from Ross and Schroeder's stability claim. Leitgeb (2013a) considers a way of reducing talk about all-or-nothing beliefs to talk about subjective probabilities, which I will not follow here (and also Chapter 4 of this book, which corresponds to it, has been revised, accordingly). Finally, Leitgeb (2014a) (and Chapter 3, which is based on that article) derives the stability of all-or-nothing belief from other principles (the axioms of subjective probability, the consistency and logical closure of belief, and the so-called Lockean thesis on belief) while in the present chapter I will move in the opposite direction: I will start from the stability of all-or-nothing belief and then derive principles such as the logical closure of belief and the Lockean thesis from it (and background assumptions).

force and stability are necessary for belief to play its intended functional role e.g. in decision-making.

In recent years, Louis Loeb has worked out this stability component of Hume's conception of belief in great detail. Loeb argues that stability is in fact *the* distinctive property of belief according to Hume—in Loeb's words,

> Tradition in Hume interpretation has it that beliefs are *lively ideas*. In my interpretation, beliefs are *steady dispositions*. (Loeb 2002, p. 65)
>
> Hume maintains that stability is the natural function of belief. (Loeb 2002, p. 81)

While liveliness is a property of occurrent beliefs, Loeb tries to show that, on a more fundamental level, Hume is concerned with dispositional belief, which Hume characterizes as a stable disposition to generate lively ideas. In Loeb's terms again:

> A *disposition to vivacity* is a disposition to experience vivacious ideas, ideas that possess the degree of vivacity required for occurrent belief. Some dispositions to vivacity are unstable in that they have a tendency to change abruptly . . . Such dispositions, in Hume's terminology, lack *fixity*. Hume in effect stipulates that a *dispositional belief* is an infixed disposition to vivacity . . .
> (Loeb 2010, p. 205)

That is: ideas are subject to degrees of vivacity, and if an idea's degree of vivacity or liveliness is high enough, then that idea counts as an occurrent belief. If a person has a stable disposition to produce such an idea with sufficiently high degree of vivacity, then Hume would also ascribe a belief to that person, but in that case—without distinguishing the two kinds of belief explicitly—a dispositional belief.

As Loeb makes clear, Hume does not just hold that stability belongs to the nature of (dispositional) belief and hence is relevant to his philosophy of mind, stability also supplies belief with defeasible justification and therefore is equally relevant to Hume's epistemology:

> there must be a property that plays a twofold role. The presence of the property must constitute a necessary condition for belief. In addition, establishing that the beliefs produced by a psychological mechanism have that property must constitute a sufficient condition for establishing justification, other things being equal. My claim is that *stability* is the property that plays this dual role, one within Hume's theory of belief, the other within Hume's theory of justification.
> (Loeb 2002, p. 65)

I will return to the question in what sense stability may belong to the nature of belief and at the same time supply belief with *pro tanto* justification in the next section.

In a nutshell, then, Loeb takes Hume to defend the following thesis on belief:

> *(Dispositional) beliefs are stable dispositions to have ideas with high 'degree of vivacity' on which acting, reasoning, and asserting are based.*

Call this the first, preliminary, and descriptive version of the *Humean thesis on belief*.

This thesis will be the starting point for my own investigations into rational belief in the rest of this chapter. I will make a normative version of the thesis more precise in section 2.2, and I will assess a sufficiently formalized version of it in terms of its consequences in section 2.3.

But before I will do so, let me illustrate—by means of three little examples—that the Humean thesis is plausible independently of (Loeb on) Hume. The point of these examples will be that in order for belief to play its characteristic functional role in decision-making (Assumption 3 in section 1.1), reasoning (Assumptions 2 and 4 in section 1.1), and asserting (Assumption 5 in section 1.1), it needs to be sufficiently stable in the course of processes such as perception, supposition, and communication:

Example 1

I am thirsty; I crave for something to drink. I believe that there is a bottle of apple spritzer in the kitchen, and I also believe that walking to the kitchen, getting me the bottle, and finishing it will quench my thirst. Based on this combination of desires and beliefs, I set out for the kitchen. Along the way, I happen to perceive various things to be so and so; nothing utterly surprising, but I do acquire some new evidence. None of it affects my belief that there is a bottle of apple spritzer in the kitchen, nor any of the other beliefs relevant in the present context. I arrive safely in the kitchen, search for the bottle (it is in the fridge), and . . . yummy!

If encountering these new pieces of perceptual evidence had resulted in abandoning my belief that there is a bottle of apple spritzer in the kitchen, or if it had led me to give up e.g. the belief that emptying the bottle will make me feel good, then I would hardly have been able to reach the kitchen and comply with my desires. Or even if I had, I would have lacked good reason for doing so, for the beliefs on which the decision for this course of action would have been based would not have been intact any more. The upshot is: beliefs need to be stable under acquiring evidence (unless the evidence is thoroughly unexpected).[116]

Example 2

I am still thirsty; I haven't walked to the kitchen as yet. I engage in an episode of suppositional reasoning: I believe that there is a bottle of apple spritzer in the kitchen, and I regard it as more likely than not that it is in the fridge. But suppose that it is not in the fridge: where is it then? I believe that after buying it in the supermarket, I did carry it home in my shopping bag which I like to take into the kitchen. So, under the supposition that the bottle is not in the fridge, and hence that I did not put it there in the first place, it must still be in the shopping bag in the kitchen. Thus, given that it is not in the fridge, I believe it is in my shopping bag.

If supposing that the bottle of apple spritzer is not in the fridge had the effect of cancelling my belief that I carried the bottle home in my shopping bag, or of removing the belief that if it is not in the fridge then I did not put it there in the first place, then

[116] I will return to the topic of stability and action in section 2.3.3 and in more detail in section 6.1, where I will also give a formal analysis of Example 1: see Example 22 in section 6.1.

that act of suppositional reasoning would not have led to the correct outcome. More generally, suppositional reasoning would be quite pointless if one were not able to supplement the assumed proposition by various background beliefs that are preserved by the act of assumption. Therefore: beliefs need to be stable under suppositions (unless they are ruled out by the supposed proposition, of course).[117]

Example 3

Still thirsty. My wife walks by: she says that she is thirsty and asks me about the bottle of apple spritzer. Being a good and altruistic husband, I answer in line with my total relevant beliefs about the situation: 'The bottle of apple spritzer is either in the fridge or in my shopping bag'. She proceeds to the kitchen. I remain thirsty.

Why didn't I just assert that the bottle is in the fridge? After all, I regarded this as likely, and by asserting it I would have been able to convey a stronger piece of information than I actually did. So, for broadly Gricean reasons, shouldn't I have gone for the more informative option? The reason why this might not be so is that what I actually asserted will remain useful to my wife independently of whether one of the disjuncts will be invalidated later; or at least that is what I believe. For if she finds out that the bottle is not in the fridge, she will apply the rule of disjunctive syllogism to my assertion of *fridge or shopping bag*, after which she will look into the shopping bag where I believe she will find the bottle then; and similarly, vice versa. As a responsible communication partner, I am foreseeing these possible developments. In contrast, merely asserting that the bottle is in the fridge would not have helped her in the possible case in which she would not have found it there. Thus: sometimes it is useful to go for the weaker disjunctive assertions, but only if the total relevant beliefs that are to be expressed in terms of a disjunction are stable under the exclusion of disjuncts (unless all disjuncts happen to be excluded).[118]

In the next section, where I will suggest a way of explicating a normative version of the Humean thesis on belief, we will see more precisely what these three examples have in common. For the moment, they should be enough to demonstrate the plausibility of the thesis that at least some kind and degree of stability is essential to belief: without it, belief could not play its functional role successfully.

2.2 Explicating the Humean Thesis

The first step of explicating a normative variant of the preliminary and informal version of the Humean thesis from section 2.1 consists in getting a better grip on the notion of 'degree of vivacity' as assigned to ideas, and on the notion of belief as expressed at the beginning of the thesis.

117 I will return to the topic of stability and suppositional reasoning in Chapter 4, where I will also give a formal analysis of Example 2: see Example 15 in section 4.3.

118 Jackson (1979) and Lewis (1986) argue for a thesis like that in a different context (leading up to their theories of indicative conditionals). I will return to this in much more detail in section 6.2, where I will also formalize Example 3: see Example 26 in section 6.2.

Fortunately, I can again rely on existing work here: Maher (1981) on Hume, and Loeb on Hume again. First of all, Maher (1981) argues that one needs to distinguish two corresponding senses of 'belief' in Hume (both in the *Treatise* and in *An Enquiry Concerning Human Understanding*):

> The belief which is characterized by a superior degree of vivacity is absolute, not admitting of degrees. To believe in this sense of to be *perswaded of the truth of what we conceive* (T97n) [*Treatise*, book I, part III, section VII]. By contrast, belief in the sense in which it is identified with vivacity must be relative, admitting of degrees, and not implying belief in the absolute sense. In order to mark this distinction, I shall restrict the term 'belief' to the absolute sense, and use 'degree of belief' for the relative notion. (Maher 1981, p. 140)

In what follows I shall apply the same terminological convention: using 'belief' for the all-or-nothing state of belief, and 'degree of belief' for the corresponding graded notion. Occasionally, I will also take 'belief' (or 'doxastic attitude') to be an umbrella term that covers both types of belief at the same time, but if so, this should become clear from the context.

The quote from Maher (1981) continues in the following way:

> In this terminology, it is belief that Hume identifies with superior vivacity, and degree of belief that he comes to identify with vivacity. Now these two identifications imply a third, namely that belief is the same thing as a superior degree of belief. This latter identification is far from trivial, and in fact is inconsistent with two other very intuitive principles about beliefs, namely: (1) One should believe the logical consequences of what one believes, and (2) One should not believe a contradiction. The inconsistency is illustrated by the well-known 'lottery paradox'.

We will see in section 2.3 that Maher is actually moving too quickly here so far as the Lottery Paradox is concerned. But what one should register indeed is that, according to Maher on Hume, belief in the absolute sense corresponds to superior degree of belief (which matches the quote from the last section: 'this different feeling I endeavour to explain by calling it a superior *force*, or *vivacity*'). With respect to Humean degrees of belief themselves, Maher (1981) defends the view—more or less, about which more later—that they coincide with our modern degrees of belief in the sense of subjective probability theory. That is: the (rational) Humean degree of belief in X corresponds to the subjective probability of X.

Taking this together with the averred correspondence between belief and superior degree of belief, one may conclude with Maher that Hume also seems to be committed to a version of what is called the Lockean thesis on belief today (see p. 140 of Foley 1993, whose formulation I will use except for replacing his 'degree of confidence' by 'subjective probability'):

> The Lockean thesis: *It is rational to believe a proposition just in case it is rational to assign a sufficiently high subjective probability to it (to have a sufficiently high degree of belief in it).*

Let us pause here for a moment. Some points above are in need of qualification or correction: some points made by Maher, and some points in my presentation of Maher.

First, if one is permitted to identify degrees of belief with subjective probabilities at all, then one may do so, strictly speaking, only for (inferentially) perfectly rational agents, as only such agents may be expected to distribute their degrees of belief over propositions in line with the axioms of probability. But this is unproblematic in the context of this book, since my own purposes are purely epistemological and normative: let us simply assume in the following that we are dealing solely with perfectly rational agents in the relevant sense of the word. One should be aware that this normative focus is quite different from Hume's more descriptive account.

Secondly, as mentioned in the last section, vivacity is meant to involve an occurrent feeling according to Hume, whereas the founding fathers of modern subjective probability theory (such as Frank P. Ramsey) made it very clear that subjective probabilities are not to be regarded as expressing feelings of conviction but rather certain dispositions to act: for instance, dispositions to bet in a certain manner. Initially, this might look like a serious discrepancy. However, if Loeb is right, then vivacity according to Hume is indeed an occurrent manifestation of a *disposition*—as explained in the last section, and as argued by Loeb after, and unknown to, Maher (1981). So the discrepancy is not that great after all: Humean degrees of belief are dispositions to vivacity that correspond to modern subjective probabilities except that they also come with a particular kind of phenomenology attached to them. As far as their functional roles are concerned, the two of them do not seem to differ at all, which is good enough for my purposes here. For example, according to standard Bayesian decision theory, a rational agent ought to choose an action that maximizes expected utility, and if e.g. the subjective probabilities of some of the possible outcomes of an action are particularly high, then the expected utility of the action will be assessed essentially in terms of the utilities of these outcomes. In this way, possibilities with high subjective probability will turn into 'governing principles of all our action', just as Hume maintained about belief in the quote from the last section.

Thirdly, Maher's (1981) proposal concerning degrees of belief actually differs slightly from the one suggested: he does not really identify Humean degrees of belief with subjective probabilities directly but rather with certain quantities that can be defined in terms of subjective probabilities. What he proposes is that the Humean degree of belief in X (relative to a person) coincides with (i) *the (person's) subjective probability of X minus that of $\neg X$* in the case in which this difference is non-negative, and with (ii) *degree* 0 otherwise. This is supposed to reflect Hume's talk of vivacities 'destroying' contrary vivacities (see Maher 1981, pp. 142 and 145) and of the mind oscillating between such contrary forces (see Maher 1981, section 3). In what follows, I will ignore this part of Hume's psychology of belief and stick to the simpler identification between Humean degrees of belief and subjective probabilities. At least in terms of what such degrees of belief are meant to *do* in an agent's cognitive life, the differences between the two analyses seem quite negligible.

Finally, following Loeb's lead again, one should not identify belief with superior degree of belief—a superior graded disposition to act—as suggested by Maher, but rather with a *stable* disposition of that kind: as quoted before, 'A *disposition to vivacity* is a disposition to experience [...] ideas that possess the degree of vivacity required for occurrent belief. Some dispositions to vivacity are unstable... Hume in effect stipulates that a *dispositional belief* is an infixed disposition to vivacity...' (Loeb 2010, p. 205). So, if anything, one ought to identify belief with stably ('infixed') superior degree of belief: stably high subjective probability.[119]

We are in the position now to return to our first and preliminary version of the Humean thesis from section 2.1. Understanding degrees of belief as subjective probabilities, and adding the intended stability component to mere 'superior degree of belief' considerations such as Maher's, my proposal will be to preserve the syntactic form of Foley's formulation of the Lockean thesis from above except for replacing 'high subjective probability' by '*stably* high subjective probability' on its right-hand side. The result is a normative principle (which is what I am actually interested in)—a more precise, though still preliminary version of the *Humean thesis on rational belief*:

> It is rational to believe a proposition just in case it is rational to assign a *stably* high subjective probability to it (or to have a *stably* high degree of belief in it).

Or formulated by reference to perfectly rational agents:

> A perfectly rational agent believes a proposition just in case she assigns a *stably* high subjective probability to it.[120]

[119] According to this analysis, Humean *degrees* of belief are interpreted dispositionally, but they are not as yet ascribed special stability properties (other than those that might be entailed already by the notion of subjective probability itself). It is all-or-nothing belief that exhibits stability. An alternative interpretation of Loeb on Hume would be to build stability into the degrees of belief themselves—maybe something like degrees of resiliency of subjective probabilities in the sense of Skyrms (1977, 1980) or subjective probabilities of high enough graded resiliency—and to identify all-or-nothing belief with superior degrees of belief in that alternative stability sense. This might be even closer to Loeb on Hume, but it would lead to the question of what mental role would be left to be played by plain subjective probabilities.

[120] If we tried to unpack this thesis in terms of some kind of rational obligation statement: how would we do it? An analogy to the possible worlds semantics for deontic operators might help here. The following is yet another reformulation of the Humean thesis (while still suppressing additional parameters, such as points in time and contexts). Pick an arbitrary proposition X. Then *for every perfectly rational agent x it holds: x realizes belief in X if and only if x realizes a stably high degree of belief in X.* If we now replace talk of *perfectly rational agents* by talk of *rationally ideal worlds*, talk of *realization by agents* by talk of *satisfaction by worlds*, and talk of *(degree of) belief states* by talk of *(degree of) belief propositions*, we get: *For every rationally ideal world w it holds: w satisfies the proposition [X is believed] if and only if w satisfies the proposition [X has a stably high degree of belief].* Semantically, this is equivalent to: *For every rationally ideal world w it holds: w satisfies the proposition [X is believed if and only if X has a stably high degree of belief].* By the semantics of the obligation operator, this is again semantically equivalent to: *It is rationally obligatory that: X is believed if and only if X has a stably high degree of belief.* So the Humean thesis may be viewed as the following state of affairs to be rationally obligatory: the equivalence of having a belief in X and having a stably high degree of belief in X. In this sense, rational obligation has a wide-scope reading in the Humean thesis. I am grateful to Catrin Campbell-Moore for urging me to address this question.

Being of great force, that is, having a sufficiently high degree of belief, is indeed necessary for a proposition to be believed; but additionally this great force also needs to be stable enough: as we have learned from Loeb, Humean belief entails not just high degree of belief but an 'infixed disposition' to having such a high degree of belief. Therefore, if it is rational for an agent to believe a proposition, it must also be rational for that agent to assign a stably high degree of belief to that proposition; which gives us the left-to-right (only if) direction of the thesis.

The right-to-left (if) direction requires additional argument: one way of arguing for it would be to understand 'Beliefs are stable dispositions to assign high degrees of belief' as 'Beliefs are *nothing but* stable dispositions to assign high degrees of belief': if being a stable disposition to have high enough degrees of belief is all there is to a belief, then if it is rational for an agent to assign a stably high degree of belief to a proposition, it must also be rational to believe the proposition. This would be a special normative version of the Reduction Option (ii) from section 1.3.2. And for all I know a descriptive variant of this Reduction Option might well be what Hume himself had in mind.

However, there are also stability views of belief which still entail the equivalence thesis but without regarding beliefs as *nothing but* dispositions to have stably high degrees of belief. Let me give you an example (which is at the same time my own preferred interpretation for reasons that I have explained at the end of section 1.3.3 and which I will make clearer later in Appendix C).

Picture a 'dual processing account'[121] of belief along the lines of the Independence Option in section 1.3.3 of Chapter 1: say, a rational agent has a system of all-or-nothing beliefs and a system of degrees of belief at the same time, where each of them is ontologically independent of the other. In principle, one could eliminate one of the systems, and the remaining system would still be able to function successfully (though maybe less successfully than before), that is, committing the agent to action (in collaboration with the agent's desires), being revisable by means of perception and inference, leading the agent to express herself in terms of assertion, and the like. Both systems maintain dispositional states of belief, though the all-or-nothing system is perhaps more closely tied to conscious reasoning and language processing in the sense that its dispositional all-or-nothing beliefs can easily be made explicit and occurrent in the conscious mind for certain periods of time. Whereas the degree-of-belief system is too complex to be accessible consciously in such an immediate manner and hence remains mostly implicit. Either way, assume that neither of the agent's systems is 'surgically removed' but that the two of them work simultaneously within the same overall cognitive system: in that case they will need to rationally 'cohere' with each other in order for the agent to behave rationally overall.

For example, although slight discrepancies between their outcomes may be forgivable—and in fact unavoidable, as all-or-nothing beliefs are simpler, more

[121] For more on dual process theories in general, see e.g. Evans (2008).

coarse-grained, and hence less sophisticated creatures than their numerical siblings—the following should not be the case: the system of degrees of belief strongly recommends a certain course of action while the system of all-or-nothing beliefs discourages the agent from taking that course of action. (Nor vice versa.)

What joint constraint on the degrees of belief and all-or-nothing beliefs of perfectly rational agents—over and above the left-to-right direction of the thesis, which we already accepted—would guarantee these agents not to face such normative dilemmas? My proposal, at the very least, is to rule out those situations in which the agent's degree-of-belief system would make the agent assign a (sufficiently) stably high probability to a proposition, while the agent's all-or-nothing belief system would *not* make the agent believe that proposition in the categorical sense: as such situations are tantamount to leading the agent into normative dilemmas of the respective kind. For instance, in the apple spritzer Example 1 from section 2.1, having stably high degrees of belief in the relevant propositions would make one's numerical belief system guide a person downstairs to the refrigerator, while if one's categorical belief system were to lack the relevant all-or-nothing beliefs, it would not give the same person any reason for doing so and maybe recommend instead to stay put; which would leave the person internally incoherent. Hence, if a perfectly rational agent with the respective dual architecture assigns a (sufficiently) stably high subjective probability to a proposition, it should be the case also that the agent believes that proposition in the all-or-nothing sense, yielding the right-to-left direction of the thesis above.

I will leave open on what grounds exactly the right-to-left direction of the Humean thesis is to be justified; as we have seen, there is more than one option here. In the following, I will simply take the thesis for granted in its full equivalence form and explore how it can be made more precise and what consequences its precisifications may have. Accordingly, I will also set aside, though only for the moment, one other concern: whether one may consistently maintain (and consistently ascribe to Hume) both this Humean thesis on rational belief and the so-called Lockean thesis on rational belief as explained before. We will see in section 2.3 that this is in fact feasible. But for the moment let us focus just on the Humean thesis, which I am not done explicating as yet.

For the next obvious question to ask—and hence the next step to take in our explication—is: what exactly is a *stably* high subjective probability or degree of belief? In particular: stable *under what*? Once again I am able to build on existing work when addressing this question. In his theory of objective chance, Skyrms (1977, 1980) emphasizes the importance of probabilistic notions of *resiliency with respect to conditionalization*: of the probabilities of certain propositions remaining approximately invariant when taking conditional probabilities given various relevant propositions. Skyrms made several such notions of resiliency formally precise and applied them in the course of his argument for the thesis that objective chances are nothing but resilient subjective probabilities. Although the goal and context of the present project differ from Skyrms's—my focus is belief, whereas he is after chance—one can gain from his work an insight into the salient role of stability or resiliency under *conditionalization*.

Accordingly, when reconsidering the three examples from the end of section 2.1, one finds that standard Bayesian probability theory would analyse each of them in terms of conditionalization. In Example 1, beliefs were meant to be stable under evidence acquired by perception, which in Bayesian terms would correspond to update on perceptual evidence by conditionalization.[122] In Example 2, beliefs were argued to be stable under supposition, and again the standard Bayesian explication of supposition would be conditionalization on the supposed proposition.[123] Finally, Example 3 concerned a communication partner's update on an asserted disjunction and her foreseeable possible further updates on the negation of one of the disjuncts, which again a Bayesian would make sense of in terms of the corresponding conditional probabilities. In other words: what the three examples have in common from the standpoint of standard subjective probability theory is that they all concern stability under conditionalization.

Last but not least, understanding stability in this way also gives us a nice answer to one of the remaining questions from the last section, that is, how Hume's theory of belief can be one of the nature of belief and of the justification of belief at the same time: the proposal is that stability, which is part of the nature of belief, is explained in terms of the fundamental operation of update or probability change that Bayesian probability theory regards as justified.[124] Update by conditionalization may in fact be said to secure an agent's resulting degree-of-belief function a state of all-else-being-equal justification, where the 'all-else-being-equal' qualification is due to remaining concerns regarding the proposition on which the agent updates, her prior subjective probability measure, and the defeasibility that arises from the possibility of further updates on propositions that might alter again the agent's degree-of-belief function. In particular, if one grants conditionalization to be warranted in the sense of Epistemic Decision Theory—and indeed e.g. Greaves and Wallace (2006) and Leitgeb and Pettigrew (2010b) have argued that conditionalization is justified in the sense of 'getting an agent as close to the truth as possible' in the face of new evidence—then the stability of belief with respect to conditionalization is actually not so far from Loeb's own considerations when he writes: 'We might put this by saying that considerations of stability *absorb* considerations of truth; the regulative disposition [in virtue of which belief aims at the truth] operates through its impact on stability' (Loeb 2002, p. 84).[125]

[122] Of course, Bayesian epistemologists might also draw on alternative methods of update instead, such as Jeffrey updating (i.e. Richard Jeffrey's probability kinematics, as encountered already in Appendix A). But normally such alternative update methods may at least be viewed as generalizations or graded versions of update by conditionalization.

[123] By this I do not want to claim that there are no relevant differences between learning a proposition and supposing it; there certainly are, and they show up most clearly in the case of learning vs supposing introspective propositions. But for most ordinary purposes, learning and supposing may indeed both be represented formally by conditionalization. I will return to this in sections 4.1.1 and 6.2.

[124] See e.g. Leitgeb and Pettigrew (2010b, p. 244) for a brief survey of arguments justifying update by conditionalization.

[125] This said, there are also important differences between this conception of stability under conditionalization and Loeb's views on stability: in particular, Loeb emphasizes the objective reliability of certain

It is about time to make things formally more precise now. Let us consider a perfectly rational agent's beliefs and degrees of belief at a certain point in time. I assume, for simplicity, that the agent's doxastic attitudes concern a finite and non-empty set W of logically possible worlds (such as the logically possible worlds for the language of propositional logic with finitely many atomic formulas); in probabilistic terms, W will function as the sample space. Intuitively, one ought to think of the members of W as coarse-grained, mutually exclusive, logically possible ways the actual world might be, so that whatever the actual world is like, it instantiates or belongs to one of these coarse-grained ways or worlds. By a proposition I simply mean a subset of W. For instance, the empty set $\varnothing$ is nothing but the (uniquely determined) contradictory proposition that is not true in any world. Let *Bel* be the class of all propositions believed by the agent at the time. I will write '$Bel(X)$' for X being a member of *Bel*. Let P be the agent's degree-of-belief function at the time by which numerical strengths of belief are assigned to propositions and which I assume to satisfy the laws of probability.[126]

I want to make precise the thesis: *a perfectly rational agent believes a proposition just in case she assigns a stably high subjective probability to it.* If believed propositions are supposed to have stably high probabilities, where 'stably high' is short for 'stably high under conditionalization', then there must be some set $\mathcal{Y}$ of propositions, such that 'stably high under conditionalization' really means: stably high under conditionalization on the members of $\mathcal{Y}$. In the following, '$\mathcal{Y}$' will function as a parameter, and I will study different ways of how it might, and ought to, be set. Presumably, one of the members of $\mathcal{Y}$ will be the set W of all worlds, so that a believed proposition, say X, will also have a sufficiently high *absolute* probability (or conditional probability given W); and the more additional members the set $\mathcal{Y}$ will have, the more stable this high absolute probability for X will be, for conditionalizing the agent's degree-of-belief function on *any* of these members will be postulated to preserve the high enough probability for X. Finally, 'high enough' or 'sufficiently high' can be made precise in terms of a threshold condition: having a probability above a certain number r, where the exact value of 'r' may be supposed to depend on context in some way, but where in any case it may be assumed to be greater than or equal to $\frac{1}{2}$ (such that the degree of belief in a believed proposition is always greater than that of its negation). So, overall, we have:

(Left-to-right direction of the thesis) If X is believed by our perfectly rational agent, that is, if $Bel(X)$, then for all propositions Y, if Y is in $\mathcal{Y}$ (and $P(Y) > 0$, so that conditionalization on Y is well-defined[127]), the conditional probability of X given Y,

processes that lead to stable belief, which differs from the mostly internalist perspective on stability that I am taking in this book.

[126] That is, I assume that: P maps subsets of W into the interval of real numbers that lie (not strictly) between 0 and 1, such that: $P(W) = 1$, and for all $X, Y \subseteq W$, if $X \cap Y$ is the empty set, then $P(X \cup Y) = P(X) + P(Y)$. Conditional probabilities are defined by: if $P(X) > 0$, then $P(Y|X) = \frac{P(Y \cap X)}{P(X)}$.

[127] Alternatively, one might develop the present theory for primitive conditional probability measures (often called Popper functions) which are not defined in terms of ratios of absolute probabilities and for

$P(X|Y)$, is above the threshold r. Or in plain words: believed propositions have a stably high subjective probability. The right-to-left direction of the thesis is the same vice versa.

This leads us to our next version of the Humean thesis on belief, which will take us one step closer to the explication (though not explicit definition) of rational belief and of the coherence between categorical and graded rational belief:

If *Bel* is a perfectly rational agent's class of believed propositions at a time, and if P is the same agent's subjective probability measure at the same time, then

$\mathrm{HT}^r_{\mathcal{Y}}$: For all X: $Bel(X)$ iff for all Y, if $Y \in \mathcal{Y}$ (and $P(Y) > 0$), then $P(X|Y) > r$.

'$\mathrm{HT}^r_{\mathcal{Y}}$' is short for: *Humean thesis with parameters* $\mathcal{Y}$ *and* r. Since $\mathrm{HT}^r_{\mathcal{Y}}$ concerns a perfectly rational agent's states of belief *at one instant of time*, it is a synchronic norm on the agent's states of belief; however, to the extent that conditional probabilities determine an agent's disposition to change her degrees of belief conditional on evidence or suppositions, $\mathrm{HT}^r_{\mathcal{Y}}$ also concerns an agent's dispositions to change her beliefs in time. And because $\mathrm{HT}^r_{\mathcal{Y}}$ is about such an agent's *beliefs and degrees of belief* simultaneously, it is nothing but a normative bridge principle for categorical and numerical belief. The principle says that the beliefs and degrees of belief of any perfectly rational agent are such that the condition is satisfied.

If we take the context to determine the value of 'r' (but high enough so that $\frac{1}{2} \leq r < 1$), the final open question about our intended explication of the Humean thesis on belief is: what exactly is that class $\mathcal{Y}$ of propositions relative to which a believed proposition ought to have a stably high probability under conditionalization? Once this question has been answered, the Humean thesis template $\mathrm{HT}^r_{\mathcal{Y}}$ will have turned into a proper thesis.

One initial observation about $\mathrm{HT}^r_{\mathcal{Y}}$ is quite obvious: smaller sets $\mathcal{Y}$ of propositions are going to yield braver (or bolder) belief sets *Bel*, whilst larger sets $\mathcal{Y}$ will return more cautious belief sets *Bel*. That is because $\mathrm{HT}^r_{\mathcal{Y}}$ with a large set $\mathcal{Y}$ imposes a heavy constraint on *Bel*—the probability of each believed proposition conditional on each of the many members of $\mathcal{Y}$ ought to be high enough—whereas for a much smaller set $\mathcal{Y}$ the constraint that is imposed by $\mathrm{HT}^r_{\mathcal{Y}}$ will be much less demanding. So: how cautious or brave should *Bel* be? Is this merely a pragmatic matter? That is: does anything go, depending only on the agent's courage and personality?

Before I address these questions, I will first discuss a couple of salient conceivable choices for $\mathcal{Y}$ in order to see what kinds of belief sets *Bel* they would determine, or, speaking more properly: what kinds of joint constraints on the agent's set *Bel* of believed propositions at a time and on the same agent's degree-of-belief function P

which also the conditionalization on zero sets is well-defined; but I will refrain from doing so here. See Pedersen and Arló-Costa (2012) for an extension of my stability theory to such generalized conditional probability measures.

at the same time they would impose. After all, as mentioned before, $\mathrm{HT}^r_{\mathcal{Y}}$ is really a bridge principle in which belief and subjective probability figure simultaneously.

Each of the proposals of how to determine $\mathcal{Y}$ will be salient in the sense of being plausible (at least prima facie), simple, and expressible solely in terms of '*Bel*' and '*P*'. And it will be helpful to view each of them as formulating a condition that concerns a set $\mathcal{Y}$ of potential *defeaters*: by that I mean here propositions Y that might potentially decrease the probability of a believed proposition X beneath the given threshold r.[128] Sticking any such 'potential defeater condition' into the placeholder '$Y \in \mathcal{Y}$' on the right-hand side of the equivalence in $\mathrm{HT}^r_{\mathcal{Y}}$ will then have the effect of demanding of the members X of *Bel* (as expressed on the left-hand side of that equivalence) *not* to be defeated by any such potential defeater Y. Defeaters of this kind are not far from rebutting defeaters in the sense of Pollock (1986), as decreasing the probability of X by means of conditionalizing on the potential defeater Y coincides with increasing the probability of $\neg X$, that is, the negation of X (or rather the complement of X, that is, the proposition $W \setminus X$ or the set of worlds in W that are not members of X).

So here are some salient proposals for how to determine $\mathcal{Y}$:

(a) $Y \in \mathcal{Y}$ iff $P(Y) = 1$: *The Lockean Thesis proposal.*
For with that set $\mathcal{Y}$ in place, $\mathrm{HT}^r_{\mathcal{Y}}$ reduces to: for all X, $Bel(X)$ iff $P(X) > r$.

(b) $Y \in \mathcal{Y}$ iff $Bel(Y)$: *A coherence theory of belief.*

(c) $Y \in \mathcal{Y}$ iff $P(Y) > r^*$:[129] *A modestly cautious proposal.*

(d) $Y \in \mathcal{Y}$ iff $\underbrace{Poss(Y)}_{\text{not } Bel(\neg Y)}$: *Another cautious proposal.*

(e) $Y \in \mathcal{Y}$ iff $P(Y) > 0$: *The Certainty or Probability 1 Proposal.*
For with that set $\mathcal{Y}$ in place, $\mathrm{HT}^r_{\mathcal{Y}}$ reduces to: for all X, $Bel(X)$ iff $P(X) = 1$.

Proposal (a) yields a version of the Humean thesis according to which believed propositions X are required to have high enough probabilities conditional on those propositions Y that are *probabilistically certain*. Obviously, at least without further assumptions in the background, this amounts to just a minimal form of stability:

[128] In epistemology, defeaters of different kinds and in different senses have been studied within the defeasibility theory of knowledge: see Chisholm (1989, the 1st edn being from 1966), Lehrer and Paxson (1969), Pollock (1974), and Lehrer (1990). Lehrer's competitor-based rule of acceptance speaks of 'competitors' that need to be 'beaten' by acceptable sentences or propositions, where competition is analysed by means of decrease of probability under conditionalization. For instance, one version of Lehrer's account has it (see Def. 1 in Olsson 1998, p. 98) that X is Lehrer-acceptable relative to the probability measure P if and only if (i) $0 < P(X) \leq 1$ and (ii) $P(X) > P(Y)$ ('X beats Y') for all Y such that $P(X|Y) < P(X)$ ('Y competes with X'). Olsson (1998) and Cantwell (1999) discuss, criticize, and further develop Lehrer's account, which they regard as too restrictive. The stability account of belief that will be developed later is similar in spirit to Lehrer's but differs in its formal details. For anyone who regards the similarity to be close enough, the stability account might count as an improvement over Lehrer's proposal. (I am grateful to an anonymous reviewer of this book for pointing this out.)

[129] That threshold r^* does not have to coincide with the threshold r in the Humean thesis, and it is not assumed either that r^* depends functionally on r.

if expressed in terms of evidence, when a proposition comes along that the agent already believes to be true to the maximal possible degree, then a believed proposition must retain its high enough probability conditional on any such piece of evidence. Indeed, it is easy to see that the Humean thesis $HT^r_{=1}$ that results from choosing $\mathcal{Y} = \{Y \mid P(Y) = 1\}$ reduces immediately to the well-known Lockean thesis with the same threshold: for all X: $Bel(X)$ iff $P(X) > r$. Effectively, conditionalization drops out, and stably high probability boils downs to mere high probability. In this sense, the Lockean thesis is a limiting case of the family of Humean theses (each member of which results from a particular choice of a potential defeater set $\mathcal{Y}$).

Proposal (b) already offers more stability: precisely those propositions X are to be believed that are likely enough given whatever *believed proposition* Y is learned or supposed by means of conditionalization. One might think that this choice of $\mathcal{Y}$ is ruled out from the start due to circularity worries: if '$Y \in \mathcal{Y}$' is replaced in $HT^r_{\mathcal{Y}}$ by '$Bel(Y)$', then 'Bel' occurs both on the left-hand side and on the right-hand side of the equivalence in $HT^r_{\mathcal{Y}}$, which might look dangerous. I will put this worry to one side for now, but I am going to return to it later. In any case, embracing option (b) would correspond to a kind of coherence theory of belief according to which every two believed propositions X and Y would have to cohere in the sense of absolute confirmation, that is, so that $P(X|Y) > r$ (though not necessarily in the sense of incremental confirmation, that is, $P(X|Y) > P(X)$).[130] By the respective Humean thesis HT^r_{Bel}, the set Bel as a whole would thus be a system of propositions that mutually support each other in such an absolute 'above a threshold' sense, which constitutes one natural way of making the traditional coherence conception of rational or justified belief precise.

Proposal (c) amounts to a purely probabilistic way of defining '$Y \in \mathcal{Y}$' again: those propositions Y are to be taken seriously as potential defeaters that *have at least some small positive probability*, that is, a probability above some given threshold r^*. Whereas, supposedly, the (contextually determined) threshold r in the Humean thesis is meant to capture some kind of practical or 'moral' certainty and hence will normally be rather high, say, 0.9, the threshold r^* that figures in proposal (c) should merely express something like 'cannot practically be ruled out' and therefore will normally be small, such as $r^* = 0.1$. The Humean thesis $HT^r_{>r^*}$ that results from this choice, therefore, makes belief safe from being defeated by any proposition that is not too unlikely or unexpected.

Proposal (d) is similar to (c), the only difference being that 'cannot practically be ruled out' is now expressed in non-probabilistic terms: the potential defeaters Y

[130] Lehrer (1990) defends a decision-theoretic account of justified belief or acceptance in which a believed proposition competes with alternative hypotheses. If one ignores the utility aspect of Lehrer's proposal, as mentioned in n. 128, it holds that Y competes with X just in case the conditional probability of X given Y is less than the absolute probability of X: so this is an example of a theory in which the explication of justified belief crucially involves the concept of *incremental* (dis-)confirmation. (See Olsson 1998 and Cantwell 1999 for further details.)

are those propositions that are *possible from the viewpoint of the agent*, that is, the negations of which are not believed by the agent; or in other words: those which are not ruled out in terms of the agent's all-or-nothing beliefs—those which are still 'live' possibilities. Hence, the corresponding Humean thesis HT^r_{Poss} makes belief safe from being defeated by any proposition that the agent does not already believe to be false. I will return to potential circularity worries again below, and I will study the properties of this type of Humean thesis in more detail later.

In the transition from (a) to (d), intuitively, our sets $\mathcal{Y}$ of potential defeaters are getting larger and larger—presumably, (b) defines a superset of the set defined by (a), and (c) and (d) define supersets of the set given by (b)—and hence the constraint that is imposed on *Bel* by the corresponding Humean theses $HT^r_{\mathcal{Y}}$ is getting more and more severe in terms of requiring of believed propositions more and more stability.

In this sequence of proposals, option (e) is the opposite extreme of (a): precisely those propositions are to be taken seriously as potential defeaters that are *probabilistically possible*, that is, which have non-zero probability of being true, or the negations of which are not certain. Clearly, sticking this condition into the Humean thesis amounts to a *lot* of stability; indeed, one can show again easily that the Humean thesis $HT^r_{>0}$ that results from choosing $\mathcal{Y} = \{Y \mid P(Y) > 0\}$ reduces immediately to the Certainty or Probability 1 Proposal for belief that one can find defended in parts of the relevant literature and which I discussed already in section 1.6 of Chapter 1:[131] for all X, $Bel(X)$ iff $P(X) = 1$. Once again, conditionalization actually drops out, and stably high degrees of belief collapse into maximal probability 1. One might call this the *Cartesian* version of rational belief: it is rational to believe X just in case X is completely certain.[132] The usual worries apply that I explained already in section 1.6: in particular, it is well-known from the debate on the Old Evidence problem (cf. Glymour 1980) that once a proposition is assigned probability 1, its probability cannot be decreased any more by means of conditionalization (on propositions with positive probability). In that sense, assigning the maximally possible probability 1 to a believed proposition yields stability to the maximally possible extent. In fact, *too much* stability: within the boundaries of certain contexts, $HT^r_{>0}$ might well be a plausible thesis, but as a *general* thesis on belief and degrees of belief, $HT^r_{>0}$ just does not sound right. Unless some of the standard Bayesian background assumptions on the interpretation of P and on the reconstruction of learning in terms of conditionalization are amended (as carried out e.g. by Levi 1980), the extreme stability of believed propositions having probability 1 would simply turn stability into dogmatic unrevisability. And checking the proposal against independent test cases does not seem to yield the right verdicts in at least some contexts either: *I believe the apple spritzer to be in the refrigerator or in the shopping bag. But if my wife offers me a bet in which I win €1 if this is the case, but where I lose*

131 E.g. Gärdenfors (1986a) defines the (all-or-nothing) belief set that is associated with P as the set of all A such that $P(A) = 1$.

132 Indeed, Loeb (2010) argues that Descartes held a stability account of rational belief or knowledge.

€1,000,000 if it is not, then I will not accept the bet. According to the standard Bayesian interpretation of degrees of belief in terms of betting quotients, while my degree of belief in the apple spritzer being in the fridge or in the bag might be high, it cannot be equal to 1, or otherwise I should have been willing to take the bet. And yet I feel perfectly rational in believing, in the all-or-nothing sense, that the apple spritzer is in the fridge or in the bag. I believe a proposition without assigning it probability 1.

In spite of such justified concerns about option (e), all of the suggestions for how to fill in the '$Y \in \mathcal{Y}$' blank in our Humean thesis template are interesting in themselves, and it is useful to observe that some of the existing proposals in the literature of how belief ought to relate to degrees of belief can be presented as different types of stability conceptions within this more general framework. All of the proposals ought to be studied in detail in order to judge their respective virtues and vices, and the same holds for many other stability conceptions of belief that have not been considered so far.

This being said, here is one reason why one of the proposals seems to stand out. (General remark: except for the more technical Chapter 4, proofs of theorems will be stated within footnotes at the very end of the respective theorems.)

Theorem 1 For every finite non-empty set W of worlds (such that propositions are subsets of W), for all Bel where not $Bel(\varnothing)$ (so the contradictory proposition is not believed), for all probability measures P, for all thresholds r with $\frac{1}{2} \leq r < 1$: if the Humean thesis HT^{r}_{Poss} that results from proposal (d) is satisfied, then there are appropriate thresholds r', r'', r''', r^*, such that the Humean theses $\mathrm{HT}^{r'}_{=1}$, $\mathrm{HT}^{r''}_{Bel}$, $\mathrm{HT}^{r'''}_{>r^*}$ are satisfied that result from proposals (a), (b), (c), respectively.[133]

[133] Here is the proof, which relies on the Representation Theorem 5 in Appendix B (which one can find proven there). According to part 2 of that theorem, the Humean thesis HT^{r}_{Poss} together with not $Bel(\varnothing)$, the axioms of probability for P, and $\frac{1}{2} \leq r < 1$ entail the following three statements: (i) there is a non-empty proposition B_W which is the least believed proposition: for all X, $Bel(X)$ iff $X \supseteq B_W$. (ii) B_W is P-stable^r (for the definition of this concept, see Appendix B). (iii) If $P(B_W) = 1$, then B_W is the least proposition with probability 1 (which must exist by W being finite). From this one gets (a)–(c).

(a) First one shows for all X, $Bel(X)$ iff $P(X) \geq s = P(B_W)$. For the $P(B_W) < 1$ case, this is just Observation 13 from subsection 'P-Stability and the First Representation Theorem' in section 4.2.2 in the special case $Z = W$ (see the notation there); the observation has (i) and (ii) above as an assumption. For the sake of self-containment, I include the proof of that special case also here. So we show that for all X, $Bel(X)$ iff $P(X) \geq s = P(B_W)$. The left-to-right direction is obvious, since if $Bel(X)$, then $X \supseteq B_W$, and the rest follows by the monotonicity property of probability: $P(X) \geq P(B_W)$. And from right-to-left: assume $P(X) \geq P(B_W)$ but not $Bel(X)$; then $X \not\supseteq B_W$, that is, $\neg X \cap B_W$ is non-empty. Thus, $[\neg X \cap B_W] \cup \neg B_W$ has non-empty intersection with B_W and its probability is greater than 0, because $1 > P(B_W) = 1 - P(\neg B_W)$ and so $P(\neg B_W) > 0$ (by the axioms of probability). But from B_W being P-stable^r it follows then that $P(B_W \mid [\neg X \cap B_W] \cup \neg B_W) > r \geq \frac{1}{2}$, that is, by the axioms of probability and the definition of conditional probability, $P(\neg X \cap B_W) > \frac{P([\neg X \cap B_W] \cup \neg B_W)}{2} = \frac{P(\neg X \cap B_W)}{2} + \frac{P(\neg B_W)}{2}$, and hence $P(\neg X \cap B_W) > P(\neg B_W)$. However, by assumption we had $P(X) \geq P(B_W)$, and by the axioms of probability again, $P(\neg B_W) \geq P(\neg X)$. So we would get $P(\neg X \cap B_W) > P(\neg X)$, which contradicts the axioms of probability. So $Bel(X)$. For the $P(B_W) = 1$ case, this follows from (iii) and (i) above and the axioms of probability. So we have found that for all X, $Bel(X)$ iff $P(X) \geq s = P(B_W)$. Now, by W—and thus also the set of propositions—being finite, there must be an r' sufficiently close to, but below, $s = P(B_W)$, such that: for all X, $Bel(X)$ iff $P(X) > r'$. But that version of the Lockean thesis is equivalent to $\mathrm{HT}^{r'}_{=1}$, as follows immediately from the axioms of probability.

In other words: let us assume that P is a probability measure over a finite set W of possible worlds, Bel is a set of propositions over W, and it is not the case that $Bel(\varnothing)$, so that our agent does not believe in the truth of the contradictory proposition $\varnothing$. Then if the Humean thesis that results from the choice $\mathcal{Y} = \{Y \mid Poss(Y)\} = \{Y \mid \text{not } Bel(\neg Y)\}$ holds for P and Bel, also instances of the Humean theses that result from the other choices of $\mathcal{Y}$ above hold true, with the possible exception of the final case (e) (about which Theorem 1 remains silent). The concluding phrase 'with appropriate thresholds' in Theorem 1 is to be understood in the way that the thresholds that figure in the entailed Humean theses need not be equal to the threshold r that figures in the entailing Humean thesis HT^r_{Poss}. At the same time, in order for HT^r_{Poss} to entail these variants of the Humean thesis, their thresholds cannot be chosen freely either. The claim is only that if HT^r_{Poss} is the case, *there exist* thresholds r', r'', r''', r^*, such that also $HT^{r'}_{=1}$, $HT^{r''}_{Bel}$, $HT^{r'''}_{>r^*}$ are the case.

The point of Theorem 1 is *unification*: HT^r_{Poss} unifies various plausible stability conceptions of belief in one fell swoop. If HT^r_{Poss} holds, then Bel is not just the very set of propositions that have a high enough degree (above r) of belief given any proposition that the agent regards as possible, but in fact Bel also coincides (a) with the set of propositions that have a high enough degree (above r') of belief given any proposition of which the agent is certain, (b) with the set of propositions that have a high enough degree (above r'') of belief given any believed proposition, and (c) with the set of propositions that have a high enough degree (above r''') of belief given any proposition to which the agent assigns a probability above a certain small threshold r^*. Returning to our previous question of 'how cautious or brave should Bel be?', a first tentative answer might thus be: cautious or brave enough in order to satisfy HT^r_{Poss}, as this will guarantee the other plausible stability principles from above all by itself.

I will therefore, tentatively, suggest the following principle to be our intended precisification of the Humean thesis on belief:

> **The Humean Thesis Explicated:** If Bel is a perfectly rational agent's class of believed propositions at a time, and if P is the same agent's subjective probability measure at the same time, then

(b) Let $r'' = r'$ for the threshold r' as in the proof of (a). I prove $HT^{r''}_{Bel}$: for all X, $Bel(X)$ iff for all Y, if $Bel(Y)$ (and $P(Y) > 0$), then $P(X|Y) > r''$. Let $X \subseteq W$. The left-to-right direction of $HT^{r''}_{Bel}$ follows from: if $Bel(X)$ and $Bel(Y)$ (and $P(Y) > 0$), then, by (i) above, $X, Y \supseteq B_W$, which is why $P(X|Y) = \frac{P(X\cap Y)}{P(Y)} \geq P(X \cap Y) \geq P(B_W) > r'$ (by a). The right-to-left direction of $HT^{r''}_{Bel}$ follows from this consideration: I show the contrapositive claim. Assume not $Bel(X)$. By (a) it follows that $P(X) = P(X|W) \not> r'$. But $Bel(W)$ is the case by HT^r_{Poss} and the axioms of probability, and $P(W) = 1 > 0$ holds by the axioms of probability again. But that means the right-hand side of $HT^{r''}_{Bel}$ fails.

(c) By (a) (with the threshold r' that was defined in its proof): for all X, $Bel(X)$ iff $P(X) > r'$. Therefore: not $Bel(\neg X)$ iff it is not the case that $P(\neg X) = 1 - P(X) > r'$; that is: $Poss(X)$ iff $1 - P(X) \leq r'$ iff $1 - r' \leq P(X)$. By the set of propositions being finite, there must be an r^* sufficiently close to, but below, $1 - r'$, such that: for all X, $Poss(X)$ iff $r^* < P(X)$. Finally, let $r''' = r$. Replacing '$Poss(X)$' by '$P(X) > r^*$' in the Humean thesis HT^r_{Poss} (the thesis was a premise) yields then precisely $HT^{r'''}_{>r^*}$.

(HT^r) For all X: $Bel(X)$ iff for all Y, if $Poss(Y)$ and $P(Y) > 0$, then $P(X|Y) > r$

where $Poss(Y)$ if and only if not $Bel(\neg Y)$ (and $\frac{1}{2} \leq r < 1$).

In the rest of this chapter I am going to study the consequences of this principle. But before I do so, let me conclude this section by a couple of preliminary remarks on this Humean thesis HT^r ($= HT^r_{Poss}$).

First of all, the two qualifying conditions on potential defeaters in the antecedent clause of the embedded conditional of HT^r match each other in being possibility conditions: if Y is *possible in the all-or-nothing sense* ($Poss(Y)$) and also *possible in the probabilistic sense* ($P(Y) > 0$), then . . . ; which is appealing.

Secondly, while the resulting constraint on *Bel* and *P* that derives from HT^r is certainly severe, it is not quite as severe as one might think. In particular, *Poss* is not meant to express *logical* possibility but only *doxastic* possibility from the viewpoint of the agent: e.g. although the proposition that there are Cartesian demons is logically possible, it is not doxastically possible for me since I happen to believe its negation, that is, the proposition that there are no Cartesian demons. Accordingly, the proposition that there are Cartesian demons is *not* amongst the potential defeaters *Y* against which my set *Bel* would be safeguarded by HT^r. In the terminology of Levi (1980): '*Poss*' expresses *serious* possibility. I would suggest to even strengthen this by adding that, at least in normal everyday situations, $Poss(X)$ should in fact require that X has a positive probability that is not too small either, or otherwise the possibility in question would not count as 'serious' enough; much like what one gets from the Lockean thesis '$Bel(X)$ iff $P(X) > r'$' of which 'not $Bel(\neg X)$ iff $P(X) \geq 1 - r'$' is a corollary.

Now, even taking all of these clarifications concerning the notion of possibility into account, one might wonder: isn't HT^r still too restrictive?[134] Here is an example:[135] say, someone believes that Hannes is a German citizen. He also regards it as possible that Hannes was born in Austria (but nevertheless has German citizenship). And his

[134] How severe is the constraint that HT^r imposes on *P* and *Bel* if considered in purely mathematical terms? I will not deal with this here in any detail, but given that *W* is finite, one can show the following to be the case: (i) for every *P* there is a *Bel*, such that HT^r holds, and vice versa, for every consistent *Bel* that is closed under logical consequence there is a *P*, such that HT^r holds; (ii) for almost all *P* there is a *Bel*, such that $HT^{\frac{1}{2}}$ holds and where there is also an *X* with $Bel(X)$ and $P(X) < 1$ (where 'for almost all' can be made precise in geometric terms by means of the so-called Lebesgue measure); (iii) for many of the concrete probability measures *P* that one can find applied in the Bayesian literature there is a *Bel*, such that HT^r holds and where there is an *X* with $Bel(X)$ and $P(X) < 1$. The cases in (i) that are not covered also by (ii) or (iii) are such cases in which the Humean thesis collapses into the Certainty or Probability 1 Proposal (e) from above; but even that does not seem too bad, as the Certainty Proposal is still one of the typical bridge postulates on belief and degrees of belief that one can find defended in the literature, and the Humean thesis collapses into it only in special circumstances, that is, for certain *P*. I will return to such worries in section 3.3 and at the end of section 4.2.2. In each of the following places one can find concrete, non-trivial, and plausible examples of how the emerging theory can be applied: section 2.3.1, Appendix B, sections 3.6, 4.2.2, 4.3, and 6.2. For remaining worries about the potential 'scarcity' of pairs $\langle P, Bel \rangle$ that satisfy the Humean thesis, see Makinson (n.d.).

[135] I owe this example to David Chalmers.

degrees of belief are distributed such that, given that Hannes was born in Austria, it is unlikely that Hannes is German. It follows that the stability of belief in the sense of HT^r rules out this type of situation. And at least at first glance it might seem that a perfectly rational agent might satisfy these assumptions by having beliefs and degrees of belief like that. If so, then this would be a counterexample to our possibility variant of the Humean thesis.

Is it? Let us take a look at its formal analysis: let G express that Hannes is a German citizen, and let A express that Hannes was born in Austria. Then the assumptions lead to: $Bel(G)$, $Poss(G \cap A)$, and $P(G|A) < \frac{1}{2}$, which implies $P(G \cap A) < P(\neg G \cap A)$ by the axioms of probability. Because of $Bel(G)$ it should also hold that $Bel(G \cup \neg A)$, that is, $Bel(\neg(\neg G \cap A))$; hence, not not $Bel(\neg(\neg G \cap A))$, which means not $Poss(\neg G \cap A)$. But intuitively that is odd: one would have to rationally rule out $\neg G \cap A$ as a serious possibility, although one does regard $G \cap A$ as possible and takes $G \cap A$ to be *less likely* than $\neg G \cap A$. Pre-theoretically, independently of considerations concerning the Humean thesis, it does not seem to be the case that a perfectly rational agent would regard a proposition as possible, another one as impossible, but assign the latter a higher degree of belief than the former. So the example does not seem to be a counterexample to the Humean thesis after all. At second glance, our pre-theoretic verdict coincides with that of the Humean thesis: a perfectly rational agent could not have the required combination of beliefs and degrees of belief.

Thirdly, returning to the circularity worries that were mentioned before: '*Bel*' occurs on the left-hand side of the equivalence in HT^r. It also occurs on the right-hand side of that equivalence, once '$Poss(Y)$' has been unpacked in terms of 'not $Bel(\neg Y)$'. Isn't that alarming? The answer is: not really, or at least not by itself. It would be so, of course, if HT^r had been put forward as an *explicit definition* of '*Bel*' in terms of '*P*' or the like, in which case HT^r would be a circular definition and hence not count as legitimate. But that is not how I suggest that one should think of HT^r: it is not a definition of '*Bel*' on the basis of '*P*' at all. More generally, it is not necessarily to be regarded as a reduction of *Bel* to *P* or anything else either. Instead, one ought to consider the Humean thesis HT^r as a postulate or an axiom—as a bridge *principle* for belief and degrees of belief. A plausible analogy would be in terms of algebraic equations: an equation in two variables expresses a joint constraint on pairs of values of the two variables to the effect that any such pair is required to satisfy the equation. HT^r is like such an equation, '*Bel*' and '*P*' are its variables, and what HT^r expresses is a joint constraint on pairs $\langle Bel, P\rangle$ to the effect that any such pair is required to be a solution to the equation, that is, to satisfy HT^r, in order for the agent in question to be rational. Whether HT^r serves that purpose well is yet to be seen, but it is certainly not ruled out by being an axiom that contains one and the same symbol on the left-hand side and the right-hand side of an embedded equivalence statement. There is no general methodological constraint that would prohibit axioms from having such a logical form.

Finally, another methodological point: one may think of this book as engaging in a bigger project on the explication of rational belief and the coherence of rational belief, where 'explication' is understood in the sense of Carnap (1950a), and where the concept of rational belief gets explicated simultaneously on different scales of measurement. Such explications can be given 'for example, by a definition' (Carnap 1950a, §2), but carrying out explications by means of systems of axioms instead is yet another possibility. Within the present project, HT^r is meant to be an axiom of precisely such kind (as are the axioms of subjective probability). Now, according to Carnap, in order to count as an adequate explication of the given imprecise concept C, the new and sharpened concept C' after explication ought to be similar to C. Additionally, C' should be exact, fruitful—that is, it should occur in various interesting general principles that can be derived on the basis of its explication and additional principles—and it should be simple (the least important of Carnap's criteria). Up to, and including, the discussion of our family of Humean theses $HT^r_{\mathcal{Y}}$, the main aim of this chapter was to argue for the similarity between, on the one hand, *Bel*, *P*, and $HT^r_{\mathcal{Y}}$, and on the other hand, the informal and pre-theoretic notions of *rational all-or-nothing belief*, *rational degree of belief*, and *coherence between the two of them*. Obviously, *Bel*, *P*, and the axioms above that govern them, such as our ultimate Humean thesis HT^r, are formally exact. Theorem 1, and the remainder of this chapter, are devoted to showing that the specific Humean thesis HT^r that I put forward is also fruitful (and reasonably simple), in which case all of Carnap's desiderata on adequate explications will be accounted for.

Let me now turn to some of these fruit-bearing consequences.

2.3 The Consequences of the Humean Thesis

2.3.1 Consequence 1: Doxastic Logic

The first salient consequence of the Humean thesis, if taken together with subjective probability theory, is doxastic logic—the closure of belief under logical consequence:

Theorem 2 If P is a probability measure, if *Bel* and P satisfy the Humean thesis $HT^r_{\mathcal{Y}}$ (relative to a class $\mathcal{Y}$ of propositions, and for $\frac{1}{2} \leq r < 1$), then the following principles of doxastic logic hold:

(i) (Whatever the $\mathcal{Y}$:) $Bel(W)$.

(ii) (Whatever the $\mathcal{Y}$:) For all propositions X, Y: if $Bel(X)$ and $X \subseteq Y$, then $Bel(Y)$.

(iii) (With $W \in \mathcal{Y}$:) For all propositions X: if $Bel(X)$ then $Poss(X)$.

(iv) (With $\mathcal{Y} = Bel$:) For all propositions X, Y, if $Bel(X)$ and $Bel(Y)$, then $Poss(X \cap Y)$.

(v) (With $\mathcal{Y} = Poss$:) For all propositions X, Y, if $Bel(X)$ and $Bel(Y)$, then $Bel(X \cap Y)$.

(vi) (With $\mathcal{Y} = Poss$:) If not $Bel(\varnothing)$, then for all propositions X, if $Bel(X)$ then $Poss(X)$.[136]

(i) says that a rational agent believes the greatest or 'tautological' proposition W that is true in every possible world (within W). (ii) expresses that rational belief is closed under one-shot logical consequence, which, for propositions, is closure under taking supersets. (iii) amounts to the consistency of a rational agent's beliefs: if X is believed, then $\neg X$ is not believed. (iv) and (v) deal with closure principles concerning the conjunction or intersection of propositions; in particular, (v) is closure of belief under conjunction. (vi) means that (iii) applies in the case where $\mathcal{Y} = Poss$, as long as the contradictory proposition $\varnothing$ is not believed. Although, for simplicity, I am assuming in this chapter that W is finite, none of the results in Theorem 2 rely on that assumption.

(i)–(ii) follow from any Humean thesis $HT^r_{\mathcal{Y}}$ whatsoever if it is taken for granted that P is a probability measure, and (iii) follows if additionally Bel is at least assumed to be stable under W. They are minimal closure conditions for belief when stability is made precise in terms of conditionalization.

136 Here is the proof:

(i) This follows from the right-hand side of $HT^r_{\mathcal{Y}}$ being satisfied (whatever the $\mathcal{Y}$).

(ii) If $Bel(X)$, then by the left-to-right direction of $HT^r_{\mathcal{Y}}$, for all Z, if $Z \in \mathcal{Y}$ (and $P(Z) > 0$), then $P(X|Z) > r$. From $X \subseteq Y$, by the axioms of probability: for all Z, if $Z \in \mathcal{Y}$ (and $P(Z) > 0$), then $P(Y|Z) \geq P(X|Z) > r$. This yields, by the right-to-left direction of $HT^r_{\mathcal{Y}}$: $Bel(Y)$.

(iii) Assume $Bel(X)$ and, for contradiction, not $Poss(X)$, that is, $Bel(\neg X)$. By applying the left-to-right direction of $HT^r_{\mathcal{Y}}$ twice, with $Y = W$, and because $Y \in \mathcal{Y}$ (by assumption) and $P(Y) = 1 > 0$ (by the axioms of probability): $P(X|Y) > r \geq \frac{1}{2}$ and $P(\neg X|Y) > r \geq \frac{1}{2}$, which is a contradiction.

(iv) Assume $Bel(X)$, $Bel(Y)$, and, for contradiction, not $Poss(X \cap Y)$, that is, $Bel(\neg X \cup \neg Y)$. If $P(\neg X \cup \neg Y)$ were 0, then $P(X \cap Y)$ would have to be 1, in which case it would be satisfied that for all Z, if $Bel(Z)$ and $P(Z) > 0$, then $P(X \cap Y|Z) > r$, which with the right-to-left direction of HT^r_{Bel} would entail $Bel(X \cap Y)$; with (i) and (iii) this would mean that $Poss(X \cap Y)$; but that had been ruled out by assumption. Hence, it holds that $P(\neg X \cup \neg Y) > 0$. By assumption, also $Bel(\neg X \cup \neg Y)$ is the case. From $Bel(X)$ and the left-to-right direction of HT^r, it follows that $P(X|\neg X \cup \neg Y) > r$. And the same holds for Y: $P(Y|\neg X \cup \neg Y) > r$. By the axioms of probability and the definition of conditional probabilities, this means: $P(X \cup Y|\neg X \cup \neg Y) = P(X \cap \neg Y|\neg X \cup \neg Y) + P(Y|\neg X \cup \neg Y) = P(X|\neg X \cup \neg Y) + P(Y|\neg X \cup \neg Y) > r + r \geq 1$ (by the assumption that $r \geq \frac{1}{2}$), which is a contradiction.

(v) The proof is very similar to the one of (iv). Suppose for contradiction that $Bel(X)$, $Bel(Y)$, but not $Bel(X \cap Y)$, that is, not $Bel(\neg(\neg X \cup \neg Y))$. So, by definition of '$Poss$': $Poss(\neg X \cup \neg Y)$. If $P(\neg X \cup \neg Y)$ were 0, then $P(X \cap Y)$ would have to be 1, in which case it would be satisfied that for all Z, if $Poss(Z)$ and $P(Z) > 0$, then $P(X \cap Y|Z) > r$, which with the right-to-left direction of HT^r (= HT^r_{Poss}) would entail $Bel(X \cap Y)$, which had been ruled out by assumption: hence, $P(\neg X \cup \neg Y) > 0$. From $Bel(X)$ and the left-to-right direction of HT^r, it follows that $P(X|\neg X \cup \neg Y) > r$. And the same holds for Y: $P(Y|\neg X \cup \neg Y) > r$. By the axioms of probability and the definition of conditional probabilities, this means: $P(X \cup Y|\neg X \cup \neg Y) = P(X \cap \neg Y|\neg X \cup \neg Y) + P(Y|\neg X \cup \neg Y) = P(X|\neg X \cup \neg Y) + P(Y|\neg X \cup \neg Y) > r + r \geq 1$ (by the assumption that $r \geq \frac{1}{2}$), which is a contradiction.

(vi) Because of not $Bel(\varnothing)$ and the definition of '$Poss$' it holds that $Poss(W)$. The rest follows from applying (iii).

The quantified conditional clause in (iv) follows for the special 'coherence with belief' option (b) from the last section, that is, HT^r_{Bel}. So if belief corresponds to stably high probability given *believed* propositions, then for every two believed propositions X and Y it must hold that their conjunction or intersection is at least possible.

Finally, the quantified conditional clause in (v), and with it that in (iv) if given also not $Bel(\varnothing)$, follow from our official Humean thesis HT^r (= HT^r_{Poss}): our Humean thesis entails that belief is closed under conjunction or intersection, and that is the case even though HT^r is partially a probabilistic thesis. While high probability, as employed in the Lockean thesis, does not by itself imply the closure of belief under conjunction, surprisingly, *stably* high belief as employed in our official version of the Humean thesis does.

(i) and the conditional parts of (ii)–(iv) correspond to axiom schemes of doxastic logic that are typically validated by means of a so-called neighbourhood semantics for the sentential belief operator (given only minor assumptions on neighbourhood sets). Neighbourhood semantics (see e.g. Chellas 1980) is a generalization of the much more common possible worlds semantics for modalities.

(i) and all of the conditionals in (ii)–(v) (or indeed (ii)–(vi)) taken together capture the full normal logic of belief as given by a standard possible worlds semantics for belief and restricted only by not allowing for nestings of the belief operator (as our present framework does not by itself provide for nestings of '*Bel*' or introspective beliefs).

So far as our official Humean thesis HT^r (= HT^r_{Poss}) from section 2.2 is concerned, Theorem 2 means:

> If P satisfies the axioms of probability, if *Bel* and P satisfy the Humean thesis HT^r (with $\frac{1}{2} \leq r < 1$), and if not $Bel(\varnothing)$ (the contradictory proposition is not believed), then *Bel* is consistent and closed under logical consequence.

It is easy to show then that there must always be a least, or logically strongest,[137] believed proposition B_W that is non-empty, finite (assuming W is finite), which coincides with the intersection of all believed propositions, and which generates the agent's belief system *Bel* in the following sense: for all propositions X, $Bel(X)$ if and only if $B_W \subseteq X$. So the summary of Theorem 2 for my own purposes is: HT^r makes a rational agent's belief system *Bel* determinable from a set B_W of doxastically accessible worlds (or 'serious possibilities', adopting the terminology of Levi 1980, 1984): $Bel(X)$ holds if and only if X is true at every doxastically accessible world, that is, at every world in B_W. $Poss(X)$ is thus the case if and only if $X \cap B_W \neq \varnothing$, that is, X being true at some doxastically accessible world. See Figure 2.1. More briefly: *Bel* has a possible worlds semantics. In some applications of the Humean thesis, *Bel* will be given, and B_W will be defined from it, while in other applications it will be the other way around.

137 In application to sets of worlds, I will use the terms 'least', 'strongest', and 'logically strongest' synonymously.

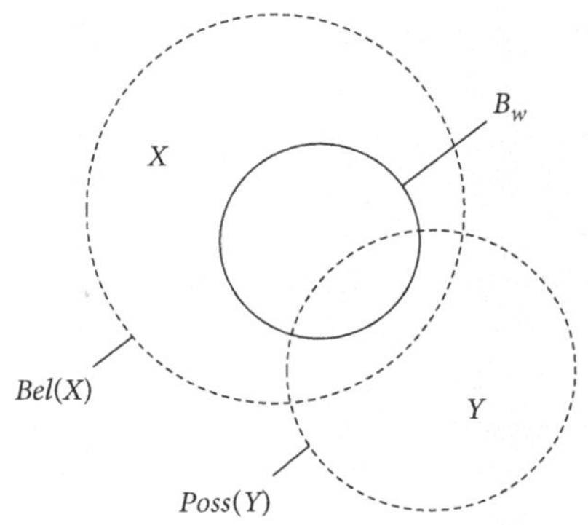

Figure 2.1. Possible worlds semantics for belief

It does not matter really, since the logical closure of belief entails (with the finiteness of W) that '*Bel*' and 'B_W' are interdefinable.

It is worth noting that variant (e) of the Humean thesis from the last section—the thesis $HT^r_{>0}$, which ultimately collapses into the Certainty Proposal for belief—satisfies the same logical closure properties as our official Humean thesis HT^r. Indeed, $HT^r_{>0}$ can be shown to entail instances of all the other Humean thesis options from the last section (for W being finite), much as we found HT^r to entail instances of the other options (other than $HT^r_{>0}$). Of course, I also argued in the last section that $HT^r_{>0}$ delivers too much stability, at least for certain purposes, and can be ruled out to hold in general on independent grounds. But what the present observation suggests is that within the 'continuum' of stability accounts of belief that range from the one extreme $HT^r_{=1}$ over the intermediate positions HT^r_{Bel}, $HT^r_{>r*}$, and HT^r (= HT^r_{Poss}) to the other extreme $HT^r_{>0}$, once belief has been postulated to be 'sufficiently' stable, then both the logical closure of belief and the unification of different versions of the Humean thesis emerge. And the cutting point of 'sufficient' stability seems to lie at HT^r.[138]

Let us take a look at a concrete example now which I take from Barber's *Bayesian Reasoning and Machine Learning* (Cambridge University Press, 2012), pp. 30–2; the example has not been altered in any way (but it itself derives from a slightly different earlier example by Judea Pearl). Consider the Bayesian network in Figure 2.2, which describes in formal terms the following situation. It is morning. Tracey has not left the house as yet, and she worries about whether her lawn is wet enough. When thinking about this, she also wonders whether it has rained, whether she has left the sprinkler on, and whether her neighbour Jack's lawn is wet. R, S, T, J are binary random variables or propositional letters that can be true or false: R expresses whether it has rained, S says whether Tracey has left the sprinkler on, T captures whether her (Tracey's)

[138] In any case, the proof of (v) in Theorem 2 really only requires the assumption that $Poss \subseteq \mathcal{Y}$. So, as long as $Poss \subseteq \mathcal{Y}$, the closure of rational belief under conjunction is going to follow. I am grateful to David Makinson for highlighting this in personal communication.

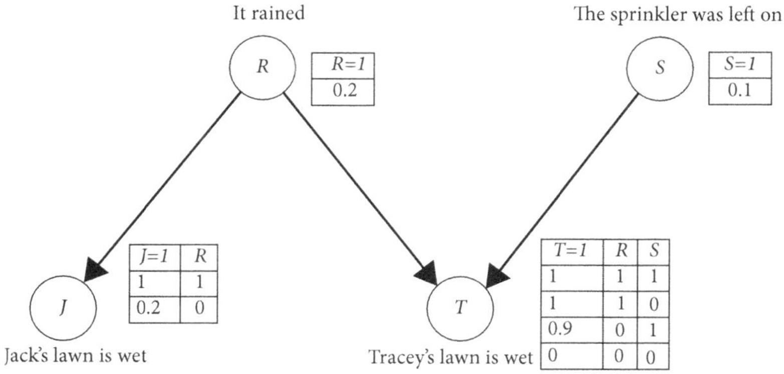

Figure 2.2. The example of Tracey's Sprinkler

lawn is wet, and J represents whether Jack's lawn is wet. In line with many typical applications of Bayesian networks, the edges can be interpreted in causal terms: if the sprinkler was on, then this will have caused her lawn to be wet, but there would not have been any effect on Jack's lawn. But if it rained, then both her neighbour's and her own lawn will have been caused to be wet.[139] Tracey is aware of these causal relationships. Finally, the tables next to R and S represent Tracey's prior probabilities in it having rained and the sprinkler having been left on, while the tables adjacent to J and T convey her respective conditional probabilities for Jack's lawn to be wet given rain/absence of rain and for her own lawn to be wet given any combination of the states of rain and of the sprinkler. For instance, even if it did not rain ($R = 0$), Tracey assigns a small probability of 0.2 to Jack's lawn being wet ($J = 1$), as Jack might e.g. own a sprinkler himself.

Once the theory of Bayesian networks has been applied to all of these components taken together (see e.g. Bovens and Hartmann 2003)—that is: the unconditional probabilities in the tables for R and S have been combined appropriately with the conditional ones in the tables for J and T—a unique probability measure P is determined which is defined on the set W of the sixteen truth value assignments to our four random variables. Eight of these worlds (or their singleton sets) can be shown to have probability 0, which is why I am going to ignore them in what follows. This leaves us with eight remaining possible worlds $w_1, \ldots, w_8$ (the interpretation of which will be explained shortly). I will take P on these eight possible worlds to be Tracey's degree-of-belief function in the context of her sprinkler considerations.[140] And given that probability measure P, my question will be: which belief sets *Bel* are such that P and

139 This causal interpretation matches the Humean pedigree of our theory quite nicely: 'as we find by experience, that belief arises only from causation, and that we can draw no inference from one object to another, except they be connected by this relation' (*Treatise*, section IX, part III, book I).

140 The respective probabilities can be shown to be: $P(\{w_1\}) = 0.576$, $P(\{w_2\}) = 0.18$, $P(\{w_3\}) = 0.144$, $P(\{w_4\}) = 0.0576$, $P(\{w_5\}) = 0.02$, $P(\{w_6\}) = 0.0144$, $P(\{w_7\}) = 0.0064$, $P(\{w_8\}) = 0.0016$.

Bel together satisfy the Humean thesis $HT^{\frac{1}{2}}$ for, say, the Humean threshold $r = \frac{1}{2}$? Or equivalently: given *P*, which sets B_W are such that if *Bel* is determined from B_W, then *P* and *Bel* satisfy $HT^{\frac{1}{2}}$? Effectively, what I am doing right now is to set the value of the one 'variable' '*P*' and then solve the 'equation' $HT^{\frac{1}{2}}$ for the remaining 'variable' '*Bel*'. But I should emphasize that this is just one way of applying the Humean thesis; e.g. alternatively, one might consider an application in which *Bel* is determined from the start and then one solves the thesis for '*P*' instead. Equations can serve many purposes.

What kinds of all-or-nothing beliefs will Tracey have? As with other 'equations', it turns out that in the sprinkler example there is actually more than just one solution: there will be a very cautious version of Tracey with very cautious all-or-nothing beliefs, a very brave version of Tracey with very brave categorical beliefs, and various in-between versions of Tracey. But all of them will satisfy the Humean thesis jointly with the given degree-of-belief function *P*.

The candidates for B_W that satisfy the Humean thesis $HT^{\frac{1}{2}}$ (given *P*) happen to be the following six sets:[141]

$\{w_1, w_2, w_3, w_4, w_5, w_6, w_7, w_8\}$
$\{w_1, w_2, w_3, w_4, w_5, w_6, w_7\}$
$\{w_1, w_2, w_3, w_4, w_5, w_6\}$
$\{w_1, w_2, w_3, w_4\}$
$\{w_1, w_2, w_3\}$
$\{w_1\}$

The last option $B_W = \{w_1\}$ corresponds to the bravest 'solution' *Bel* according to which Tracey believes w_1 to be the actual world, where the intended interpretation of each world w_i can be read off the following table (about which more below):

6. w_8: $T = 0, J = 1, R = 0, S = 1$
5. w_7: $T = 0, J = 0, R = 0, S = 1$
4. w_5: $T = 1, J = 1, R = 1, S = 1$; w_6: $T = 1, J = 1, R = 0, S = 1$
3. w_4: $T = 1, J = 0, R = 0, S = 1$
2. w_2: $T = 1, J = 1, R = 1, S = 0$; w_3: $T = 0, J = 1, R = 0, S = 0$
1. w_1: $T = 0, J = 0, R = 0, S = 0$

141 These findings were determined by Krombholz (2013) by means of a suitable computer programme. The six sets are characterized by the following property: they satisfy what will be called in Appendix B the 'Outclassing Condition' for B_W relative to *P* and $r = \frac{1}{2}$. That is: if B_W is any of these six sets, then each single world in B_W exceeds the probability of the set of worlds outside of B_W. The six sets are also characterized by a probabilistic stability property that I am going to mention for the first time in Appendix B and which will play an important role also in later chapters: the sets are *P*-stable^r (where *r* equals $\frac{1}{2}$ here). The representation theorem of Appendix B will state the relationship between the Humean thesis, the Outclassing Condition, and this stability property in exact terms. There is also a simple algorithm that determines sets that satisfy these conditions: it will be sketched in sections 3.2 and 4.2.2.

So e.g. w_1 is the world in which neither Tracey's nor Jack's lawn is wet, it did not rain, and she did not leave the sprinkler on. That is what she believes to be true if *Bel* is generated from $B_W = \{w_1\}$, and the so-determined *Bel* satisfies $\mathrm{HT}^{\frac{1}{2}}$ together with *P*.

The most cautious option is the other extreme, that is, the set $B_W = \{w_1, w_2, w_3, w_4, w_5, w_6, w_7, w_8\}$: in that case, since that set B_W has probability 1, $Bel(X)$ holds just in case $P(X) = 1$, and hence the Humean thesis $\mathrm{HT}^{\frac{1}{2}}$ would collapse into the Certainty Proposal (e) from the last section. There are more doxastic possibilities here than in the $B_W = \{w_1\}$ case, and hence it is harder to keep the probability of one's believed propositions above $\frac{1}{2}$ conditional on whatever proposition the corresponding version of Tracey would regard as possible. But $\mathrm{HT}^{\frac{1}{2}}$ does not *necessitate* that Certainty Proposal—as there are five alternative options—it only allows for it.

The other four 'solutions' are in between the two extremes. In fact, 'in between' can be taken quite literally here: as one can check above, the six candidate sets B_W that satisfy the Humean thesis given *P* are nested like Russian dolls or like spheres in David Lewis's semantics for counterfactuals, and one can prove that this is always so: for *every* given *P* (and for every Humean threshold $\frac{1}{2} \leq r < 1$), the class of sets B_W that validate the Humean HT^r with *P* is well-ordered in terms of the subset relation.[142] Consequently, for every world w_i that is a member of some such set B_W there must be a first 'time' at which it enters this hierarchy of B_W sets: e.g. w_1 is in there from the start, w_2 and w_3 join the sets from the second stage (at which $B_W = \{w_1, w_2, w_3\}$), and so on, up to w_8 which becomes a member of these sets at the sixth and final stage. In other words: the Humean thesis induces a ranking of possible worlds, which explains the hierarchical manner in which I have denoted possible worlds above (with the 'more plausible' worlds further down in the hierarchy). For instance, while the set $\{w_1\}$ of worlds of the first rank is an option, and so is the set $\{w_1, w_2, w_3\}$ of worlds of the first or the second rank, the set $\{w_1, w_2\}$ (in which w_2 is included but the equally ranked w_3 is not) could *not* be used as B_W since it would not yield the required stability: if $B_W = \{w_1, w_2\}$, then $\{w_1, w_2\}$ would have to be believed, $\{w_2, \ldots, w_8\}$ would have to be possible (as it has non-empty intersection with $\{w_1, w_2\}$), but $P(\{w_1, w_2\} \mid \{w_2, \ldots, w_8\}) = 0.42\ldots < \frac{1}{2}$.

As one can see from this example, even given a uniquely determined probability measure *P*, the Humean thesis does not always determine an agent's categorical belief set *Bel* uniquely. In the present case, there are six distinct belief sets that would do the trick of satisfying the Humean thesis together with *P* (and for a Humean threshold of $\frac{1}{2}$).

In order to see what all of that means for Tracey's beliefs, let us consider the intermediate $B_W = \{w_1, w_2, w_3\}$ option in more detail. With that set of doxastically possible worlds in place, Tracey's beliefs are as follows, where I am going to ascribe belief contents by means of propositional connectives for the sake of readability (but

[142] This follows from combining Theorem 5 in Appendix B with Theorem 18 in Chapter 4.

where all of this could be rephrased in terms of set-theoretic complement, intersection, and union again):

> $Bel(S = 0)$: Tracey believes that she has not left the sprinkler on (each of w_1, w_2, w_3 satisfies $S = 0$).
>
> $Bel(T = 1 \leftrightarrow R = 1)$: Tracey believes that her lawn is wet just in case it has rained (which makes sense: after all, she believes not to have left the sprinkler on).
>
> $Bel(\neg(J = 0 \wedge R = 1))$: Tracey believes that it is not the case that Jack's lawn is dry and it rained (the rain would have made it wet).
>
> $Bel(S = 0 \wedge (T = 1 \leftrightarrow R = 1) \wedge \neg(J = 0 \wedge R = 1))$: since she believes each of the three conjuncts, she also believes their conjunction by (v) of Theorem 2, that is, the closure of belief under conjunction.
>
> $Poss(R = 1)$ (that is, not $Bel(R = 0)$): Tracey regards it as possible that it has rained (as w_3, according to which $R = 1$, is one of her serious possibilities).

Taking the last two facts together, the Humean thesis entails that $P(S = 0 \wedge (T = 1 \leftrightarrow R = 1) \wedge \neg(J = 0 \wedge R = 1) \mid R = 1) > \frac{1}{2}$: even given that it has rained, which is a possibility from Tracey's point of view, Tracey's degree of belief in the respective conjunction remains high enough. That is the *Poss*-variant of our stability conception of belief in action.

All of these features of Tracey's beliefs and degrees of belief sound reasonable, I believe.

2.3.2 *Consequence 2: The Lockean Thesis*

As I said before (see Theorem 1 in section 2.2), our Humean thesis HT^r entails an instance of the alternative Humean thesis $\mathrm{HT}^{r'}_{=1}$ that is equivalent to an instance of the Lockean thesis,

> *The Lockean thesis*: For all X: $Bel(X)$ iff $P(X) > r'$ $(\geq \frac{1}{2})$.

Or, turning to the greater-than-*equals* version of the Lockean thesis instead, which will prove to be more convenient for present purposes: an instance of

> *The Lockean thesis*: For all X: $Bel(X)$ iff $P(X) \geq s$ $(> \frac{1}{2})$

is derivable from HT^r (always given not $Bel(\emptyset)$ and the axioms of probability for P).[143]

At the same time, Theorem 2 states that the very same assumptions entail also the logical closure of belief, in particular, the closure of belief under conjunction. And now one might think: how can HT^r support both the Lockean thesis and the closure of

143 Since W, and thus also the set of propositions, is finite by assumption, the two versions are equivalent by choosing the respective threshold appropriately (and appropriately close to the threshold in the other version). Compare n. 163 in Chapter 3.

belief under conjunction? Isn't that ruled out by Lottery-Paradox-like considerations (as famously deriving from Kyburg 1961)?

The answer to this seeming contradiction is that not any old instance of the Lockean thesis follows from HT^r but only an instance of the Lockean thesis with a very special Lockean threshold, s, which depends on both P and Bel:

Theorem 3 (The Lockean Part of Theorem 1 Reconsidered)
If P is a probability measure, if Bel and P satisfy the Humean thesis HT^r, and if not $Bel(\varnothing)$, then the following instance of the Lockean thesis holds:

$$\text{For all } X\text{: } Bel(X) \text{ iff } P(X) \geq s = P(B_W)\ (> r \geq \tfrac{1}{2}).$$ [144]

So the Lockean threshold in question is simply the agent's degree of belief $P(B_W)$ in the logically strongest believed proposition B_W that must exist by the Humean thesis. It is with *this* Lockean threshold that an instance of the Lockean thesis follows from the Humean thesis. Once one has given up the idea that the threshold in the Lockean thesis can be set arbitrarily—in particular, independently of what P is like—then there is nothing contradictory any more about the logical closure of belief and *an* instance of the Lockean thesis being satisfied simultaneously.[145]

That is also why I said in section 2.2 that Maher was a bit too quick in claiming Hume's 'superior degree of belief' considerations to be inconsistent with the logical closure and the consistency of belief. It is not contradictory to ascribe to Hume both a stability account of belief that is made precise in terms of HT^r, and which implies the logical closure of belief, and also a 'superior degree of belief' account of belief that is precisified in terms of the very instance of the Lockean thesis in the theorem, and which is also entailed by HT^r.

For example: in the story of Tracey's sprinkler, the relevant Lockean thresholds s that correspond to the probabilities of the six sets B_W that satisfy the Humean thesis $HT^{\frac{1}{2}}$ (given Tracey's degree-of-belief function P) are:

$\{w_1, w_2, w_3, w_4, w_5, w_6, w_7, w_8\}$ $(\geq s = 1.0)$

$\{w_1, w_2, w_3, w_4, w_5, w_6, w_7\}$ $(\geq s = 0.9984)$

$\{w_1, w_2, w_3, w_4, w_5, w_6\}$ $(\geq s = 0.992)$

$\{w_1, w_2, w_3, w_4\}$ $(\geq s = 0.9576)$

$\{w_1, w_2, w_3\}$ $(\geq s = 0.9)$

$\{w_1\}$ $(\geq s = 0.576)$

144 For the proof, see Theorem 1 in section 2.2, part (a). The Lockean threshold $s = P(B_W)$ can be seen to be strictly greater than the Humean threshold r in HT^r_{Poss}, by the left-to-right direction of HT^r with $X = B_W, Y = W$.

145 Much more will be said about this in Chapter 3. Section 3.4 will be devoted to the Lottery Paradox in particular. I will deal with the related case of the Preface Paradox in sections 3.5 and 6.4.

In particular, choosing $B_W = \{w_1, w_2, w_3\}$, which satisfies $HT^{\frac{1}{2}}$ with P, implies an instance of the Lockean thesis of the form: for all X, $Bel(X)$ iff $P(X) \geq 0.9$ $(= P(B_W))$. Thus, Tracey believes precisely those propositions to be true to which she also assigns a sufficiently high subjective probability, as long as 'sufficiently high' means in this context: 0.9. Call the version of Tracey for whom this is the relevant Lockean threshold *Cautious Tracey*.

But for yet another version of Tracey, say, *Brave (or Bold) Tracey*, B_W might be equal to $\{w_1\}$, in which case the Lockean threshold in question would be 0.576. Clearly, Brave Tracey believes propositions in the categorical sense of the term that Cautious Tracey does not.

Cautious Tracey and Brave Tracey are epistemically on par in so far as both of them satisfy an instance of the Humean thesis; they are both rational in that respect. In fact, I aim to show in the course of this book that by satisfying the Humean thesis they seem to tick *all* boxes that are mandatory for rational belief. They do, however, differ in some pragmatic respects that a rational agent is free to choose. When there is more than one epistemically permissible set of beliefs available—more than one possible belief set for a perfectly rational agent—answering pragmatic questions such as 'How brave or cautious do I want to be?' may break the epistemic tie between these belief sets. And Cautious Tracey answers this question, whether implicitly or explicitly, differently than Brave Tracey does.

What does this tell us about the Lottery Paradox? I will not go into any detail here, since I will turn to that question in detail in Chapter 3 (especially in section 3.4), but the short answer is: it depends. Say, one is dealing with a fair lottery of 1,000,000 tickets, and one is aware of this. If one is interested in *which ticket will be drawn?*, the corresponding space of possible worlds will be the set $W = \{w_1, \ldots, w_{1000000}\}$ in which each world w_i represents that ticket #i wins. The resulting subjective probability measure is 'flat' or uniform:

$$P(\{w_1\}) = \frac{1}{1000000}, P(\{w_2\}) = \frac{1}{1000000}, \ldots, P(\{w_{1000000}\}) = \frac{1}{1000000}$$

It is easy to see then that the only way of satisfying the Humean thesis HT^r is by B_W being identical to W, that is, by a Lockean threshold of $s = 1 = P(W)$: one believes that *some ticket will be drawn* but one cannot rationally believe of any ticket that it will not win.

On the other hand, if one is interested in *will ticket #i be drawn or not?*,[146] then the relevant space of possibilities boils down to the set $W' = \{w_i, \overline{w_i}\}$ where $\overline{w_i}$ represents that ticket i will *not* win; $\overline{w_i}$ is, as it were, the 'fusion' of all worlds w_j above where $j \neq i$. In such a context, the corresponding probability measure is of course not 'flat' any more:

146 This kind of question-sensitivity had already been exploited for Lottery-Paradox-like situations by Levi (1967).

$$P'(\{w_i\}) = \frac{1}{1000000}, \; P'(\{\overline{w_i}\}) = \frac{999999}{1000000}$$

Accordingly, one can show that there are now two candidates for sets B_W (and hence sets *Bel*) so that the Humean thesis HT^r is satisfied given P': one is the cautious option $B_W = W$ with a Lockean threshold of 1 again, but another one is the brave option $B_W = \{\overline{w_i}\}$ that corresponds to a Lockean threshold of $s = \frac{999999}{1000000} = P'(\{\overline{w_i}\})$ (where I assume the Humean threshold r to be less than s). In the brave case $B_W = \{\overline{w_i}\}$, one believes that *ticket #i will not be drawn.* This is stable now, since the coarse-grained space $W' = \{w_i, \overline{w_i}\}$ of possibilities does not make any proposition entertainable that could drag the probability of $\{\overline{w_i}\}$ below the threshold by means of conditionalization.

As becomes apparent from examples like that, our Humean thesis comes with a price: a strong context-sensitivity of rational belief. For if the Humean thesis is right, then what one believes rationally co-depends on the context, where by 'context' I do not so much mean the context of the ascriber of belief—as it would be the case in proper *contextualism* about belief—but the context of reasoning of the *rational agent who has the beliefs*. Such a context will then include the agent's 'question' or partitioning of possibilities (W vs W'), the agent's degree-of-belief function (P vs P'), and the ranges of permissible Lockean thresholds that are restricted to numbers $P(B_W)$ for those sets B_W that satisfy the Humean thesis (the single option 1 vs the two options 1 and $\frac{999999}{1000000}$).

Section 3.3 in Chapter 3 will address contexts in this sense of the term in much more detail. Let me just point out one feature here that comes with a certain kind of 'Humean' flavour: one can prove that if HT^r holds, then the more fine-grained a rational agent's partitioning of possibilities is, the more cautious her rational beliefs must be. For instance, if W is a *very* fine-grained infinite set of possible worlds, and if one assumes that there is an infinitely descending chain of smaller and smaller propositions $B_W = X_1 \supsetneq X_2 \supsetneq X_3 \supsetneq \ldots$ that are subsets of the agent's least believed proposition B_W, then HT^r will actually entail the Certainty or Probability 1 Proposal again:[147] for all X, $Bel(X)$ iff $P(X) = 1$.[148] Roughly: even in the case of a biased lottery, if the agent does not just care which ticket will win, but also who will own the winning ticket, what the ticket's weight will be, at what point of time it will be drawn, and so on and so forth, then rational belief must end up being 'elusive' (using Lewis's 1996 term from his discussion on the elusiveness of knowledge). The reason is: in any such context in which B_W is composed of a great number of very 'fine-grained' ways the world might be, there is also a great range of potential defeaters; for instance, for each of the infinitely many $w \in B_W$ the proposition of the form $\{w\} \cup \neg B_W$ will be doxastically possible from the viewpoint of the agent (because of the $\{w\}$ subset), and

[147] This follows from Theorem 18 in section 4.2.2. But it is also a special case of more general considerations that one can find in Smith (2010).

[148] Additionally, P will be entailed to be such that there is a least set of probability 1 included in its underlying algebra of propositions.

hence the Humean thesis would require believed propositions to have a stably high probability conditional on $\{w\} \cup \neg B_W$. And the result mentioned before means that in any such context in which one sees potential defeaters everywhere, the only way of achieving stability of belief is by becoming a sceptic in the sense of believing only those propositions of which one can be completely certain—a 'quasi-Humean' scepticism.[149] These properties of context-sensitivity and elusiveness of rational belief are probably the most worrying features of the Humean thesis on belief as explicated in section 2.2. For a detailed exposition of these worries, and for a defence of the theory in the face of them, see section 3.3.

This feature of context-sensitivity also means that the stability of rational belief according to the Humean thesis is bounded by context, by which I mean here: bounded by the *partition* of the agent's context of reasoning. *Within a context* (and its partition), rational all-or-nothing belief in X is stable; but if the context (the partition) changes, then the Humean thesis does not guarantee that the belief in X is retained.[150] If the time span of such a context coincides with the time it takes me to walk down the stairs from the second to the ground floor (Example 1), or to go through an episode of suppositional reasoning (Example 2), or to have a conversation with my wife (Example 3), then the stability is rather short-term. But still the stability may pay off as it was the case in my three examples. If what is required is *long-term* stability over the period in which, say, a house is being planned and built, or in which a scientific research programme is carried out, then either the degrees of belief that are assigned to the believed propositions would need to be really close to 1, or something would need to be 'done' to the agent's degree-of-belief function so that these degrees 'become' really close to 1 or even equal to 1. That is where an agent's *acceptance* of propositions will enter the picture, to which I will turn later in this book in section 6.3. Acceptance, which I am going to distinguish from belief, will also be based on stability.

2.3.3 Consequence 3: Decision Theory

The final consequence of the Humean thesis that I will consider is a practical one.

One attraction of the Bayesian approach to belief is Bayesian decision theory: given an agent's rational degree-of-belief function P and a utility measure u that represents the agent's desires, standard decision theory suggests precisely those actions that maximize expected utility to be (pragmatically) permissible in the Bayesian sense.

While there is no comparably developed formal decision theory for all-or-nothing belief,[151] functionalists about belief have always assumed some kind of belief–desire

[149] If granted sufficient liberty in terms of Hume interpretation, this is much like Hume becoming a sceptic about belief in the course of the *Treatise*: 'When I reflect on the natural fallibility of my judgment, I have less confidence in my opinions, than when I only consider the objects concerning which I reason' (*Treatise*, section I, part IV, book I).

[150] But there are also some cross-context invariance laws: see section 3.3 again.

[151] For a recent exception, see Lin (2013), which also includes some further references to qualitative decision theories.

model of rational decision-making also on the qualitative side; the only question is how to make that model precise.

Here is a very simple way of doing so: let us assume that O is the set of possible outcomes that our agent's actions might have; formally, let O be a set with at least two members. An action is understood very liberally as *any* function from the set of worlds to that set of outcomes: if A is an action, then $A(w)$ is the outcome in O of carrying out A in w. Moreover, let us presuppose a utility measure u that is just as simple and coarse-grained as *Bel* is: either an action's outcome $A(w)$ is *useful* to the agent (relative to her present desires), in which case $u(A(w))$ is identical to the 'good' value, say, u_{max}; or the action's outcome $A(w)$ is *not useful* to the agent, in which case $u(A(w))$ equals the 'bad' value, say, u_{min}; and of course $u_{max} > u_{min}$. So I will assume $u : O \to \{u_{max}, u_{min}\}$ to be an onto utility function that takes precisely these two real values u_{max} and u_{min}. In the simplest case, u_{max} might be 1 and u_{min} might be 0, although it is not necessary to make this additional assumption.

With that being in place, for every action A one can collect those worlds w in W in which A is useful (or 'good'): $Use(A) = \{w \in W \mid u(A(w)) = u_{max}\}$, which is just the set of worlds in which carrying out A serves the agent's desires as given by u. $Use(A)$ may be considered the proposition that is expressed by the sentence 'Action A is useful', which is true precisely in the worlds that are members of $Use(A)$. Finally, let us count precisely those actions A as pragmatically permissible in the all-or-nothing belief sense that the agent believes to be useful: so A is permissible in that sense if and only if $Bel(Use(A))$. Since $Use(A)$ is a proposition, that is, a subset of W, this is well-defined. For example: *Shall I walk downstairs to the kitchen and get myself a bottle of apple spritzer? Yes, I believe that to be useful, given my desire to drink.* Let this be our simple all-or-nothing belief counterpart to Bayesian decision theory.

There will be no surprises on the Bayesian side: an action A is permissible in the Bayesian sense if and only if A maximizes expected utility. The expected utility $E_P(u(A))$ of A, in which $u(A)$ acts a random variable that takes values (the utilities u_{max} and u_{min}) at worlds, may be defined as: $\sum_{w \in W}[P(\{w\}) \cdot u(A(w))]$. And a perfectly rational Bayesian agent's decision-making can at least be described as if she were making her decisions by maximizing their expected utilities in that sense.

The Humean thesis can now be shown to entail that the simple decision theory based on all-or-nothing beliefs from before is compatible with standard Bayesian decision theory (given such a binary utility measure):

Theorem 4 If P is a probability measure, if *Bel* and P satisfy the Humean thesis HT^r (with $\frac{1}{2} \leq r < 1$), and if not $Bel(\varnothing)$, then:

- for all actions A, B: if $Bel(Use(A))$ and not $Bel(Use(B))$ then

$$E_P(u(A)) > E_P(u(B)),$$

- for all actions A: if

$$E_P(u(A)) \text{ is maximal,}$$

then $Bel(Use(A))$, and for all actions B with $Bel(Use(B))$ it holds that

$$E_P(u(A)) - E_P(u(B)) < (1 - P(B_W))\,(u_{max} - u_{min}) < (1 - r)\,(u_{max} - u_{min}),$$

where B_W is the least believed proposition (which must exist by Theorem 2 in section 2.3.1).[152]

In words: first of all, if an action A is permissible in the all-or-nothing belief sense, so that $Bel(Use(A))$, while action B is not, that is, not $Bel(Use(B))$, then by HT^r it follows that Bayesian decision theory confirms the all-or-nothing recommendation of A over B: the expected utility $E_P(u(A))$ of any such A (relative to the given coarse-grained utility measure u) will always exceed the expected utility $E_P(u(B))$ of any such B.

Secondly, consider an action A that is permissible in the Bayesian sense, that is, where $E_P(u(A))$ is maximal amongst all actions. Then HT^r implies that any such A is also permissible in the all-or-nothing belief sense: $Bel(Use(A))$. Hence, the best probabilistic options are included amongst the best qualitative ones.[153]

Thirdly: clearly, Bayesian decision theory is more sophisticated than our simple decision theory in terms of all-or-nothing beliefs, just as probability measures can be more fine-grained than belief sets. Accordingly, there must be some potential drawback for an agent to follow the all-or-nothing recommendations. If we reconsider any action A from before that was permissible in the Bayesian sense, and if we compare it, with regards to expected utility, to any action B that is permissible in the all-or-nothing belief sense: what is the worst possible discrepancy? By Theorem 4, the difference cannot be that bad: it is always less than $(1 - P(B_W))\,(u_{max} - u_{min}) < (1 - r)\,(u_{max} - u_{min}) \leq \frac{u_{max}-u_{min}}{2}$. This means that any action B permitted by

152 Here is the proof: First of all, by calculation: $E_P(u(A)) = \sum_{w \in W}[P(\{w\}) \cdot u(A(w))] = \sum_{w \in Use(A)}[P(\{w\}) \cdot u_{max}] + \sum_{w \notin Use(A)}[P(\{w\}) \cdot u_{min}] = u_{max} \cdot \sum_{w \in Use(A)}[P(\{w\})] + u_{min} \cdot \sum_{w \notin Use(A)}[P(\{w\})] = P(Use(A)) \cdot u_{max} + [1 - P(Use(A))] \cdot u_{min}$. Similarly, $E_P(u(B)) = P(Use(B)) \cdot u_{max} + [1 - P(Use(B))] \cdot u_{min}$. Because of $Bel(Use(A))$, it follows that $Use(A) \supseteq B_W$, while the failure of $Bel(Use(B))$ entails that there is a world w, such that w is in B_W but not in $Use(B)$. By Theorem 5 in Appendix B, part 3 (about the Outclassing Condition), $P(\{w\}) > P(W \setminus B_W)$. This implies (with the axioms of probability): $P(Use(A)) \geq P(B_W) = P(B_W \setminus \{w\}) + P(\{w\}) > P(B_W \setminus \{w\}) + P(W \setminus B_W) = P(W \setminus \{w\}) \geq P(Use(B))$. But that means that the convex combination $P(Use(A)) \cdot u_{max} + [1 - P(Use(A))] \cdot u_{min}$ (where $u_{max} > u_{min}$) is strictly greater than the convex combination $P(Use(B)) \cdot u_{max} + [1 - P(Use(B))] \cdot u_{min}$. (It is easy to see by plain calculation that their difference is positive.) In other words: $E_P(u(A)) > E_P(u(B))$.

Secondly, if $E_P(u(A))$ is maximal, then by our liberal definition of an action and u being onto, it must be the case that $Use(A) = W$. So $Bel(Use(A))$, because $Bel(W)$.

Thirdly, let A maximize expected utility again, and let $Bel(Use(B))$: $E_P(u(A)) - E_P(u(B)) = P(Use(A)) \cdot u_{max} + [1 - P(Use(A))] \cdot u_{min} - (P(Use(B)) \cdot u_{max} + [1 - P(Use(B))] \cdot u_{min}) =$, since $Use(A) = W$, $= u_{max} - (P(Use(B)) \cdot u_{max} + [1 - P(Use(B))] \cdot u_{min}) \leq$, by $Use(B) \supseteq B_W$ and reasoning about convex combinations as before, $\leq u_{max} - (P(B_W) \cdot u_{max} + [1 - P(B_W)] \cdot u_{min}) = u_{max} - u_{min} - P(B_W)(u_{max} - u_{min}) = (1 - P(B_W))(u_{max} - u_{min}) <$, since $P(B_W) = P(B_W|W) > r$ by HT^r, $< (1 - r)(u_{max} - u_{min})$.

153 This relies on my liberal assumption that *any* function from worlds to outcomes counts as an action. By that assumption, any best Bayesian action must produce 'good' outcomes of utility u_{max} in every possible world, and it is such actions that are then amongst the actions that are permissible also in the all-or-nothing sense. The other claims in Theorem 4 do not rely on this liberal conception of actions. I will switch to a more restrictive notion of action *as a member of a given repertoire of actions* in section 6.1. (I am grateful to Alexandru Baltag for a discussion of this.)

all-or-nothing beliefs (and u) will always be closer in expected utility to any best Bayesian option A than to any worst Bayesian option.[154]

Summing up: Theorem 4 tells us that if the Humean thesis holds, then rational qualitative decisions are probabilistically reliable.[155]

For instance: let us extend the story of Tracey's sprinkler by assuming that Tracey's goal is for her lawn to be wet; furthermore, she does not care at all about wasting water, but she does not want to lose time by engaging with the sprinkler in those cases in which it is on already. Her probability measure P is as described before, the Humean thesis holds with a Humean threshold of $r = \frac{1}{2}$ again, and her utility measure u is such that the two actions

A_1: Turning Tracey's sprinkler on (or in any case attempting to).

A_2: If Tracey's sprinkler is off, turning it on; else leaving it on.

are useful in the following worlds: $Use(A_1) = \{w_1, w_2, w_3\}$, $Use(A_2) = \{w_1, \ldots, w_8\}$. w_1, w_2, w_3 are the three worlds amongst $w_1, \ldots, w_8$ in which $S = 0$, that is, where Tracey's sprinkler is off; consequently, unconditional attempting to turn on the sprinkler in these worlds will have the intended effect of watering Tracey's lawn without her losing time pointlessly by fiddling with a sprinkler that was on already. However, A_1 will not be useful in the same sense in any of the other worlds. On the other hand, A_2 is the perfect action, as it achieves precisely what is to be done in every world in $\{w_1, \ldots, w_8\}$. Admittedly, A_2 is not the most natural action to consider, but let us assume it is available to Tracey, too. (She is able to check with a mere glance whether the sprinkler is off, and only then turns it on.)

If B_W is equal to $\{w_1, w_2, w_3\}$ again—satisfying the Humean thesis and corresponding to a Lockean threshold of 0.9—this means: $Bel(Use(A_1))$, $Bel(Use(A_2))$, and hence both actions are permissible in the all-or-nothing belief sense. With, say, $u_{max} = 1 > 0 = u_{min}$, the difference between the two actions in terms of their expected utilities is just $E_P(u(A_2)) - E_P(u(A_1)) = 1 - 0.9 = 0.1 < \frac{1}{2} = \frac{u_{max} - u_{min}}{2}$: Tracey is not much worse off by acting according to A_1 than according to the top Bayesian option A_2.

If the Humean thesis holds, and if utilities are just as categorical as all-or-nothing beliefs, then deciding on the basis of all-or-nothing beliefs is coherent with deciding on the basis of subjective probabilities, where the exact meaning of 'coherent' is captured in this case by Theorem 4. That kind of practical coherence is itself a consequence of the coherence between categorical belief and numerical belief in the sense of the Humean thesis.

Much more about such practical aspects of the Humean thesis will be said in Chapter 6.

154 Once again, by our assumptions, any such worst Bayesian action must produce 'bad' outcomes of utility u_{min} in every possible world.

155 In Humean terms again: '[belief] renders them [the ideas] the governing principles of all our actions' (Hume, *Treatise*, section VII, part III, book I).

2.4 Conclusions

In this chapter I have developed a joint theory of rational belief and rational degrees of belief based on what I called the Humean thesis of belief: *rational belief corresponds to resiliently high (enough) probability*. The theory has attractive consequences: all-or-nothing belief is stable, closed under logical consequence, corresponds to high enough probability, and supports reliable qualitative decisions. The price to be paid for this is a strong context-sensitivity of belief.

Let me round out the emerging picture by some concluding remarks.

First: in section 2.3, I pointed out three consequences of our Humean thesis HT^r. As it happens, one can also prove *converses* of these results: if the closure of belief under logic, which was consequence 1, is combined with either of the consequences 2 or 3 (and the consistency of belief and the axioms of subjective probability), then this combination in turn entails HT^r for some r. This will be shown in Chapter 3 and in section 6.1, respectively. That is the kind of argument structure highlighted in a different context by Koellner (2013) in which a thesis has fruitful consequences, and where additionally that thesis can be recovered from combinations of some of its fruitful consequences. In Koellner's words, it is such recovery theorems that might 'seal the case'.

Secondly, the theory has lots of applications: one such application—the description of a Bayesian network in all-or-nothing terms—was sketched in section 2.3. There are many more, which, however, I do not have space to develop here (such as to theory choice, belief revision, assertion, a new Ramsey test for conditionals, pragmatic acceptance, and more). But I will turn to some of these applications in later chapters.

Finally, here is a little postscript for the (radical Bayesian) sceptic who might still wonder why it might be useful to invoke the concept of all-or-nothing belief in the first place—why not leave things with degrees of belief alone? Here is why.

In some situations it may be useful to determine *Bel* from *P* in line with the Humean thesis: e.g. to put things qualitatively and simply. Consider the following little example: say, data from a probabilistic database need to be conveyed to the layperson who does not understand, or is not willing to digest, statements about probabilities. How can complex probabilistic data be broken down so that categorical answers to queries of the form 'Shall I believe *A*?' are given in a rational manner? And what if the layperson happens to insist on the satisfaction of certain 'quality criteria', such as the set of answers being closed under logical consequence? The Humean thesis suggests a method for achieving this.

In some situations, however, it may be useful to operate in the converse direction—to determine *P* from *Bel* in line with the Humean thesis: e.g. this might be the case when the exact probabilities are not available initially, but when they would be helpful to have. This is much like the situation in measurement theory in which one starts from a qualitative empirical structure and aims to map that structure to a numerical one in the course of measurement. Reconsider the case of Tracey's sprinkler: say, she has not as yet assigned degrees of belief to the relevant propositions, but she only has certain all-or-nothing beliefs. For simplicity, let us restrict attention just to the three possible worlds w_1, w_2, w_3, let the Humean threshold be $r = \frac{1}{2}$ again, and suppose Tracey believes w_1 to be the actual world: $Bel(T = 0 \wedge J = 0 \wedge R = 0 \wedge S = 0)$. Given *Bel*, the task is now to determine *P* so that the Humean thesis $\mathrm{HT}^{\frac{1}{2}}$ is satisfied. In geometric terms, one can show that this constrains *P* to a convex set of probability measures: any probability measure will do the job that assigns to $\{w_1\}$ a probability greater than the sum of the probabilities of $\{w_2\}$ and $\{w_3\}$; and every probability measure that lies 'between' two probability measures that do the job will also do the job. It was merely a choice that we 'solved' the Humean thesis in section 2.3 for *Bel* rather than for *P*; in other cases it might well be the other way around.

In yet other situations it may be useful to do neither: each of *P* and *Bel* have a life of their own, too. E.g. they can be updated on the same *X* separately; and one can show that if this is done according to the standard theories on both sides (conditionalization on the side of *P*, and so-called AGM belief revision on the side of *Bel*), then if *P* and *Bel* satisfy the Humean thesis, also their updates will do so. I will explain this in detail in Chapter 4 (see section 4.1.3).

Finally, in some situations it may be useful to do a little bit of both: e.g. when one is given a constraint on *P* that does not pin down *P* uniquely, and a constraint on *Bel* that does not determine *Bel* uniquely either. In such a case, one would first have to check whether the two constraints are even 'consistent' with each other. Initially, it would not even seem clear what this should mean exactly, but the Humean thesis suggests an answer: one needs to check whether *P* and *Bel* jointly satisfy an instance of the Humean thesis. If this is so, then the thesis can be used to translate the given constraint on *P* into an additional constraint on *Bel*, and vice versa; in this way, the overall constraint on *P* and *Bel* may well be more than the sum of the parts. I will give an example like that later in section 6.2 (see Example 25).

There is hope for a joint formal epistemology of belief and degrees of belief that will meet the so-called Bayesian challenge (cf. Kaplan 1996) by demonstrating 'how an account of rational human activity will be the poorer if it has no recourse to talk of belief' (Kaplan 1996, p. 100).

Appendix B

Where Does Stability Come from? Stability through Repetition

I argued in Chapter 2 that rational all-or-nothing belief corresponds to stably high rational degree of belief. But where might the required stability of degrees of belief come from?

The short answer is: from any source from which also all-or-nothing belief might emerge.

First of all, stability might result from the representation of causal relationships. The Bayesian network from Chapter 2, which concerned Tracey's sprinkler and the causal nexus of which it is a part, was an example of that kind.

Secondly, stability might follow from prior presumptions about worldly uniformities: I will give an example of this kind in Chapter 4 (Example 18) when I turn to conditional beliefs (which, in the example in question, will correspond to dispositional patterns of inductive reasoning).

Thirdly, stability might arise from the evidence itself in a very immediate manner. One might say, with Hume, and Loeb on Hume: stability can emerge from repetition. 'Belief is an act of the mind arising from custom' (*Treatise*, section IX, part III, book I), and custom 'proceeds from a past repetition' (*Treatise*, section VIII, part III, book I; cf. Loeb 2002, p. 75).

Here is an example. Let P be our rational agent's degree-of-belief function at a time t_0. Let us assume that W consists of eight possible worlds which represent the logically possible combinations of three propositions E_1, E_2, E_3; so w_1 is $E_1 \cap E_2 \cap E_3$, and so on, up to w_8, which is $\neg E_1 \cap \neg E_2 \cap \neg E_3$. Or more concretely: say, $E_1 = \{w_1, w_3, w_4, w_7\}$, $E_2 = \{w_1, w_2, w_4, w_6\}$, $E_3 = \{w_1, w_2, w_3, w_5\}$. For simplicity, I assume that P is uniform over these eight worlds, that is, for each world w_i: $P(\{w_i\}) = \frac{1}{8}$. Consequently, E_1, E_2, E_3 are mutually independent relative to P: each one is probabilistically independent of the intersection of the others.

Now let us suppose that the agent faces a stream of evidence that consists of, say, precisely the three propositions

$$E_1, E_2, E_3,$$

which reach the agent at times $(t_0 <)\ t_1 < t_2 < t_3$, respectively. Obviously, the evidence is consistent overall, since there is a world (w_1) that satisfies all of the propositions

E_j simultaneously. Let us assume also that each piece of evidence is probabilistically uncertain: when E_j comes along at time t_j, the agent learns it only with probability α_j, where $0 < \alpha_j < 1$. For simplicity again, let $\alpha_1 = \alpha_2 = \alpha_3 = \alpha$. Intuitively, the greater α, the more 'entrenched' E_i will be after the update.

The standard Bayesian method of learning for such a type of situation is Jeffrey update (or probability kinematics).[156] After taking E_1 on board with a target degree of α, the agent's subjective probability measure is given by (with X being an arbitrary proposition):

$$P_1(X) = \alpha\, P(X|E_1) \;+\; (1-\alpha)\, P(X|\neg E_1).$$

So, by the axioms of probability, indeed $P_1(E_1) = \alpha$. The mutual independence of E_1, E_2, E_3 is preserved by this transition to P_1, as follows just like the independence preservation observation that was shown already in section A.3 of Appendix A (where we encountered Jeffrey update for the first time).

Accordingly, updating P_1 with E_2 in the analogous manner yields (using the independence of E_2 from E_1 with respect to P_1):

$$P_2(X) = \alpha\, P_1(X|E_2) \;+\; (1-\alpha)\, P_1(X|\neg E_2) =$$
$$\alpha^2\, P(X|E_1 \cap E_2) \;+\; \alpha(1-\alpha)\; [P(X\,|\neg E_1 \cap E_2) + P(X\,|E_1 \cap \neg E_2)]$$
$$+\; (1-\alpha)^2\, P(X\,|\neg E_1 \cap \neg E_2).$$

Finally, when P_2 is the agent's degree-of-belief function, learning E_3 in the analogous way (and applying analogous independence considerations) leads to:

$$P_3(X) =$$
$$\alpha^3\, P(X|E_1 \cap E_2 \cap E_3) +$$
$$\alpha^2\,(1-\alpha)\; [P(X|\neg E_1 \cap E_2 \cap E_3) + P(X|E_1 \cap \neg E_2 \cap E_3) + P(X|E_1 \cap E_2 \cap \neg E_3)] +$$
$$\alpha\,(1-\alpha)^2\; [P(X|\neg E_1 \cap \neg E_2 \cap E_3) + P(X|\neg E_1 \cap E_2 \cap \neg E_3) + P(X|E_1 \cap \neg E_2 \cap \neg E_3)] +$$
$$(1-\alpha)^3\, P(X|\neg E_1 \cap \neg E_2 \cap \neg E_3).$$

For instance: if $\alpha = 0.8$, then P_3 assigns probabilities to (singleton sets of) worlds as follows:

w_1 (i.e. $E_1 \cap E_2 \cap E_3$):	0.512
w_2 (i.e. $\neg E_1 \cap E_2 \cap E_3$):	0.128
w_3 (i.e. $E_1 \cap \neg E_2 \cap E_3$):	0.128
w_4 (i.e. $E_1 \cap E_2 \cap \neg E_3$):	0.128
w_5 (i.e. $\neg E_1 \cap \neg E_2 \cap E_3$):	0.032
w_6 (i.e. $\neg E_1 \cap E_2 \cap \neg E_3$):	0.032
w_7 (i.e. $E_1 \cap \neg E_2 \cap \neg E_3$):	0.032
w_8 (i.e. $\neg E_1 \cap \neg E_2 \cap \neg E_3$):	0.008

156 The method was suggested first in Jeffrey (1965, ch. 11).

I turn to the same question now as in the case of Tracey's Sprinkler in Chapter 2: given the probability measure P_3, which belief sets *Bel* are such that P and *Bel* taken together satisfy the Humean thesis with a threshold of, say, $r = \frac{1}{2}$?

Here is the answer stated first in more general terms—a representation theorem for the Humean thesis:

Theorem 5 (Representation Theorem for the Humean Thesis)
Let W be finite and non-empty. Let P be a probability measure on W. Let r be a threshold, such that $\frac{1}{2} \leq r < 1$. Assume that not $Bel(\varnothing)$. Then the following three statements are equivalent:

1. *Bel* and P satisfy the Humean thesis HT^r (that is, HT^r_{Poss}) from Chapter 2.
2. There is a (non-empty) proposition B_W, such that: (i) for all X: $Bel(X)$ iff $B_W \subseteq X$. (ii) B_W has the following stability property with respect to P and r:

B_W is P-stable^r: for all Y with $Y \cap B_W \neq \varnothing$ and $P(Y) > 0$ it holds that $P(B_W|Y) > r$.

(iii) If $P(B_W) = 1$, then B_W is the least proposition with probability 1.

3. There is a (non-empty) proposition B_W, such that: (i) for all X: $Bel(X)$ iff $B_W \subseteq X$. (ii′) B_W satisfies the following condition with respect to P and r:

Outclassing Condition:[157] for all w in B_W it holds that $P(\{w\}) > \frac{r}{1-r} \cdot P(W \setminus B_W)$.

('$W \setminus B_W$' denotes the set W without B_W, or: $\neg B_W$.)[158]

[157] David Makinson suggested this name in personal communication.

[158] Here is the proof. '2 $\leftrightarrow$ 3': This somewhat technical observation follows from (ii) + (iii) being equivalent to (ii'), which in turn follows from Observation 11 in subsection 'P-Stability and the First Representation Theorem' in section 4.2.2. That subsection proves the main properties of P-stable^r sets that are required for the technical observations in this book.

'1 $\rightarrow$ 2': The existence of the (uniquely determined) least believed set B_W, which therefore has property (i), follows from Theorem 2 in section 2.3.1. For the sake of self-containment, I repeat the relevant steps here.

First of all, there is some believed set, since $Bel(W)$: this follows from the right-hand side of HT^r being satisfied.

Secondly, for all propositions X, Y, if $Bel(X)$ and $X \subseteq Y$, then $Bel(Y)$: if $Bel(X)$, then by the left-to-right direction of HT^r (= HT^r_{Poss}), for all Z, if $Poss(Z)$ (and $P(Z) > 0$), then $P(X|Z) > r$. From $X \subseteq Y$, by the axioms of probability: for all Z, if $Poss(Z)$ (and $P(Z) > 0$), then $P(Y|Z) \geq P(X|Z) > r$. This yields, by the right-to-left direction of $\mathrm{HT}^r_{\mathcal{Y}}$: $Bel(Y)$.

Thirdly, *Bel* is closed under intersections: Suppose for contradiction that $Bel(X)$, $Bel(Y)$, but not $Bel(X \cap Y)$, that is, not $Bel(\neg(\neg X \cup \neg Y))$. So, by definition of '*Poss*': $Poss(\neg X \cup \neg Y)$. If $P(\neg X \cup \neg Y)$ were 0, then $P(X \cap Y)$ would have to be 1, in which case it would be satisfied that for all Z, if $Poss(Z)$ and $P(Z) > 0$, then $P(X \cap Y|Z) > r$, which with the right-to-left direction of HT^r (= HT^r_{Poss}) would entail $Bel(X \cap Y)$, which had been ruled out by assumption: hence, $P(\neg X \cup \neg Y) > 0$. From $Bel(X)$ and the left-to-right direction of HT^r, it follows that $P(X|\neg X \cup \neg Y) > r$. And the same holds for Y: $P(Y|\neg X \cup \neg Y) > r$. By the axioms of probability and the definition of conditional probabilities, this means: $P(X \cup Y|\neg X \cup \neg Y) = P(X \cap \neg Y|\neg X \cup \neg Y) + P(Y|\neg X \cup \neg Y) = P(X|\neg X \cup \neg Y) + P(Y|\neg X \cup \neg Y) > r + r \geq 1$ (by the assumption that $r \geq \frac{1}{2}$), which is a contradiction.

Fourthly, let B_W be the intersection of all believed propositions (there are such by $Bel(W)$): so $Bel(B_W)$, by what we have shown under 'Thirdly' (and by W, and hence the set of propositions, being finite). Moreover: for all X, $Bel(X)$ iff $B_W \subseteq X$, by the definition of B_W and by what we have shown under 'Secondly'.

That B_W is also P-stable^r follows from an application of the left-to-right direction of HT^r_{Poss} with: $Bel(B_W)$; $Poss(Y)$ iff $Y \cap B_W \neq \varnothing$. Finally, assume $P(B_W) = 1$, but suppose that the least proposition

This is a representation theorem in the sense that every pair $\langle Bel, P\rangle$ that satisfies condition 1 can be represented as a pair $\langle B_W, P\rangle$ that satisfies a certain probabilistic property ('B_W is P-stable^r', 'B_W satisfies the Outclassing Condition relative to P and r'). Representation theorems like that will continue to play a crucial role for the rest of this book.[159]

Condition 2 in Theorem 5 shows that the stability of believed propositions in the sense of the Humean thesis from Chapter 2 can be 'condensed' into a special stability property that applies just to the least believed proposition, that is, B_W. I call this stability property: P-stability^r (stability with respect to P and r). The property of being P-stable^r in the special case in which $r = \frac{1}{2}$ will be a big topic in Chapter 3, and P-stability^r for general r with $\frac{1}{2} \leq r < 1$ will be analysed formally in Chapter 4, especially in section 4.2.2. Given (ii) and the formal properties of P-stable^r sets, clause (iii) may be seen to be interpretable in the way: if a perfectly rational agent regards a proposition X as certain (so $P(X) = 1$), then by the Humean thesis the agent also believes X in the all-or-nothing sense ($B_W \subseteq X$).

So far as condition 3 is concerned, in the special case of $r = \frac{1}{2}$, the Outclassing Condition for B_W relative to P and r takes the simpler form:

> Outclassing Condition for $r = \frac{1}{2}$: for all w in B_W it holds that $P(\{w\}) > P(W \setminus B_W)$.[160]

X with $P(X) = 1$ (which exists by W being finite) is a proper subset of B_W. Then because of $P(X) = 1$ it holds for all Y with $Poss(Y)$ and $P(Y) > 0$ that $P(X|Y) = 1 > r$; but because of $X \not\supseteq B_W$, it must also hold that not $Bel(X)$; which contradicts HT^r_{Poss}. So $X = B_W$.

'2 → 1': Let $X \subseteq W$. The left-to-right direction of HT^r_{Poss} follows from: assume $Bel(X)$, $Poss(Y)$, $P(Y) > 0$. By $Bel(X)$, it holds that $X \supseteq B_W$. Because of $Poss(Y)$, it is the case that $Y \cap B_W \neq \varnothing$. By B_W being P-stable^r, $P(B_W|Y) > r$. But since $X \supseteq B_W$, it follows that $P(X|Y) \geq P(B_W|Y) > r$. The right-to-left direction of HT^r_{Poss} follows from: assume for all Y, if $Poss(Y)$ and $P(Y) > 0$, then $P(X|Y) > r$. Suppose not $Bel(X)$: then $Poss(\neg X)$, that is, $\neg X \cap B_W \neq \varnothing$. If $P(B_W) = 1$, then B_W is the least proposition of probability 1, which cannot have any non-empty subset of probability 0, for otherwise B_W without that subset would still have probability 1 but would be a proper subset of B_W, which would contradict B_W being the least such set. If $P(B_W) < 1$, then B_W cannot have any non-empty subset of probability 0 either, by Observation 10 in subsection 'P-Stability and the First Representation Theorem' of section 4.2.2. Either way it follows with $\neg X \cap B_W \neq \varnothing$ that $P(\neg X) > 0$. So we have $Poss(\neg X)$, $P(\neg X) > 0$, and thus, by assumption, it has to be the case that $P(X|\neg X) > r$. But of course $P(X|\neg X) = 0$ by the axioms of probability, which is a contradiction. Therefore, $Bel(X)$.

159 Generally, representation theorems are of the form: every structure S in a class C is isomorphic to a structure S' in a subclass C' of C (so $C' \subseteq C$). Or in other words: every structure S in a class C can be represented (up to isomorphism) as a structure S' in a subclass C' of C. (As in, for instance, Stone's Representation Theorem for Boolean algebras: every Boolean algebra is isomorphic to, or can be represented as, a Boolean field of sets.) In my case, the isomorphism in question will always be the identity map: $S = S'$. What will make the representation theorems still interesting and non-trivial will be that the manner in which 'S'' will present the structure in question will differ substantially from how 'S' will present it. As in, for example, part '1 ↔ 2' of Theorem 5: every $\langle Bel, P\rangle$ that satisfies the Humean thesis HT^r can be represented as a pair $\langle Bel', P\rangle$, such that Bel' is generated from a set B_W (that is, for all X, $Bel'(X)$ iff $B_W \subseteq X$), where B_W is P-stable^r and if $P(B_W) = 1$ then B_W is the least proposition of probability 1.

160 In the computer science literature, a compatibility condition on probability measures and strict total orders of worlds has been formulated that is similar to this equivalent reformulation of P-stability^r with $r = \frac{1}{2}$: compare the 'big-stepped probabilities' of Benferhat et al. (1999) and Snow's (1998) 'atomic bound systems'. I will discuss this literature in more detail in section 4.1.4.

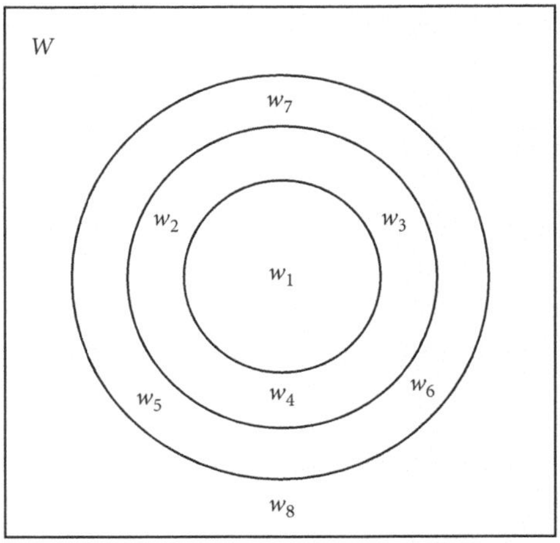

Figure B.1. Jeffrey update with $\alpha = 0.8$

Applying this to P_3 from our example finally determines the following answer to our question: given P_3, there are exactly four belief sets *Bel*, such that *P* and *Bel* satisfy the Humean thesis $HT^{\frac{1}{2}}$, and these belief sets are generated by the following four choices of B_W, respectively: either $B_W = \{w_1\}$ or $B_W = \{w_1, w_2, w_3, w_4\}$ or $B_W = \{w_1, \ldots, w_7\}$ or $B_W = W$. It is easy to check that these sets, and only these sets, satisfy the Outclassing Condition with respect to *P* and $r = \frac{1}{2}$. Graphically, these options are depicted as spheres in Figure B.1.

In words: one way for belief to cohere with P_3 in the sense of the Humean thesis would be for the agent to believe in each of E_1, E_2, E_3 (and hence, by logical closure, in their conjunction or intersection): that is the $B_W = \{w_1\}$ option. That should not be particularly surprising, since the evidence had presented w_1 as a candidate of what the actual world might be like three times in a row (though each time the evidence was certain only to a degree of $\alpha = 0.8$): w_1 had been a member of E_1 and E_2 and E_3.

Or the agent is more cautious and only believes at least two out of three pieces of evidence to be true: that is, $B_W = (E_1 \cap E_2) \cup (E_1 \cap E_3) \cup (E_2 \cap E_3) = \{w_1, w_2, w_3, w_4\}$, which is the set of possibilities presented by the evidence at least twice. Or at least one out of three, that is, $B_W = \{w_1, \ldots, w_7\}$. Or she is maximally cautious and only believes the tautological proposition *W*.

In each of the four cases, belief is stable: for instance, in the $B_W = \{w_1, w_2, w_3, w_4\}$ case, the agent believes $(E_1 \cap E_2) \cup (E_1 \cap E_3) \cup (E_2 \cap E_3)$, she holds $\neg E_3$ to be possible (since she does not believe E_3), and indeed

$$P((E_1 \cap E_2) \cup (E_1 \cap E_3) \cup (E_2 \cap E_3) \mid \neg E_3) = \frac{0.128}{0.2} = 0.64 > r = \frac{1}{2}.$$

Summing up: iterated Jeffrey update on uncertain evidence may lead to stably high degrees of belief. In such cases it is the update itself that entrenches information by

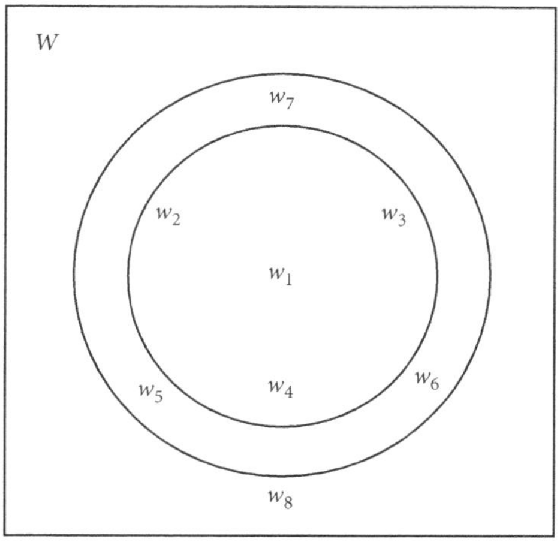

Figure B.2. Jeffrey update with $\alpha = 0.78$

means of repetition of possibilities so that the information becomes stable enough to be believed rationally.[161]

It is easy to see that, as long as $\alpha > \sqrt[3]{\frac{1}{2}} \approx 0.7937\ldots$ is the case in the present kind of situation, there will always be the same four belief sets that do the job as described before. However, if $\alpha \leq \sqrt[3]{\frac{1}{2}}$, then given the probability measure that results from updating P by E_1, E_2, E_3 with that α, the number of possibilities for *Bel* to satisfy the Humean thesis decreases. For example, if $\alpha = 0.78$, the resulting probabilities are these:

w_1 (i.e. $E_1 \cap E_2 \cap E_3$):	0.474552
w_2 (i.e. $\neg E_1 \cap E_2 \cap E_3$):	0.133848
w_3 (i.e. $E_1 \cap \neg E_2 \cap E_3$):	0.133848
w_4 (i.e. $E_1 \cap E_2 \cap \neg E_3$):	0.133848
w_5 (i.e. $\neg E_1 \cap \neg E_2 \cap E_3$):	0.037752
w_6 (i.e. $\neg E_1 \cap E_2 \cap \neg E_3$):	0.037752
w_7 (i.e. $E_1 \cap \neg E_2 \cap \neg E_3$):	0.037752
w_8 (i.e. $\neg E_1 \cap \neg E_2 \cap \neg E_3$):	0.010648

In this case, there are only three sets *Bel* that cohere with the probabilities as demanded by the Humean thesis, and their corresponding sets B_W are depicted in Figure B.2. For yet smaller values of α it can happen that W remains as the only possible choice for B_W that yields enough stability.

161 Before the iterated Jeffrey update took place, the only set B_W that together with the prior uniform measure P would have satisfied the Humean thesis was W itself: the set of all worlds. So at t_0 a perfectly rational agent in the Humean thesis sense of the word would have believed $\{w_1, \ldots, w_8\}$ to be true and nothing else.

3

Logical Closure and the Lockean Thesis

The last chapter determined a bridge principle for rational (all-or-nothing) belief and degrees of belief which I called the Humean thesis on belief. It unified different stability conceptions of rational belief, and various plausible conclusions turned out to be derivable from it.

This chapter develops a joint theory of rational (all-or-nothing) belief and degrees of belief again, but this time the starting point will be different. The theory will be based on three assumptions: the logic of rational belief; the axioms of probability for rational degrees of belief; and the so-called Lockean thesis in which the concepts of rational belief and rational degree of belief figure simultaneously. Contrary to what is commonly believed, I will show that this combination of principles is satisfiable and indeed non-trivially so, and that the principles are jointly satisfied if and only if the Humean thesis from Chapter 2 holds with a Humean threshold of $\frac{1}{2}$ (where it is assumed additionally that the contradictory proposition is not believed).

Although the logical closure of belief and the Lockean thesis are attractive postulates in themselves, this result may seem initially like a formal 'curiosity'. However, as I am going to argue in the rest of the chapter, a very reasonable theory of rational belief can be built around these principles that is not ad hoc but which has various philosophical features that are plausible independently. The downside of the theory will be that rational belief will turn out to be context-sensitive in a sense that I will explain in the chapter. But I will also give reasons for believing that we should be able to live with that kind of context-sensitivity.[162]

3.1 The Lockean Thesis and Closure of Belief under Conjunction

Each of the following three postulates on belief (*Bel*) and degrees of belief (*P*) for perfectly rational agents seems tempting, at least if taken just by itself:

[162] Hanti Lin and Kevin Kelly, and Sonja Smets and Alexandru Baltag, kindly commented on an extended abstract of the journal article (Leitgeb 2014a) on which this chapter is based. Their comments, together with the extended abstract and my replies to their comments, appeared in Van Benthem and Liu (2013).

P1 The logic of belief, in particular, the closure of belief under conjunction, that is: for all propositions A, B,

$$\text{if } Bel(A) \text{ and } Bel(B) \text{ then } Bel(A \wedge B).$$

P2 The axioms of probability for the degree-of-belief function P.

P3 The Lockean thesis (cf. Foley 1993, pp. 140–1) that governs both *Bel* and P: there is a Lockean threshold s that is greater than $\frac{1}{2}$ and less than or equal to 1, such that for every proposition B, it holds that B is believed if and only if the degree of belief in B is not less than s, or more briefly,

$$Bel(B) \text{ if and only if } P(B) \geq s.$^{[163]}$

P1 is entailed by the doxastic version of any normal system of modal logic for the operator '*Bel*'. It was included in Assumption 8 from section 1.5 of Chapter 1. P2 is at the heart of Bayesianism. It is part of Assumption 9 from section 1.5. P3 expresses the natural thought that it is rational to believe a proposition if and only if it is rational to have a sufficiently high degree of belief in it. (More will be have to said soon about how to disambiguate the relative positions of the tacit universal quantifier over P and the explicit existential quantifier over s.)

Yet this combination of rationality postulates is commonly rejected. And the standard reason for doing so is that, given P1–P2, there does not seem to be any plausible value of 's' available that would justify the existence claim in P3.

Here is why: the first possible option, s being equal to 1, seems too extreme; '$Bel(B)$ if and only if $P(B) \geq s$' would turn into the trivializing '$Bel(B)$ if and only if $P(B) = 1$' condition, by which all and only propositions of which one is probabilistically certain are to be believed. But that cannot be right, as explained already in section 1.6 (and reiterated in Chapter 2), at least if it is taken as a requirement on believed propositions that is meant to hold in each and every context. For example: it is morning; I rationally believe that I am going to receive an email today. However, I would not regard it as rational to buy a bet in which I would win €1 if I am right, and in which I would lose €1,000,000 if I am wrong. But according to the usual interpretation of subjective probabilities in terms of betting quotients, I should be rationally disposed to accept such a bet if I believed the relevant proposition to the maximal degree of 1. Hence, I rationally believe the proposition even though I do not believe it with probability 1.

[163] In many formulations of the Lockean thesis, a 'greater-than' symbol is used instead of 'greater-than-equals', but since I am going to assume the underlying set of possible worlds to be finite, nothing will really hang on this choice of formulation. In the finite case, for any $\geq$-version of the Lockean thesis with threshold s, there is a $>$-version of the Lockean thesis with a threshold r' that is slightly below s. Vice versa, for any $>$-version of the Lockean thesis with threshold r', there is a $\geq$-version of the Lockean thesis with a threshold s that is slightly above r'. However, the 'greater-than-equals' formulation will prove to be a bit more convenient for the purposes of this chapter.

The remaining option for how to argue for the existence claim in P3 would be to turn to some value of 's' that is less than 1; and as long as one considers '$Bel(B)$ if and only if $P(B) \geq s$' just by itself, this looks more appealing and realistic. But then again, if taken together with P1 and P2, this option seems to run into the famous Lottery Paradox (cf. Kyburg 1961) to which I will return later.[164]

Therefore, in spite of the prima facie attractiveness of each of P1–P3, it just does not seem to be feasible to have all of them at the same time. Which is why a large part of the classical literature on belief (or acceptance) can be categorized according to which of the three postulates are being preserved and which are dropped—as Levi (1967, p. 41) formulates it, 'either cogency [our P1] or the requirement of high probability as necessary and sufficient for acceptance [our P3] must be abandoned'. For instance, putting P2 to one side for now, Isaac Levi keeps P1 but rejects P3, while Henry Kyburg keeps P3 and rejects P1.[165] Hempel (1962) still had included both P1 and P3 as plausible desiderata, although he was already aware of the tension between them.

In the following I want to show that this reaction of dropping any of P1–P3 is premature; it is in fact not clear that one could not have all of P1–P3 together and the existence claim in P3 being true in virtue of some threshold $s < 1$. Indeed, we have already seen these postulates to follow from a combination of the axioms of probability and the Humean thesis in Chapter 2, and we have found them to be consistent with each other there in view of an example. In the following I will address similar points again, but now in a context in which P1–P3 are the axioms and in which the stability of belief will turn out to be a corollary.

The first step is to note that P3, as formulated, is *ambiguous* with respect to the position of the 'there is a threshold s' quantifier in relation to the implicit universal quantification over degree-of-belief functions P.[166] According to one possible disambiguation, there is indeed no value of 's' less than 1 so that 'for all B, $Bel(B)$ if and only if $P(B) \geq s$' could be combined consistently with P1 and P2. But according to a second kind of disambiguation, taking all of these assumptions together will in fact be logically possible, and it will be that manner of understanding P3 on which my stability theory of belief in this chapter will be based.

Here is the essential point: we need to distinguish a claim of the form 'there is an $s < 1 \ldots$ for all $P \ldots$' from one of the form 'for all $P \ldots$ there is an $s < 1 \ldots$'. As we are going to see, the difference is crucial: while it is *not* the case that

there is an $s < 1$, such that for all P (on a finite space of worlds)

[164] A similar point can be be made in terms of the equally well-known Preface Paradox; see Makinson (1965). I will discuss the Preface Paradox in two parts: first in section 3.5 and later in section 6.4.

[165] Both Levi and Kyburg also reject P2, but I will not discuss this here.

[166] For simplicity, I disregard additional quantifiers here, such as those ranging over belief sets Bel or over their underlying spaces of possible worlds.

the logical closure of *Bel*, the probability axioms for P, and for all B, $Bel(B)$ if and only if $P(B) \geq s$, are jointly satisfied, it *is* the case that

for all P (on a finite space of worlds), there is an $s < 1$

such that the same conditions are jointly the case.

Let me explain why. I will start with what will be interpreted later on in sections 3.3 and 3.4 as a typical lottery example:

Example 4

Assume that $s = \frac{999999}{1000001}$. Consider W to be a set $\{w_1, \ldots, w_{1000000}\}$ of one million possible worlds, and let P be the uniquely determined probability measure that is given by $P(\{w_1\}) = P(\{w_2\}) = \ldots = P(\{w_{1000000}\}) = \frac{1}{1000000}$. A fortiori, the axioms of probability are satisfied by P, as demanded by P2 above. At the same time, by the corresponding instance of the Lockean thesis (P3), it would follow that for every $1 \leq i \leq 1,000,000$, it is rational to believe the proposition $W - \{w_i\}$ (that is, W without $\{w_i\}$), as $P(W - \{w_i\}) = \frac{999999}{1000000} \geq \frac{999999}{1000001} = s$. Therefore, by P1, the conjunction (that is, intersection) of all of these propositions would rationally have to be believed as well; but this conjunction is nothing but the contradictory proposition $\varnothing$, which has probability 0 by P2, and which for that reason is not rationally believed according to P3. We end up with a contradiction. That is: for s as being picked before, we can determine a probability measure P, such that the logical closure of *Bel*, the probability axioms for P, and for all B, $Bel(B)$ if and only if $P(B) \geq s$, do not hold jointly. By the same token, for every $\frac{1}{2} < s < 1$ a uniform probability measure can be constructed, such that these conditions are not satisfied simultaneously.

Example 5

Let W be the set $\{w_1, \ldots, w_{1000000}\}$ again, and assume the probability measure P to be given again by $P(\{w_1\}) = \ldots = P(\{w_{1000000}\}) = \frac{1}{1000000}$. But now set $s = \frac{1000000}{1000001}$: in that case, the only proposition that is to be believed according to the Lockean thesis is W itself, which has probability 1. Trivially, then, the set of believed propositions is closed under logic (including closure under conjunction); which is why the logical closure of *Bel*, the probability axioms for P, and for all B, $Bel(B)$ if and only if $P(B) \geq s$, hold jointly. For P as being chosen before, we can determine a threshold s, such that all of our desiderata are satisfied.

It is evident that in Example 5 we were able to circumvent the contradiction from Example 4 by another trivializing method (just as opting for $s = 1$ and '$Bel(B)$ if and only if $P(B) = 1$' had been trivializing before): given a P with a finite domain, one can push the threshold in the Lockean thesis sufficiently close to (though short of) 1 so that only those propositions that have probability 1 end up believed.

While the same method enables us to determine for every probability measure (over a finite set of worlds) a suitable threshold $s < 1$ and *Bel* such that P1, P2, and for all B, $Bel(B)$ if and only if $P(B) \geq s$, are jointly the case, this is hardly satisfying;

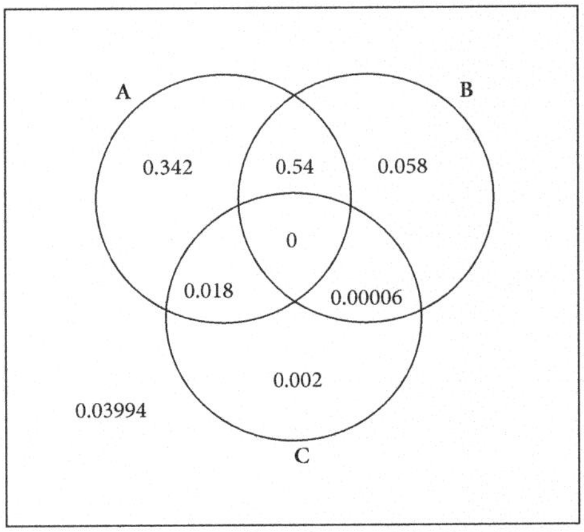

Figure 3.1. Example 6

for once again, rational belief would be restricted to propositions of which one is probabilistically certain.

The much more exciting observation is that in many cases one can do much better: it is possible to achieve the same result without trivializing consequences, in the sense that at least some proposition of probability *less than 1* happens to be believed.

Here is an example (to which I will return also in subsequent sections and which will be given a concrete interpretation in section 3.6):[167]

Example 6

Let $W = \{w_1, \ldots, w_8\}$ be a set of eight possible worlds; one might think of these eight possibilities as coinciding with the 'state descriptions' that can be built from three propositions A, B, C: w_1 corresponds to $A \wedge B \wedge \neg C$, w_2 to $A \wedge \neg B \wedge \neg C$, w_3 to $\neg A \wedge B \wedge \neg C$, w_4 to $\neg A \wedge \neg B \wedge \neg C$, w_5 to $A \wedge \neg B \wedge C$, w_6 to $\neg A \wedge \neg B \wedge C$, w_7 to $\neg A \wedge B \wedge C$, and w_8 to $A \wedge B \wedge C$. ('A', 'B', 'C' will receive a proper interpretation in section 3.6.) Let P be the unique probability measure that is defined by: $P(\{w_1\}) = 0.54$, $P(\{w_2\}) = 0.342$, $P(\{w_3\}) = 0.058$, $P(\{w_4\}) = 0.03994$, $P(\{w_5\}) = 0.018$, $P(\{w_6\}) = 0.002$, $P(\{w_7\}) = 0.00006$, $P(\{w_8\}) = 0$. Figure 3.1 depicts what this probability space looks like. Now consider the following six propositions,

$$\{w_1\}, \{w_1, w_2\}, \{w_1, \ldots, w_4\}, \{w_1, \ldots, w_5\}, \{w_1, \ldots, w_6\}, \{w_1, \ldots, w_7\},$$

only the last one of which has probability 1. Pick any of them, call it 'B_W', and let *Bel* be determined uniquely by stipulating that B_W is the least or strongest proposition that is believed, so that a proposition is believed if and only if it is entailed by (is a superset of) B_W. In other words: for all propositions $X \subseteq W$,

$$Bel(X) \text{ if and only if } B_W \subseteq X.$$

[167] It is the probability measure that has made an appearance already in section 1.5.

Finally, take $s = P(B_W)$ to be the relevant threshold. One can show that the so-determined *Bel*, *P*, and *s* satisfy the logical closure of *Bel*, the probability axioms for *P*, and for all *B*, *Bel*(*B*) if and only if $P(B) \geq s$. Once again, for our given *P*, there is a threshold *s*, such that all of our desiderata hold simultaneously. But this time, as far as the first five choices of B_W are concerned, there is in fact a proposition of probability less than 1 that is being believed. E.g. if B_W is $\{w_1, w_2\}$, then $\{w_1, w_2\}$ is believed even though it has a probability of $0.882 < 1$.

What should we conclude from these examples? *Maybe it is possible to have one's cake and to eat it, too:* to preserve the logic of belief and the axioms of probability while at the same time assuming consistently that the beliefs and degrees of belief of perfectly rational agents relate to each other as expressed by an instance of the Lockean thesis even for a threshold of less than 1.

The price to be paid for this proposal will be that not any old threshold in the Lockean thesis will do; instead the threshold must be chosen suitably depending on what the agent's beliefs and her degree-of-belief function are like. Whether that price is affordable or not, I will discuss later, but first I will turn to a different question: given a degree-of-belief function *P*, what are the belief sets *Bel* and thresholds *s* like which, together with *P*, satisfy all of our intended conditions? The answer will be given by section 3.2 in which P1–P3 will be made formally precise, and in which the intended belief sets and thresholds will be characterized by means of a probabilistic notion of stability or resiliency. Based on this, we will see that P1–P3 taken together are equivalent to the stability theory of belief from Chapter 2 given a *Humean* threshold (*not* a Lockean one) of $r = \frac{1}{2}$. So P1–P3 will turn out to constitute a stability theory of belief, too: the stability theory of belief from the last chapter, only presented differently. Afterwards, in section 3.3, I will outline the costs of accepting this theory: a strong form of context-sensitivity of belief, where the context in question involves both the agent's degree-of-belief function *P* and the partitioning or individuation of the underlying possibilities. Section 3.4 explains what the theory predicts concerning the Lottery Paradox; the observed context-sensitivity of belief will actually work to the theory's advantage there. In section 3.5 I will give a first and preliminary analysis of the Preface Paradox (which will be continued later in section 6.4 of Chapter 6). In section 3.6 I will present an example of how the theory can be applied in other areas, in that case, to a problem in formal epistemology or general philosophy of science. Section 3.7 summarizes what has been achieved and, on these grounds, makes the case for the theory.

3.2 *P*-Stability

I begin by stating P1–P3 from the last section in full detail.

Let us consider a perfectly rational cognitive agent and her beliefs and degrees of belief at a fixed point of time. By 'perfectly rational' I only mean '*inferentially* perfectly rational'—so that the usual logical and probabilistic principles of rational belief can be taken for granted for any such agent—but of course I do not assume e.g. that any such agent would be perfectly rational in the sense of believing all and only truths, or the like.[168]

Let W be a (non-empty) set of possible worlds. Throughout the chapter I will keep W fixed, and I will assume that W is *finite*; the theory that I am going to develop will work also in the infinite case (subject to some constraints, as explained in Chapter 4), but I want to keep things as simple as possible here. Like in the previous chapters, W may be regarded again as the set of logically possible worlds for a simple propositional language with finitely many (atomic) propositional letters.

Given W, by a proposition I mean any subset of W; so propositions will be regarded as sets of possible worlds. I will apply the standard terminology that is normally used for sentences also to propositions: when I say that a proposition is consistent I mean that it is non-empty, and accordingly $\varnothing$ is the unique contradictory proposition. When I say that a proposition A is consistent with another proposition B, then this is: $A \cap B \neq \varnothing$. When I say that A entails B, this amounts to A being a subset of B. When I refer to the negation of A, I actually refer to A's complement $(W \setminus A)$ relative to W (which I will also denote by '$\neg A$'). The conjunction $A \wedge B$ of A and B is their intersection, and their disjunction $A \vee B$ is their union.

I represent the agent's beliefs at the relevant time by means of a set *Bel* of propositions: the set of propositions believed by the agent in question at the time in question. Instead of '$A \in Bel$', I will usually write: $Bel(A)$.

This being in place, P1 from the last section was really a shorthand for the standard laws of doxastic logic adapted to the current propositional context (and disregarding introspective belief which will not play any role here and which I leave aside throughout the book):

P1 For all propositions $A, B \subseteq W$:
- $Bel(W)$;
- not $Bel(\varnothing)$;
- if $Bel(A)$ and $A \subseteq B$, then $Bel(B)$;
- if $Bel(A)$ and $Bel(B)$, then $Bel(A \wedge B)$.

The first two clauses express that the agent believes that one of the worlds within her total set W of worlds is the actual world, and she does not believe the empty set to include the actual world. The other two clauses express the closure of belief under logical consequence.

Since W is finite by assumption, there can be only finitely many members of *Bel*; by P1, the conjunction of all of them, say, B_W, must also be a member of *Bel*, B_W must be

[168] Ultimately, we should be concerned with *real-world* agents, but methodologically it seems like a good strategy to sort out the tension between belief and degrees of belief first for ideal agents—whom we strive to approximate—and only then for agents such as ourselves. Compare the discussion of this point in section 1.4.

consistent, and by the definition of B_W and by P1 again, the following must hold for every proposition B: $Bel(B)$ if and only if $B_W \subseteq B$.

Vice versa, assume there to be a consistent proposition B_W in *Bel*, such that for every proposition B: $Bel(B)$ if and only if $B_W \subseteq B$. Then it follows that P1 above is satisfied.

In other words, we can reformulate P1 equivalently as follows:

P1 [Reformulated] There is a consistent proposition $B_W \subseteq W$, such that for all propositions B:

$$Bel(B) \text{ if and only if } B_W \subseteq B.$$

So P1 really amounts to a possible worlds model of belief: the agent implicitly or explicitly divides the set W of possible worlds into those which are 'serious possibilities' for the agent at the time (using the terminology of Levi 1980, 1984)—that is, serious candidates for what the actual world might be like—and those which are not. B_W is that set of serious possibilities, and it is determined uniquely given the belief set *Bel* and our assumptions.

Now I turn to P2 from the last section: at the relevant point of time, let P be the agent's degree of belief or credence function which I take to be defined for all subsets of W; in probabilistic terms, W is the sample space for P. Indeed, P2 assumes that P is a probability measure, and accordingly it states that:

P2 For all propositions $A, B \subseteq W$:

- $P(W) = 1$;
- $P(A) \geq 0$;
- if A is inconsistent with B, then $P(A \vee B) = P(A) + P(B)$;
- finally, I extend the previous (substantial) assumptions on P by the following definition: conditional degrees of belief are introduced by

$$P(B|A) = \frac{P(B \cap A)}{P(A)}$$

whenever $P(A) > 0$.

Since W was assumed to be finite, we may think of probabilities this way: they are assigned first to the singleton subsets of W—or, if one prefers, to the worlds in W themselves—and then the probabilities of larger sets are determined by adding up the probabilities of its singleton subsets. Because W is finite, we do not need to deal at all with the probabilities of infinite unions or intersections of propositions.

Finally, the Lockean thesis:

P3 There is an s with $\frac{1}{2} < s \leq 1$, such that for all propositions $B \subseteq W$:

$$Bel(B) \text{ if and only if } P(B) \geq s.$$

Now drop the existential quantifier 'there is an s' for a moment so that 's' becomes a free variable, and call the resulting open formula 'P3[s]': read this as 'the (instance of the)

Lockean thesis with threshold *s*'. If the interpretations of '*P*' and '*Bel*' are fixed, then, depending on the value of '*s*', *P*3[*s*] might turn out to be either true or false. I will be interested in characterizing those values of '*s*' for which it is true. I do allow for $s = 1$, but I will be particularly interested in choosing the value of '*s*' so that also propositions of probability less than 1 will be believed by the agent.

For the moment, I will focus especially on the right-to-left direction of the Lockean thesis with threshold *s*:

$LT_{\leftarrow}^{\geq s > \frac{1}{2}}$: For all B, $Bel(B)$ if $P(B) \geq s$.

That is because, with the right background assumptions, $LT_{\leftarrow}^{\geq s > \frac{1}{2}}$ will actually turn out to be equivalent to P3[*s*], which is interesting in itself to observe. Other than that, in what follows, I could have worked just with P3[*s*] directly.

Note that assuming P1, and hence the existence of a least believed proposition B_W, and also assuming P2, it follows that $LT_{\leftarrow}^{\geq s > \frac{1}{2}}$ for the special threshold $s = P(B_W)$ is equivalent to the very plausible 'Monotonicity Principle'

for all B, C, if $Bel(B)$ and $P(C) \geq P(B)$, then $Bel(C)$,

which says that if B is believed by a perfectly rational agent, and the agent gives C a degree of belief that is at least as great as that of B, then also C must be believed by the agent. Given P1 and P2, if $LT_{\leftarrow}^{\geq P(B_W) > \frac{1}{2}}$ is the case, then it is easy to see that also this Monotonicity Principle holds, and vice versa.[169] So $LT_{\leftarrow}^{\geq P(B_W) > \frac{1}{2}}$ is especially plausible. I am going to turn to that plausible special case $LT_{\leftarrow}^{\geq P(B_W) > \frac{1}{2}}$ of the right-to-left direction of the Lockean thesis shortly.

Now we are almost ready to spell out under what conditions P1, P2, and $LT_{\leftarrow}^{\geq s > \frac{1}{2}}$ (or P3[*s*]) are jointly satisfied. In order to formulate the corresponding theorem I will need one final probabilistic concept which is closely related, though not identical, to the notions of resiliency introduced by Skyrms (1977, 1980) within his theory of objective chance. It corresponds to the notion of *P*-stability^r from Appendix B for the special case $r = \frac{1}{2}$:

Definition 6 With P being a probability measure on the sample space W, I define for all $A \subseteq W$:

A is P-stable if and only if for all $B \subseteq W$, such that B is consistent with A and $P(B) > 0$:

$$P(A \mid B) > \frac{1}{2}.$$

169 From $LT_{\leftarrow}^{\geq P(B_W) > \frac{1}{2}}$ derive the Monotonicity Principle by concluding from $Bel(B)$ and P1 that $B \supseteq B_W$ and hence from $P(C) \geq P(B)$ and P2 that also $P(C) \geq P(B) \geq P(B_W)$; then $Bel(C)$ follows from $LT_{\leftarrow}^{\geq P(B_W) > \frac{1}{2}}$. In the other direction, assume that $P(B) \geq P(B_W)$ and then apply the Monotonicity Principle using $Bel(B_W)$ (as follows from P1).

Thus, a proposition is P-stable just in case it is sufficiently probable given any proposition with which it is compatible.

In order to get a feel for this definition, consider a consistent (non-empty) proposition A that is P-stable: one of the suitable values of 'B' above is the total set W of worlds—as W is consistent with A, and $P(W) = 1 > 0$—which is why P-stability entails that $P(A|W) = P(A) > \frac{1}{2}$. Therefore, any consistent P-stable proposition A must have a probability greater than that of its negation. What *P-stability* adds to this is that *this is going to remain so* under the supposition of any proposition B that is consistent with A and for which conditional probabilities are defined: A's high enough probability is resilient or robust.

It follows immediately from the axioms of probability that every proposition of probability 1 must be P-stable. For trivial reasons also the empty proposition is P-stable. And it might seem that this might actually exhaust the class of consistent P-stable sets, since P-stability might seem pretty restrictive; but things will turn out to be quite different.

The relevance of P-stability is made transparent by the following representation theorem:

Theorem 7 (Representation Theorem for the Logic of Belief and the Lockean Thesis) Let W be a finite non-empty set, let *Bel* be a set of subsets of W, and let P assign to each subset of W a number in the interval $[0, 1]$. Then the following two statements are equivalent:

I. *Bel* satisfies P1, P satisfies P2, and P and *Bel* satisfy $\mathrm{LT}_{\leftarrow}^{\geq P(B_W) > \frac{1}{2}}$.

II. P satisfies P2, and there is a (uniquely determined) $A \subseteq W$, such that
- A is a non-empty P-stable proposition (and hence $P(A) > \frac{1}{2}$),
- if $P(A) = 1$ then A is the least (with respect to $\subseteq$) subset of W with probability 1; and:
- for all $B \subseteq W$:

$$Bel(B) \text{ if and only if } A \subseteq B$$

(and hence, $B_W = A$).[170]

170 Here is the proof.

- 'II $\rightarrow$ I': *Bel* satisfies P1 by the proposition A being non-empty, and for all B, *Bel*(B) if and only if $B_W = A \subseteq B$. P satisfies P2 by assumption. Finally, one cannot just derive $\mathrm{LT}_{\leftarrow}^{\geq P(B_W) > \frac{1}{2}}$ but really the full Lockean thesis with threshold $P(B_W)$. There are two cases here: the $P(B_W) < 1$ case and the $P(B_W) = 1$ case.

The $P(B_W) < 1$ case follows by Observation 13 in Chapter 4 (where Z needs to be set to W and where 'P-stable$^{\frac{1}{2}}$' corresponds to 'P-stable' in the present chapter). I have given the required proof already also in the course of the proof of Theorem 1 in Chapter 2. But for the sake of self-containment I include the proof of that instance of the Lockean thesis here as well. So I show now that, given $P(B_W) < 1$, it holds that for all X, *Bel*(X) iff $P(X) \geq s = P(B_W)$.

The left-to-right direction of the Lockean thesis is obvious, since if *Bel*(X), then $X \supseteq B_W$, and the rest follows by the monotonicity property of probability: $P(X) \geq P(B_W)$. And about the right-to-left

This is a (universally quantified) equivalence statement: its left-hand side (I) summarizes all of our desiderata, if for the moment we restrict ourselves just to one direction of the Lockean thesis, and if we use $P(B_W)$ as the corresponding threshold. The right-hand side (II) expresses that B_W is P-stable, and if B_W has probability 1 then it is the least proposition of probability 1 (which must always exist for finite W).

Summing up: if P and Bel are such that P1, P2, and the right-to-left direction of the Lockean thesis with threshold $P(B_W)$ are satisfied, where B_W is the least proposition that is believed and which exists by P1: then B_W must be P-stable. And if given P and a P-stable proposition (which, if it has probability 1, is the least of that kind), then one can determine Bel from that P-stable proposition, so that P and Bel satisfy all of the desiderata, and the given P-stable proposition is the strongest believed proposition B_W. Or once again in other terms: assuming that P and Bel make the left-hand-side condition (I) true carries exactly the same information as assuming that P is a probability measure and the least believed proposition is P-stable (and, if it has probability 1, is the least proposition of probability 1).

One can show even more: either side (I or II) of the equivalence statement that is embedded in the theorem above actually implies the *full* Lockean thesis with threshold $P(B_W)$, that is, for all propositions B: $Bel(B)$ if *and only if* $P(B) \geq P(B_W) > \frac{1}{2}$. This follows from the proof of Theorem 7. Consequently, one can replace '$LT^{\geq P(B_W) > \frac{1}{2}}_{\leftarrow}$' in condition I by P3[$P(B_W)$] (the Lockean thesis with threshold $P(B_W)$), and still the

direction of the Lockean thesis: assume $P(X) \geq P(B_W)$ but not $Bel(X)$; then $X \not\supseteq B_W$, that is, $\neg X \cap B_W$ is non-empty. Thus, $[\neg X \cap B_W] \cup \neg B_W$ has non-empty intersection with B_W and its probability is greater than 0, because $1 > P(B_W) = 1 - P(\neg B_W)$ and so $P(\neg B_W) > 0$ (by the axioms of probability). But from B_W being P-stable it follows then that $P(B_W \mid [\neg X \cap B_W] \cup \neg B_W) > \frac{1}{2}$, that is, by the axioms of probability again, $P(\neg X \cap B_W) > \frac{P([\neg X \cap B_W] \cup \neg B_W)}{2} = \frac{P(\neg X \cap B_W)}{2} + \frac{P(\neg B_W)}{2}$, and hence $P(\neg X \cap B_W) > P(\neg B_W)$. However, by $P(X) \geq P(B_W)$ and the axioms of probability again, $P(\neg B_W) \geq P(\neg X)$. So we get $P(\neg X \cap B_W) > P(\neg X)$, contradicting the axioms of probability. So $Bel(X)$.

For the $P(B_W) = 1$ case one uses the assumption that if $P(A) = P(B_W) = 1$ then $A = B_W$ is the least (with respect to $\subseteq$) subset of W with probability 1. The antecedent is satisfied in this case, hence B_W is the least subset of W with probability 1, and therefore (with the axioms of probability): for all X, $Bel(X)$ iff $P(X) \geq s = P(B_W) = 1$.

- 'I $\rightarrow$ II": P satisfies P2 by assumption. Let $A = B_W$: by P1, B_W is non-empty, and it holds that for all $B \subseteq W$, $Bel(B)$ if and only if $A = B_W \subseteq B$. ($A = B_W$ is also determined uniquely by this condition.) Finally, we need to show that $A = B_W$ is P-stable, and if $P(B_W) = 1$ then B_W is the least subset of W with probability 1.

First, one proves that the Outclassing Condition from Theorem 5 in Appendix B applies to B_W with respect to P and $r = \frac{1}{2}$: for all w in B_W, $P(\{w\}) > P(W \setminus B_W)$.

For assume otherwise: then there is a w in B_W, such that $P(\{w\}) \leq P(W \setminus B_W)$. But then $P(B_W) \leq P([B_W \setminus \{w\}] \cup [W \setminus B_W]) = P(W \setminus \{w\})$, which by $LT^{\geq P(B_W) > \frac{1}{2}}_{\leftarrow}$ would imply that $Bel(W \setminus \{w\})$. But it is not the case that $Bel(W \setminus \{w\})$, since $B_W \not\subseteq W \setminus \{w\}$. Contradiction. So the Outclassing Condition holds.

This implies what needs to be shown (B_W is P-stable, and if $P(B_W) = 1$ then B_W is least with probability 1), by part '3 $\rightarrow$ 2' of Theorem 5 in Appendix B, or by Observation 11 in subsection 'P-Stability and the First Representation Theorem' in section 4.2.2. (That subsection proves the main properties of P-stable^r sets that are required for the technical observations in this book. P-stability is again P-stability^r with $r = \frac{1}{2}$.)

equivalence holds. This means: one might have thought that one could do just with the right-to-left half of the Lockean thesis, but once one throws in enough of the logic of belief, there is no such halfway house—one always ends up with the full Lockean thesis.

The threshold term '$P(B_W)$' as employed in the Lockean thesis above is, more or less, the only choice really given the logic of belief: by P1 there must be a least believed proposition B_W; therefore, if one also wants the Lockean thesis with threshold s to be satisfied, the threshold s cannot exceed $P(B_W)$. While s may well be a bit smaller than $P(B_W)$, it cannot be so small that some proposition ends up believed on grounds of the Lockean thesis that is not at the same time a superset of B_W, or otherwise the definition of B_W would be invalidated. Hence, in the present context, if one wants an instance P3[s] to be satisfied at all, one may just as well use P3[$P(B_W)$] from the start—for given P1, any such P3[s] must determine the same beliefs as P3[$P(B_W)$] anyway.

The additional 'if $P(B_W) = 1$ then B_W is the least proposition of probability 1' clause is a consequence of P1 and P3, too: by P1 (and W being finite), there must be a logical strongest believed proposition B_W. And if $P(B_W) = 1$, then B_W must be least amongst the propositions with probability 1, for otherwise there would have to be a least proposition B with $P(B) = 1$ ($= s = P(B_W)$) that would have to be believed by the Lockean thesis but which would not be believed by B_W not being a subset of B; which is impossible.

By the theorem from above, in a context in which P1 and P2 have already been presupposed, we can therefore reformulate postulate P3 from before (as the two formulations are equivalent given P1 and P2):

P3 [Reformulated] B_W is P-stable, and if $P(B_W) = 1$ then B_W is the least proposition $A \subseteq W$ with $P(A) = 1$.

From the theorem it also follows that, if one has complete information about what the P-stable sets for a given probability measure P are like, then one knows exactly how to satisfy P1–P3 from above for this very P: either one picks a P-stable set of probability less than 1—if there is such—and uses it as B_W; or one uses the least proposition of probability 1 for that purpose.

Fortunately there is an algorithm that makes it very easy to compute precisely those P-stable sets over which the right-hand side (condition II) in our theorem quantifies. I will (tacitly) apply that algorithm to some examples soon in this section.[171]

[171] For the record, here is at least a sketch of the algorithm: assume that $W = \{w_1, \ldots, w_n\}$, and $P(\{w_1\}) \geq P(\{w_2\}) \geq \ldots \geq P(\{w_n\})$. If $P(\{w_1\}) > P(\{w_2\}) + \ldots + P(\{w_n\})$ then $\{w_1\}$ is the first, and least, non-empty P-stable set, and one moves on to the list $P(\{w_2\}), \ldots, P(\{w_n\})$; e.g. if $P(\{w_2\}) > P(\{w_3\}) + \ldots + P(\{w_n\})$, then $\{w_1, w_2\}$ would be the next P-stable set. On the other hand, if $P(\{w_1\}) \leq P(\{w_2\}) + \ldots + P(\{w_n\})$ then consider $P(\{w_2\})$: if it is greater than $P(\{w_3\}) + \ldots + P(\{w_n\})$ then $\{w_1, w_2\}$ is the first P-stable set, and one moves on to the list $P(\{w_3\}), \ldots, P(\{w_n\})$; but if $P(\{w_2\})$ is less than or equal to $P(\{w_3\}) + \ldots + P(\{w_n\})$ then consider $P(\{w_3\})$: and so forth. The procedure is terminated when the least subset of W of probability 1 is reached. More details can be found in section 4.2.2.

As I explained in Appendix B (see Representation Theorem 5), given not *Bel*(∅), a finite W, and the axioms of probability for P, it holds that P and *Bel* satisfy the Humean thesis $HT^{\frac{1}{2}}$ with a Humean threshold of $r = \frac{1}{2}$ just in case the following is satisfied: there is a logically strongest believed proposition B_W; B_W is P-stable^r, which in the case $r = \frac{1}{2}$ means that B_W is P-stable; and if B_W has probability 1 then it is the least proposition of probability 1.

In other words, there is yet another reformulation of our postulates, but this time of P1 and P3 taken together (given our original P2):

P1&P3 [Reformulated] *Bel* and P satisfy the Humean thesis $HT^{\frac{1}{2}}$, and not *Bel*(∅).

In Chapter 2, the Humean thesis HT^r characterized believed propositions as those stably having a probability greater than r, where in the present case r equals $\frac{1}{2}$. So the theory of this chapter, which is based on the logic of belief, the axioms of subjective probability, and the Lockean thesis, turns out to be equivalent to the one of Chapter 2, which was based on the axioms of subjective probability, the Humean thesis (where the Humean threshold needs to be set to $\frac{1}{2}$), and it not being the case that *Bel*(∅). In Chapter 2, the logic of belief and the Lockean thesis on belief were corollaries of the Humean thesis (given the axioms of probability and not *Bel*(∅)). Now we have determined that the Humean thesis with Humean threshold $\frac{1}{2}$ is actually *equivalent* to the conjunction of these corollaries (given the axioms of probability and not *Bel*(∅) again). Or yet another way of putting this is: what the Humean thesis from Chapter 2 adds to the Lockean thesis is precisely the amount of stability that is required to yield also the logical closure of rational belief—not more.

If one applies the algorithm mentioned before to Examples 4 and 5 from section 3.1, the only P-stable set B_W so constructed is W itself, which is at the same time the least proposition of probability 1. The Lockean threshold $P(B_W)$ $(= P(W))$ is 1, but one might just as well choose some number that is less than but sufficiently close to 1 instead. On the other hand, e.g. $\{w_2, \ldots, w_{1000000}\}$, that is, the proposition of ticket 1 not winning, would not be P-stable: $\{w_1, w_2\}$ is consistent with $\{w_2, \ldots, w_{1000000}\}$, $P(\{w_1, w_2\}) > 0$, but

$$P\left(\{w_2, \ldots, w_{1000000}\} \mid \{w_1, w_2\}\right) = \frac{1}{2} \not> \frac{1}{2}.$$

In the case of Example 6, as promised, the algorithm determines (starting at the bottom):

- $\{w_1, w_2, w_3, w_4, w_5, w_6, w_7\}$ $(s = 1.0)$
- $\{w_1, w_2, w_3, w_4, w_5, w_6\}$ $(s = 0.99994)$
- $\{w_1, w_2, w_3, w_4, w_5\}$ $(s = 0.99794)$
- $\{w_1, w_2, w_3, w_4\}$ $(s = 0.97994)$
- $\{w_1, w_2\}$ $(s = 0.882)$
- $\{w_1\}$ $(s = 0.54)$

These are all the P-stable sets for P from Example 6. For instance, $\{w_1, w_2\}$ is P-stable: e.g. $\{w_1, w_3\}$ is consistent with $\{w_1, w_2\}$, $P(\{w_1, w_3\}) > 0$, and indeed $P(\{w_1, w_2\} \mid \{w_1, w_3\}) = 0.903\ldots > \frac{1}{2}$; similarly for all other propositions of positive probability that are consistent with $\{w_1, w_2\}$. On the other hand, e.g. $\{w_1, w_2, w_3\}$ is not P-stable: $\{w_3, w_4, w_5, w_6\}$ is consistent with $\{w_1, w_2, w_3\}$, $P(\{w_3, w_4, w_5, w_6\}) > 0$, but $P(\{w_1, w_2, w_3\} \mid \{w_3, w_4, w_5, w_6\}) = 0.491775\ldots < \frac{1}{2}$.

Each of these P-stable sets, and only these, can be turned into logically strongest believed propositions, such that P1–P3 are the case. For instance, if $\{w_1, w_2\}$ is taken to be the least believed proposition B_W, then all of P1–P3 are satisfied, and the same holds for $\{w_1, w_2, w_3, w_4\}$; in contrast, neither $\{w_1, w_2, w_3\}$ nor $\{w_1, w_2, w_4\}$ will do. To the right of the list of P-stable sets, I have stated the corresponding Lockean thresholds $s = P(B_W)$ that are to be used in P3. The bravest option would be to use $s = 0.54$ as a threshold, in the sense that it yields the greatest number of believed propositions: all the supersets of $\{w_1\}$. The other extreme is $s = 1$ (or something just a bit below that), which is the most cautious choice: the only propositions believed by the agent will then be $\{w_1, w_2, w_3, w_4, w_5, w_6, w_7\}$ and W itself. All the other thresholds lie somewhere in between these two extremes; e.g. the Lockean threshold $P(B_W)$ for $B_W = \{w_1, w_2\}$ is 0.882.

The six P-stable sets taken together look very much like one of David Lewis's 'spheres systems' in his semantics for counterfactuals (cf. Lewis 1973): for every two of them, one is a subset of the other or vice versa. (We have already discussed the same phenomenon in section 2.3.1 of Chapter 2.) And indeed one can prove in general, including the infinite case: if there is a P-stable proposition A with $P(A) < 1$ at all, then the set of all such propositions A is well-ordered with respect to the subset relation; and the least P-stable proposition of probability 1 (if it exists) is a proper superset of all of them.[172] Clearly, this induces a 'ranking' of worlds according to the first 'time' at which a world enters this hierarchy of P-stable sets or 'spheres': in the example, the rank of w_0 is 0, since it enters the hierarchy right at the start (or bottom) and remains there throughout all spheres; the rank of w_2 is 1, as it enters at the next state; the rank of both w_3 and w_4 is 2; and so forth. In this sense, the postulates for unconditional belief of the present chapter are already en route to those of Chapter 4 in which postulates for conditional belief or AGM belief revision (Alchourrón et al. 1985, Gärdenfors 1988) will be seen to correspond to such sphere systems or total pre-orders of worlds.

One final example: Figure 3.2 shows the equilateral triangle that represents geometrically all probability measures on the set $\{w_1, w_2, w_3\}$ of worlds. E.g. the w_1-corner represents the measure that assigns a degree of belief of 1 to $\{w_1\}$ and 0 to the other two singletons; the centre point represents the uniform measure that assigns $\frac{1}{3}$ to each singleton set; the closer one moves from the centre towards the w_1-corner, the

172 See section 4.2.2, Theorem 18. (Once again, 'P-stable' in the present chapter corresponds to 'P-stable$^{\frac{1}{2}}$' in Chapter 4.)

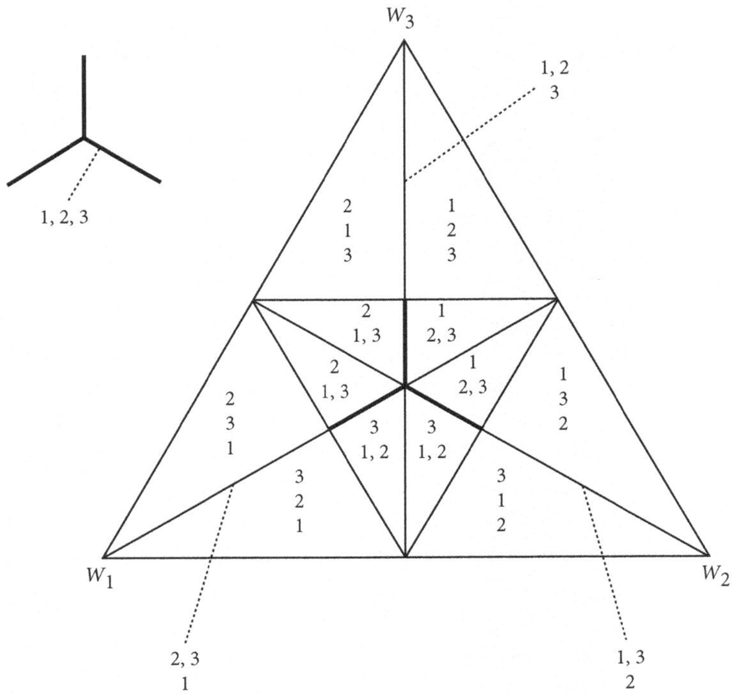

Figure 3.2. P-stable sets for $W = \{w_1, w_2, w_3\}$

greater the probability of $\{w_1\}$; and so forth. The ordered numbers in the interior small triangles encode the P-stable sets for the probability measures that are represented by points within the respective triangles: e.g. all P that are represented by points in the lower of the two small triangles adjacent to the w_1-corner have $\{w_1\}$, $\{w_1, w_2\}$, $\{w_1, w_2, w_3\}$ as P-stable sets; the ordered numbers $\substack{3\\2\\1}$ are the indices of worlds which, in this order, generate the P-stable sets if read from below. So worlds whose indices appear further down in a numerical array carry more probabilistic weight than the worlds whose indices appear higher up. Accordingly, every measure that is represented by a point in the upper of the two small triangles adjacent to the w_1-corner has $\{w_1\}$, $\{w_1, w_3\}$, and $\{w_1, w_3, w_2\}$ as its P-stable sets. Intuitively, all of this makes sense: in both of the small triangles, w_1 counts as the most plausible world, because geometrically all of the corresponding measures are close to the w_1-corner. w_2 is more plausible than w_3 in the lower triangle, because, from the viewpoint of w_1, this triangle belongs to the w_2-half of the whole equilateral triangle. Things are just the other way round in the upper of the two small triangles. If one moves closer to the centre again, the resulting systems of P-stable sets become more coarse-grained, that is, the number of P-stable sets decreases; e.g. no singleton set is P-stable any more. Furthermore, probability measures that are represented by points that are close to each other in the triangle have similar sets of P-stable propositions.

The only points in the full equilateral triangle that represent probability measures for which *there are no P-stable propositions of probability less than 1 at all* are: the vertices; the midpoints of the edges of the full equilateral triangle; and all points on the bold line segments that meet at the centre of the triangle. In particular, the uniform probability measure P at the centre only allows for W to be P-stable. This gives us: *almost all* probability measures P have a least P-stable set of probability less than 1.[173] Hence, for *almost all* probability measures P there exists an $s < 1$ and a *Bel*, such that *Bel* is closed logically, where for all B it holds that $Bel(B)$ iff $P(B) \geq s$, and where there is a B, such that $Bel(B)$ and $P(B) < 1$. The same can be shown to be true if there are more than three, but still finitely many, possible worlds.

Returning to the discussion in section 3.1 (but using the notation for postulates that was used in the present section), we find that: for all P (on a finite space of worlds), there is an $s < 1$ such that P1, P2, P3[s] are jointly satisfied. And, additionally, almost always there is a non-trivializing way of satisfying P1, P2, P3[s] with $s < 1$ so that at least some proposition of probability less than 1 is believed.

3.3 The Theory and its Costs

The results from the last section suggest a theory of belief and degrees of belief for perfectly rational agents that consists of the following three principles:

P1 There is a consistent proposition $B_W \subseteq W$, such that for all propositions B: $Bel(B)$ if and only if $B_W \subseteq B$.

P2 The axioms of probability hold for the degree-of-belief function P.

P3 B_W is P-stable, and if $P(B_W) = 1$ then B_W is the least proposition $A \subseteq W$ with $P(A) = 1$.

In a nutshell: *Belief is determined by a proposition of resiliently or stably high subjective probability (in the sense of P-stability)*. As it were, the 'grounds' (that is, the set B_W) of a perfectly rational agent's belief system must satisfy a probabilistic stability property. Call P1–P3 the *stability theory of belief* as presented in this chapter. As we have seen in the last section, the stability theory of belief as presented is actually provably equivalent to the stability theory of belief as presented in the last chapter, that is, to the conjunction of the axioms of subjective probability, the Humean thesis on belief

[173] The term 'almost all' can be made precise by means of the so-called Lebesgue measure that one finds defined in typical textbooks in measure theory. It means: all points in the triangle represent probability measures P that have a least P-stable set of probability less than 1 *except for a set of points of Lebesgue measure 0*. I should add that things change if one insists on the existence of, for instance, at least *two* distinct P-stable sets of probability less than 1: for three worlds, the Lebesgue measure of (points representing) probability measures P that allow for this is then $\frac{3}{4} < 1$.

(with a Humean threshold $r = \frac{1}{2}$), and the thesis that the contradictory proposition is not rationally believed (that is, not $Bel(\varnothing)$). So alternatively, up to equivalence, I may simply speak of *the* stability theory of belief, which is then going to refer simultaneously to the theory presentations of Chapter 2 and the present chapter.

By what I have shown in the previous section, it follows from this that rational belief is closed under logic, the rational degree-of-belief function obeys the axioms of probability, and the Lockean thesis relates belief and degrees of belief, which is what I started from in the first section. In fact, if taken together, P1–P3 as stated in this section are equivalent to the postulates stated in section 3.1. And we also found in the last section that for almost all P it is possible to satisfy P1, P2, P3[s] by means of a P-stable proposition B_W for which $s = P(B_W) < 1$. If measured by these consequences, P1–P3 seem to make for a very nice normative theory of theoretical rationality as far as belief and degrees of belief are concerned—normative, as the theory deals with the beliefs and degrees of belief of perfectly rational agents.

To be sure, P1–P3 is not a *complete* theory of rational belief. For instance, it lacks diachronic norms on how to *change* belief; and it lacks norms on the interaction between belief and *decision-making*.[174] Let me briefly comment on this.

So far as belief change is concerned, one would have to supply P1–P3, which are synchronic in nature, by *diachronic* principles. P1–P3 are meant to hold for all *Bel* and P at arbitrary times t. In order to add an account of how to proceed from one time t to another time t' between which all that the agent learns is some piece of evidence E for which $P(E) > 0$, one would extend P2 by maintaining that P ought to be updated by conditionalizing it on E: for all B, $P_{new}(B) = P(B|E)$. Accordingly, as recommended by belief revision theory (see AGM—that is, Alchourrón et al. 1985—and Gärdenfors 1988), one would add to P1 the principle that, given some piece of evidence E that is consistent with B_W and which is therefore also consistent with every proposition believed by the agent, *Bel* ought to be updated so that: Bel_{new} is the set of supersets of the new strongest believed proposition $B_W^{new} = B_W \cap E$. All of that would be consistent with P1–P3, in the sense that, if *Bel* and P satisfy P1–P3, then also Bel_{new} and P_{new} satisfy the corresponding conditions that are imposed by P1–P3 on them: B_{new} can be shown to be P_{new}-stable again.[175] Over and above that, one would also have to add principles of update on pieces of evidence E that have probability 0 or which contradict some of the agent's present beliefs or both. I will return to such a

[174] There is more that is lacking here: e.g. the theory also lacks norms on *introspective* belief which, as mentioned before, I will not deal with in this book at all. Or one might want to extend the theory to one of *social* belief. And so forth.

[175] This is explained in more detail at the end of section 4.1.3 and it corresponds to Observation 17 in section 4.2.2. Note that it is not the case for all P that if B_W is the *least* P-stable set, then B_W^{new} so defined is always the *least* P_{new}-stable set again. This is very easy to see directly, but it can also be derived from a much more general result proven by Lin and Kelly (2012b); see their Corollary 1. I will return to their result again in Appendix C (the appendix to Chapter 4), but that will be in a different context.

case briefly in section 3.6. All of these diachronic principles will be covered in detail by Chapter 4.

If, finally, the resulting theory were extended also by adequate principles of *practical* rationality—a belief–desire model of action on the all-or-nothing side, Bayesian decision theory on the other side—the resulting package would get closer to a theory of rationality for belief and degrees of belief more generally. Section 2.3.3 was my first shot at such practical aspects, and section 6.1 will be devoted to them completely. But for the rest of the present chapter I will restrict my discussion to P1–P3.

Before I turn to the potential downsides of the theory, let me make clear that P1–P3 leave a lot of room for, even substantially diverging, interpretation. In particular: because of the centrality of the probabilistic notion of P-stability, one might think that the stability theory necessarily amounts to a *reductive* account of belief in terms of probability; however, such a view would be misguided.

First of all, while P3 demands that the strongest believed proposition B_W is P-stable, it does not determine with *which* P-stable set the proposition B_W ought to be identified, and as we know already, there might be more than one choice.[176] Only if P3 were strengthened appropriately—e.g. by postulating B_W to be the *least* P-stable set (which must always exist for finite W)—would one be able to explicitly define 'B_W' and hence '*Bel*' in terms of 'P', and thus belief would be reducible to degrees of belief.[177] But I did not presuppose any such strengthening of P3.

Secondly, although the Lockean thesis is very often understood in the way that *Bel* can be determined from P by applying the thesis from the right to the left, *and therefore P is prior to Bel*, the latter 'therefore'-part is not actually contained in the thesis itself. After all, the Lockean thesis with threshold s is merely a universally quantified material equivalence statement, which says: for all B, either $Bel(B)$ and $P(B) \geq s$, or not $Bel(B)$ and $P(B) < s$. This *allows* for probability to be prior to belief, but it does not necessitate it.

For instance, one might want to defend the Lockean thesis in conjunction with the view that *belief is prior to probability*: then the thesis is a constraint on P of the form that, given *Bel* and s, the measure P ought to be such that all and only the believed propositions are to be assigned a probability greater than or equal to s.

Or it might be that *neither side of the Lockean thesis is taken to be prior to the other*: in that case, the thesis is a simultaneous constraint on *Bel*, P, and s, which might e.g. be regarded as a normative principle of coherence or harmony between two ontologically and conceptually distinct systems of belief, that is, the system of all-or-nothing belief and the system of quantitative belief. In order for an agent

176 If the sample space W is infinite, then one can prove that there are even probability measures P for which there exist infinitely many P-stable propositions of probability less than 1. See the last example at the very end of section 4.2.2.

177 That is precisely the route that I followed in Leitgeb (2013a) and which I will discuss later in Appendix C: I will also explain there why I do not support this view any more.

to be rational, the two systems would be demanded to cohere with each other as expressed by the Lockean thesis. This interpretation would be the appropriate one if the Independence variant of option (iii) on belief and degree of belief in section 1.3.3 of Chapter 1 turned out to be correct.

I will leave open which of these interpretations is the most plausible one.[178] But all of these interpretations are consistent with P1–P3 from above. And in all of these interpretations, belief ends up as some kind of coarse-graining of probability, for, by the Lockean thesis, believing a proposition is always equivalent to abstracting away all the different degrees of belief that a proposition might have as long as its degree is not less than s. For the same reason, all of the uncountably many probability measures represented by points within one and the same of the little triangles in Figure 3.2 yield one and the same system of finitely many P-stable sets. In other words: in the transition from P to *Bel* information is being lost, which was to be expected, as 'P' expresses a quantitative concept whilst '*Bel*' expresses a qualitative one. But none of this entails that rational belief is reduced to subjective probability by the theory.

This stability theory of belief and degrees of belief looks almost too good to be true. Where have the paradoxes gone? Why is it that, all of a sudden, closure of belief under conjunction does not work against the Lockean thesis any more? There must be a catch.

And there is: for the rest of the present section I will discuss the two kinds of costs that follow from the principles of stability theory. That is: on the one hand, (C1) the sensitivity of the threshold in the Lockean thesis to P, and on the other, (C2) the sensitivity of *Bel* to partitionings of the set W of worlds, where additionally thresholds that are not particularly close to 1 demand there to be a small number of worlds or partition cells in B_W. In a nutshell: *a serious sensitivity of belief to the context* (in a sense of 'context' that is to be explained in more detail). Afterwards, in the two subsequent sections, I will explain what this means for the Lottery and the Preface Paradoxes. Ultimately the goal will be to evaluate whether the benefits of the theory outweigh its costs.

I should add that nothing in my theory will force an agent's rational *degrees* of belief to be context-dependent in the same way: they may be so, but if so then this is not entailed by the norms that I am defending. So I will take them to be context-*in*dependent in everything that follows.[179]

[178] I should add that in those interpretations in which one type of belief is said to be prior to the other, one would also need to specify the *kind* of priority that one has in mind; and of course it is perfectly possible e.g. that probability is claimed to be *ontologically* prior to belief, while at the same time belief is regarded as *epistemologically* prior to probability (since beliefs seem more easily accessible than subjective probabilities). Hence, much more would have to be said about the kind of priority in question.

[179] This is in contrast e.g. with Clarke (2013), who takes also numerical degrees of belief to be context-sensitive (where the context in question may be viewed to consist in a contextually restricted set of worlds or a contextually determined set of accepted propositions). So far as sensitivity or insensitivity of belief to *partitionings* of possibilities is concerned—to which I will turn very soon—one might think that the whole point of subjective probability theory is to avoid any such sensitivity. By finite additivity, in whatever way X is partitioned into subsets, the sum of the probabilities of these subsets is always the same: the probability of X. This said, every probability measure is defined only on some algebra of events or propositions, and there will

According to the stability theory, only particular thresholds s (equal to $P(B_W)$ or slightly below) are permissible to be used in the Lockean thesis, as follows from the results in the last section. Which thresholds one is permitted to choose depends on what 'P' and 'B_W' in '$P(B_W)$' refer to, that is, the probability measure P and the belief set *Bel*. Furthermore, B_W is itself constrained to be P-stable. So overall, if one grants the stability theory, one must learn to live at least with the fact that:

> C1 The range of permissible choices of threshold in the Lockean thesis co-depends on the agent's degree-of-belief function P (that is, it depends on P but it does not necessarily only depend on P). Not every combination of Lockean threshold and degree-of-belief function is permissible.[180]

Let us take a step back, for a moment. What determines the choice of threshold in the Lockean thesis more generally? The usual answer is: *the context*. Compare:

> The level of confidence an agent must have in order for a statement to qualify as *believed* may depend on various features of the context, such as the subject matter and the associated doxastic standards relevant to a given topic, situation, or conversation. (Hawthorne 2009, p. 73)

What this means exactly depends on whether the Lockean thesis is meant to govern the ascription of belief—in which case the choice of threshold will depend on features of the situation in which belief is ascribed—or whether the Lockean thesis is meant to govern the belief states themselves—in which case the threshold will be determined only by features of the believer. In the first case, it is possible that the agent, say, x, who ascribes beliefs to an agent, y, is distinct from y. In the second case, only one agent, y, is relevant: the agent whose belief states are in question. Either way, the respective threshold s in the Lockean thesis functions as a 'level of cautiousness', since demanding a greater lower boundary of the probabilities of believed propositions is more restrictive than demanding a smaller lower boundary. But in the first interpretation in terms of belief *ascription*, with P being fixed, it might well be that the belief *ascriber* (x) determines the value of 's': the greater that value is, the more demanding the resulting contextually determined concept of belief and hence the more cautiously x must ascribe beliefs to y. The context in question is then x's *context of belief ascription*, and it comprises everything that determines x's own standards of belief ascription at a time. Whereas in the second interpretation, the value of 's' would be due to the *believing* agent (y): the greater the value is, the more restrictive the constraint on y's *belief set* in the sense that y is more cautious about believing: the context in question

always be *some* proposition that cannot be expressed by means of the members of such an algebra. So even a rational agent's degree-of-belief function comes with a restriction to a particular class of propositions to which degrees of belief are assigned, and one might think of that class to be given contextually again. I will say more about this in section D.5 of Appendix D.

180 Once again, this does not mean that degrees of belief must be determined prior to the choice of any such threshold s: for instance, for given s, a measure P might be determined so that s is the probability of some P-stable set.

is then what might be called y's own *context of reasoning*, and it comprises everything that determines y's own standards of belief at a time.

Here is another way of explaining the difference between the two interpretations: in the first case, when x ascribes beliefs to y at time t, she might express by the term 'belief' a rather tolerant concept of belief—the threshold in the Lockean thesis might be just a bit above $\frac{1}{2}$, and y might happen to have lots of beliefs in that sense of the word. At a later time t', the context of belief ascription might have changed: in that new context, x might express a much more demanding concept of belief by using the same term 'belief' as before. Perhaps x's attention is focused now on some sceptical scenario from the epistemology classroom, the Lockean threshold moves closer to 1, and in *that* sense of the word, y might cease to believe some propositions that she would have been said to believe in the sense of 'belief' from time t. In order for that to be so, y's mental state would not have had to change at all between t and t', but which proposition is expressed by 'y believes that A' would have changed between t and t' (for some 'A').

In contrast, in the second interpretation, the concept that is expressed by the term 'belief' would remain the same in each and every context. But of course y's mental state might change between t and t': initially, at t, some aspects of y's mental state might have determined the threshold in the Lockean thesis to be close to $\frac{1}{2}$. But then, say, y's context of reasoning might change—perhaps y's attention becomes focused on the high stakes of some decisions that will be based on her all-or-nothing beliefs—and y might no longer believe something at t' that she had believed before at t when her attention had been focused on something else. The proposition that is expressed by 'y believes that A' would be the same at t and t' (for whatever 'A'), but the (time-dependent) truth value of such a proposition might be different at t from what it is at t'.

The stability theory is open to both interpretations: in the first case it would be meant to govern the joint ascription of belief and degrees of belief to perfectly rational agents. Presumably, the two kinds of ascriptions ought to cohere with each other, and the suggestion would be that the coherence in question can be made precise in terms of the Humean thesis from Chapter 2 or by means of the postulates in the present chapter. So this would be about coherence between concepts. In the second case the stability theory would be supposed to govern a perfectly rational agent's beliefs and degrees of belief. Belief states and degree-of-belief states would have to cohere with each other in order for the agent to avoid serious dilemmas in the course of acting upon her beliefs and degrees of belief simultaneously. So this would be about coherence between mental states.

As far as I can see, no argument against the first interpretation emerges from the stability theory of belief, and I do not want to argue against contextualism about 'belief' in this first semantic sense here either.[181] But in what follows I will go for the second

[181] Appendix C will argue against the Reduction Option (ii) from section 1.3.2 of Chapter 1 by explaining why a reduction of rational belief to rational degrees of belief alone is not plausible given the rest of

interpretation, which will allow me to use the term 'belief' with the same content independently of the context of assertion. In the terms of the corresponding debate on knowledge: I aim at something closer to a sensitive moderate invariantism in the sense of Hawthorne (2004) (or interest-relative invariantism in the sense of Stanley 2005, or pragmatic encroachment in the sense of Fantl and McGrath 2009) rather than a contextualist understanding in the sense of proper *contextualism*.[182] Indeed, if in the following quotation 'knowledge' is replaced by 'belief', then I will subscribe to the resulting statement:

> the kinds of factors that the contextualist adverts to as making for ascriber-dependence—attention, interests, stakes, and so on—[have] bearing on the truth value of knowledge claims only insofar as they [are] the attention, interests, stakes, and so on of the subject.
>
> (Hawthorne 2004, p. 157)

According to this second non-contextualist interpretation that I will focus on now, even if the agent's degree-of-belief function is kept fixed, if what is salient to an agent changes, then her beliefs might change; the more that is (perceived to be) at stake for the agent, the more it might take her to believe; and so on. The question is really: how much risk is the agent willing to take whose beliefs are in question? And according to the stability theory, the subject's degree-of-belief function P must be counted amongst the factors that co-determine the answer at the relevant time; it is the subject's attention, interest, stakes, . . . , *and her degree-of-belief function* that are relevant for determining the threshold in the Lockean thesis.[183]

For the same reason, the term 'context' as I will understand it here might be misleading to anyone who associates it immediately with contextualism about knowledge or, here, about rational belief: the view according to which the *content* of terms such as 'knowledge' or 'rational belief' may vary with the context of ascription (as when the belief ascriber determines different values for 's' in the Lockean thesis).[184] The notion of context in that sense is a semantic one: context in a similar sense in

my assumptions. However, Appendix C will leave open whether rational belief might still be reducible to rational degrees of belief plus x, where x might comprise certain practical features of either the belief ascriber or the agent of belief (or both), such as attention, interest, what is regarded as important, and the like. I do not have an argument against any such refined reductionist treatment of rational all-or-nothing belief. But according to such a refined reductionist proposal, belief ascription still concerns 'almost' the same phenomenon as degree-of-belief ascription (the difference being just those practical features). That might be enough of a reason to expect that successful belief ascription will have to cohere with successful degree-of-belief ascription, and the present stability theory might be exactly what is needed for making that notion of coherence precise enough.

182 See Coffmann (2012) for an overview of the corresponding debate so far as knowledge is concerned.

183 When I say that the Lockean threshold is co-determined by the subject's attention, interests, and the like, I do not mean necessarily that the subject *consciously* decides on the threshold. I only mean that the Lockean threshold is a function of certain parameters, including the subject's attention, interests, and the like.

184 See e.g. Hawthorne (2004) and Stanley (2005) for a discussion and criticism of contextualism about knowledge in that sense.

which e.g. the reference of indexicals is supposed to be determined by the context.[185] As mentioned before, any such context of belief ascription would involve or depend on what the belief *ascriber* is attentive to, what she is interested in, what is at stake for her, and so on.

On the other hand, in different parts of philosophy, e.g. Thomason (1986, 2007), Nozick (1994), Bratman (1999), and Cresto (2010), the term 'context' is used as I want to understand it here: in order to denote and highlight certain aspects of the circumstances in which the agent who has the beliefs in question is reasoning. In particular, these will be practical aspects, such as what the agent attends to, what is relevant to her, the practical pressures that she is facing, and the like; but perhaps, as I am arguing, also epistemic aspects, such as the agent's degree-of-belief function. I will normally regard all of these contextual features as internal to the agent, though there may be an externalist version of this theory, too, in which some aspects of the context would be permitted to be beyond the agent's epistemic reach. In any case, the content of the term 'rational belief' will not be affected by changing the context in that sense, but what an agent *believes* might be affected by it. This seems to be the sense in which Nozick (1994, p. 96) takes rational belief to be context-dependent: in his example, he believes that his new junior colleague is not a child molester, but when the context changes from an ordinary one to one in which stakes are high—he needs someone to watch his young child for two weeks—he might not retain that belief. It is the believing agent that makes this shift. This is much like in Example 6 from before in which shifting the Lockean threshold from 0.882 to the more cautious 0.97994 corresponds to a switch from the set $\{w_1, w_2\}$ of doxastically accessible worlds to the set $\{w_1, w_2, w_3, w_4\}$ in which further possibilities are taken into account.[186] Thomason (1986, 2007) gives similar examples in which perceived risk, interest, and topic affect an agent's beliefs. The morals that he draws from this are:

> There are occasions when we can't act without a belief, and in which high standards for belief prevent us from having an appropriate belief. In these cases, an urgent need to act can cause us to lower our standards....
>
> There are occasions when we have a belief that is well justified, but the consequences of acting on this belief if we are wrong are very harmful. In these cases, we can destroy the belief by changing our standards. In a theory of practical reasoning where actions are determined by beliefs and desires (rather than by probabilities and utilities) mechanisms of this sort are essential in order to deal with uncertainty and risk.

(Both quotations are from Thomason 2007, section 3.) The notion of context that Thomason (2007, section 3) invokes to analyse situations like that is the one that I use:

[185] See Gauker (2010) for a survey on contexts in that semantic sense. For the additional semantic difference between contexts of utterance and contexts of assessment (as in what John MacFarlane calls 'assessment relativism'), see Fantl and McGrath (2009, s. 2.2) and MacFarlane (2011b) for overviews. Furthermore, there is also a non-indexical form of semantic context-sensitivity according to which *extension* (rather than content) depends on context; see MacFarlane (2009) for the details.

[186] In other respects, Nozick's theory of belief differs from mine. In some ways, what he calls belief seems to be closer to what I will call *accepted belief* in section 6.3.

'The context-dependence . . . belongs to a theory of philosophical psychology, or to an agent architecture, rather than to semantics'.[187]

Unfortunately, I will not able to put forward a definition of the term 'context' in that second non-semantic sense, but I hope that the examples will be clear enough to get some preliminary understanding of what is at issue here. So far as my formal treatment of belief and degrees of belief is concerned, it will at least be clear enough what a context is supposed *to do*: to determine a threshold in the Lockean thesis and (about which more later) to determine the underlying partition of possibilities. Since, according to this theory, the threshold in the Lockean thesis depends also on the agent's degree-of-belief function, one may just as well count the agent's degree-of-belief function among the components of the agent's context of reasoning in this non-semantic sense of context. Alternatively, one may reserve the term 'context of reasoning' just for the agent's attention, interest, perceived stakes, and the like, and keep the agent's degree-of-belief function out of it. In the latter case, the present theory would say that the Lockean threshold depends on the agent's context of reasoning *and* the agent's degree-of-belief function taken together.

In any case: the sensitivity of Lockean thresholds in our theory to an agent's degree-of-belief function should not be too surprising. Why should the choice of threshold in the Lockean thesis be allowed to be sensitive to the agent's attention and interests but not to the agent's degrees of belief? After all, all of them are salient components of the agent's state of mind. Or from the viewpoint of Bayesian decision theory: assume that the Lockean thesis is taken for granted but only the choice of the corresponding threshold is left unresolved. How would a good Bayesian determine the right threshold in the corresponding context? She would view the whole situation as a decision problem: should I choose the threshold in the Lockean thesis to be s_1, or should I choose it to be s_2, or . . . ? The outcome of each such choice of threshold would be a particular set of beliefs, which would be determined by plugging in that threshold in the Lockean thesis. These possible outcomes would be evaluated in terms of their utilities, and ultimately, by the tenets of standard decision theory, a threshold ought to be chosen that maximizes the expected utility of these outcomes. Hence: given the relevant utility measure *and her subjective probability measure*, she would choose a threshold so that the expected utility of the choice is maximal. In this way, obviously, P would co-determine the threshold s in the Lockean thesis, simply because the

[187] Bratman (1999, ch. 2) employs the same notion of context, but he argues against Thomason's (1986) views and for the context-*in*sensitivity of belief. Cresto (2010) uses the same notion of context in her formal model of belief and acceptance. In the epistemological literature on the 'pragmatic encroachment' of knowledge, authors use terms such as 'one's circumstances' or 'practical facts about one's environment' instead of 'context' as understood by e.g. Thomason or myself. But the idea is the same or at least close enough (barring differences to do with externalism vs internalism about circumstances). Compare: 'How strong your epistemic position must be—which purely epistemic standards you must meet—in order for a knowledge-attributing sentence, with a fixed content of use, to be true of you varies with *your circumstances*' (Fantl and McGrath 2009, p. 35, my emphasis). Or: 'Bare Interest-Relativist Invariantism . . . is simply the claim that whether or not someone knows that p may be determined in part by *practical facts about the subject's environment*' (Stanley 2005, p. 85, my emphasis). While Stanley only deals with knowledge, Fantl and McGrath also deal with belief (see their ch. 5).

expected utility of choosing one threshold rather than another co-depends on P: with the utility measure being fixed, different probability measures P might well determine different ranges of permissible thresholds that all maximize expected utility relative to P. This is just like in the stability theory developed here, where different probability measures P may determine different sets of permissible thresholds that all correspond to the probabilities of sets that are stable relative to P. So the dependency of s on P should not be particularly problematic in itself.

Still one might wonder: in the case of Example 6 as discussed in the first two sections, why is one allowed to choose $s = 0.882$ or $s = 0.97994$ as a threshold—corresponding to B_W being either of the P-stable sets $\{w_1, w_2\}$ and $\{w_1, w_2, w_3, w_4\}$, respectively—but not, say, $s = 0.94$, which is the probability of the P-*un*stable set $\{w_1, w_2, w_3\}$?

An analogy might help here. It is well-known that for some purposes, we conceive of *properties* in the way that every set of individuals whatsoever is guaranteed to be the extension of some property. But then again, for other purposes, we may want to restrict properties just to 'natural' ones, so that not every set of individuals may count as an extension of a property in this restricted sense—a standard move in semantics, metaphysics, philosophy of science, and other areas (see e.g. Lewis 1983). What 'natural' means exactly may differ from one area to the next, but in each case natural properties ought to 'carve nature at its joints', in some sense.

Now let us apply the same thought in the present context. For some purposes, for which the logic of belief is not relevant, we may conceive of the threshold in the Lockean thesis in the way that every threshold whatsoever can be combined with every probability measure whatsoever. But then again, for other purposes for which the logic of belief is an issue, we may want to restrict thresholds just to 'natural' ones, so that not every threshold can be combined with every probability measure. Natural thresholds ought to 'carve probabilities at their joints', and

s is natural with respect to P if and only if
there exists an A, such that $s = P(A)$ and for all $w \in A$, $P(\{w\}) > P(W \setminus A)$

may be just the kind of 'probability cutting' that is appropriate here. As Theorem 5 in Appendix B registered (with r chosen now to be $\frac{1}{2}$),[188] if $P(A) < 1$, then this so-called 'Outclassing Condition' to the effect that for all $w \in A$, $P(\{w\})$ exceeds the probability of W without A, is equivalent to A being P-stable. So the naturalness of this kind of 'probability cutting' would manifest itself in the stability of A, that is, the stability of B_W, the strongest believed proposition.

Or analogously: if one is interested only in the logic of belief, then every proposition whatsoever may be a candidate for the strongest believed proposition B_W. However, in a context in which both belief and degrees of belief are of interest, only

[188] I will also return to this in Chapter 4: see Observation 11 in subsection 'P-Stability and the First Representation Theorem' of section 4.2.2.

'probabilistically natural' propositions may count as candidates for B_W, and P-stability may be just the right notion of naturalness, since it belongs to a similar ballpark as other 'natural' notions of stability or resiliency or robustness in statistics (see Skyrms 1977, 1980), economics (cf. Woodward 2006), metaphysics (cf. Lange 2005), epistemology (cf. Rott 2004 and Stalnaker 2006 on stability analyses of knowledge), and beyond. Hence, the fact that P1–P3 as formulated in section 3.1 impose more constraints on the value of 's' than P2–P3 would do just by themselves, and the fact that P1–P3 impose impose more constraints on B_W than P1 would do just by itself, should not be thought to speak against the theory.

Now for the second, and more substantial, worry: according to the stability theory of belief, it turns out that

C2(i) belief is partition-dependent, and

C2(ii) generally, the smaller the probabilities of the partition cells, the greater the probabilities of believed propositions must be in order for P1–P3 to be satisfied.

Let me explain this in detail (still presupposing W to be finite). It is quite common in applications of probability theory that even when initially P had been defined for all subsets of W, there might be a context in which not all subsets of W are actually being required for the purposes in question. E.g. if one is interested only in the possible outcomes of a lottery, then only the propositions of the form *ticket* 1 *wins*, *ticket* 2 *wins*, . . . together with their logical combinations will be relevant; accordingly only the probabilities of such propositions will count. Formally, this can be achieved by introducing a *partition* Π on W: a set of pairwise disjoint non-empty subsets u_i of W, such that the union of these sets u_i is just W again. E.g. in the lottery case, initially W might have been the set of all, say, metaphysically possible worlds, but then a set of partition cells u_i might have been introduced, such that any such set u_i would be the set of all worlds in which ticket i wins.[189] Such partition cells u_i might then be viewed themselves as 'coarse-grained' possible worlds in which all differences between two distinct metaphysically possible worlds within one and the same cell would be ignored; the probabilities of these 'pseudo-worlds' would be given by $P(u_i)$, and only unions of such sets u_i would be considered propositions in the relevant context. 'Coarse-grained' possible worlds in that sense are similar to what are called 'small' worlds in decision theory (cf. section 2.6 of Joyce 1999).

If one wants to make all of that completely precise, one needs to build up a new probability space that has the set Π of all partition cells as its sample space, where propositions are now subsets of Π, and where a new probability measure P_Π is defined based on P. The probability space in Examples 4 and 5 from section 3.1 could be seen as arising from precisely that procedure, with each 'coarse-grained world' in W corresponding to a particular ticket winning in a fair lottery of 1,000,000 tickets.

189 Let us disregard the question of whether any such class u_i would actually be a set in that case or rather a proper class of worlds.

If the context changes again, and one needs to draw finer distinctions than before—for example: it is not just relevant which ticket wins but also *who bought the ticket*—one may refine the partition accordingly, so that what had been one partition cell u_i before is broken up into several new and smaller partition cells. Or one can afford to draw coarser distinctions—like, it is not relevant any more which ticket wins but only *whether ticket 1 wins or not*—and hence the partition is made coarser, so that what had been several partition cells before are now fused into just one large partition cell.

In each case, the probabilities of the partition cells and of their unions are determined from the original probability measure P that is defined for all subsets of W. Or equivalently: where the original probability measure is given with respect to the maximally fine-grained partition whose partition cells are just the singleton sets $\{w\}$ for $w \in W$. For it does not really matter whether W is $\{w_1, \ldots, w_n\}$ or whether the set of 'worlds' considered is the maximally fine-grained partition $\Pi = \{\{w_1\}, \ldots, \{w_n\}\}$ of W; whether the probability measure is P or whether it is the measure P_Π that assigns to the singleton set $\{\{w_i\}\}$ the same number that P assigns to the singleton set $\{w_i\}$. More generally, it does not matter whether P_Π assigns to $X \subseteq \Pi$ a number, or whether P assigns the same number to $\bigcup X$, that is, to the set of members of members of X. And in terms of the intended interpretation of propositions, it does not matter whether the proposition that ticket 1 is drawn is $\{w_1\}$ or $\{\{w_1\}\}$; and so forth. Accordingly, in the following, I will move back and forth between such numerically distinct but formally equivalent constructions of worlds, propositions, and probability measures, without much additional comment.

Since operating with partitions is such a natural and useful doxastic procedure, it is important to determine what happens to an agent's beliefs when partitions are introduced and changed. If P1–P3 are taken for granted, the answer is: [C2(i)] refining a partition may lead to a *loss* of beliefs, that is, belief may be turned into suspension of judgement. [C2(ii)] Whatever the partition, in order for P1–P3 and $P(B_W) < 1$ to be satisfied, the probability of every singleton subset of B_W must be greater than the probability of $W \setminus B_W$, whether the members of B_W are some 'maximally fine-grained worlds' in W or some more or less coarse-grained partition cells on W; which has some worrisome consequences. I will illustrate C2(i) by means of an example, and I will demonstrate C2(ii) and its consequences by a little calculation.

Examples 4 and 5-Reconsidered:
Let $W = \{w_1, \ldots, w_{1000000}\}$ be a set of 1,000,000 possible worlds again, where each world w_i corresponds to ticket i being drawn in a fair lottery. Accordingly, let P be the uniform probability measure that is given by $P(\{w_1\}) = \ldots = P(\{w_{1000000}\}) = \frac{1}{1000000}$ again.

Now introduce the partition

$$\Pi = \{\{w_1\}, \{w_2, \ldots, w_{1000000}\}\}$$

of W, or in other words: the agent is interested only in whether ticket 1 wins or not. Consider the partitions cells $\{w_1\}$ and $\{w_2, \ldots, w_{1000000}\}$ as new coarse-grained worlds and Π as the resulting new set of such worlds. Based on our original P, we can then define a new probability measure P_Π, for which Π serves as its sample space, and where P_Π assigns probabilities to subsets of Π as expected: $P_\Pi(\{\{w_1\}\}) = \frac{1}{1000000}$, $P_\Pi(\{\{w_2, \ldots, w_{1000000}\}\}) = \frac{999999}{1000000}$, $P_\Pi(\{\{w_1\}, \{w_2, \ldots, w_{1000000}\}\}) = 1$, $P_\Pi(\emptyset) = 0$. The new probability for a set X results from applying the original probability measure P to $\cup X$ (the set of members of W that are members of the partition cells in X); in particular, $P_\Pi(\{\{w_1\}\}) = P(\{w_1\})$ and $P_\Pi(\{\{w_2, \ldots, w_{1000000}\}\}) = P(\{w_2, \ldots, w_{1000000}\})$.

The algorithm from section 3.2 (as sketched in n. 171) tells us then that the corresponding P_Π-stable sets are

$$\{\{w_2, \ldots, w_{1000000}\}\} \text{ and } \{\{w_1\}, \{w_2, \ldots, w_{1000000}\}\},$$

the first one of which has a probability slightly less than 1, while the second one has a probability of exactly 1.

Finally, let $B_W^\Pi = \{\{w_2, \ldots, w_{1000000}\}\}$ and $s = P(\{\{w_2, \ldots, w_{1000000}\}\})$: then all of P1–P3 are satisfied, and since $\{\{w_2, \ldots, w_{1000000}\}\}$ is nothing but the negation of the proposition $\{\{w_1\}\}$, this means that the agent believes that ticket 1 will not win (relative to Π).

In order to drive the point home, let us now maximally refine Π to Π' so that one is interested again in which ticket will be drawn; or equivalently: simply use the original W and P again. Then, as observed already in section 3.2, W is the only P-stable set, and my theory demands that $B_W = W$: consequently, the agent does *not* believe that ticket 1 will not win (relative to the most fine-grained available partition). That is: refining a partition can lead to a loss of beliefs.

In section 3.4 I will return to this example, when I will evaluate its consequences for the Lottery Paradox. So much concerning C2(i), for the moment.

And about C2(ii): this is just the alternative characterization of P-stable sets in terms of the Outclassing Condition that I had mentioned before (and also in Appendix B), but we will see that it makes best sense to address its consequences in a context in which one discusses the workings of partitions.

By P1–P3, the strongest believed proposition B_W is P-stable. Let us assume that we are dealing with the non-trivial case in which $P(B_W) < 1$: by the Outclassing Condition, for all $w \in A$, $P(\{w\}) > P(W \setminus B_W)$. So every singleton subset of B_W must have a probability greater than $\neg B_W = W \setminus B_W$, whether the worlds in question are the 'given' worlds in W or some more or less coarse-grained 'pseudo-worlds' as determined from some partition Π of W. Either way, consequently, for all $w \in B_W$, $P(\{w\}) > 1 - P(B_W)$, and hence, for all $w \in B_W$, $P(B_W) > 1 - P(\{w\})$. In words: if the probability of some serious candidate for the actual world is really small, then $P(B_W)$, and hence the probability of *every* believed proposition, must be really high, or otherwise P1–P3 could not hold jointly. Or contrapositively: if $P(B_W)$, or for that matter the probability of *some* believed proposition, is not particularly high, then the probabilities of all worlds or partition cells in B_W cannot be particularly low either. For instance: if one wants P1–P3 to hold, and the agent ought to believe some proposition of probability 0.91, then all worlds or partition cells in B_W need to have a probability of at least 0.09. That is: B_W cannot contain more than 10 worlds or partition cells. Or: if P1–P3 are meant to be satisfied, and the agent ought to believe some proposition of probability 0.98, then all worlds or partition cells in B_W must have a probability of at least 0.02. Therefore: B_W cannot contain more than 49 worlds or partition cells. And if we let the number of members of B_W go to infinity, then the probability of B_W, and thus of every believed proposition, must tend to 1 in the limit.[190]

Observations C2(i) and C2(ii) should make the limitations of the stability theory of belief quite clear. How serious are they, and what, if anything, can one say in defence of the theory? C2(i) suggests that P1–P3 taken together make belief dependent on, or relativized to, partitions. If, as is plausible, we count the agent's choice of partition as belonging to the context of reasoning in which the agent's beliefs 'take place', or if the partition is at least determined from such a context, then we might say that belief ends up relativized to contexts. But that should not take us by surprise any more: we have already seen that, according to P1–P3, the threshold in the Lockean thesis—and thus what the agent believes—depends on the context (comprising the agent's attention, interests, stakes, degrees of belief, and so forth). I have also made clear already that this does not entail any kind of priority of probability over belief.

What we have established now is that also the agent's manner of partitioning W ought to be included in the context on which the agent's beliefs depend. But in view of the general impact that the context has on belief according to the stability theory, this is hardly a big deal at this point of argumentation. If formulated in the terms of the semantics of questions (cf. Groenendijk and Stokhof 1984), in which questions are reconstructed as partitions of worlds and where answers correspond to partition cells: if the present theory is right, rational belief ends up being sensitive to *the agent's underlying question*.[191]

190 This follows also from Theorem 18 in section 4.2.2.

191 Other than representing them formally with partitions, I will have to leave open what exactly *an agent's* underlying question is. That is, which aspects of a real-world agent's cognitive system correspond to a context or partition as understood by my theory? Or in other words: in virtue of what is a certain partition of logical space the partition relative to which an agent believes or does not believe various propositions at a given time? How can one tell what an agent's partition is like at a given time? At least partly, these are empirical questions: if I am right, then an agent must *somehow* mentally represent partitions, whether by symbolically representing a problem in a certain manner, or by focusing attention on a certain kind of question, or by developing and maintaining a certain kind of cognitive interest, or by something 'in the mind' that all of these have in common. Cognitive scientists might be able to operationalize these mental counterparts to partitions in empirical terms, and they might determine what these counterparts are like, and how they are causally related to beliefs. Or they find *nothing* like them, which might even put my normative theory under pressure. Unfortunately, I will not be able to say more about this here. (I am grateful to an anonymous referee of this book for urging me to comment on that.)

Furthermore, there are quite a few well-known and successful theories around that presuppose probabilities of some sort and for which the same relativization to partitions can be observed. Take Levi's (1967) theory of acceptance, in which partition cells are again regarded as the 'relevant answers' to a question posed by the agent. Or consider Skyrms's (1984) theory of objective chance, in which partition cells are 'natural hypotheses' that derive from the causal-statistical analysis of a subject matter. In the former theory, what an agent *accepts* at a time depends on what she regards as relevant answers, and in the latter theory the *chance* of an event depends on what hypotheses the agent regards as natural candidates and how the agent distributes her subjective probabilities over them.

Thirdly, C2(i) does not just affect my stability theory but really a broad class of theories of belief/acceptance and probability, as proven by Lin and Kelly (2012a, sections 13–14). Stability theory's partition-dependence of belief is just a special case of the general phenomenon that Lin and Kelly refer to as lack of 'question-invariance' of acceptance. Roughly, what they prove is that given some pretty general background assumptions on belief and degree of belief, assuming the logical closure of belief will always necessitate belief to be partition-sensitive.[192]

Fourthly, there are even some empirical findings on the so-called 'Alternative-Outcomes Effect' which seem to support the view that belief is in fact partition-sensitive (see e.g. Windschitl and Wells 1998): if possible scenario outcomes are presented to people in terms of different partitions (e.g. you hold three raffle tickets and seven other people each hold one vs you hold three and another single person holds seven), then the participant's *numerical* probability estimates of the focal outcomes remain unaffected, while their corresponding *non-numerical* or qualitative certainty estimates turn out to be sensitive to the partitions. I do not claim that I could rationally reconstruct these experimental results on the basis of my stability account. And, as always, it is not so clear what kind of bearing empirical results like these should have on a normative theory of rational belief. But at least such findings may be taken to indicate that actual beliefs of actual people *are* indeed partition-dependent (where the partitions in that case would be given by the different linguistic presentations of the problem space).

Fifthly, while the stability theory has it that an agent's beliefs may change from one partition to another, there are also some invariances: the same logical closure conditions apply to believed propositions relative to every partition whatsoever. The agent's degrees of belief in propositions are never required to change numerically when the partition is changed (as long as the propositions are entertainable before and after the change). Relative to every partition, the probability of every believed proposition must exceed that of its negation (by the Lockean thesis). And one can also derive a

[192] I should add that my theory is not itself covered by the background assumptions of their theorem, since Lin and Kelly presuppose that an agent's belief set is a function of the agent's subjective probability measure, which I do not. This will become important later in Appendix C.

couple of cross-partition laws for all-or-nothing belief: take a partition to be given. A set B_W has been determined to be the strongest believed proposition. Now coarsen the partition inside of B_W (or do not change anything there) and repartition any way you want outside of B_W. However, do not repartition so that a member of any original partition cell from inside of B_W ends up in the same cell in the new partition as a member of an original partition cell from outside of B_W. If you abide by these constraints on repartitioning, then the original B_W still determines a set which is P-stable also on the new partition. Only if a partition is altered on B_W without making it coarser there, may previously P-stable sets no longer have stable counterparts after repartitioning, which may force a rational agent to give up some of her belief in the transition from the one context to the other. However, even in that case, it can never happen that an agent is forced to turn a belief into disbelief or vice versa: at worst, a belief may need to be changed to suspension of judgement. Summing up, it is not as if the theory entailed that changing partitions would always affect an agent's beliefs in some completely erratic and unpredictable manner.[193]

Sixthly, one might still wonder how repartitioning possibilities could affect a rational agent's belief in a proposition X at all. As long as X is entertainable before and after changing the partition, nothing seems to have changed really as far as the content of the agent's belief in X is concerned. So how come X is, say, believed by the agent before the partition change but no longer afterwards? One way of making sense of this situation is to remind oneself that belief operators in natural language are actually known to create hyperintensional contexts rather than intensional ones (cf. Cresswell 1975, Nolan 2014). In spite of the attractions of the standard (possible worlds or neighbourhood) semantics of belief, perhaps belief contents ought not really to be identified with sets of possible worlds after all but instead with more fine-grained entities, so that one might rationally believe that A, not believe that B, while 'A' and 'B' are true in precisely the same worlds. A standard choice at this point would be, for instance, to invoke structured propositions or the like. But here is another possibility: perhaps one should not identify belief contents with sets of possible worlds but rather with sets of possible worlds *relative to a conceptual framework or partition*. In that case, changing a partition *would* affect a belief content: X under partition π would differ in content from X under partition π', if the partition co-determines content. One should not then expect there to be any guarantee that belief in the one content would entail belief in the other. In the terminology of Yablo (2014): what one believes also depends on what the belief is about, and subject matters can be analysed as partitions of possibilities (see Yablo 2014, section 2.2, and Lewis 1988a). But I will not work this out in any more detail here.

Finally, it is illuminating to compare the situation with another one in which partition invariance or lack thereof has been an issue: Bayesian decision theory. For

[193] I am grateful to an anonymous referee here whose comments on the corresponding part of Leitgeb (2014a) were invaluable.

instance, it is well-known (see sections 3.4 and 4.2 of Joyce 1999) that Leonard Savage's classic decision theory is partition-sensitive while Richard Jeffrey's is not: expected utilities of actions may vary with the underlying partitioning of possibilities according to the former but not according to the latter. The sense in which Jeffrey's theory is partition-insensitive is this: given a probability space on a sample space W (and also a utility measure), *coarse-graining* W and/or the algebra on W on which the probability measure is defined will never change the expected utility of actions. But rational belief according to the stability account is *also* partition-*insensitive* in that sense: given *Bel* relative to a set of worlds or partition cells, *coarse-graining* that set of worlds will never force one to change any of one's beliefs (as long as worlds within B_W are not fused with any outside of B_W). So in that sense my theory is just as partition-*insensitive* as Jeffrey's. It is only if one assumes P to be defined on the 'maximally fine-grained' space W of possibilities, whereas *Bel* is given relative to a coarse-graining Π of the possibilities in W, then *fine-graining* Π again (against the backdrop of W) may force a rational agent to abandon some of her beliefs and to suspend judgement instead.[194]

Taking these points together, my overall diagnosis is: while belief certainly becomes more strongly dependent on contexts than one might have hoped for, no decisive argument against P1–P3 emerges from C2(i).[195]

Now for C2(ii), that is, if $P(B_W) < 1$ then the probability of every world or partition cell in B_W must be greater than the probability of $W \setminus B_W$. Given P1–P3, this leaves the agent with the following options. Either (A) she only believes propositions that have probabilities very close to, or identical with, 1, in which case she is flexible about drawing fine-grained distinctions within B_W. Or (B) she believes some proposition with a probability that is not particularly close to 1, in which case she can only make very few distinctions in terms of serious possibilities in B_W. Or (C) she opts for a position in between the two extremes. Let us assume that W itself is very fine-grained in the sense of containing a lot of worlds: then, by means of partitioning, the obvious manner of realizing (A) is to introduce a partition that is very fine-grained with respect to B_W; for (B) a very coarse-grained partition with respect to B_W will be the right choice; and (C) will be the case if the agent opts for a partition that lies somewhere in between.

Option (A) should be appealing to all those who defend a view according to which believed propositions ought to have a degree of belief of 1 *in all contexts*; for option (A) approximates that kind of position at least in contexts with fine-grained partitions. Examples in the relevant literature would be Levi (1980) (though for 'knowledge' and 'credal probability' instead of 'belief' and 'degree of belief'), Gärdenfors (1986a),

[194] There are also arguments in decision theory to the effect that certain individuations of outcomes are rational, while others are not, and that super-fine individuation can be problematic (cf. Broome 1991, Dreier 1996). I am grateful to Seamus Bradley for pointing me to this literature.

[195] However, see Staffel (2016) for an argument to the contrary: Staffel's argument partially relies on Buchak's (2014) arguments for the thesis that all-or-nothing belief has certain roles to play that degrees of belief do not.

van Fraassen (1995), Arló-Costa (2001), Arló-Costa and Parikh (2005); and a closely related position is held by Williamson (1998) if 'belief' is replaced by 'knowledge' and 'subjective probability' by 'epistemic probability'. All of these proposals also share with the present theory the assumption that ideal belief (or ideal knowledge) is closed under logic. By invoking further resources—for instance, by starting from a primitive conditional probability measure (or Popper function), as van Fraassen, Arló-Costa, and Parikh have done—one might even finesse P1–P3 so that option (A) would get even closer to some of these proposals, e.g. by singling out only particular sets of probability 1 or only particular sets of very high probability as believed propositions. In any case, P1–P3 cannot be much worse off than these proposals, as P1–P3 allow for them or for something close to them to be realized. But P1–P3 are also less restrictive than these proposals by not turning option (A) into a general requirement *in all contexts*.

Option (B) ought to be attractive to anyone who favors the Lockean thesis with a 'realistic' threshold that is not particularly close to 1; examples include Kyburg (1961, 1970), or more recently, Kyburg and Teng (2001), Foley (1993), Hawthorne and Bovens (1999), and Hawthorne and Makinson (2007). Of course, in contrast with the current theory, these proposals do not include the closure of belief under conjunction, but that might be because they think *one could not have* it in the presence of the Lockean thesis anyway, which is not right by what we found in the first two sections of this chapter. The downside of P1–P3 if compared to these other Lockean proposals is the additional constraint that in order to realize (A) one needs to reason relative to sufficiently likely serious possibilities only (of which there cannot be too many).

But how severe is this constraint really? Is it really plausible to assume that, when we have beliefs and when we reason on their basis, we always take into account *every maximally fine-grained possibility whatsoever*? Instead, in typical everyday contexts, we might reason relative to some contextually determined partition of *salient* and *sufficiently likely* alternatives. Say, for some reason in some context we are interested only whether the three propositions A, B, C are the case or not. Hence, the possible worlds or partition cells on which we concentrate are precisely all of the logical combinations of these three propositions,

$$A \wedge B \wedge C,\quad A \wedge B \wedge \neg C,\quad A \wedge \neg B \wedge C, \ldots,\quad \neg A \wedge \neg B \wedge \neg C,$$

and we only take into account the propositions that can be built from them. The same thought applies to the rational reconstruction of reasoning patterns as carried out in epistemology or philosophy of science. For instance, in formal epistemology, when one rationally reconstructs confirmation or coherence or learning, one typically does so by means of 'small' probability spaces which may well correspond to what is required by option B. Indeed, when I will turn to a concrete application of my theory in section 3.7 of this chapter, I will deal with precisely such a situation in which only the logical combinations of three propositions happen to be relevant. More generally, when we represent an argument from natural language in logical terms, we usually follow Quine's (1960, p. 160) *maxim of shallow analysis* and end up with a formalization in terms of, say, just a couple of propositional letters.[196] When people draw inferences in everyday situations, according to what is perhaps the empirically most successful theory of reasoning in cognitive psychology—Johnson-Laird's (1983) theory of mental models[197]—they do not do so by representing infinitely many super-fine-grained possibilities but rather by representing the, usually very few, distinctions that are required in order to build a model of the situation. It is not clear at all why this should be a sign of irrationality. And so on. In all of these cases, it seems that satisfying P1–P3 along the lines of option B should be perfectly viable. In other words: there is no reason why a perfectly rational agent who faces an everyday question or problem would be forced to rely on a very fine-grained partition of possibilities. It is only when one's attention got directed simultaneously towards a great number of case distinctions that belief would have to get ever closer to having a probability of 1, according to the present theory. Adapting the title of Lewis (1996), also rational belief turns out to be elusive then, not just knowledge. But *mostly and normally* rational belief would not have to be elusive like that (which matches also the qualifications that figured in the defining definite description for belief in section 1.1: the propositional attitude the function of which is to realize . . . *at least to a great extent and in normal circumstances*).

This also means that if an agent aimed at stating each and every proposition that she believes—one after the other: $A_1, A_2, A_3, \ldots$—then there would have to be respective contexts in which each of them would be believed. For instance: A_1 might be believed with respect to the partition A_1 vs $\neg A_1$ and one Lockean threshold; A_2 might be believed with respect to the partition A_2 vs $\neg A_2$ and another Lockean threshold; and so on. Or maybe $A_1, \ldots, A_{10}$ are believed in one context, $A_5, \ldots, A_{15}$ in another, and so on. But there would not be a joint context in which all of these different propositions would be believed simultaneously, since the corresponding partition would have to be too fine-grained in order to support all of these beliefs at one and the same time (unless the probabilities of $A_1, A_2, A_3, \ldots$ would be super-high or equal to 1).[198] However, as long as the conjunction of all of these propositions were to have a probability above one-half, the agent might still be able to believe that conjunction (or intersection) holophrastically as one 'building block': for the corresponding set of worlds might well be believed by the agent relative to yet another context (and hence, partition and threshold). It is just that she could not normally *logically derive* that conjunction from its conjuncts, since the logical closure of belief is only guaranteed

[196] 'A *maxim of shallow analysis* prevails: *expose no more logical structure than seems useful* for the deduction or other inquiry at hand' (Quine 1960, p. 160, original emphasis).

[197] Of course, verdicts like this about the relative success of psychological theories may vary with the psychologists to whom one talks; for instance, some psychologists will regard the more recent Bayesian accounts of psychology to be more successful.

[198] I should add that if the agent asserted one sentence after the other in order to express her beliefs in $A_1, A_2, A_3, \ldots$, then it would still not be the case that any of these sentences would have to express one proposition in one context and a different proposition in another: rather, which proposition is *entertainable by the agent* may shift from one context to the next.

to hold *within* a context, not *across* contexts. I should also stress again that the agent's degrees of belief in any of these propositions need not be affected by such context shifts at all. When one compares an agent's categorical belief set *Bel* with her degree-of-belief function *P*, it is useful to restrict *P* to the partition cells on which *Bel* is defined (in the relevant context), but one may always think of *P* as being given really on some more fine-grained space of possibilities.

Finally: the stability theory of belief allows for continuous transitions between options (A) and (B) and hence for the compromise option (C). All of these options are still governed by the same set of general principles, that is, P1–P3.

Let us take stock. If P1–P3 are satisfied, and thus their consequences C1, C2(i), and C2(ii) are true as well, the following picture of our perfectly rational agent emerges: the agent must hold her beliefs, and reason upon them, always relative to a context of reasoning that involves the agent's attention, interests, stakes, the degree-of-belief function *P*, and more. The context must include or determine a partition of the underlying set of presumably very fine-grained worlds into more or less coarse-grained partition cells that figure as 'pseudo-worlds' in the subsequent reasoning processes. Additionally, the context restricts the permissible thresholds in the Lockean thesis to a range of natural candidates that are given by the probabilities of *P*-stable sets. From these thresholds, whether implicitly or explicitly, the agent needs to choose the one that is to be used for the Lockean thesis. The greater the threshold is, the more cautious the agent will be about her beliefs; but also the greater the threshold is, the greater the number of serious possibilities that the agent is potentially able to distinguish. Different contexts of reasoning are available to an agent at a time, but, presumably, at each time only one context is chosen to be 'active' (implicitly or explicitly) and will thus ground the agent's rational all-or-nothing beliefs at the time. That context, or at least certain aspects of it (most importantly, the partition), will be maintained for a certain period of time in which the stability of the agent's beliefs will (hopefully) pay off. But at some point the context (and in particular its crucial aspects, such as the partition) will change again due to changing questions, perceived stakes, interests, and the like.

In this way, the agent is able to maintain the logic of belief, the axioms of probability, and the Lockean thesis simultaneously. The price to be paid is this very dependency of belief on contexts. Accordingly, while the logic of beliefs does hold locally within every context, logical inferences across contexts are not licensed unrestrictedly. The same holds for the Lockean thesis: the Lockean threshold, and the set of propositions to which the thesis applies, vary with the context, and one cannot always export the consequences that the Lockean thesis has from one context to the next one. Although the results from Chapter 2 secure the agent's all-or-nothing beliefs in *Bel* to be stable under new and doxastically possible evidence—and even though the Humean threshold *r* from Chapter 2 does not need to change under the impact of evidence—the Lockean threshold *s* may well have to change given new evidence, because the agent's degree-of-belief function changes. But P1–P3 also guarantee some doxastic invariances across contexts. In particular, as long as the partitioning of possibilities remains intact, all-or-nothing belief will have the Humean stability property from Chapter 2 (with a Humean threshold of $r = \frac{1}{2}$). Moreover, in a lot of everyday and scientific contexts an agent may restrict herself to coarse-grained possibilities without loss, and the corresponding Lockean threshold may thus be rather low (though above one-half). Finally, the fallback position of reasoning in terms of the most fine-grained partition is available to her, too, in which case P1–P3 amount to the more conservative Probability 1 Proposal of belief (or something close to it), which would not be crazy either, and which would only be required by such fine-grained contexts.

While it is always hard to weigh the benefits of a theory against its limitations, so far, the logic of belief, the axioms of probability, and the Lockean thesis seem to do quite well against the drawbacks of contextualization.

Before I put the theory to the test again by considering how well it does in the face of paradox, let me conclude this section by pointing out what in this chapter I do *not* assume contexts to do: to *eliminate possibilities* merely by the agent ignoring or disregarding them, as e.g. Lewis (1996) on elusive knowledge would have it. The reason is that doing so might well go against belief's aiming at the truth (and thus against my Assumption 2 from section 1.1 of Chapter 1). Let me explain this just for the degree-of-belief side: one way of explicating the act of ignoring or disregarding possibilities is in terms of conditionalizing one's subjective probability measure on some propositions that, for the time being, are to be taken for granted or presupposed. For instance: say, I am ignoring all the *brain-in-a-vat* possibilities. I am operating on the assumption that $\neg$*brain-in-a-vat* is the case: my degree-of-belief function has been conditionalized on $\neg$*brain-in-a-vat*, and it is the resulting function that guides my reasoning and acting now. But then these resulting probabilities are no longer 'pure' degrees of belief any more but rather degrees of *acceptance*: degrees of belief modified by the acceptance of certain propositions, where in this case the acceptance results from my *ignoring* certain possible cases. My original degree of belief in some proposition X might have been, say, $P(X) = 0.8$, while $P(X \mid \neg \textit{brain in a vat})$ might only be 0.4. The original value of 0.8 was my best possible shot at the truth value of X—the result of my degree-of-belief function aiming at the truth—while 0.4 may at best be said to *conditionally* aim at the truth *given certain premises* (namely, $\neg$*brain-in-a-vat*). Accepting propositions is an important kind of mental act, and I will deal with acceptance in precisely that sense in section 6.3 of Chapter 6 (while relying on the related notion of conditional belief from Chapter 4), but I will also distinguish its functional role there from that of belief. Unlike belief, acceptance does not just by its nature aim at the truth; in certain cases, the acceptance of propositions may still aim at the truth in a sense, but there is no guarantee for that to be so just in virtue of these propositions being accepted. Combining the context-sensitive elimination of possibilities with truth-aiming would require special additional assumptions. One such assumption might be that the propositions accepted are also *believed*: cases of accepted belief, as I am going to call them in section 6.3. But I will turn to these matters in more detail there.[199]

Changing the manner in which one partitions the set of all logical possibilities, or in which one individuates them, does not affect belief's aiming at the truth, since only the 'resolution' of one's contents of beliefs is thereby changed. It is like: the 'pixel size' may be changed, but nothing is blanked out. But suppressing possibilities for the 'wrong' reason, e.g. merely because I do not want to deal with them now—*not* because they have been ruled out by the evidence, or the like—may well be incompatible with belief aiming at the truth, as a part of reality would simply be blanked out. That is why I allow my rational belief contexts to individuate possibilities differently but not to throw any of them away.

3.4 Application to the Lottery Paradox

'Solving' a paradox by a theory usually involves the following ingredients: the theory should avoid the absurd conclusion of the paradox. It should preserve some, or many, of the original premises of the paradox. The theory should explain why some of the premises need to be given up; and it should explain why those premises that are given up appeared to be true initially, by explaining—and maybe explaining away—the intuitions that seemed to warrant these premises.

I want to argue that the theory from the last section does solve the Lottery Paradox, which is the topic of this section. To the extent to which the Preface Paradox resembles the Lottery Paradox, similar considerations apply, and I will briefly point this out in the next section. But the Preface story involves additional complications that I do not want to get into there; I will discuss the Preface Paradox in full detail in section 6.4. (Of course, I hope that section 6.4 will also 'solve' the paradox in the way sketched before.)

The main task for now will be to interpret and evaluate two of the formal examples that we have already encountered before: Examples 4 and 5 from section 3.1 and Examples 4 and 5-Reconsidered from the last section.

There will be a fair lottery of a million tickets. By the Lockean thesis, a rational agent ought to believe of each ticket that it will not win, because each ticket is very likely to lose. But it is also plausible that belief is closed under conjunction and that the agent's degrees of belief should reflect the fairness of the lottery. Taking these together leads to contradiction, along the lines of what was pointed out in Example 4 in section 3.1.

What does our joint stability theory of belief and degrees of belief predict concerning this paradox? First, for $W = \{w_1, \ldots, w_{1000000}\}$ and P being uniform over W again, it suggests that a partition of the underlying set of worlds needs to be determined first. The salient options are:

[199] For the same reason, I interpret e.g. Clarke's (2013) proposal of 'Belief is Credence One (in Context)'—which in the case of perfectly rational agents involves conditionalizing a global credence function on contextually determined propositions—as a proposal that concerns the *acceptance* of propositions.

- In a context in which the agent is interested in *whether ticket i will be drawn*; e.g. for $i = 1$:

  Let Π be the corresponding partition $\{\{w_1\}, \{w_2, \ldots, w_{1000000}\}\}$. The resulting probability measure P_Π is given by P so that:

  $$P_\Pi(\{\{w_1\}\}) = \frac{1}{1000000}, P_\Pi(\{\{w_2, \ldots, w_{1000000}\}\}) = \frac{999999}{1000000}.$$

  As determined in Examples 4 and 5-Reconsidered in the last section, there are two P-stable sets, and one of the two possible choices for the strongest believed proposition B_W^Π is $\{\{w_2, \ldots, w_{1000000}\}\}$. If B_W^Π is chosen as such, our perfectly rational agent believes of ticket $i = 1$ that it will not be drawn, and of course P1–P3 are satisfied.

  This might be a context in which a single ticket holder—the person holding ticket 1—would be inclined to say of her ticket: 'I believe it won't win.' Π will be the natural partition to consider since the person is, presumably, primarily interested in whether her own ticket will win or not.

- In a context in which the agent is interested in *which ticket will be drawn*:

  Let Π' be the corresponding partition that consists of all singleton subsets of W. Or equivalently: keep W as it is. Consequently, the probability measure $P_{\Pi'}$ can be identified with P again, and it is distributed uniformly over the 1,000,000 alternatives.

  As mentioned in Examples 4 and 5-Reconsidered in the last section, the only P-stable set—and hence the only choice for the strongest believed proposition B_W—is W itself: our perfectly rational agent believes that some ticket will be drawn but she does not believe of any ticket that it will not win.[200] Of course, P1–P3 are satisfied again.

  This might be a context in which a salesperson of tickets in a lottery would be inclined to say of each ticket: 'It might win' (that is, it is not the case that I believe that it won't win). That is also what many epistemologists these days would say concerning the knowledge version of the Lottery case: no ticket is known not to win. Π' will be the natural partition to consider for a salesperson who is primarily interested in selling their tickets and who wants to present each of the tickets as having the same chance of winning the lottery.

  If formulated with regard to stability in the sense of the Humean thesis HT$^{\frac{1}{2}}$ from Chapter 2: the salesperson could not believe anything more specific than W, since that would not be stable enough. For assume otherwise: suppose the cardinality of any maximally specific proposition believed by the salesperson is i

[200] Douven and Williamson (2006) prove on very general grounds that if a probability space is 'quasi-equiprobable' (their term)—a generalization of uniform or equiprobable probability measures—the corresponding belief set must either consist only of propositions of probability 1 or it must include a proposition of probability 0. B_W coinciding with W falls under the first disjunct, of course.

where $1 \leq i < 1{,}000{,}000$; for instance, without loss of generality, she might believe $\{w_1, \ldots, w_i\}$. Because she does not believe any proposition that is more specific, she must regard $\{w_i, \ldots, w_{1000000}\}$ as possible (as she does not believe its negation). But the probability of $\{w_1, \ldots, w_i\}$ conditional on $\{w_i, \ldots, w_{1000000}\}$ is $\frac{1}{1000000-(i-1)}$, which is in fact less than or equal to $\frac{1}{2}$, but which by P-stability should be strictly greater than $\frac{1}{2}$. Contradiction.

The same relativization to partitions as in these two contexts had been exploited already by Levi (1967, p. 40), in order to analyse Lottery-Paradox-like situations.

Thomason's (1986) analysis of the Lottery Paradox does not invoke partitions explicitly, but he certainly gets very close to them when he is arguing for the context-sensitivity of belief in the lottery story:

> If belief is context-sensitive, we can say that the context is switched in the course of the paradoxical argument. Suppose that we are dealing with a fair lottery that has 10,000 entries. I have before me a list of entries' names: Bas van Fraassen is the first entry, David Lewis is the second. When I think of van Fraassen, the first entry, I believe that he won't win. When I think of Lewis, I believe that *he* won't win; and the same is true when I think of any other entry. But the proposition that none of the entries will win is not among my beliefs, even though this proposition is equivalent to the conjunction of my beliefs about each entry. Here, the change of context seems to be determined by topic. Limiting my attention to a single individual provides a circumscribed arena of relevant suppositions and interests that somehow help to condition the belief...
>
> Whether or not this is a matter of dependence on topic, I think there is no doubt that it involves dependence on a context. If I am deciding whether to bother trying to sell a Florida condominium to Lewis, I will suppose that he won't win; there is no point in deciding to convince him to buy on the strength of his one lottery ticket. But if I am wondering whether to send prospectuses on the condominium to all of the 10,000 entries, I will *not* suppose that he won't win... a change of belief that is not conditioned by any gain or loss of information. And it is precisely the switch of topic accompanying this change that amounts for the paradox.
>
> (Thomason 1986, p. 352)

With respect to Thomason's reference to *topics* here, one might note that Lewis (1988a) has argued that the proper way of formalizing subject matters or topics is in terms of partitions of a set of worlds.[201] The subject matter of, say, a particular geographic region in a particular period of time corresponds to a partition of the set of possible worlds according to which worlds end up in the same partition cell if and only if that geographic region in that particular period of time is the same in each of them. Or the subject matter of the number of planets in the solar system corresponds to the partition in which each partition cell collects all possible worlds in which the earth has one and the same number of planets; and so on. Hence it is plausible that different subject matters correspond to different partitions of the underlying set of possible worlds.

[201] See also Yablo (2014) who further develops this thought.

In either of the two lottery contexts from before, the theory avoids the absurd conclusion of the Lottery Paradox; in each context, it preserves the closure of belief under conjunction; and in each context, it preserves the Lockean thesis for some threshold ($s = \frac{999999}{1000000}$ in the first case, $s' = 1$ in the second case). All of this follows from P-stability and the theorem from section 3.2. In the first Π-context, the intuition is preserved that, in some respect, one believes of ticket i that it will lose, since it is so likely to lose. In the second Π'-context, the intuition is preserved that, in a different respect, one should not believe of any ticket that it will lose, since the situation is symmetric with respect to tickets (as expressed by the uniform probability measure), and some ticket must win. Finally, by disregarding or mixing the contexts, it becomes apparent why one might have regarded all of the premises of the Lottery Paradox as true. But according to the present theory, contexts should not be disregarded or mixed: partitions Π and Π' differ from each other, and different partitions may lead to different beliefs, as observed in the last section and as exemplified in the Lottery Paradox. Accordingly, the thresholds in the Lockean thesis may have to be chosen differently in different contexts, and once again that is what happens in the Lottery Paradox—which makes good sense: in the second Π'-context, by uniformity, the agent's degrees of belief do not give her much of a hint of what to believe. That is why the agent ought to be super-cautious about her beliefs in that context; hence the maximally high threshold. In contrast, in the first Π-context, the agent's degrees of belief are strongly biased against ticket i being drawn. That is why the agent may afford to be brave in terms of her beliefs about i not winning in that context. No contradictory conclusion follows from that, since, according to the stability theory, it is not permissible to apply the closure-under-conjunction rule across different contexts.

This seems to be a plausible rational reconstruction and solution (in the sense specified before) of the Lottery Paradox, based on the theory from the last section. I conclude that the stability theory handles the Lottery Paradox quite successfully. The context-sensitivity of belief that was observed in the previous section actually works to the theory's advantage here, since one can analyse the different reasons for assuming the various premises in the paradox in terms of different contexts, without running into contradictions. And the contexts in question arise naturally—from the interest in a particular ticket winning or not, or the interest in which ticket will win.

3.5 A First Shot at the Preface Paradox

To some extent, similar conclusions apply in the case of the Preface Paradox (cf. Makinson 1965). The story is this: a (non-fiction) book is published. The author seems to believe each statement A_i that is made in the main part of the book (in any of its proper chapters), at the same time the author apologizes in the preface for the mistakes that inevitably will be contained in it. By the logical closure of rational belief, the author seems to be committed to believe the conjunction of all statements in the main part, but what the author says in the preface seems to commit her also to believe the negation of that conjunction: $\neg(A_1 \wedge \ldots \wedge A_n)$. So it looks as if her beliefs overall are inconsistent.

But now we can apply a similar analysis as before: each single statement A_i in the main part of the book corresponds to a ticket-losing proposition $\neg\{w_i\}$ in the Lottery Paradox. Both are likely from the viewpoint of the agent, and if the author focuses her attention on any such single statement A_i, then in such a context with a partition $\{A_i, \neg A_i\}$ she will believe that statement to be true. However, in the context of the preface, in which a different partition of possibilities might be salient, she may well believe the negation of the conjunction of all statements in the main part (or equivalently $\neg A_1 \vee \ldots \vee \neg A_n$) which she also regards as likely. This is just as in the lottery case where in the million-tickets context the agent believed that at least some ticket will win but where she did not believe anything more specific than that: $B_W = \{w_1, \ldots, w_{1000000}\}$ $(= \{w_1\} \cup \ldots \cup \{w_{1000000}\})$. As in the discussion of Thomason, we might even be able to relate the different partitions in question to different subject matters: whatever the statements in the main part of the book may be about, the preface is normally about something else, that is, the *book* itself, which would explain the different partitions. All of that is in line with the reconstruction of the Lottery Paradox before, and all of it is compatible with the stability theory of belief.

But there are also some differences between the two paradoxes:[202] first, $\{w_1, \ldots, w_{1000000}\}$ has probability 1 in the Lottery Paradox, while $\neg A_1 \vee \ldots \vee \neg A_n$ may just have a really high probability. Secondly, each proposition $\neg\{w_i\}$ is negatively relevant to each other ticket-losing proposition $\neg\{w_j\}$, since assuming $\neg\{w_i\}$ makes the ticket-winning proposition $\{w_j\}$ a bit more likely. In contrast, we may expect each A_i in the Preface Paradox to be positively relevant to at least some other statement A_j made in the main part of the book. Thirdly, and perhaps most importantly, by writing and publishing the book the author seems to express some sort of commitment to all of the statements in the book *taken as a whole*. There is no similar act of 'mass assertion' involved in the lottery story. Moreover, in the course of her arguments in the book, the author is likely to express a network of inferences from various statements in the main part to various other such statements, which is not part of the lottery story either. And neither of these more holistic features can be captured by the multiple piecemeal $\{A_i, \neg A_i\}$ partitions that we invoked before. What kind of commitment does the author express by asserting all of the statements in the book as a whole? What kind of mental state corresponds to the author's presentation of her theory in its entirety? I will turn to these questions in section 6.4 of Chapter 6.

In the next section I will consider another application of the theory that does not involve paradoxical circumstances.

202 See also Pollock (1994, pp. 396–7) and Foley (1993, s. 4.2) for a discussion of some of the differences between the Lottery and the Preface cases.

3.6 An Application in Formal Epistemology

Sometimes, when we analyse a concept, problem, or question on the basis of subjective probabilities, we still want to be able to express our findings also in terms of beliefs. Or the other way round. Or we want to refer to both belief and probability right from the start. In all of these cases a joint theory of belief and degrees of belief is required.

In this section, I will present an example of the first kind by applying the stability theory of belief in the context of Bayesian formal epistemology (or Bayesian philosophy of science).

By the *secular acceleration of the moon* one refers to the phenomenon that the movement of the moon around the earth appears to accelerate slowly. Astronomers had been aware of this for a long time, and in the nineteenth century they wanted to explain the phenomenon by means of the physics at the time, that is, Newtonian mechanics, which turned out to be a non-trivial problem.

In logical terms, when T is the relevant part of Newtonian mechanics, H is a conjunction of auxiliary hypotheses including the assumption that tidal friction does not matter, and E is the observational evidence for the moon's secular acceleration, then T and H together logically imply $\neg E$. In other words: T, H, and E are not jointly satisfiable. So given E, either T or H needs to be given up, and it is not clear which—a classical Duhem–Quine case of underdetermination of theories by evidence (as discussed in every textbook in philosophy of science), or so it seems.

That is where the Bayesian story begins: Dorling (1979) argues that this apparent instance of underdetermination vanishes as soon as one takes into account subjective probabilities. For that purpose, he reconstructs what might be called the 'ideal' astrophysicist's degrees of belief at the time. Obviously, this is all fictional, but that is how it goes with rational reconstructions, and Dorling does a sophisticated job of deriving the probability measure on systematic grounds. He ends up with precisely the probability measure from Example 6 as discussed in the first two sections, with 'T' replacing 'A', 'H' replacing 'B', and 'E' replacing 'C'; compare Figure 3.1 from section 3.1. Hence, $T = \{w_1, w_2, w_5, w_8\}$, $H = \{w_1, w_3, w_7, w_8\}$, $E = \{w_5, w_6, w_7, w_8\}$. Since '$T$', '$H$', '$E$' are treated like propositional letters here, the probability of $T \wedge H \wedge E$ needs to be set to 0 'by hand', for the logically omniscient ideal astrophysicist at the time already knew that this conjunction could be ruled out. Accordingly, in Example 6, the probability of $\{w_8\}$ had been set to 0. The probability space as a whole is a typical case of a Bayesian philosopher of science abstracting away from all further complications, such as the precise propositional contents of the single axioms of T, of the various conjuncts of H, and of the various data that are summarized by E. In terms of coarse-graining, when we will introduce beliefs into this Bayesian model further below, we will thus be heading for option B from section 3.3.

Now what is the Bayesian response to the Duhem–Quine case from before? The prior probability measure P assigns a high degree of belief to Newtonian mechanics, it assigns a degree of belief to the conjunction of the auxiliary hypotheses that is greater than what it assigns to its negation, and it assigns initially a tiny probability to E:

$$P(T) = 0.54 + 0.342 + 0.018 = 0.9,\ P(H) = 0.54 + 0.058 + 0.00006 = 0.59806,$$
$$\text{and } P(E) = 0.002 + 0.00006 = 0.02006.$$

A perfectly rational Bayesian agent would then update her degrees of belief by the relevant evidence E: the resulting new degrees of belief are

$$P_{new}(T) = P(T|E) = 0.8976,\ P_{new}(H) = P(H|E) = 0.003,\ P_{new}(E) = P(E|E) = 1.$$

This means that, after taking into account the observational data: the ideal astrophysicist at the time still ought to assign a high degree of belief to Newtonian mechanics; she has become certain about the evidence; but she should assign only a tiny degree of belief to the conjunction of the auxiliary hypotheses. And that is pretty much what happened in actual history: physicists gave up some of the auxiliary assumptions, including the one of tidal friction being negligible, but of course they continued to support Newtonian mechanics. No Duhem–Quine problem emerges: a success story of Bayesianism.

This said, Dorling (1979, p. 179) mentions that 'while I will insert definite numbers so as to simplify the mathematical working, nothing in my final *qualitative* interpretation ... will depend on the precise numbers'. And that had better be right: because of the fictional character of P, it would be ridiculous if any of Dorling's findings depended on his precise choice of numbers. Dorling (1979, p. 180) also states that 'scientists always conducted their serious scientific debates in terms of finite *qualitative* subjective probability assignments to scientific hypotheses', the idea being that scientists never put forward numerical degrees of belief in their academic debates: instead they argue that some hypothesis is highly plausible, that given some hypothesis some other hypothesis is not very plausible at all, or the like.[203]

However, Dorling does not seem to have the resources available to derive the intended *qualitative* interpretation of his probabilistic results in any systematic manner, nor to prove the *robustness* of his interpretation under slight modifications of numbers, nor to offer any precise account of *qualitative* subjective probability assignments.[204] There is an obvious way of filling this gap: by expressing Dorling's findings by means of the qualitative concept of belief, based on a joint theory of belief

[203] In Dorling's (1979, p. 180) own terms: scientists use expressions such as 'more probable than not', 'very probable', 'almost certainly correct', 'so probable as to be almost necessary', and so on.

[204] Sometimes by 'qualitative probability' one means *comparative* probability: probability theory based on the primitive predicate 'is at least as likely as'. And that is certainly available to Dorling. But at the same time that is not how Dorling (1979, p. 179) understands 'qualitative probability': as he points out, in order for his example to work, 'H should have been regarded at the time as more probable than not and T should have been regarded as substantially more probable than H'. In order to make locutions such as 'substantially more probable' precise, he concludes, 'something semi-quantitative is necessary' for which comparative probability is not sufficient. For a comparison between 'qualitative probability' in the sense of comparative probability and 'qualitative probability' in the sense of the present theory, see Leitgeb (2014f).

and subjective probability. The stability theory of belief seems to be an obvious choice for this purpose, for the following reasons.

First of all, Dorling's argument seems to rely, if only tacitly, on the following inference step: he determines that, after taking account of evidence, the probability of T is high and the probability of H is tiny, *from which he concludes that T ought to be maintained but H ought to be abandoned*. After all, he wants to justify why scientists gave up on H but not on T, and giving up is still a binary act. It is hard to see anything other than a version of the Lockean thesis to be in operation here, which is what P3 offers.

Secondly, according to the stability theory, and as I argued in section 3.3, belief turns out to be a coarse-grained version of subjective probability due to the presence of the Lockean thesis again. So when we translate facts about P into facts about Bel by means of the Lockean thesis, we know that a lot of information is being abstracted away; infinitely many probability measures will correspond to one and the same belief set. What is more, we have seen in Figure 3.2 of section 3.2 that probability measures whose geometric representations are close to each other also yield similar P-stable sets and hence similar candidates for B_W. Therefore, if we can confirm Dorling's diagnosis about underdetermination in terms of the ideal astrophysicist's *beliefs* as determined by the stability theory, we can be quite certain that he was right when he claimed that his interpretation did not 'depend on the precise numbers'.

Thirdly, scientists do seem to express their own (all-or-nothing) beliefs, and criticize the beliefs of others, when they 'conduct their serious scientific debates'; and they also apply the standard logical rules, including closure under conjunction, when they do so: picture a scientist writing A on a blackboard and then later B, arguing that both are satisfied, and then imagine another scientist stopping her colleague from writing $A \wedge B$ below—this would certainly seem at odds with scientific practice. Which gives us P1.

So the all-or-nothing concept of belief, with P1 and P3 from section 3.3 in the background, seems to be precisely what is required to supply Dorling with the lacking theoretical resources. Since P2 is a given anyway by Bayesian lights, the stability theory of belief is what emerges.

In sections 3.1 and 3.2, I already determined the six P-stable sets that result from Dorling's choice of numerical values. According to the stability theory, a perfectly rational agent's beliefs at the time need to be given by one of these P-stable sets. I settle for the bravest possible choice in light of the fact that the probability of H is not particularly high; this gives us:

$$B_W = \{w_1\} \quad (s = 0.54)$$

At this point, the agent believes Newtonian mechanics, the conjunction of the auxiliary hypotheses, and the negation of E—that is: $Bel(T)$, $Bel(H)$, $Bel(\neg E)$—as well as all of their logical consequences: e.g. $Bel(T \wedge H \wedge \neg E)$. Bel and P taken together satisfy the Lockean thesis with $s = 0.54$ as a threshold. We also know from the previous sections that if that Lockean threshold had not been identical to (or, more precisely, had not been identical to, or sufficiently close to and below) the probability of a P-stable set, then belief would not have been closed under conjunction; e.g. it might have been the case then that $Bel(T)$, $Bel(H)$, $Bel(\neg E)$ without $Bel(T \wedge H \wedge \neg E)$ being the case at the same time.

Just as in the probabilistic story from before, the next step for the agent is to update her beliefs by means of $E = \{w_5, w_6, w_7, w_8\}$. Since E contradicts B_W, that is, since the agent had expected $\neg E$ to be true beforehand, this is a case of proper belief *revision* in the sense of AGM (1985) and Gärdenfors (1988). Given a sphere system of doxastic fallback positions, the standard method of revision in such a case (cf. Grove 1988) is for the agent to move to the least sphere that is consistent with the evidence, to intersect it with the evidence, and to use the resulting set B_W^{new} of worlds as the new strongest believed proposition. Formally, this is just like a Lewis–Stalnaker semantics for conditionals in which one considers the least sphere that is consistent with the antecedent proposition: one intersects the two, and then one determines which consequent propositions are supersets of that intersection.[205] I will make all of that formally precise in Chapter 4, and the mechanics of spheres or 'doxastic fallback positions' in particular will be explained in intuitive terms in section 4.1.2.

If we use the total set of P-stable propositions as the obvious choice of sphere system (recall section 3.2) in the present case, then the least P-stable set that is consistent with E is

$$\{w_1, \ldots, w_5\}.$$

Intersecting it with E yields

$$B_W^{new} = \{w_5\}.$$

Therefore, the propositions that the agent believes after the update are precisely the supersets of $\{w_5\}$.

This means that after taking into account the observational data, the ideal astrophysicist at the time still ought to believe Newtonian mechanics, she takes on board the evidence, but she should also believe the *negation* of the conjunction of the auxiliary hypotheses. In short:

$$Bel_{new}(T),\ Bel_{new}(\neg H),\ Bel_{new}(E);$$

and, accordingly,

$$Bel_{new}(T \wedge \neg H \wedge E).$$

Once again, that is exactly what happened in actual history. And all of this is consistent with stability theory and with the previous purely probabilistic considerations, since

205 I take what Lewis (1973) calls the 'Limit Assumption' for granted here: the assumption that if the evidence is possible (a possible antecedent) there is always a least sphere that is consistent with it.

B_W^{new} turns out to be P_{new}-stable again (where $P_{new}(.) = P(.|E)$).[206] We can thus confirm Dorling's intended qualitative conclusions by applying the stability theory of belief to what would otherwise be a purely Bayesian, and hence quantitative, theory. The pure Bayesian theory lacked the resources for doing so itself, or it might have only done so in a manner that might have been unsystematic and ad hoc.

3.7 Summary

In this chapter I have presented a theory of belief and degrees of belief that combines three parts, P1–P3, which are usually thought to lead jointly to trivialization or inconsistency; in particular, the theory includes the closure of rational belief under conjunction and the Lockean thesis on rational belief. In the first two sections I made it clear that, actually, neither trivialization nor inconsistency follows from these assumptions. In section 3.3, I gave a reformulation of the theory of this chapter, which I called the stability theory because of the central notion of P-stability that figures in it. The theory was found to be equivalent with that of Chapter 2 which had been based on the Humean thesis on belief. I also discussed the main cost of the theory: a strong form of sensitivity of belief to context. In particular, the theory entails that what an agent believes rationally will depend crucially on how the underlying space of possibilities is partitioned. However, I argued that the benefits of the theory seemed to outweigh its limitations. In section 3.4 I showed that the theory is able to handle the Lottery Paradox (and, to a first approximation, also the Preface Paradox, as considered in section 3.5). Finally, section 3.6 dealt with a concrete application of the theory to a problem in formal epistemology, which demonstrated that this joint theory of belief and degrees of belief is more than just the sum of doxastic logic and subjective probability theory taken together. All of this seems to speak in favour of the theory.

There are several important questions that this theory does not answer. The most important ones, it seems to me, concern *contexts*. When does the context change? How should the corresponding partition be determined? Is there a rational way of achieving this? Unfortunately, I will have to leave these questions open. As far as I can see, they are the 'small-world' versions of deeper philosophical questions that have been investigated on a much larger scale, e.g. by Friedman (2001) (in the wake of Rudolf Carnap's and Thomas Kuhn's work), but that have not been answered completely either: when does a scientific paradigm change? How should the corresponding linguistic framework be determined? Is there a rational way of achieving this? Let me conclude this chapter by briefly explaining this analogy.

According to Friedman, Kuhnian scientific paradigms are nothing but Carnapian linguistic frameworks: for instance, the paradigm of Newtonian mechanics differs

[206] This is not just a random coincidence. From the principles of stability theory, one can derive such correspondence results for conditionalization and belief revision *in general*. See section 4.1.3 for further details.

from that of relativistic mechanics in so far as the conceptual resources of these theories differ. But what such linguistic frameworks do is just to determine a space of possibilities: ways the world might be that can be expressed in the corresponding framework. So my first analogy is between partitions (as considered in this chapter) and linguistic frameworks. Secondly, on the background of a scientific paradigm, certain scientific hypotheses are put forward: for example, in Newtonian mechanics, the Newtonian Law of Gravitation. That is like an agent having certain beliefs relative to certain contexts (that involve partitions), as described in this chapter. Within a context, belief change follows the preservative logic of belief revision (as will be discussed in Chapter 4), just as theory change is cumulative in normal science. So the second analogy is between beliefs in a context and scientific hypotheses in a scientific paradigm. Finally, belief contexts may shift, just as scientific paradigms do, such as from Newtonian mechanics to relativistic mechanics. Carnap (1950b) argued famously that the question of which linguistic framework to choose for one's scientific theory is ultimately a pragmatic one. This is much like what we saw before when the choice of context for rational all-or-nothing belief turned out not to be separable from the agent's interests, attention, questions, and other pragmatic features of her total state of mind. So the third analogy is between pragmatic aspects of choosing a partition for belief and those of choosing a linguistic framework. Synchronically, normally, just one scientific paradigm dominates a field, and similarly an agent's beliefs at a time might normally be those that she holds relative to one and only one context of reasoning: the one that occupies her mind at the time. In other cases, paradigms might coexist, and perhaps, in some way, it might also be the case that two or more belief contexts can be salient to an agent at the same time. There are lots of options here.

In any case, this has led me to the analogy between questions: questions about contexts, partitions, and belief on the one hand and questions about scientific paradigms and hypotheses on the other. How are belief contexts and their corresponding partitions determined? How are scientific paradigms determined? Is there a rational way of achieving this? I do not have answers to these questions, whether in their small-world or large-scale versions.

4

Conditional Belief and Belief Dynamics

In this chapter I will turn to notions of rational *conditional* belief: rationally believing a proposition *given another proposition*.

Among other things, conditional belief plays a role for belief revision by entailing a disposition for how to revise one's belief given new evidence.[207] The numerical version of conditional belief is subjective conditional probability, and in section 4.1.1 I will first call to mind how the concept of conditional probability figures in the standard diachronic norm on *degree-of-belief* change. Section 4.1.2 will then do the same for conditional all-or-nothing belief and the standard diachronic norm on *all-or-nothing* belief change, all of which will be based on the so-called AGM (i.e. Alchourrón et al. 1985) postulates for belief revision. Before I start developing my own stability account of rational conditional belief, I will give a preview of its main outcomes in section 4.1.3: subjective *conditional* probability and rational *conditional* all-or-nothing belief will be found to cohere with each other just as their absolute or unconditional versions did in Chapters 2 and 3. From this, together with the standard diachronic norms on belief change, it will follow that also rational degree-of-belief *change* and categorical belief *change* must cohere with each other once the stability theory of conditional belief is in place. Section 4.1.4 will relate the theory to some existing literature. This will conclude the first, and more informal, part of this chapter.

From section 4.2.1 onwards I will develop the theory in full formal detail. The only bridge postulate that will be introduced for rational conditional degrees of belief and rational conditional all-or-nothing belief will be the left-to-right direction of a conditional version of the Lockean thesis: conditional belief in B given A implies a high enough conditional degree of belief in B given A. Taking this together with subjective probability theory and the AGM postulates for belief revision (or, rather, conditional belief) will entail a conditional version of the stability theory of Chapters 2 and 3 again, as will become clear from two further representation theorems (Theorems 12 and 23)

[207] We already encountered an instance of belief revision in this sense in section 3.6.

that I will prove in this chapter. The theory as a whole will thus amount to a stability theory of conditional belief.[208]

Other than extending the stability account of Chapters 2 and 3 to conditional belief, this chapter will also serve another purpose: to supply some of the mathematical machinery on which the theory in this book is based. In particular, the chapter will include the proofs of some lemmata that were presupposed in the proof of Representation Theorem 5 in Appendix B. In contrast with the other chapters, all proofs of theorems will be stated completely (in the main text). The chapter will also be the only one in this book in which I will allow for rational belief to be given relative to a space of infinitely many worlds and infinitely many propositions, both on the quantitative and on the qualitative side of belief. In the previous chapters I wanted to keep things as simple as possible, but it will follow from the results in this chapter that there are also infinitary versions of the theories presented in the previous chapters. For these reasons, and also because conditional belief is in itself more complex than unconditional belief, this chapter will be more intricate mathematically than the rest of this essay. But its first four sections (4.1.1–4.1.4), which summarize the philosophical essentials, should be easy enough to digest. And so is section 4.3 in which I will work out some concrete toy examples of rational conditional belief (reconsidering also some examples from previous chapters).

4.1 A Stability Theory of Conditional Belief and Belief Dynamics: Introduction and Synopsis

4.1.1 *Conditional Probability and Conditionalization*

The axioms of subjective probability, which I will state again in proper detail later in this chapter (see section 4.2.1), are *synchronic* coherence postulates about a perfectly rational agent's degrees of belief. For instance, the postulate of finite additivity demands of such an agent's degree-of-belief function P *at a time* t: if propositions A and B are logically inconsistent with each other, then $P(A \cup B) = P(A) + P(B)$. That is: the agent's degree of belief in the union or disjunction $A \cup B$ at time t ought to be the sum of the degree of belief in A at t and the degree of belief in B at t, given that $A \cap B$ is empty.

But the standard Bayesian account of degrees of belief, to which I have committed myself in Assumption 9 of Chapter 1, also includes diachronic norms. By the meaning of 'degree of belief'—or, more generally, 'belief', as expressed by Assumptions 1–5 from Chapter 1—any degree-of-belief assignment carries with it certain dispositions: dispositions to act in certain ways given certain circumstances. Some of these acts are

[208] I am particularly grateful to Stanislav Speranski for various very helpful suggestions and corrections concerning this second part of Chapter 4. Stanislav was also the first to note that countable additivity (σ-additivity) for subjective probability measures P is not actually required for any of the results in the chapter, while I had still presupposed countable additivity in the journal article (Leitgeb 2013a) on which parts of this chapter are based.

doxastic themselves: the two most salient examples being the disposition to change one's degrees of belief in certain ways given a new piece of evidence E, and the disposition to hypothetically change one's degrees of belief within the suppositional context of adopting an assumption E. Standard subjective probability theory derives both of these dispositions from one and the same conditional quantity:[209] the conditional probability or conditional degree of belief $P(A|E)$ of a proposition A given the proposition E (relative to an agent's subjective probability measure P).

While the corresponding conditional notion of degree of belief could, and perhaps should,[210] be taken as primitive—in which case an ideal agent's degree-of-belief function would be conditional from the start—standard probability theory actually derives such conditional probabilities from absolute or unconditional ones. That is achieved by means of the so-called ratio formula:

$$\text{Ratio Formula: } P(A|E) = \frac{P(A \cap E)}{P(E)} \text{ if } P(E) > 0.$$[211]

It is characteristic of virtually all forms of (subjective) Bayesianism that, so long as only plain factual evidence about the world is concerned, perfectly rational agents update their degrees of belief in line with their corresponding prior conditional probabilities:[212]

> Diachronic Norm on Degrees of Belief: suppose that an agent's degree-of-belief function at t is P_t. Suppose that, between t and t', the agent learns proposition $E \subseteq W$ with certainty and nothing more. And suppose further that $P_t(E) > 0$. Then her degree-of-belief function $P_{t'}$ at time t' ought to be such that, for each $A \subseteq W$,
>
> $$P_{t'}(A) = P_t(A|E) = \frac{P_t(A \cap E)}{P_t(E)}.$$

So the agent's prior degree-of-belief function P_t gives rise to a disposition which, if triggered with a new piece of evidence E of positive prior probability $P_t(E)$, leads to the posterior degree-of-belief function $P_{t'}(\cdot) = P_t(\cdot|E)$ that is given by conditionalizing P_t on E.

[209] As mentioned before, treating learning and supposing formally in precisely the same manner is actually too coarse-grained for some purposes: learning a proposition and assuming a proposition may well affect an agent's degrees of belief differently in certain cases. But normally these differences show up only when degrees of belief in *introspective* propositions are concerned. For instance (this is a variation of a so-called 'Thomason conditional', as mentioned in Van Fraassen 1980b): *suppose* that my (clever) business partner is cheating on me. Then, presumably, I will never know it. But given that I *learn* that my business partner is cheating on me, I will normally know this to be so. Since I will not deal with introspection or introspective propositions here at all (such as *I will know this to be so*), I will take the liberty of neglecting these differences. See Leitgeb (2011) for more on this topic, both on the numerical and on the categorical belief side.

[210] See Hájek (2003) for an argument.

[211] If probability theory is based instead on primitive conditional probability measures (often called Popper functions), then, in turn, absolute or unconditional probabilities would be derived from conditional ones. See e.g. Makinson (2011) for an overview of this alternative approach to probability theory.

[212] The formulation of this norm can be found e.g. in Leitgeb and Pettigrew (2010b) in which also some defences of the norm are discussed.

We have already seen this diachronic norm in action in Appendix A and in section 3.6 of Chapter 3. It can be generalized to so-called Jeffrey conditionalization in order to deal with those cases in which the evidence is not learned with certainty: that is what we dealt with in both Appendices A and B.

Now I will turn to the all-or-nothing counterparts of conditional degrees of belief and of update by conditionalization.

4.1.2 Conditional Belief and Belief Revision

Just as there are both absolute (unconditional) probability and conditional probability on the quantitative side, there are also both absolute (unconditional) belief and conditional belief on the qualitative side. In the previous chapters of this book I only dealt with unconditional categorical belief, for which consistency and logical closure were the synchronic coherence postulates. It is about time to turn to conditional all-or-nothing belief and the disposition for all-or-nothing belief *change* that it induces.[213]

Even when an agent does not (absolutely or unconditionally) believe *A*—not *Bel*(*A*)—it might be the case that she does so *conditional on* some further proposition *E*. Imitating the standard notation for conditional probability, I will say: *Bel*(*A*|*E*). The two most salient ways in which conditional belief in that sense may manifest itself are again in cases of learning new evidence *E* and in cases of supposing a proposition *E*: if *Bel*(*A*|*E*), then learning *E* (and nothing else) leads rationally to belief in *A*; and furthermore, if *Bel*(*A*|*E*), then supposing *E* (and nothing else) leads rationally to the 'hypothetical belief' or 'offline belief' in *A* within the corresponding suppositional context.[214] The case of learning will be more important for this chapter, while suppositional reasoning will be on the agenda when I will develop the stability account of the assertability of indicative conditionals that will constitute the main part of section 6.2 in Chapter 6. But in order to get a better feel for conditional all-or-nothing belief in general, let me start with at least a couple of remarks about the suppositional consequences of conditional belief first.

Let us assume that an agent does not believe *A* unconditionally. It might still be that *on the supposition that E is the case* the agent does believe that *A* is the case, which is one typical way in which the agent's conditional belief in *A* given *E* can become transparent. This is very much like applying the first step in the natural deduction rule of conditional proof in classical logic: when assuming *E*, in combination with whatever else has been assumed or derived before, *A* is derivable; which does not mean either that *A* is derivable without the assumption of *E*. The only difference between

[213] For detailed treatments of conditional all-or-nothing belief in this sense, see e.g. Stalnaker (1987, ch. 6) and Leitgeb (2007).

[214] For a logical-philosophical analysis of the mental process of supposing a proposition, see Leitgeb (2012c). The only type of supposition that I will be concerned with here is supposition *as a matter of fact*, that is, suppositions that are usually expressed in the indicative, rather than the subjunctive mood: 'Suppose that *X* *is* the case. Then I believe that *Y* is the case.' For the difference between supposition in the indicative and in the subjunctive mode, see e.g. Joyce (1999, ch. 6).

modern formal accounts of belief revision under suppositions and the classical rule of conditional proof is that according to the former it is even possible to suppose a (consistent) statement E that is *inconsistent* with what one believes absolutely or unconditionally—inconsistent with whatever premises were given outside of the suppositional context—*without a contradiction following from this*. Instead, the act of supposing E may be viewed as functioning as follows: at first, enough of one's other beliefs are withdrawn in order to 'make room' for E in one's belief system; secondly, E is added as an assumption or, as it were, as a hypothetical belief; and thirdly the resulting hypothetical belief set is closed under logic again. It is well known that there is not necessarily a unique way of 'making room' for a supposition E in this way, in view of the different possibilities of how to withdraw propositions from a belief set so that the resulting set is consistent with E. But all ways of going through the three-step procedure from before have at least been argued to plausibly satisfy one and the same set of general rationality postulates. These postulates constitute the so-called AGM theory of all-or-nothing belief revision (cf. Alchourrón et al. 1985, Gärdenfors 1988) which, however, is not usually interpreted as a theory of rational supposition but rather as a theory of how to rationally take in evidence. In Assumption 8 from section 1.5 I have already committed myself to that theory, which is the standard theory of qualitative belief revision in the relevant part of the literature.

Traditionally, the AGM postulates are not formulated by means of the '$Bel(\cdot|\cdot)$' notation[215] that I am going to use, and they do not quantify over propositions in my sense either (that is, over sets of possible worlds). Instead, they are typically spelled out in terms of a so-called belief revision operator $*$ for statements or formulas. Such an operator $*$ takes a perfectly rational agent's prior (deductively closed) set K of believed statements and an input formula E as arguments and maps them to a posterior (deductively closed) set $K * E$ that is the result of revising K by E. In typical interpretations of the theory, that input formula E is not considered as assumed but rather as a piece of evidence that is to be learned, although the axioms of belief revision allow for both interpretations.[216] With the learning interpretation, $K * E$ is the agent's set of believed statements once learning E has taken place.

Here is thus the classical AGM axiomatization of belief revision (where A and B are now arbitrary formulas in a given object language):

K*1 Closure: $K * A = Cn(K * A)$ (where Cn is the deductive closure operator).

K*2 Success: $A \in K * A$.

[215] Sometimes I use '$\cdot$', such as in '$Bel(\cdot|\cdot)$', in order to signal an argument place. For instance, the concept of conditional belief comes with two argument places for propositions: hence, '$Bel(\cdot|\cdot)$'.

[216] Some of the axioms of belief revision are in fact more easily defendable if read suppositionally. E.g. the so-called Success postulate '$A \in K * A$' is not unproblematic if given the learning interpretation ('evidence A is included in the revision of K by A'): sometimes one's 'evidence' might be regarded as flawed or misleading and should be rejected rather than taken on board. But the same postulate is perfectly plausible if A is regarded as *supposed*: once A has been assumed, A becomes something that is believed hypothetically in that suppositional context and will not be rejected as long as one remains within the boundaries of that suppositional context.

K*3 Inclusion: $K * A \subseteq K + A$ (where $K + A$ is the result of adding A to K and then closing deductively: so $K + A = Cn(K \cup \{A\})$).

K*4 Preservation: If $\neg A \notin K$, then $K + A \subseteq K * A$ (so, with Inclusion, $K + A = K * A$).

K*5 Consistency: If A is consistent, so is $K * A$.

K*6 Equivalence: If $(A \leftrightarrow B) \in Cn(\varnothing)$, then $K * A = K * B$.

K*7 Superexpansion: $K * (A \wedge B) \subseteq (K * A) + B$.

K*8 Subexpansion: If $\neg B \notin K * A$, then $(K * A) + B \subseteq K * (A \wedge B)$.

Since I will turn to these rationality postulates in more detail later—though spelled out by means of '$Bel(\cdot|\cdot)$' and for propositions rather than formulas—I will not explain them here in much detail. Just to get a feel for them, take for instance Preservation (K*4): the if-part of that principle expresses that the negation of A is not included in one's deductively closed belief set K, or equivalently, A is consistent with everything that the agent believes at this point. The then-part says that in this case $K + A$, that is, the set of all sentences that follow logically from the members of K taken together with A, is a subset of $K * A$, the belief set that results from revising K by A.[217] If taken together with Inclusion (K*3), this means that if A is consistent with all of the agent's (unconditional) beliefs, then $K * A$, the result of revising K by A, coincides with $K + A$, the result of expanding K by A, that is, adding A to K and closing deductively again. It is easy to see that in the presence of K*2 and K*1 one might in fact weaken AGM's original postulate K*4 to the following principle that would bear the name 'Preservation' even more appropriately: if $\neg A \notin K$, then $K \subseteq K * A$. That is: if A is consistent with the agent's present beliefs, then these beliefs are preserved under revision.[218]

One justification for the set of AGM postulates is given by Grove's (1988) representation theorem: an operator $*$ satisfies all of these postulates if, and only if, it can be represented in terms of what is called a sphere system (cf. Lewis 1973) of 'doxastic fallback positions' (cf. Lindström and Rabinowicz 1990), or, equivalently, a total pre-order of possible worlds (as in Lehmann and Magidor's 1992 semantics for nonmonotonic reasoning) by which worlds might be said to get ranked in terms of their 'plausibility'. A sphere system is just a set of 'nested' non-empty sets of worlds, that is, for every two spheres in a sphere system one must be a subset of the other or vice versa. A pre-order of worlds is like a partial order except that it allows for ties between worlds: for two numerically distinct worlds having the same rank or being equally plausible. The totality or linearity of such a pre-order means that for every two worlds one is strictly more plausible than the other, or the other way around, or they are equally plausible. The total pre-order that corresponds to a sphere system is given

[217] As I am going to say soon: preservation deals with belief revision by *expansion*—simply adding the evidence and closing under logic.

[218] I am grateful to Hans Rott for urging me to add a remark on this.

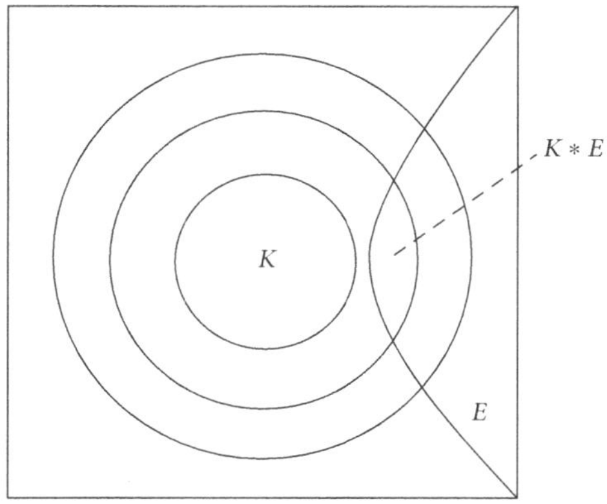

Figure 4.1. Spheres semantics for AGM belief revision

by: a world w is at least as plausible as w' ($w \preccurlyeq w'$) just in case every sphere that includes w' also includes w. K from above coincides then with the set of formulas that are true in the innermost sphere or equivalently in all those possible worlds that are most plausible overall (have least rank overall). That innermost sphere is precisely the set of worlds that I have called 'B_W' in the previous chapters: the set of doxastically accessible worlds. So K corresponds to B_W. $K * E$ is the set of formulas that are true in all worlds in the intersection of E with the least E-permitting sphere, that is, the least sphere that is consistent with what is learned or supposed (E): see Figure 4.1. Or equivalently: $K * E$ is the set of formulas that are true in all those possible worlds that are the most plausible ones amongst those that satisfy E, that is, in which what is learned or supposed is the case: see Figure 4.2.[219] And vice versa every sphere system or total pre-order of possible worlds determines an operator $*$ in this way that satisfies all of the AGM postulates from above.

Sphere models like this are well-known from Lewis's (1973) standard semantics for counterfactuals, even though their intended interpretation differs: Lewis's spheres are meant to capture an objective ordering of worlds by means of similarity or closeness to the actual world, while Grove's spheres may be taken to correspond to a subjective ordering of worlds based on plausibility or, perhaps, closeness to an agent's present unconditional belief set.

More importantly for present purposes, models of the same kind are used also in nonmonotonic reasoning: an area in theoretical computer science in which so-called nonmonotonic consequence or inference relations $\mid\!\sim$ are studied (see e.g. Lehmann and Magidor 1992). Where the belief revision theorist would say

$$A \in K * E$$

[219] The question whether for every non-empty set E of worlds in any such a model there always *exists* a least sphere that intersects E (equivalently, there always exists a non-empty subset of maximally plausible members of E) needs special care; but I will not need to discuss this here.

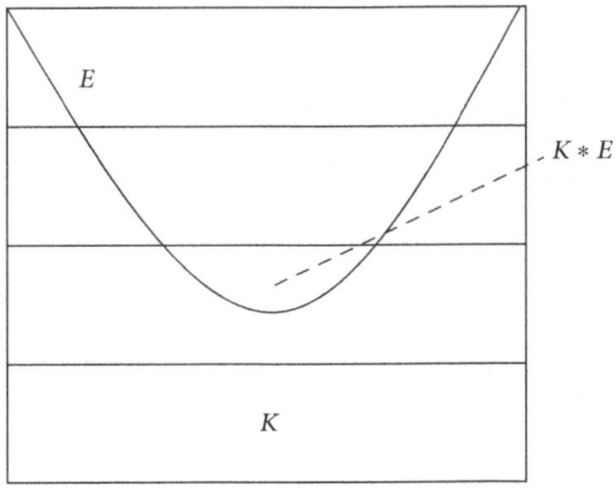

Figure 4.2. Order semantics for AGM belief revision

('A is believed after revising one's beliefs by E'), the same state of affairs would be described in nonmonotonic reasoning by

$$E \mid\sim A$$

('A can be defeasibly inferred from E'). The agent's prior belief set K may be taken to be encoded by $\mid\sim$: K is the set of formulas which can be inferred from the trivial or tautological information $\top$ (the logical verum), that is: $K = \{B\colon \top \mid\sim B\}$. So $\mid\sim$ really corresponds to K and $*$ taken together.[220] The sphere-systems-like models for $\mid\sim$ are usually called 'ranked models' and the axioms for $\mid\sim$ that correspond to the AGM postulates are then said to characterize *rational* consequence relations. Semantically, '$E \mid\sim A$' expresses again: the most plausible E-worlds are A-worlds (the worlds in E with least rank are worlds in A).[221]

In the present chapter, I am going to consider a perfectly rational agent's conditional beliefs, and I will express them by means of my '$Bel(\cdot|\cdot)$' notation. Such an agent's conditional beliefs will include the agent's absolute or unconditional beliefs as a special case, by an analogous move as that regarding K and $\mid\sim$ before: $Bel(B)$, that is, the agent believes proposition B (unconditionally), holds just in case $Bel(B|W)$, that is, when the agent believes B conditional on the trivial or tautological set W of all worlds. Hence, absolute or unconditional belief corresponds to belief conditional on trivial information. For that reason, B_W will also be the least or strongest set of worlds believed conditional on W. All of this is exactly analogous to the case of absolute vs conditional probability, which are related by: $P(B) = P(B|W)$. One of the functional

220 Even on the belief revision side, K may be thought of as given by $*$ from the start.

221 Lehmann and Magidor (1992) is the standard reference that includes all formal details on this. See Makinson and Gärdenfors (1991) and Gärdenfors and Makinson (1994) for more on mutual interpretations between nonmonotonic reasoning and belief revision theory. Versions of some of the axioms for rational consequence relations $\mid\sim$ will reappear later in section 6.2 as postulates governing the assertability of indicative conditionals.

roles of conditional belief is to dispose the agent to revise her beliefs in a certain manner in case new evidence arrives. Therefore, much the same state of affairs that was described before in belief revision terms and in nonmonotonic inference terms can also be formulated in terms of

$$Bel(A|E)$$

where 'A' and 'E' denote sets of possible worlds now (while they had denoted statements or formulas above). Indeed, it will always be possible to determine (propositional versions of) a belief revision operator $*$ and its left argument K, as well as a nonmonotonic consequence relation $\mid\sim$, from a conditional belief set $Bel(\cdot|\cdot)$ by means of: $A \in K * E$ iff $E \mid\sim A$ iff $Bel(A|E)$.[222] And the postulates that I am going to impose on conditional *Bel* later in this chapter will imply that the so-determined belief revision operator $*$ will satisfy (propositional variants of) the AGM postulates, just as the so-determined consequence relation $\mid\sim$ will follow to satisfy (propositional variants of) the axioms of rational consequence relations in the sense of Lehmann and Magidor (1992). The exact details concerning $Bel(\cdot|\cdot)$ and its rationality postulates will be supplied later in this chapter from section 4.2.2 onwards.[223]

This leads us to the diachronic norm that governs the impact that learning a piece of evidence ought to have on all-or-nothing beliefs:

> Diachronic Norm on All-or-Nothing Belief: Suppose that an agent's set of conditional beliefs at t is $Bel_t(\cdot|\cdot)$ (which will be assumed to satisfy analogues of the AGM postulates). Suppose that, between t and t', the agent learns proposition $E \subseteq W$ in the all-or-nothing sense, and nothing more. Then her absolute or unconditional belief set $Bel_{t'}(\cdot)$ at time t' ought to be such that, for each $A \subseteq W$,
>
> $$Bel_{t'}(A) \text{ iff } Bel_t(A|E).$$

So the agent's prior conditional belief set $Bel_t(\cdot|\cdot)$ gives rise to a disposition which, if triggered with a new piece of evidence E, leads to the posterior absolute or unconditional belief $Bel_{t'}(\cdot) = Bel_t(\cdot|E)$. The transition from the prior absolute belief set $Bel_t(\cdot)$, that is, $Bel_t(\cdot|W)$, to the posterior absolute belief set $Bel_{t'}(\cdot)$ (which is determined such that it is given by $Bel_t(\cdot|E)$) corresponds to the transition from the prior unconditional belief set K to the posterior unconditional belief set $K * E$ in belief revision theory.

Four remarks on this diachronic norm are in order.

222 In Chapter 6 I will add another equivalent reformulation to this: the indicative conditional $E \Rightarrow A$ is (subjectively) assertable for the agent in question.

223 When I said before that '$Bel(A|E)$' describes 'much the same state of affairs' as '$A \in K*E$' and '$E \mid\sim A$' do, I meant that the conditional belief in A given E *entails* a disposition to react on evidence E by revising one's beliefs so that A is believed after the revision (and similarly if expressed in terms of a nonmonotonic-inference ascription). But the conditional belief in A given E may not *reduce* to that kind of disposition: just as its unconditional counterpart, also conditional belief has multiple functional roles to play (e.g. the role to commit the agent to action given the acceptance of certain premises—see section 6.3).

First, the norm is obviously analogous to the one on degree-of-belief change from the last section. This said, if one compares their two formulations, one finds that there is no qualitative analogue to the qualification '$P_t(E) > 0$' that was part of the probabilistic norm from section 4.1.1. A conceivable all-or-nothing analogue would have been to restrict E to cases in which '$Poss_t(E)$' holds, that is, 'not $Bel_t(\neg E)$'.[224] By omitting any such restriction, the norm above does not just amount to an analogue but, in a sense, to a generalized analogue of the Diachronic Norm on Degrees of Belief.

Let me explain this point also in the terms of belief revision theory: there are really two cases of belief revision. Either E is consistent with the agent's prior (unconditional) beliefs, that is, $Poss_t(E)$: then, as we have seen before, revising one's unconditional beliefs should correspond to expanding one's stock of beliefs by E and closing the result under logical consequence. This is called revision by belief *expansion*.[225] Or E contradicts the agent's prior beliefs, that is, $Bel_t(\neg E)$, in which case the agent needs to drop at least some of her prior beliefs before she can add E to her belief set—the case of belief *revision* proper. The Diachronic Norm on All-or-Nothing Belief covers both cases in one fell swoop. In contrast, the probabilistic norm from section 4.1.1 only dealt with the $P_t(E) > 0$ case, but not with the other possible case $P_t(E) = 0$ in which the agent's prior degree-of-belief function would have ruled out E completely at the time when E arrived as a new piece of evidence. In order to deal also with that case by means of conditionalization, standard probability measures would have to be generalized to the primitive conditional probability measures that were mentioned briefly in section 4.1.1 and which even allow for conditionalization on propositions E of unconditional probability 0. But I want to hold on to standard probability theory here, which is why I added the respective qualification in the probabilistic case.

Secondly: there is another difference between our two diachronic norms on belief. In the Diachronic Norm on Degrees of Belief, $P_{t'}(\cdot)$ $(= P_t(\cdot|E))$ is a subjective probability measure from which, by means of the ratio formula, conditional probabilities can be defined again: probabilities conditional on propositions that have positive probability relative to $P_{t'}$. These conditional probabilities will then determine how the agent will continue to update if new evidence of positive probability comes along. But the Diachronic Norm on Belief only constrains an agent's posterior *absolute* or *unconditional* belief set $Bel_{t'}(\cdot)$, without saying anything on how the agent's prior conditional belief set $Bel_t(\cdot|\cdot)$ ought to be changed into a posterior *conditional* belief set $Bel_{t'}(\cdot|\cdot)$ that would determine her dispositions again for further belief revision. In other words: the norm underdetermines what the dynamics of all-or-nothing belief change ought to be like. The reason why I did not state any norm stronger than the one

[224] I regard this as a merely structural analogue. I do not want to claim that $Poss_t(E)$ holds if and only if $P_t(E) > 0$, which would entail the Certainty Proposal from section 1.6 that I think applies only in special contexts, not in general.

[225] In this case, belief (revision by) expansion for consistent belief sets K is covered completely by the first four axioms of AGM: Closure, Success, Inclusion, and Preservation. (The fifth axiom Consistency holds trivially in that case.)

above is that AGM belief revision is in fact just a theory of one-shot belief revision—it does not cover iterated belief revision, except for the special case of iterated belief *expansion*: if $E_1, E_2, \ldots, E_n$ is a sequence of pieces of evidence the conjunction of which is consistent with the agent's prior unconditional belief set $Bel(\cdot)$, then AGM does determine the corresponding iteratively revised unconditional belief sets to be (using my notation): $Bel(\cdot|E_1), Bel(\cdot|E_1 \cap E_2), \ldots, Bel(\cdot|E_1 \cap E_2 \cap \ldots \cap E_n)$. But AGM remains silent about iterated belief revision *proper* in which several pieces of evidence may contradict the agent's unconditional belief set at different points in time. There are various ways of extending AGM to iterated belief revision in that sense, but none has become 'canonical' so far,[226] and there are arguments for why there might not be a unique epistemically mandatory scheme of iterated belief revision at all.[227] In any case, I will avoid further discussion of these difficulties by restricting myself to the diachronic norm from above. If, and when, I am going to consider iterated belief revision at all, it will concern only the simple case of iterated belief *expansion* that is indeed covered by AGM. Accordingly, on the probabilistic side, I will only deal with cases of iterated conditionalization on a sequence $E_1, E_2, \ldots, E_n$ of pieces of evidence when $P(E_1 \cap E_2 \cap \ldots \cap E_n) > 0$, so that the standard ratio formula for conditional probabilities can be applied iteratively.

Thirdly, if stated in semantic terms, classical AGM belief revision demands that in any spheres model every *logically* possible world or truth value assignment for the object language in question must occur somewhere within the spheres (or the respective plausibility ordering). For instance, if the underlying object language is the language of propositional logic, then, since $p \wedge q \wedge r$ is a consistent formula, there must be a world in the ranking that corresponds to $*$, such that $p \wedge q \wedge r$ holds at that world; the same is the case for all other consistent formulas of the object language. If the agent's prior belief set K is revised by a consistent formula, say, $p \wedge q \wedge r$, then the resulting posterior belief set $K * (p \wedge q \wedge r)$ will follow to be consistent again. Semantically, that is because there are worlds in the ranking for $*$ so that these worlds make $p \wedge q \wedge r$ true, and $K * (p \wedge q \wedge r)$ is just the set of formulas true in the most plausible of these worlds. With respect to the AGM postulates from above, this consistency assumption corresponds to the Consistency postulate K*5: revision by consistent evidence always leads to a consistent belief set (by K*5), only inconsistent evidence yields an inconsistent belief set (by K*2). In contrast, my theory in section 4.2 will allow for cases in which the revision of K by a consistent proposition E—say, the one expressed by $p \wedge q \wedge r$—leads to an inconsistent belief state. However, that will

226 The paper on iterated belief revision that comes closest to canonicity is Darwiche and Pearl (1997).

227 Rott (2008) suggests that the choice of any such scheme might ultimately be a matter of 'personality' (and corresponding features such as personal style of reasoning or personal risk-aversity). See Rott (2009) for a survey and formal reconstruction of twenty-seven iterated belief change operators. Spohn (1988, 2012) argues that the problem of iterated all-or-nothing belief revision can only be resolved by strengthening AGM: from a theory for ordinal-scale belief revision operators (belief revision based on ordering of worlds), such as AGM, to a theory of belief revision that operates on the ratio scale (as occupied by Spohn's so-called ranking functions).

be so only, and indeed precisely, in those cases in which the agent's degree of belief in the evidence E is 0: so the role of inconsistent evidence in AGM will be taken over in my theory by evidence of probability 0. This will follow from an auxiliary postulate (BP2* in section 4.2.3) that I will adopt for reasons of simplicity, as I will explain in the corresponding section. But it will still mean that one of the original AGM postulates will not be contained in my own postulates on rational belief. This said, the change will only concern beliefs conditional on propositions that are disbelieved unconditionally (the case of revision proper), while Consistency will still hold below for beliefs conditional on propositions that the agent regards as possible. Furthermore, one can show that even full Consistency could be restored by changing parts of my formal framework (dropping BP2* in section 4.2.3, and reformulating Theorem 23 a bit). I merely found the way in which I am going to proceed to be the formally most convenient one.[228] All other AGM postulates will be contained in the theory, but of course stated for propositions rather than for formulas. I will first deal with restricted conditional belief (from section 4.2.2)—or, dynamically: belief expansion—which will involve versions of K*2–K*5. Afterwards (from section 4.2.3), I will deal with conditional belief in general—dynamically: belief revision—which will be built on versions of K*2, K*7–K*8. The AGM postulates K*1 and K*6 will be implicit in my propositional treatment of belief expansion and belief revision.

Fourthly, the theory does not just build on the AGM theory of belief revision. By adding a probabilistic component, it will also throw some new light on it. For instance: belief revision relies on the existence of doxastic 'fallback' positions. If new evidence comes along that contradicts one's present beliefs, then one withdraws to a more generous fallback position (a sphere in Grove's semantics), adds the evidence, and closes under logic. But where do these fallback positions 'come from'? The theory will be able to give at least a partial answer to that question: as we have seen in the last chapter, even when P is held fixed, there is generally more than just one P-stable set available so that all of our rationality postulates would be satisfied if the agent's rational *unconditional* belief system were generated from that set. (The notion of P-stability from Chapter 3 will be generalized again to that of P-stability^r with an additional numerical parameter r in section 4.2.2 of this chapter: the same notion of P-stability^r that was introduced already in Appendix B.) One of these P-stable sets is the agent's *actual* set B_W: the one that actually generates the agent's unconditional beliefs. What are the fallback positions if new recalcitrant evidence comes along? Whatever sets of worlds they may be exactly, the theory at least predicts that they will be *among the P-stable sets that would have been available also beforehand as permissible candidates for* B_W. So AGM fallback positions are nothing but possible but non-actualized choices for the agent's prior set B_W, where these choices are more cautious than (that is, supersets of) the very set B_W that is actually generating the agent's unconditional beliefs at the time. Fallback positions are ways the agent's unconditional beliefs might have been

228 There are accounts of nonmonotonic reasoning in which Consistency is not assumed either: Kraus, Lehmann, and Magidor (1990) is an example.

like if the agent had been more cautious.[229] Finally, as Theorem 18 in section 4.2.2 will demonstrate, the class of P-stable sets of probability less than 1 is ordered by the subset relation in the same way as Grove's sphere systems for AGM belief revision operators are: the notion of probabilistic stability from the previous chapters demands the formal structure of a sphere system or of a total pre-ordering of worlds all by itself.

4.1.3 Conditionalization vs Belief Revision: A Preview

The theory in this chapter will consist of: standard subjective probability theory for conditional probability, versions of the AGM postulates formulated for conditional all-or-nothing belief, and one rather weak-looking bridge principle that postulates that having a sufficiently high conditional degree of belief is a *necessary* condition for conditional belief:

$$\text{if } Bel(Y|X) \text{ then } P(Y|X) > r,$$

where 'r' will denote a threshold value again that is determined contextually in some way (but which will be at least $\frac{1}{2}$ and below 1). So this is just the left-to-right direction of a conditional version of the very Lockean thesis that occupied us in Chapter 3.[230]

In Chapter 2, adding the right-to-left (P to Bel) direction of the unconditional Lockean thesis to subjective probability theory and to the assumptions of consistency and logical closure for categorical belief was found to be equivalent to a combination of subjective probability theory and the special case of the Humean thesis on belief with threshold $r = \frac{1}{2}$. In the present chapter, we will find that adding the other—left-to-right (Bel to P)—direction of the conditional Lockean thesis to subjective probability theory and the AGM postulates on conditional belief will have the same effect of determining a stability account of belief, but this time of *conditional* belief. In this sense, the theory of the present chapter will generalize that from the previous chapters to the conditional case. Moreover, all of the postulates of this chapter taken together will entail for each X, for which $Poss(X)$ and $P(X) > 0$ are the case, an instance of the *full* Lockean thesis for conditional belief of the form

$$Bel(Y|X) \text{ if and only if } P(Y|X) \geq s_X$$

in which the Lockean threshold s_X will depend not just on Bel and P but also on X. So one will also need to count the supposed or learned proposition X to the threshold-determining context in that case, which sounds plausible: *in the light of* X, the agent might be disposed to change the Lockean threshold (e.g. X might make the agent worry more about the conceivable outcomes of some actions based on her beliefs, or the

229 Formally, this will correspond to: Representation Theorem 23 in section 4.2.3 will show that all spheres in a belief-revision sphere system are P-stable^r, and Representation Theorem 12 in section 4.2.2 will show that each of these P-stable^r sets is a potential candidate for B_W.

230 The reason why I denote the threshold here by a 'Humean' 'r' rather than a 'Lockean' 's' will become clearer later, when we will find that the Humean thesis with threshold r will be entailed by the conjunction of 'for all X, Y, if $Bel(Y|X)$ then $P(Y|X) > r$' with the other postulates of this chapter (compare Observation 16 in section 4.2.2).

like).[231] In any case: one might have thought that one could satisfy just the left-to-right direction of the conditional Lockean thesis, but as soon as one adds the AGM postulates for conditional belief into the mix (and subjective probability theory), the full Lockean thesis is a consequence. See Observation 13 in section 4.2.2 for the details. It is also possible to derive an instance of the full Lockean thesis for conditional belief in cases in which the learned or supposed proposition X contradicts the agent's present unconditional beliefs (so not $Poss(X)$, that is, $Bel(\neg X)$), but I will restrict myself to the derivation of the Lockean thesis in the $Poss(X)$ case later.

Here is a little example that may serve as an illustration of what is to come. It concerns the probability measure that we encountered already in Chapter 3 (see Example 6) and a concrete interpretation of which was given in section 3.6.

Example 7

Let $W = \{w_1, \ldots, w_8\}$ be a set of eight possible worlds. Let P be a probability measure on the power set algebra on W (the set of all propositions, that is, subsets of W), such that P is given by: $P(\{w_1\}) = 0.54$, $P(\{w_2\}) = 0.342$, $P(\{w_3\}) = 0.058$, $P(\{w_4\}) = 0.03994$, $P(\{w_5\}) = 0.018$, $P(\{w_6\}) = 0.002$, $P(\{w_7\}) = 0.00006$, $P(\{w_8\}) = 0$. Assume P to be our perfectly rational agent's degree-of-belief function at time t. Conditional degrees of belief are given by P in terms of the ratio formula.

Let the agent's set Bel of conditional beliefs at t be given as follows. Consider the following sphere system $\mathcal{X}$ of propositions:

$$\mathcal{X} = \{\{w_1\}, \{w_1, w_2\}, \{w_1, \ldots, w_4\}, \{w_1, \ldots, w_5\}, \{w_1, \ldots, w_6\}, \{w_1, \ldots, w_7\}\}$$

As we have seen already in section 3.2, these six sets are precisely those propositions X for which it holds that: X is P-stable, and if X has probability 1 with respect to P then X is the least proposition of probability 1. In section 4.2.2 I will generalize the notion of P-stability from Chapter 3 to the notion of P-stability^r that we encountered already in Appendix B, where $\frac{1}{2} \leq r < 1$. r is the threshold that will figure in the bridge principle 'for all X, Y, if $Bel(Y|X)$ then $P(Y|X) > r$' and that will be seen to coincide with the threshold r in our Humean thesis from Chapter 2. Note that it is accidental that I am putting *all* P-stable sets of probability less than 1 into $\mathcal{X}$ here: we will find that conditional belief will always be given by a sphere system of P-stable sets, but the system will not necessarily have to include *every* P-stable set of probability less than 1. It is just in the present example that I include all of them in $\mathcal{X}$.

Now, if X is a subset of W, let B_X be the intersection of X with the least member of $\mathcal{X}$ that has non-empty intersection with X if there is one; and if there is none, let B_X be the empty set. For instance, if $X = \{w_1, w_2\}$, then $B_X = X \cap \{w_1\} = \{w_1\}$, and if $X = \{w_3, w_4, w_6\}$, then $B_X = X \cap \{w_1, \ldots, w_4\} = \{w_3, w_4\}$. Intuitively, B_X should be thought of as the set of most plausible X-worlds, that is, the set of worlds that

[231] In the terminology that will be introduced later in the chapter, that number s_X will be equal to $P(B_X|X) = P(X \cap B_W|X) = P(B_W|X)$.

are most plausible amongst those that satisfy X. Accordingly, B_W is the set of worlds that are most plausible overall: in the present case, that set is $\{w_1\}$. Now let $Bel(\cdot|\cdot)$ be determined from $\mathcal{X}$ in the way that for all subsets X, Y of W,

$$Bel(Y|X) \text{ if and only if } Y \supseteq B_X.$$

In words: the agent believes Y conditional on X if and only if all most plausible X-worlds are Y-worlds.

Therefore, for example: $Bel(Y \mid \{w_1, w_2\})$ iff $Y \supseteq \{w_1\}$, and $Bel(Y \mid \{w_3, w_4, w_6\})$ iff $Y \supseteq \{w_3, w_4\}$. In words: conditional on $\{w_1, w_2\}$, the agent believes $\{w_1\}$ and all of its supersets, and given $\{w_3, w_4, w_6\}$ the agent believes $\{w_3, w_4\}$ and all of its supersets. Or formulated differently again: $B_{\{w_1,w_2\}} = \{w_1\}$ is the logically strongest believed proposition given $\{w_1, w_2\}$, while $B_{\{w_3,w_4,w_6\}} = \{w_3, w_4\}$ is the logically strongest believed proposition given $\{w_3, w_4, w_6\}$. $B_W = \{w_1\}$ is the logically strongest believed proposition overall, and if we think of absolute or unconditional belief to be given by $Bel(\cdot|W)$ again, then $Bel(Y)$ holds just in case $Y \supseteq \{w_1\}$. Moreover, reconsidering the case of a piece of evidence $E = \{w_5, \ldots, w_8\}$ as described in section 3.6, it holds that $B_E = E \cap \{w_1, \ldots, w_5\} = \{w_5\}$, so $Bel(Y|E)$ if and only if $Y \supseteq \{w_5\}$. This is just as was promised in section 3.6.

With P and (conditional) *Bel* determined in that way, the following can be shown: P satisfies the axioms of subjective probability. *Bel* satisfies the versions of the AGM postulates of belief that will be formulated later (in terms of conditional belief in propositions). *Bel* and P taken together satisfy the left-to-right direction of the conditional version of the Lockean thesis with a threshold of $r = \frac{1}{2}$: if $Bel(Y|X)$ then $P(Y|X) > r = \frac{1}{2}$. Finally, the unconditional belief set $Bel(\cdot)$ that is determined by $Bel(\cdot|W)$ satisfies, jointly with P, the Humean thesis HT^r from Chapter 2, and the two of them satisfy all of our postulates in Chapter 3 (which, taken together, had turned out to be equivalent to the conjunction of subjective probability theory and the Humean thesis HT^r with a threshold of $r = \frac{1}{2}$).

In turn, it will follow from the representation results in this chapter that if P and *Bel* satisfy the axioms of probability, AGM, and the left-to-right direction of the Conditional Lockean thesis, then *Bel* can always be represented by a sphere system of P-stable^r sets like $\mathcal{X}$ above. In the present example, $\mathcal{X}$ happened to be the set of *all* P-stable^r sets with the *least possible* threshold $r = \frac{1}{2}$, but more generally such a sphere system may consist of only *some* P-stable^r sets, and the threshold may be chosen to lie *strictly between* $\frac{1}{2}$ and 1.

It is possible to derive all of these claims from the theory that will be developed later in this chapter, and I will return to this Example 7 at various places in the chapter.

With rational conditional degrees of belief and rational conditional all-or-nothing beliefs being in place, the two diachronic norms from the last two sections translate these rational conditional beliefs into dispositions for how to change one's beliefs rationally. For instance, in the example before (and as required by section 3.6), if the agent were confronted with $E = \{w_1, \ldots, w_5\}$ as a new piece of evidence, then the agent's posterior unconditional belief state after updating on E would be given by $P_{new}(\cdot)$ ($= P(\cdot|E)$) and $Bel_{new}(\cdot)$ (where $Bel_{new}(\cdot)$ iff $Bel(\cdot|E)$). For instance, reconsidering the auxiliary hypothesis $H = \{w_1, w_3, w_7, w_8\}$ from section 3.6, it would follow that $P_{new}(H) = 0.003$ on the one hand, and not $Bel_{new}(H)$ on the other; indeed, $Bel_{new}(\neg H)$. That would be so, since $P(H|E) = 0.003$, and $B_W^{new} = B_E = \{w_5\}$ is a subset of $\neg H$. So far as the all-or-nothing side is concerned, updating with E in this example amounts to a case of proper belief revision, because the agent believed $\neg E$ initially: $Bel(\neg E)$ (that is, $Bel(\neg E|W)$). But of course the theory in section 4.2 will also cover belief revision by *expansion*, such that the agent updates on a proposition that is consistent with the agent's logically strongest believed proposition, that is, with B_W (and hence with all believed propositions).

The following will follow from the results (see Observation 17 in section 4.2.2): if $P(\cdot)$ and $Bel(\cdot)$ jointly satisfy the Humean thesis HT^r, then updating on a piece of evidence that has positive prior probability and that is consistent with B_W leads to a posterior degree-of-belief function $P_{new}(\cdot)$ and a posterior unconditional belief set $Bel_{new}(\cdot)$ that will jointly satisfy the Humean thesis HT^r again. In a nutshell: *updating rationally on each side preserves the Humean thesis*. Call this *Robustness Persistence*.[232] If degrees of belief and beliefs cohere with each other initially, and some new piece of evidence is being learned—by conditionalization on the one hand and by belief expansion on the other—then the resulting degrees of belief and categorical beliefs will also cohere afterwards.

It is also possible to derive a version of Robustness Persistence even for the case in which *the evidence contradicts the agent's present all-or-nothing beliefs* (but where the evidence still has positive probability). The corresponding result would thus belong in section 4.2.3 concerning conditional belief in general (or belief revision proper). But I will content myself later just with deriving robustness persistence for the case of belief *expansion*, which will belong in section 4.2.2 (a section devoted just to the simpler case of restricted conditional belief, or, dynamically, to belief expansion).

By iterating updates on both sides, one also gets an iterative form of Robustness Persistence: let $E_1, E_2, E_3, \ldots$ be a stream of evidence, such that $P(E_1 \cap E_2 \cap E_3 \cap \ldots) > 0$ and $(E_1 \cap E_2 \cap E_3 \cap \ldots) \cap B_W \neq \varnothing$ (so the evidence taken together is consistent with what the agent believes). Assume that update proceeds probabilistically by iterated conditionalization: $P_{E_1} = P(\cdot|E_1)$, $[P_{E_1}]_{E_2} = P(\cdot|E_1 \cap E_2)$,... And suppose on the categorical side that update proceeds by means of iterated belief revision (here, by belief expansion): $[Bel * E_1](Y)$ iff $Bel(Y|E_1)$, $[[Bel * E_1] * E_2](Y)$ iff $Bel(Y|E_1 \cap E_2)$,...[233] In total, there will be two learning processes going on simultaneously, one for quantitative belief and the other one for qualitative belief:

[232] I am grateful to Chris Gauker for suggesting this term.

[233] In such a case it holds by AGM (but stated in my terminology): $Bel(Y|E_1)$ iff $Y \supseteq B_{E_1} = B_W \cap E_1$, $[[Bel * E_1] * E_2](Y)$ iff $Bel(Y|E_1 \cap E_2)$ iff $Y \supseteq B_{E_1 \cap E_2} = B_W \cap (E_1 \cap E_2)$, and so on.

$$\begin{array}{lllll} P & \mapsto P_{E_1} & \mapsto [P_{E_1}]_{E_2} & \mapsto [[P_{E_1}]_{E_2}]_{E_3} & \mapsto \cdots \\ Bel & \mapsto [Bel * E_1] & \mapsto [[Bel * E_1] * E_2] & \mapsto [[[Bel * E_1] * E_2] * E_3] & \mapsto \cdots \end{array}$$

It will then also follow from Observation 17 in section 4.2.2 that each pair $\langle [P_{E_1}] \ldots, [Bel * E_1] * \ldots \rangle$ will satisfy the Humean thesis if the initial pair $\langle P, Bel \rangle$ does. In addition to synchronic coherence, the stability theory of belief as developed in this chapter also entails a diachronic form of coherence between rational belief and rational degrees of belief.

For how long may such chains of iterated parallel updates of degrees of belief and beliefs go on? *Until the context changes*, where the context includes (or determines) the relevant Humean threshold r and the underlying partitioning of possibilities (about the latter see section 3.3). Once the context has changed, new streams of evidence will trigger new chains of iterated update by conditionalization and belief revision, respectively. And so forth. But *within* each such context, updating rational degrees of belief and rational beliefs simultaneously will always preserve coherence as being given by the Humean thesis for some Humean threshold r.[234]

[234] What if evidence comes along that does not correspond to any set of worlds in W on which *Bel* is defined? In particular: what if the context has imposed a certain coarse-grained partitioning π of the worlds in W, such that conditional all-or-nothing belief is given on that partition π, but where the evidence E that the agent is facing is a subset of W that does not coincide with any union of partition cells in π? (For a detailed discussion of partitions of W and of the sensitivity of belief to partitioning, see section 3.3.)

I see two possible ways of coping with this kind of problem. (i) Either the proposition E is approximated by an E' that does live in the more coarse-grained algebra of propositions that is given by the partition π in question. So E is adapted to π. (ii) Or the agent's present partition π of possibilities gets refined to a new partition π'. That is, π is adapted to E. (Which means that the agent's context of reasoning changes.) (i) is much like approximating a (black-and-white) high-resolution image by a low-resolution image. For instance, E' might be defined in the way that in precisely those cases in which E has non-empty intersection with a partition cell, the cell is counted as belonging to E' as a whole; or in precisely those cases in which E overlaps with more than a certain fraction of the partition cell (being measured in terms of P), the cell is counted as belonging to E'; or in precisely those cases in which E covers a partition cell completely, the cell is counted as belonging to E'; or the like. The exact approximation scheme may itself depend on context, e.g. on how cautious or brave the agent wants to be. It might even depend on 'where' in W it is to be applied: e.g. within the logically strongest believed proposition B_W or outside of it. Sometimes the approximation will need to take into account the agent's degree-of-belief function and not just the algebras in question (as in the second approximation strategy mentioned before); in other cases it might not be necessary to invoke P in the course of approximation. In any case, approximating E by E' will normally distort the evidence to some extent, but there might also be bounds on the error depending on the approximation strategy selected. Much the same is true of any kind of rounding or digitization process, or of any instance of describing a real-world situation in a simple language with restricted vocabulary. In contrast, (ii) corresponds to changing one's 'resolution' in order to match that of the given image (the evidence), or to increase the expressiveness of one's language for that purpose. Formally, this might be done by defining π to be the most coarse-grained partition that is more fine-grained than both π and $\{E, \neg E\}$. According to the present theory, if this act of refinement takes place within B_W, then some of the agent's beliefs might have to give way to suspension of judgement (as discussed in section 3.3).

It is worth mentioning that even a purely Bayesian agent may face the very same type of problem: how should a probability measure be updated on evidence that is not a member of the algebra on which P is defined? The possible answers to this question are analogous to (i) and (ii) above. With respect to (i), the same considerations on approximation strategies, errors of approximation, and so on will apply. With respect to (ii), if the original algebra is refined in order to make room for the evidence, it will be underdetermined what the probability measure ought to look like on the more fine-grained sample space

4.1.4 Some Closely Related Theories

Let me conclude this introductory part of Chapter 4 with a couple of remarks on some closely related theories.

I have already mentioned Skyrms's work on resilience in Chapter 3. Indeed, amongst philosophical theories, it is the one that comes closest to the theory that is to be developed in section 4.2. Skyrms (1977, 1980) investigates the notion of objective chance and its applications in terms of a more fundamental concept: the *resiliency* of the (subjective) probability of a statement. In the simplest case—Skyrms's theory is actually much more general than that—the degree of resiliency of a (non-probabilistic) statement A is the infimum of the set of conditional probabilities of the form $P(A|B)$ where 'B' varies over all statements that are consistent with A in a given language. As we will see in section 4.2.2, the concept that will be fundamental for my own theory is a categorical notion of stability with respect to a probability measure P and a threshold r: a proposition X will be defined P-stable^r if $P(X|Y) > r$ for all Y that are consistent with X and have positive probability. Clearly, the two concepts are closely related, even though the underlying aims of the two theories differ: Skyrms's is to explicate objective chance, mine is to explicate (the coherence of) rational belief. It will follow from the postulates of my theory that the logically strongest proposition that is believed by a perfectly rational agent must be P-stable^r. While the results below will be new—in particular the two Representation Theorems 12 and 23—some of Skyrms's results overlap with mine: in particular, Note (2) on pp. 712–13 of Skyrms (1977), in which Skyrms is dealing with *acceptance*, and his theorem on p. 713 go some way towards Theorem 12 below.

Here is how the notion of P-stability^r figures in Example 7 from before:

Example 8 (Example 7 and P-Stability^r)
For P as before and $r = \frac{1}{2}$, it will turn out that the non-empty P-stable$^{\frac{1}{2}}$ sets will be:

$$\{w_1\}, \{w_1, w_2\}, \{w_1, \ldots, w_4\}, \{w_1, \ldots, w_5\}, \{w_1, \ldots, w_6\}, \{w_1, \ldots, w_7\}, \{w_1, \ldots, w_8\},$$

only the last two of which have probability 1. For instance, $\{w_1\}$ is P-stable$^{\frac{1}{2}}$ since for every Y with $Y \cap \{w_1\} \neq \varnothing$ and $P(Y) > 0$, it holds that $P(\{w_1\} \mid Y) > \frac{1}{2}$. On the other hand, $\{w_1, w_2, w_3\}$ is not P-stable$^{\frac{1}{2}}$, because e.g. $\{w_3, \ldots, w_8\} \cap \{w_1, w_2, w_3\} = \{w_3\} \neq \varnothing$ and $P(\{w_3, \ldots, w_8\}) > 0$, but $P(\{w_1, w_2, w_3\} \mid \{w_3, \ldots, w_8\}) = 0.49152\ldots < \frac{1}{2} = r$. The proposition $\{w_1, \ldots, w_7\}$ is the least P-stable$^{\frac{1}{2}}$ sets of probability 1. $\{w_1, \ldots, w_8\}$ has probability 1, too, but it is not least among the sets of probability 1—it includes a 'zero set' as a subset, that is, $\{w_8\}$ (which has probability 0 as given by P). The sphere system in Example 7 was such that it included only P-stable$^{\frac{1}{2}}$ sets, and it

(relative to which the original sample space is but a coarse-grained partition). So questions like these are important for both *Bel* and *P*, and much more should be said about them. I am grateful to Teddy Seidenfeld for urging me to address them at all, if only too briefly.

included precisely one P-stable$^{\frac{1}{2}}$ set of probability 1, that is, the least set of probability 1 overall ($\{w_1, \ldots, w_7\}$). It will follow from the postulates in section 4.2.3 about general conditional belief that a perfectly rational agent's sphere system of doxastic fallback positions only includes one P-stable$^{\frac{1}{2}}$ set of probability 1: the least set of probability 1 (the existence of which will follow from my overall postulates for P and *Bel*).

With $r = \frac{3}{4}$, the non-empty P-stable$^{\frac{3}{4}}$ sets will turn out to be:

$$\{w_1, w_2\}, \{w_1, \ldots, w_4\}, \{w_1, \ldots, w_5\}, \{w_1, \ldots, w_6\}, \{w_1, \ldots, w_7\}, \{w_1, \ldots, w_8\}.$$

For instance, $\{w_1, w_2\}$ is P-stable$^{\frac{3}{4}}$ since for every Y with $Y \cap \{w_1, w_2\} \neq \varnothing$ and $P(Y) > 0$, it holds that $P(\{w_1, w_2\} \mid Y) > \frac{3}{4}$.

The logically strongest believed proposition in Example 7 was $\{w_1\}$, and as we have seen it is indeed P-stable^r for $r = \frac{1}{2}$.

In the literature in computer science, Snow's (1998, 1999, 2003) account of *atomic bound systems* and Benferhat et al.'s (1999) related account of so-called *big-stepped probabilities* deal with a special case of the theory here. Both of them consider special probability measures P on a finite set of atoms (singletons of worlds) for which the following *atomic bound condition* is satisfied: there is a strict total order $<$ so that the probability of an atom a is greater than the sum of probabilities of all atoms b that lie above a in that strict total order. Probability measures that satisfy this condition may be seen to induce rational nonmonotonic inference relations (which are equivalent to AGM belief revision operators, and which will correspond to our conditional beliefs below). One can also prove easily that for all sets X, Y of atoms, if X is non-empty, the uniquely determined $<$-least member of X is a member of Y if and only if $P(Y|X) > \frac{1}{2}$. Probability measures that satisfy the atomic bound condition can be generated by distributing probabilities over a given strict total order in exponential steps: e.g. the $<$-greatest atom receives probability $\frac{1}{2^n-1}$, the next one below gets probability $\frac{2}{2^n-1}$, the next one probability $\frac{4}{2^n-1}$, and so on, down to the $<$-least atom overall, which receives the largest probability, $\frac{2^{n-1}}{2^n-1}$.

In a similar manner, my second representation theorem for conditional belief—Theorem 23 in section 4.2.3—will show that, given a probability measure P, the postulates on conditional belief in section 4.2.3 will always determine a conditional belief set that will correspond to a unique total pre-order of worlds or to a sphere system that is given by a sequence of P-stable^r sets: for one can show that every sequence of P-stable^r sets of probability less than 1 is a well-ordering with respect to the subset relation, from which it is going to follow that every world of positive probability can be assigned an ordinal rank according to its first 'appearance' in a P-stable^r set in that well-ordering (compare Theorem 18). Furthermore, in the finite case, the probability of (a singleton set of) a world w will be greater than the sum of the probabilities of (the singleton sets of) all worlds w' that lie above w in that ordinal ranking, much as in the atomic bound condition above. This property, which I will call the 'Sum Condition', is a generalization of what was called the 'Outclassing Condition' in Appendix B and section 3.2. See Observation 24 at the end of section 4.2.3 for the details.

The essential difference between my approach in this chapter and Snow's and Benferhat et al.'s is that I will be able to do this *for all probability measures P whatsoever*, and not just for special ones. This will be possible because the theory will not demand total partial orders of worlds but total *pre-orders* or *rankings* of worlds, which allow for ties between worlds. That is: different worlds whose singletons have different probabilities may nevertheless have the same rank on the conditional belief side according to my theory.

For instance:

Example 9 (Example 7 and Total Pre-Orders of Worlds)
For P as in Example 7, $r = \frac{1}{2}$ again, and conditional *Bel* as in Example 7, it will turn out that the ranking in question is given by:

$$w_1 \prec w_2 \prec w_3, w_4 \prec w_5 \prec w_6 \prec w_7.$$

(A world such as w_8 is omitted from any such ranking because of the probability of its singleton set being zero.) So here one can observe a tie between the worlds w_3 and w_4. The rank of a world is determined by means of the least P-stable^r set (in the sphere system that corresponds to *Bel*) in which the world occurs:[235] e.g. the rank of w_1 is 0 because, as pointed out before, $\{w_1\}$ is the least non-empty P-stable$^{\frac{1}{2}}$ set in the sphere system; accordingly, the rank of w_2 is 1, the rank of w_3 and w_4 is 2, and so forth.

And for $r = \frac{3}{4}$, with the sphere system for conditional belief being given by the set of all P-stable$^{\frac{3}{4}}$ sets (other than W itself), we will have:

$$w_1, w_2 \prec w_3, w_4 \prec w_5 \prec w_6 \prec w_7.$$

Note that in the case $r = \frac{1}{2}$, $P(\{w_3\})$ is greater than $P(\{w_5\}) + P(\{w_6\}) + P(\{w_7\}) + P(\{w_8\})$, and also $P(\{w_4\})$ is greater than $P(\{w_5\}) + P(\{w_6\}) + P(\{w_7\}) + P(\{w_8\})$, but neither $P(\{w_3\}) > P(\{w_4\}) + P(\{w_5\}) + P(\{w_6\}) + P(\{w_7\}) + P(\{w_8\})$ nor $P(\{w_4\}) > P(\{w_3\}) + P(\{w_5\}) + P(\{w_6\}) + P(\{w_7\}) + P(\{w_8\})$ is the case. That is exactly why neither of w_3 and w_4 happens to be ranked strictly below the other. Section 4.2.2 will make precise why and how P-stable^r sets determine total pre-orders of worlds, and the section will also present the simple algorithm by which P-stable^r sets can be computed efficiently in terms of inequalities between probabilities. None of these results on P-stability^r nor the representation theorems below are contained in Snow's or Benferhat et al.'s work.

There are also related theories and notions of robustness or stability in other areas, such as in statistics (which partially inspired Skyrms's theory) or in game theory; but

[235] I have mentioned such rankings of worlds in earlier chapters: see sections 2.3.1 and 3.2.

to the best of my knowledge, all of them differ from what will follow here. E.g. in game theory, the concept of *strong belief* (see Battigalli 2002) is nothing but certainty of a hypothesis under all histories consistent with the hypothesis (where this is spelled out in the context of primitive conditional probability measures). So an agent strongly believes X if she is certain that X is the case at the beginning of a game, and continues to do so as long as X is not falsified by the evidence. The theory of belief in section 4.2 will be similar to that; however, belief will not entail certainty, that is, probability 1, but merely a probability above r.

4.2 A Stability Theory of Conditional Belief and Belief Dynamics: The Formal Details

For the rest of this chapter, the goal will be to enumerate, and to study, a couple of postulates about quantitative and qualitative conditional belief. I will assume that we are dealing with a perfectly rational agent again who, at a fixed point in time, has conditional belief states of both the qualitative (*Bel*) and the quantitative (P) types available, so that these states obey the given postulates. Each postulate will express a constraint either on the referent of 'P' or on the referent of '*Bel*' or on the referents of 'P' and '*Bel*' simultaneously. When I will state theorems, 'P' and '*Bel*' will be variables, so that I will be able to say: 'for all P, *Bel*, it holds that P and *Bel* satisfy so-and-so if and only if so-and-so is the case.'

Let W be our given non-empty set of possible worlds, which one may think of again as a set of logically possible worlds for some kind of language (as in the previous chapters).

Assume that at a given point in time our agent is capable in principle of entertaining all and only propositions (sets of worlds) in a class $\mathfrak{A}$ of subsets of W. For simplicity, and just as in the previous chapters, I will assume that class $\mathfrak{A}$ to be the power set of W: the set of *all* propositions over W, that is, all subsets of W. But one can show that all of what I will be doing below would also go through if $\mathfrak{A}$ were merely assumed to be a so-called σ-algebra over W, that is, a class of subsets of W such that: W and $\varnothing$ are members of $\mathfrak{A}$; if $X \in \mathfrak{A}$ then the relative complement of X with respect to W, $W \setminus X$, is also a member of $\mathfrak{A}$; and finally if all of $X_1, X_2, \ldots, X_n, \ldots$ are members of $\mathfrak{A}$, then $\bigcup_{n \in \mathbb{N}} X_n \in \mathfrak{A}$. (It follows then for any σ-algebra that $\mathfrak{A}$ is closed under finite unions and under countable intersections, too.) σ-algebras are the standard choice in probability theory so far as the domains of probability measures are concerned, and the theory here is prepared for being applied also in cases in which $\mathfrak{A}$ is only assumed to be a σ-algebra. It is merely a matter of convenience that I will restrict myself to the special case in which $\mathfrak{A}$ is the power set algebra of all subsets of W (which is also a σ-algebra, of course). At some points I will add some footnotes in which I will point out that certain formal constructions will generate propositions in $\mathfrak{A}$ even in cases in which $\mathfrak{A}$ had merely been assumed to be a σ-algebra.[236] When, in the following, I speak of $\mathfrak{A}$ as the *algebra of propositions* over W, it should be kept in mind that that algebra will simply be the set of all subsets of W again but that the formal results below would go through also in the more general case of an arbitrary σ-algebra $\mathfrak{A}$. That is why I will keep referring to an algebra $\mathfrak{A}$ rather than quantifying over all subsets of W directly.

As in previous chapters, I will extend the standard logical terminology that is normally defined only for formulas or sentences to the propositions in $\mathfrak{A}$: so when I speak of a proposition as a logical truth I actually have in mind the unique proposition W (the top element of $\mathfrak{A}$); the empty set (the bottom element of $\mathfrak{A}$) is the unique logical falsity; when I say that a proposition is consistent I mean that it is non-empty; and one proposition X logically entailing another one Y coincides with X being a subset of Y, that is, every world that satisfies X (or is a member of X) also satisfies Y (is a member of Y). When I refer to the negation of a proposition X, I refer to its complement relative to W (and I will sometimes denote it by '$\neg X$'); the conjunction of two propositions is of course their intersection; and so on. I shall speak of conjunctions and disjunctions of propositions even in cases of *infinite* intersections or unions of propositions.

4.2.1 *Probabilistic Postulates*

Let me start by recapitulating the postulates on rational degrees of belief.

Let P be our ideal agent's degree-of-belief function at the given point in time. Following Assumption 9 from section 1.5 in Chapter 1 (the Bayesian take on quantitative belief), I postulate:

> P1 (Probability) P is a probability measure on $\mathfrak{A}$, that is, P has the following properties:
> $P: \mathfrak{A} \to [0,1]$; $P(W) = 1$; and P is finitely additive: if X_1, X_2 are pairwise disjoint members of $\mathfrak{A}$, then $P(X_1 \cup X_2) = P(X_1) + P(X_2)$.
> Conditional probabilities are introduced by: $P(Y|X) = \frac{P(Y \cap X)}{P(X)}$ whenever $P(X) > 0$.

In particular, the conditional probability $P(Y|W)$ is nothing but the absolute probability $P(Y)$ of Y again.

I will *not* add *Countable Additivity* (or σ-additivity) as another postulate. Countable Additivity would say: if $X_1, X_2, \ldots, X_n, \ldots$ are pairwise disjoint members of $\mathfrak{A}$, then $P(\bigcup_{n \in \mathbb{N}} X_n) = \sum_{n=1}^{\infty} P(X_n)$.[237] While countable additivity is indispensable in mathematical theories such as measure theory (or the theory of integration), it is often not assumed in Bayesian epistemology.[238] But of course it is still permissible to think the agent's degree-of-belief function P to be some σ-additive probability measure, and

[236] These notes are: nn. 251, 252, 253, 264.

[237] In this respect, the present chapter will differ from Leitgeb (2013a) on which it is based, since in Leitgeb (2013a) I did assume Countable Additivity.

[238] For more on the role(s) of σ-additivity, see e.g. Schurz and Leitgeb (2008).

accordingly to assume $\mathfrak{A}$ to be a σ-algebra, if one likes to apply the theory below to such a case. All of the results will go through also for a σ-additive P on a σ-algebra $\mathfrak{A}$ as long as the required additional assumptions in the subsequent sections are satisfied.

4.2.2 Restricted Conditional Belief and a Bridge Postulate

Let me now turn to conditional all-or-nothing belief. At first I will only deal with conditional belief restricted to cases in which the 'given' proposition is consistent with everything that the agent believes unconditionally: cases of $Bel(Y|X)$ in which $\neg X$ is not believed unconditionally by the agent. I am calling them cases of *restricted conditional belief*. Hence, if the agent were to receive evidence X, and if she were to revise her beliefs in line with $Bel(Y|X)$ and thus end up believing Y, then this would be a case of belief revision by expansion. Afterwards, in section 4.2.3, I will deal with *conditional belief in general*: cases of $Bel(Y|X)$ without any further qualitative restrictions on X. Belief revision in that case will be either belief expansion again or belief revision proper (revising one's belief given evidence that one *dis*believed beforehand). Accordingly, the postulates in section 4.2.3 will include those of the present section 4.2.2 as a special case.

BELIEF POSTULATES

Each belief that our agent holds at a given point in time is assumed to have a set in $\mathfrak{A}$ as its propositional content. In other words: quantitative and qualitative beliefs take their contents from the same space of propositions. As usual, by '*Bel*' I am going to denote the class of propositions that our perfectly rational agent believes to be true at the time. Instead of writing '$Y \in Bel$', I will rather say: $Bel(Y)$; and I call *Bel* the agent's *belief set*.

In line with the synchronic part of Assumption 8 from section 1.5, *Bel* is assumed to satisfy the following postulates:

1. $Bel(W)$.
2. For all $Y, Z \in \mathfrak{A}$: if $Bel(Y)$ and $Y \subseteq Z$, then $Bel(Z)$.
3. For all $Y, Z \in \mathfrak{A}$: if $Bel(Y)$ and $Bel(Z)$, then $Bel(Y \cap Z)$.

So the agent's belief set is closed under logic. Actually, I am going to strengthen the principle on finite conjunctions of believed propositions to the case of the conjunction of *all* believed propositions *whatsoever*:

4. For $\mathcal{Y} = \{Y \in \mathfrak{A} \mid Bel(Y)\}$ $(= Bel)$, $\bigcap \mathcal{Y}$ is a member of $\mathfrak{A}$,[239] and $Bel(\bigcap \mathcal{Y})$.

$\bigcap \mathcal{Y}$ is simply the intersection of all members of $\mathcal{Y}$.

Principle (4) involves a good deal of idealization even in the case of a perfectly rational agent: (4) is much like assuming a system of arithmetic to be closed under

[239] $\bigcap \mathcal{Y}$ being a member of $\mathfrak{A}$ is trivial for our power set algebra $\mathfrak{A}$, but it would amount to a proper constraint if e.g. $\mathfrak{A}$ had merely been assumed to be some σ-algebra.

the infinitary ω-rule,[240] which may well yield a set of theorems that is not recursively axiomatizable. On the other hand, if $\mathfrak{A}$ is finite, then (4) simply reduces to the case of the closure of belief under finitely many conjuncts again. In any case, (4) has the following obvious and pleasant consequence: there is a *least set* (a strongest proposition) Y, such that $Bel(Y)$. Y is of course just the conjunction of all propositions believed by the agent. As in the previous chapters, I will denote this very proposition by 'B_W'. The main reason I presuppose (4) is that it will enable me in this way to represent the totality of the agent's beliefs in terms of a unique proposition or a unique set of possible worlds. In the semantics of doxastic or epistemic logic, this set B_W would correspond to the set of doxastically accessible worlds from the viewpoint of the agent. Accordingly, using the terminology that is quite common in belief revision theory or nonmonotonic reasoning, one might think of the members of B_W as the most plausible candidates for what the actual world might be like, from the viewpoint of the agent at the given time.

Finally, I add

5. (Consistency) $\neg Bel(\varnothing)$,

which was also contained in Assumption 8 of Chapter 1. So much for belief if taken *unconditionally*.

But I will require more than just categorical belief in that sense—indeed, that is the key move in this chapter. Let us assume that our perfectly rational agent also holds *conditional* beliefs, that is, beliefs conditional on certain propositions in $\mathfrak{A}$. In this extended context, *Bel* itself should now be regarded as a class of *ordered pairs* of members of $\mathfrak{A}$, rather than as a set of members of $\mathfrak{A}$ as before. Instead of '$\langle Y, X\rangle \in Bel$', we may simply say again: $Bel(Y|X)$. And we may identify the agent's absolute or unconditional belief set from before with the class of propositions that the agent believes to be true *conditional on the tautological proposition* W, just as this is the case with absolute vs conditional probability.

As mentioned before, in the present section 4.2.2 I will be interested only in conditional beliefs in Y given X where X is consistent with everything that the agent believes absolutely (or conditionally on W) at that time; equivalently: where X is consistent with B_W; equivalently: where $\neg X$ is not believed by the agent. (We will see later that all of these conditions are pairwise equivalent.) That case is much easier to handle than the case of conditional belief in general. Accordingly, the postulates in the present section will be weaker than the postulates in section 4.2.3, and it will be important to observe that even these weaker postulates will allow us to derive a substantial representation theorem for belief.

For every $X \in \mathfrak{A}$ that is consistent with what the agent believes, the set of propositions conditional on X will be assumed to satisfy postulates that impose

240 The ω-rule says: from the infinite set of premises $A[0], A[1], A[2], \ldots$ taken together, the universally quantified formula $\forall n A[n]$ is derivable.

constraints of the same type as (1)–(4) did before for absolute beliefs. In order to make clear that I am dealing only with suppositions that are consistent with what the agent believes unconditionally, I will add conditions of the form '$\neg Bel(\neg X | W)$' (or '$Poss(X)$' or '$\neg Bel(\neg X)$', if one prefers) antecedently, when I state these postulates:[241]

B1 (Reflexivity) If $\neg Bel(\neg X | W)$, then $Bel(X|X)$.

B2 (One Premise Logical Closure)

If $\neg Bel(\neg X | W)$, then for all $Y, Z \in \mathfrak{A}$: if $Bel(Y|X)$ and $Y \subseteq Z$, then $Bel(Z|X)$.

B3 (Finite Conjunction)

If $\neg Bel(\neg X | W)$, then for all $Y, Z \in \mathfrak{A}$: if $Bel(Y|X)$ and $Bel(Z|X)$, then $Bel(Y \cap Z|X)$.

B4 (General Conjunction)

If $\neg Bel(\neg X | W)$, then for $\mathcal{Y} = \{Y \in \mathfrak{A} \mid Bel(Y|X)\}$, it holds that $\bigcap \mathcal{Y}$ is a member of $\mathfrak{A}$, and $Bel(\bigcap \mathcal{Y}|X)$.

Or in other words: for every X, such that $\neg Bel(\neg X | W)$, the set $Bel_X = \{Y \mid Bel(Y|X)\}$ is a belief set in the sense of conditions (1)–(4) from before.

However, I will still assume the Consistency postulate to apply only to absolute beliefs or beliefs conditional on W at this point (but a version of Consistency conditional on any X with $\neg Bel(\neg X | W)$ will turn out to be derivable later). So, just as in the case of (5), I only demand:

B5 (Consistency) $\neg Bel(\varnothing | W)$.

Assuming B1 is unproblematic at least under a suppositional manifestation of conditional belief: under the supposition of X, our ideally rational agent must hold X true *in the context of the supposition of* X. This is much like in conditional proofs in which the statement that was first assumed may then also be concluded. B3 above is really redundant in view of B4 and B2, but I shall keep it for the sake of continuity.

As before, B4 now entails that for every $X \in \mathfrak{A}$ for which $\neg Bel(\neg X | W)$ is the case there is a *least set* (a strongest proposition) $Y \in \mathfrak{A}$, such that $Bel(Y|X)$, which by B1 must be a subset of X. For any such given X, I will denote this very proposition by: B_X. For all $Y \in \mathfrak{A}$ it holds then that: $Bel(Y|X)$ iff $Y \supseteq B_X$. From left to right this is by the definition of 'B_X', and from right to left it is in view of the definition of 'B_X', B4 (hence $Bel(B_X|X)$), and B2 combined. So determining B_X suffices in order to pin down completely our agent's beliefs conditional on X.

By B5, W itself is such that $\neg Bel(\neg W | W)$ (since $\neg W$ equals $\varnothing$), hence all of B1–B4 apply to $X = W$ unconditionally, and by B5 again it holds that B_W must be non-empty.

[241] I am abusing notation here a bit. In '$\neg Bel(\neg X | W)$', I use the formal negation symbol '$\neg$' both outside and inside of the *Bel* context, where really the outer occurrence of the negation symbol should be an informal 'not'. This is just for brevity and readability. But it should be understood that only the second negation symbol expresses negation or complement for propositions, that is, subsets of W.

With $X = W$, the notation 'B_X' is consistent with the notation 'B_W', and for all $Y \in \mathfrak{A}$ it holds that: $Bel(Y|W)$ iff $Y \supseteq B_W$.

So far there have not been any postulates on how belief sets conditional on *different propositions* relate to each other logically. At this point I demand just one such condition to be satisfied that corresponds to the standard AGM postulates K*3 and K*4 (Inclusion and Preservation) on belief revision taken together. B_W will be the propositional counterpart of AGM's syntactic belief set K, and belief revision in the sense of AGM (which reduces to expansion in this case) will be replaced by conditional belief:

B6 (Restricted $Bel(\cdot|\cdot)$ / Expansion)
For all $Y \in \mathfrak{A}$ such that $\neg Bel(\neg Y|W)$:
for all $Z \in \mathfrak{A}$, $Bel(Z|Y)$ if and only if $Z \supseteq Y \cap B_W$.

In words: if the negation of the proposition Y is not believed, then the agent believes Z conditional on Y if and only if Z is entailed by the conjunction of Y with B_W.

B6 is not independent of the previous postulates—in fact, it entails some of them, but that should not concern us.

There is an easy but helpful reformulation of B6. As we have seen before, because 'B_W' denotes the least (unconditionally) believed proposition, it holds that for all $Y \in \mathfrak{A}$: $Bel(Y|W)$ iff $Y \supseteq B_W$, and therefore also for all $Y \in \mathfrak{A}$: $Bel(\neg Y|W)$ iff $\neg Y \supseteq B_W$. Thus, $\neg Bel(\neg Y|W)$ iff $\neg Y = [W \setminus Y] \not\supseteq B_W$, which yields immediately: $\neg Bel(\neg Y|W)$ iff $Y \cap B_W \neq \varnothing$. For that reason, instead of qualifying the postulates by means of '$\neg Bel(\neg X|W)$', we may just as well do so by means of '$X \cap B_W \neq \varnothing$' (and indeed I will do so). And for the same reason, we can reformulate B6 as follows:

B6 (Restricted $Bel(\cdot|\cdot)$ / Expansion)
For all $Y \in \mathfrak{A}$ such that $Y \cap B_W \neq \varnothing$:
for all $Z \in \mathfrak{A}$, $Bel(Z|Y)$ if and only if $Z \supseteq Y \cap B_W$.

In line with the sphere semantics for AGM, B6 can be justified based on total plausibility pre-orders or rankings of possible worlds. Say, a conditional belief expresses that the most plausible of the antecedent-worlds are among their consequent-worlds. Then if some of the most plausible worlds overall (the worlds in B_W) are Y-worlds, these worlds must be precisely the most plausible Y-worlds, and therefore in that case the most plausible Y-worlds are Z-worlds if and only if all the most plausible worlds overall that are Y-worlds are Z-worlds. Which is B6 as formulated above.

From our previous considerations on $Bel(Z|W)$ being equivalent with $Z \supseteq B_W$, it is clear that this is yet another equivalent way of stating B6:

B6 (Restricted $Bel(\cdot|\cdot)$ / Expansion)
For all $Y \in \mathfrak{A}$, such that for all $Z \in \mathfrak{A}$, if $Bel(Z|W)$ then $Y \cap Z \neq \varnothing$:
for all $Z \in \mathfrak{A}$, $Bel(Z|Y)$ if and only if $Z \supseteq Y \cap B_W$.

$\neg Y$ not being believed (or Y being possible) as well as Y being consistent with B_W are each equivalent to Y being consistent with *every* proposition that is believed by the agent unconditionally. That is what was applied in order to derive this reformulation of B6.

However formulated, B6 is the crucial postulate for AGM belief revision in the case in which revision reduces to expansion on propositional information that is not ruled out by what the agent believes already.

Let me mention some consequences of B6. As we know, we have anyway that for all X with $\neg Bel(\neg X|W)$ and for all $Y \in \mathfrak{A}$:

$$Bel(Y|X) \text{ if and only if } Y \supseteq B_X.$$

Now with B6 it also follows that for such X and for all $Y \in \mathfrak{A}$:

$$Bel(Y|X) \text{ if and only if } Y \supseteq X \cap B_W.$$

Taking the two together entails that if $\neg Bel(\neg X|W)$ (or $X \cap B_W \neq \varnothing$):

$$B_X = X \cap B_W.$$

That is, we may reformulate B6 one more time in the handy form:

B6 (Restricted $Bel(\cdot|\cdot)$ / Expansion) For all $Y \in \mathfrak{A}$ such that $Y \cap B_W \neq \varnothing$: $B_Y = Y \cap B_W$.

Supplying conditional belief with a suppositional interpretation again: if Y is consistent with everything the agent believes absolutely, then supposing Y (in the sense of matter-of-fact supposition) amounts to nothing else than adding Y to one's stock of absolute beliefs and closing under logical consequence; or in propositional terms: taking the intersection of Y and B_W (see Figure 4.3) and believing every superset of that intersection.

We have also already shown in the course of our reformulation of B6 that $\neg Bel(\neg X|W)$ iff $X \cap B_W \neq \varnothing$. So for all X with $\neg Bel(\neg X|W)$ it is the case that the proposition $B_X = X \cap B_W$ is non-empty. This means that we can derive from B1–B6 a Consistency postulate that is more general than B5: if $\neg Bel(\neg X|W)$, then $\neg Bel(\varnothing|X)$. This corresponds to the part of AGM's K*5 (Consistency) postulate that deals with belief expansion.

It also follows that if $\neg Bel(\neg X|W)$ then $B_X \cap B_W$ $(= [X \cap B_W] \cap B_W = X \cap B_W)$ is non-empty. Consequently, we can apply B6 in its last version above to the proposition B_X itself and derive from $\neg Bel(\neg X|W)$ that

$$B_{B_X} = B_X \cap B_W,$$

which yields, since $B_X = X \cap B_W$ is a subset of B_W,

$$B_{B_X} = B_X.$$

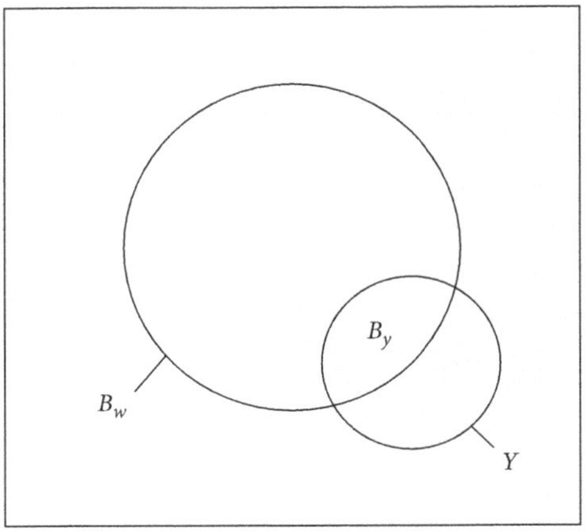

Figure 4.3. The expansion operation

Or formulated differently: for all $Y \in \mathfrak{A}$ it holds that

$$Bel(Y|B_X) \text{ if and only if } Bel(Y|X) \text{ if and only if } Y \supseteq B_X.$$

Hence what is believed by the agent conditional on X may always be determined just by considering all and only those members of $\mathfrak{A}$ that the agent believes conditional on the subset B_X of X. In the literature on nonmonotonic reasoning, the corresponding property of nonmonotonic consequence relations is called Cumulativity:[242] importing plausible conclusions into one's set of premises neither weakens nor strengthens what can be plausibly inferred from these premises. Accordingly, for absolute or unconditional belief: $Bel(Y|B_W)$ iff $Bel(Y|W)$ iff $Y \supseteq B_W$. I will use equivalences like that at several points, and when I do so I will not state this explicitly any more.

Although AGM's K*3 (Inclusion) and K*4 (Preservation) determine expansion to be something like an obvious qualitative counterpart of probabilistic conditionalization—supposing Y means restricting the space B_W of doxastically possible worlds to the doxastically possible worlds in Y, that is, $B_W \cap Y$—they have not remained unchallenged, of course. One typical worry is that revising by some new evidence or suppositional information Y may be thought to lead to *more* beliefs than what one would get deductively by adding Y to one's current beliefs: that might be so because there might be *inductively* strong inferences that the presence of Y might warrant. One line of defence of AGM concerning that point is: if the agent's current beliefs (as given by B_W) are themselves already the result of the application of rational inductive reasoning methods to the sum of the agent's evidence, so that the agent's beliefs are really what she *expects* to be the case on inductive grounds, then revising her

[242] Cumulativity was suggested first by Gabbay (1984). It corresponds to the combination of the rules of Cautious Monotonicity and Cautious Cut in Kraus et al. (1990). I will return to these rules in section 6.2.

beliefs by consistent information might well reduce to merely adding this information to her beliefs and closing off deductively. Another line of defence would be: a postulate such as B6 might be true of conditional belief simply because without it qualitative belief would not have the simplifying power that is essential to it. Inductive reasoning based on *quantitative* belief is yet another matter, and the mentioned criticism of the conjunction of K*3 (Inclusion) and K*4 (Preservation) might simply result from mixing up considerations on qualitative and quantitative belief. (I will turn to joint principles for conditional belief and degrees of belief soon.)

Lin and Kelly (2012b, section 8) criticize Preservation on the basis of a Gettier-style example: who owns the Ford? The possible relevant answers are *Nogot*, *Havit*, and *Nobody* (corresponding to a set of three possible worlds). One's total relevant unconditional belief is supposed to be given by *Nogot-or-Havit*, but by the story that they are telling it is really the *Nogot* option that makes *Nogot-or-Havit* likely (*Havit* is meant to be just a little bit more likely than *Nobody*). However, Nogot by itself would not be likely enough to be believed on its own. Given the new information $\neg$*Nogot*, Lin and Kelly would like to conclude *Havit-or-Nobody*, while Preservation would predict *Havit* (which preserves the prior belief in *Nogot-or-Havit*). My defence of Preservation in that case would be: if one's total relevant prior belief is *Nogot-or-Havit*, then each of the two options should be sufficiently more likely than *Nobody* (where the meaning of 'sufficiently more likely' is given by the context). If that is so, then if *Nogot* is eliminated by the evidence, one should still believe *Havit*. If it is not so, that is, if it is not the case that each of the two options is sufficiently more likely than *Nobody*, then one should not have had the total belief in *Nogot-or-Havit* in the first place. One argument for assuming that total belief in *Nogot-or-Havit* ought to entail that each of the two options is sufficiently more likely than *Nobody* is a stability argument again: if this were not so, then belief in *Nogot-or-Havit* would not be stable under updating on what is a serious possibility in all-or-nothing terms, that is, $\neg$*Nogot*. (Compare the Outclassing Condition in Representation Theorem 5 for the Humean thesis in Appendix B.[243])

Indeed, as its name says already, one may think of Preservation as a kind of stability principle very much in the spirit of the Humean thesis from Chapter 2, but formulated solely for all-or-nothing conditional belief, or, with the diachronic norm from section 4.1.2 in place: for all-or-nothing belief revision. As long as the evidence is consistent with everything that the agent believes, revising her beliefs on the basis of the evidence will not affect her prior beliefs. They will be stable under

[243] What counterargument might there be *against* the thesis that total belief in *Nogot-or-Havit* ought to entail that each of *Nogot* and *Havit* is sufficiently more likely than *Nobody*? Perhaps one might want to argue from the Lockean thesis: the probability of *Nogot-or-Havit* is above the Lockean threshold, which is why *Nogot-or-Havit* is believed, and one might think that this might be so without each of *Nogot* and *Havit* being sufficiently more likely than *Nobody*. However, Lin and Kelly (2012b) do not actually endorse the Lockean thesis on belief. If they did, we know already from Chapter 3 that combining the logical closure of belief with the Lockean thesis would have the consequence that the Lockean threshold would have to be $P(B_W)$ (or a number slightly below that), B_W would be P-stable, and at least the unconditional belief version of the present stability theory of belief would follow.

revision in that case, and that kind of stability might be just as rational to have as that supplied by the Humean thesis in Chapter 2. Putnam (1963) calls the corresponding feature of inductive methods 'tenacity' and argues for it by observing that it guarantees that a true hypothesis, once accepted, is not subsequently rejected on the basis of (true) evidence.[244] Preservation also keeps belief revision simple in the case in which the evidence is compatible with what one believes unconditionally: simply 'throw' in the evidence and close deductively! As argued at the end of section 1.5, it belongs to the distinguishing features of all-or-nothing belief to be simpler than numerical belief, and logical closure supplies belief with a lot of simplicity; the same might be the case for Preservation and all-or-nothing *conditional* belief.

In the spheres semantics or, equivalently, the total 'plausibility' pre-order semantics for AGM belief revision, Preservation follows from the nestedness of spheres or the totality of the pre-order. I will say more about this in section 4.2.3 when I introduce the postulate B6* that extends B6 to the case of conditional belief in general. I will also return to Preservation in sections 6.2 and 6.3 of Chapter 6, when I will discuss the assertability of indicative conditionals and the relationship between belief and acceptance. In both cases we will encounter further arguments for Preservation.

Example 10 (Example 7 from section 4.1.3 Reconsidered)
Let W be as in Example 7, but let us assume that *Bel* has not been determined as yet. By choosing B_W to be some arbitrary non-empty subset of W, it becomes possible to determine *Bel* so that all of our postulates B1–B6 in this section are satisfied.

For instance, let $B_W = \{w_1\}$, and turn B6 from above into a definition of *Bel* for all $Y \cap \{w_1\} \neq \varnothing$ and for all $Z \subseteq W$: $Bel(Z|Y)$ if and only if $Z \supseteq Y \cap B_W$. It follows then that all of our belief postulates hold true, and e.g. $Bel(\{w_1\}|W)$ and $Bel(\{w_1\}|\{w_1, w_2\})$ are the case.

But we might just as well choose $B_W = \{w_1, w_2\}$ and define for all $Y \cap \{w_1, w_2\} \neq \varnothing$ and for all $Z \subseteq W$: $Bel(Z|Y)$ if and only if $Z \supseteq Y \cap B_W$. Once again all of our belief postulates are satisfied, and e.g. $Bel(\{w_1, w_2\}|W)$ (but not $Bel(\{w_1\}|W)$), and not $Bel(\{w_1\}|\{w_1, w_2\})$ are the case.

Since I have not introduced any bridge principles for belief and degrees of belief up to this point, so far the choice of *Bel* or B_W has been unconstrained by P (e.g. by the P in Example 7); the same holds vice versa. This is going to change now.

THE MAIN BRIDGE POSTULATE

Finally, I turn to the promised necessary condition for having a conditional belief—the left-to-right direction of a conditional version of the Lockean thesis. Again, I will

[244] 'tenacity: if h is once accepted, it is not subsequently abandoned *unless* it becomes inconsistent with the data' (Putnam 1963, p. 772). Putnam (1963, p. 776) also relates this to the conservativeness that 'is an important and essential part of the actual procedure of science: a hypothesis once accepted is not easily abandoned'.

first formulate this condition for beliefs conditional on propositions that are consistent with all the agent believes. This will make the agent's conditional degrees of belief at a time t and (some of) her conditional beliefs at t compatible, in a sense. The resulting bridge principle between qualitative and quantitative belief will involve a numerical constant 'r' again the value of which I will leave indeterminate at this point—for the moment, just assume that r is some real number in the half-open interval $[0, 1)$. So I will even leave open for now whether $r \geq \frac{1}{2}$ (though ultimately this will be assumed). As in the previous chapters, this bridge principle is not meant to give us anything like a definition of '*Bel*' or 'B_W' on the basis of 'P'. It only expresses a joint constraint on the references of '*Bel*' and 'P', that is, on both our agent's actual conditional beliefs and her actual subjective probabilities at the given time.

The principle says (where '*BP*' signals that this is for *B*elief and *P*robability simultaneously):

$\mathrm{BP1}^r$ (Likeliness) For all $Y \in \mathfrak{A}$ such that $Y \cap B_W \neq \varnothing$ and $P(Y) > 0$: for all $Z \in \mathfrak{A}$, if $Bel(Z|Y)$, then $P(Z|Y) > r$.

$\mathrm{BP1}^r$ is just the obvious generalization of the left-to-right direction of the Lockean thesis to the case of beliefs conditional on propositions Y that are consistent with all of the agent's absolute beliefs. The antecedent clause '$P(Y) > 0$' in $\mathrm{BP1}^r$ makes sure that the conditional probability $P(Z|Y)$ is well-defined. By using W as the value of 'Y' and B_W as the value of 'Z' in $\mathrm{BP1}^r$, and then applying the definition of B_W (a subset of W which exists by B1–B4 and which is non-empty by B5) and postulate P1, it follows that $P(B_W|W) = P(B_W) > r$. Therefore, from the definition of 'B_W' and P1 again, having a subjective probability of more than r is a necessary condition for a proposition to be believed *absolutely*, although it will become clear later that this is not a sufficient condition (but having a probability greater than $P(B_W)$ will be a sufficient condition for absolute belief).

As we will see from Observation 16 later in the present section 4.2.2, it is perfectly permissible to think of 'r' as the corresponding 'Humean threshold' in the Humean thesis HT^r from Chapter 2. But taken by itself $\mathrm{BP1}^r$ just says: conditional beliefs (given the relevant Ys) entail having corresponding conditional probabilities of more than r. One might wonder why there should be *one* such threshold r *for all* propositions Y and Z as stated in $\mathrm{BP1}^r$ at all, rather than having *for all* Y (or for all Y and Z) *some* threshold value that might depend on Y (or on Y and Z). But without any further qualification, this revised principle would be almost empty, because as long as for Y and Z it is the case that $P(Z|Y) > 0$, there will always be an r such that $P(Z|Y) > r$: just take r to be equal to 0, or let r be any number just a little bit below $P(Z|Y)$. Further qualifications might consist in, e.g. postulating additionally some instances of the right-to-left (P-to-*Bel*) direction of the Lockean thesis, as it was the case in Chapter 3 (where we postulated the full right-to-left direction); but this would have to be worked out first in full detail. We are going to see later that a claim of that form will actually be *derivable* from all of our postulates taken together, including $\mathrm{BP1}^r$:

compare Observation 13, which will derive a conditional version of the Lockean thesis below. But BP1^r itself postulates a conditional probabilistic boundary from below that is *uniform* for all conditional beliefs whatsoever—such an r should thus derive from contextual considerations on *belief* itself rather than from contextual considerations on the *contents* of belief.[245]

For further illustration, think of r in BP1^r as being equal to $\frac{1}{2}$ for the moment. If conditional probabilities and conditional beliefs ought to be compatible in some sense at all, then the resulting BP1$^{\frac{1}{2}}$ is pretty much the weakest possible expression of any such coherence that one could think of: if the agent believes Z (conditional on one of the Ys referred to above), then she assigns a subjective probability to Z (conditional on Y) that exceeds the subjective probability that she assigns to the negation of Z (conditional on Y). If BP1$^{\frac{1}{2}}$ were invalidated, then there would be Z and Y, such that our agent believes Z conditional on Y, but where $P(Z|Y) \leq \frac{1}{2}$: if $P(Z|Y) < \frac{1}{2}$, then the agent would be in a position in which she regarded $\neg Z$ as more likely than Z, conditional on Y, even though she believes Z, but not $\neg Z$, conditional on Y. On the other hand, if $P(Z|Y) = \frac{1}{2}$, then she would be in a position in which she regarded $\neg Z$ as equally likely as Z, conditional on Y, even though she believes Z, but not $\neg Z$, conditional on Y. At least the former should be very difficult to accept—and the more difficult the lower the value of $P(Z|Y)$.

Instead of defending BP1$^{\frac{1}{2}}$ or any other particular instance of BP1^r at this point, I will simply move on, taking for granted one such instance BP1^r has been chosen. Within the theory, choosing $r = \frac{1}{2}$ will in fact be the right choice for the *least possible* threshold value that would give us an account of 'believing that'; but taking any greater threshold value less than 1 will be permissible, too. However, for weaker forms of subjective commitment than *belief*, such as *suspecting that Z* or *hypothesizing that Z*, r might well be chosen to be less than $\frac{1}{2}$, and some of the formal results below (though not all of them) will still be applicable, since they do not depend on r being greater than or equal to $\frac{1}{2}$.

For the moment this exhausts our list of postulates, with more principles to come later.

P-STABILITY AND THE FIRST REPRESENTATION THEOREM

For now let us pause as regards the introduction of postulates. Instead I will focus on finding jointly necessary and sufficient conditions for the conjunction of our postulates up to this point being satisfied. This will lead me to the first representation theorem in this chapter: the theorem will characterize in transparent terms those pairs $\langle P, Bel\rangle$ whose coordinate entries jointly satisfy all of the postulates so far.

In order to formulate the theorem, I will need the following probabilistic concept which will turn out to be crucial for the whole theory in this chapter; the concept was

245 It would be possible to weaken '$>$' in '$P(Z|Y) > r$' in BP1^r to '$\geq$'; in what follows, not much will depend on this, except that whenever I am going to use BP1^r with a threshold $r \geq \frac{1}{2}$ below, one would rather have to choose some threshold $r' > \frac{1}{2}$ instead and then demand that '... $P(Z|Y) \geq r'$' is the case.

defined before in previous chapters (see Appendix B), and it generalizes the concept of P-stability that was central to Chapter 3:

Definition 8 (P-Stability^r) Let P be a probability measure on $\mathfrak{A}$ (so P1 is satisfied), let $0 \leq r < 1$. For all $X \in \mathfrak{A}$ we define:

X is P-stable^r if and only if for all $Y \in \mathfrak{A}$ with $Y \cap X \neq \varnothing$ and $P(Y) > 0$: $P(X|Y) > r$.

If one thinks of $P(X|Y)$ as the degree of X under the supposition of Y, then a P-stable^r proposition X has the property that whatever proposition Y one supposes, as long as Y is consistent with X, and when probabilities conditional on Y are well-defined, it will be the case that the degree of X under the supposition of Y exceeds r. For any such non-empty P-stable^r set X, one of the Ys that we could choose is of course the full set W of possible worlds (which is a member of $\mathfrak{A}$ with probability 1): such a non-empty P-stable^r X must therefore have an absolute probability greater than r. What P-stability^r adds to this is that this will *remain* so as long as one supposes propositions that are consistent with X and on which conditionalization is defined at all. So once again a P-stable^r proposition has a special *stability property*: it is characterized by its stably high probabilities under all suppositions of a particularly salient type. Trivially, the empty set is P-stable^r. W is P-stable^r, too, and more generally all propositions X in $\mathfrak{A}$ with probability $P(X) = 1$ are P-stable^r. More importantly, and perhaps surprisingly, as we will see later (e.g. in this section 4.2.2), there are in fact lots of probability measures for which there are lots of *non-trivial* P-stable^r propositions which have a probability *strictly between 0 and 1.*[246] Finally: it is clear that the greater the value of 'r' is, the more severe is the constraint that is expressed by P-stability^r.

Obviously, the right-hand side of Definition 8 looks a lot like the right-hand side of the Humean thesis HT^r from section 2.2 in Chapter 2, and both of them concern probabilistic stability. But one should also see the differences: while the Humean thesis is a *joint constraint on Bel and P*, P-stability^r is a *purely probabilistic property of propositions X*. According to the Humean thesis, a believed proposition X needs to retain a high enough probability conditional on propositions Y *the negations of which are not believed*. In contrast, a P-stable^r proposition X needs to retain a high enough probability conditional on propositions Y *that are consistent with the P-stable^r proposition X itself*.

A different way of thinking of P-stability^r is the following one. Let X be non-empty and P-stable^r: for all Y being such that $Y \cap X \neq \varnothing$ and $P(Y) > 0$, it holds then that $P(X|Y) = \frac{P(X \cap Y)}{P(Y)} > r$, which is equivalent to: $P(X \cap Y) > r \cdot P(Y)$. (This means also that $P(X \cap Y)$ must be greater than 0.) But by P1 this is again equivalent with $P(X \cap Y) > r \cdot [P(X \cap Y) + P(\neg X \cap Y)]$, which yields $P(X \cap Y) > \frac{r}{1-r} \cdot P(\neg X \cap Y)$. By letting the value of 'Y' vary over all members of $\mathfrak{A}$ that have non-empty intersection with X and which also have non-zero probability, the value of '$X \cap Y$' actually varies over precisely

246 Of course, this is not so surprising any more given the results and examples from the previous chapters.

the subsets of X that are members of $\mathfrak{A}$ and which have non-zero probability. And the value of '$\neg X \cap Y$' varies over precisely the subsets of $\neg X$ that are members of $\mathfrak{A}$. So we have the following equivalent characterization of P-stability^r:

Observation 9 Let P satisfy P1. For all $X \in \mathfrak{A}$ with X non-empty:

> X is P-stable^r if and only if for all $Y, Z \in \mathfrak{A}$, such that $Y \subseteq X$ with $P(Y) > 0$, and where $Z \subseteq \neg X$, it holds that:
>
> $$P(Y) > \frac{r}{1-r} \cdot P(Z).$$

Of course, we could also reformulate this equivalence by only considering $Z = \neg X$ on the right-hand side from the start.

In the special case in which $r = \frac{1}{2}$, the factor $\frac{r}{1-r}$ is just 1, and hence X is P-stable$^{\frac{1}{2}}$ if and only if the probability of any subset of X that has positive probability at all is greater than the probability of any subset of $\neg X$. So P-stability^r is also a *separation property* that divides the class of sub-propositions of a proposition from the class of sub-propositions of its negation or complement in terms of their probabilities.

Here is another property of non-empty P-stable^r propositions X that I will need on various occasions, which is why it is worth stating explicitly:

Observation 10 Let P satisfy P1. For all $X \in \mathfrak{A}$ with X non-empty and P-stable^r:

> if $P(X) < 1$, then it is not the case that there is a non-empty $Y \in \mathfrak{A}$ with $Y \subseteq X$ and $P(Y) = 0$.

For assume this is not so: then $Y \cup \neg X$ has non-empty intersection with X since Y has, and at the same time $P(Y \cup \neg X) > 0$ because $P(\neg X) > 0$ (by P1). By X being P-stable^r, it would therefore have to be the case that $P(X | Y \cup \neg X) = \frac{P(X \cap Y)}{P(Y \cup \neg X)} > r$, in contradiction with (using P1): $P(X \cap Y) \leq P(Y) = 0$.

From the last two observations taken together, it follows that in certain simple cases one can simplify the formulation of the right-hand side of the separation property from Observation 9 by dropping the assumption '$P(Y) > 0$'. In previous chapters (see Appendix B in particular) I have called this the 'Outclassing Condition' for X relative to P and r:

Observation 11 Assume W is finite, and let $\mathfrak{A}$ again be the power set algebra on W. Let P satisfy P1, and let $X \in \mathfrak{A}$ be such that X is non-empty. Then the following two statements are equivalent:

- X is P-stable^r, and if $P(X) = 1$ then X is the least member of $\mathfrak{A}$ with probability 1 (which must exist in that kind of situation).
- Outclassing Condition: For all w in X it holds that:

$$P(\{w\}) > \frac{r}{1-r} \cdot P(\neg X).$$

Proof. From left-to-right: assume X is P-stable^r. If $P(X) < 1$, then every singleton subset of X must have positive probability by Observation 10. If $P(X) = 1$, then X is the least proposition of probability 1 by assumption, and therefore once again every singleton subset of X must have positive probability: for if not then X without that singleton would still have probability 1 (by P1) but would be a proper subset of X, which would contradict X being the least such set. So for all w in X it is the case that $P(\{w\}) > 0$. The rest follows from Observation 9. From right-to-left: X being P-stable^r follows immediately from Observation 9 and P1. If $P(X) = 1$, then X must be the least member of $\mathfrak{A}$ with probability 1, for otherwise X would have a singleton subset $\{w\}$ of probability 0 (by P1), which would contradict $P(\{w\}) > \frac{r}{1-r} \cdot P(\neg X)$. ■

From this point onwards, I will not mark each and every application of P1 in proofs explicitly; instead I will sometimes apply simple probabilistic inference steps in proofs with further comment.

Using the concept of P-stability^r, we can now formulate the following rather simple representation theorem for restricted conditional belief. (There will be another more intricate one in section 4.2.3—Representation Theorem 23—which will extend the present one to conditional belief in general.)

Theorem 12 (Representation Theorem for Restricted Conditional Belief)
Let *Bel* be a class of ordered pairs of members of $\mathfrak{A}$, let $P : \mathfrak{A} \to [0,1]$, and let $0 \leq r < 1$. Then the following two statements are equivalent:

I. P and *Bel* satisfy P1, B1–B6, and BP1^r.
II. P satisfies P1, and there is a (uniquely determined) $X \in \mathfrak{A}$, such that X is a non-empty P-stable^r proposition, and:
   - for all $Y \in \mathfrak{A}$ such that $Y \cap X \neq \varnothing$, and for all $Z \in \mathfrak{A}$:

$$Bel(Z \mid Y) \text{ if and only if } Z \supseteq Y \cap X$$

   (and hence, $B_W = X$).

Proof. From left to right (I to II): P1 is satisfied by assumption. Now we let $X = B_W$, where B_W exists by B1–B4 and has, by definition, the property of being the strongest believed proposition. First of all, as derived before by means of B5, B_W is non-empty. Secondly, B_W is P-stable^r: let $Y \in \mathfrak{A}$ with $Y \cap B_W \neq \varnothing$, $P(Y) > 0$; since $B_W \supseteq Y \cap B_W$, it follows from B6 that $Bel(B_W|Y)$, which with BP1 and $P(Y) > 0$ entails that $P(B_W|Y) > r$, which was to be shown. Thirdly, let $Y \in \mathfrak{A}$ be such that $Y \cap B_W \neq \varnothing$, let $Z \in \mathfrak{A}$: then it holds that $Bel(Z|Y)$ if and only if $Z \supseteq Y \cap B_W$ by B6, as intended. Finally, uniqueness: assume $X' \in \mathfrak{A}$, X' is non-empty, P-stable^r, and for all $Y \in \mathfrak{A}$ with $Y \cap X' \neq \varnothing$, for all $Z \in \mathfrak{A}$, it holds that $Bel(Z \mid Y)$ if and only if $Z \supseteq Y \cap X'$. But from the latter it follows that $Bel(B_W \mid W)$ if and only if $B_W \supseteq W \cap X' = X'$, and hence, with $Bel(B_W \mid W)$ from B1–B4 and the definition of B_W, we may conclude $B_W \supseteq X'$. On the other hand, by choosing X' as the value of 'Z' and W as the value of 'Y', we have

$Bel(X' \mid W)$ if and only if $X' \supseteq W \cap X'$, and thus $Bel(X' \mid W)$; but by the definition of B_W again this entails: $X' \supseteq B_W$. Therefore, $X' = B_W$.

From right to left: suppose P satisfies P1, and there is an X, such that X and Bel have the required properties. Then, first of all, all the instances of B1–B5 for beliefs conditional on W are satisfied: for it holds that $W \cap X = X \neq \emptyset$ because X is non-empty by assumption, so, again by assumption, $Bel(Z|W)$ if and only if $Z \supseteq W \cap X = X$, therefore B5 is the case, and the instances of B1–B4 for beliefs conditional on W follow from the characterization of beliefs conditional on W in terms of supersets of X. Indeed, it follows: $B_W = X$. So, for arbitrary $Y \in \mathfrak{A}$, $\neg Bel(\neg Y|W)$ is really equivalent to $Y \cap X \neq \emptyset$, as I have already shown after the introduction of B1–B5, and hence all instances of B1–B4 are satisfied by the assumed characterization in II of beliefs conditional on any Y with $Y \cap X \neq \emptyset$ in terms of supersets of $Y \cap X$. B6 holds trivially, by assumption and because of $B_W = X$. About BP1^r: let $Y \cap X \neq \emptyset$ and $P(Y) > 0$. If $Bel(Z|Y)$, then by assumption $Z \supseteq Y \cap X$, hence $Z \cap Y \supseteq Y \cap X$, and by P1 it follows that $P(Z \cap Y) \geq P(Y \cap X)$. From X being P-stable^r and $Y \cap X \neq \emptyset$ and $P(Y) > 0$ we also have $P(X|Y) > r$. Taking these two facts together, and by the definition of conditional probability in P1, this implies $P(Z|Y) > r$, which was to be shown. ■

This simple theorem will prove to be fundamental for all subsequent arguments in this chapter.

Note that the following is *not* entailed by Theorem 12: *every* believed proposition is P-stable^r. In fact, given our postulates, P-stability^r is only required for *precisely one* believed proposition: the logically strongest proposition that is believed at all, or equivalently, the conjunction of all believed propositions. This said, the theorem still shows very clearly that we are on the way to extending the stability theory of belief from Chapters 2–3 to the conditional case: by the postulates in this chapter and the representation theorem above it holds that restricted conditional belief can be represented by the same P-stable^r (or P-stable) sets by which unconditional rational belief turned out to be representable in Appendix B (the appendix to Chapter 2) and in Chapter 3. Over and above unconditional belief, also belief *conditional on propositions that are possible from the viewpoint of the agent* turns out to be stable. In the present case, this finding does not rely on a joint stability principle for unconditional belief and degree of belief (as in Chapter 2) nor on the right-to-left direction of the Lockean thesis for unconditional belief (as in Chapter 3) but rather on the left-to-right direction of the Lockean thesis for conditional belief (BP1^r). More or less the same stability theory of belief follows from different sets of, more or less, natural assumptions—the theory itself happens to be quite robust.

But there is more to come, and I will take things slowly for now. Let me start by exploiting Theorem 12 in a rather trivial fashion: let us concentrate on its right-hand side, that is, condition II. Disregarding for the moment any considerations on qualitative belief, let us just assume that we are given a probability measure P over $\mathfrak{A}$. We know already that one can in fact always find a non-empty set X, such that X is a P-stable^r proposition: just take any proposition with probability 1. For now, let us assume the simplest case: take X to be W itself. Non-emptiness (indeed $P(W) > 0$) and P-stability^r are then seen to be the case immediately. Now consider the very last equivalence clause of II and turn it into a (conditional) *definition* of $Bel(.|Y)$ for all those cases in which $Y \cap W = Y \neq \varnothing$: that is, for all $Z \in \mathfrak{A}$, define $Bel(Z \mid Y)$ to hold if and only if $Z \supseteq Y \cap W = Y$. In particular, $Bel(Z \mid W)$ holds if and only if $Z \supseteq W$, which obviously is the case if and only if $Z = W$. $B_W = W$ follows, all the conditions in II of Theorem 12 are satisfied, and thus by Theorem 12 all of our postulates from above must be true as well. What this shows is that given a probability measure, it is always possible to define belief *simpliciter* in a way such that all of our postulates turn out to be the case. What would be believed thereby by our agent would be maximally cautious: having such beliefs, the agent would only believe W unconditionally, and therefore, trivially, every proposition that is believed unconditionally would have probability 1. Furthermore, she would believe conditional on the respective Ys just what is logically entailed by them, that is, all supersets of Y.

But we actually find a much more general pattern to emerge from Theorem 12: let P be given again as before. Now choose *any* non-empty P-stable^r proposition X, and partially define conditional belief just for all those cases in which $Y \cap X \neq \varnothing$ by: $Bel(Z \mid Y)$ if and only if $Z \supseteq Y \cap X$. Then $B_W = X$ follows again, and all of our postulates hold by Theorem 12—including B3 (Finite Conjunction) and B4 (General Conjunction)—even though it might well be that $P(X) < 1$. If so, *there will be beliefs the propositional contents of which have a subjective probability of less than 1 as being given by P*. Such beliefs are not maximally cautious any more—as seems to be the case for many of the beliefs of real-world human agents in normal contexts.

These are ways of turning the right-to-left direction of Theorem 12 into a method of generating *Bel* from P, such that all of the postulates P1, B1–B6, and BP1^r are satisfied. But as we have seen already in the previous chapters, the present theory does not *have* to be interpreted or applied in such a reductive manner.

Let me now relate the theorem above to the stability theory of belief as developed in the previous chapters. Theorem 12 implies: if P and *Bel* satisfy P1, B1–B6, and BP1^r, then the logically strongest believed proposition B_W is P-stable^r, and if $Z \cap B_W \neq \varnothing$ then $B_Z = Z \cap B_W$ is the strongest proposition believed conditional on Z. In the case in which $\frac{1}{2} < r < 1$ and also $P(B_W) < 1$, this implies a conditional version of the Lockean thesis: the conjunction of what one believes conditional on Z sets the Lockean threshold for what one believes conditional on Z.

Observation 13 Let P satisfy P1, let X be non-empty and P-stable^r, $\frac{1}{2} \leq r < 1$, $P(X) < 1$, such that $X = B_W$ is the strongest proposition believed, let $Z \cap X \neq \varnothing$, let B_Z $(= Z \cap B_W = Z \cap X)$ be the strongest proposition believed conditional on Z, and $P(Z) > 0$. Then it holds:

(Conditional) Lockean Thesis:
for all $Y \in \mathfrak{A}$: $Bel(Y|Z)$ iff $P(Y|Z) \geq P(B_Z|Z)$ $(= P(Z \cap B_W|Z) = P(B_W|Z))$.

Proof. The left-to-right direction is obvious, since if $Bel(Y|Z)$, then $Y \supseteq B_Z$, hence $Y \cap Z \supseteq B_Z \cap Z$, and the rest follows by the monotonicity property of probability. And from right-to-left: assume $P(Y|Z) \geq P(B_Z|Z) = P(B_W|Z)$ but not $Bel(Y|Z)$; then $Y \not\supseteq B_Z = Z \cap B_W$, that is, $\neg Y \cap Z \cap B_W$ is non-empty. Thus, $[\neg Y \cap Z \cap B_W] \cup \neg B_W$ has non-empty intersection with B_W and its probability is greater than 0, because $1 > P(B_W) = 1 - P(\neg B_W)$ and so $P(\neg B_W) > 0$. But from B_W being P-stable^r it follows then that $P(B_W \mid [\neg Y \cap Z \cap B_W] \cup \neg B_W) > r \geq \frac{1}{2}$, that is, $P(\neg Y \cap Z \cap B_W) > \frac{P([\neg Y \cap Z \cap B_W] \cup \neg B_W)}{2} = \frac{P(\neg Y \cap Z \cap B_W)}{2} + \frac{P(\neg B_W)}{2}$, and hence $P(\neg Y \cap Z \cap B_W) > P(\neg B_W)$. Therefore, with $P(Z) > 0$, $\frac{P(\neg Y \cap Z \cap B_W)}{P(Z)} > \frac{P(\neg B_W)}{P(Z)} \geq \frac{P(\neg B_W \cap Z)}{P(Z)}$, and so $P(\neg Y \cap B_W|Z) > P(\neg B_W|Z)$. However, by $P(Y|Z) \geq P(B_W|Z)$, it also holds that $P(\neg B_W|Z) \geq P(\neg Y|Z)$. So we get $P(\neg Y \cap B_W|Z) > P(\neg Y|Z)$, which contradicts P1. So $Bel(Y|Z)$. ■

With $Z = W$, the previous proof is essentially the proof of the right-to-left direction of Theorem 7 from Chapter 3.[247] Indeed, the previous observation can be used to derive the following observation concerning the unconditional Lockean thesis that was discussed already in Chapter 3 (where r was always set to $\frac{1}{2}$):

Observation 14 Let P satisfy P1, let X be non-empty and P-stable$^{\frac{1}{2}}$ (that is, P-stable in the terminology of Chapter 3), such that $X = B_W$ is the strongest proposition believed, and where additionally if $P(B_W) = 1$ then B_W is the least proposition (in the sense of $\subseteq$) of probability 1. Then it holds:

(Unconditional) Lockean Thesis: for all $Y \in \mathfrak{A}$: $Bel(Y)$ iff $P(Y) \geq P(B_W)$.

Proof. If $P(B_W) < 1$, this follows directly from Observation 13 by setting $Z = W$.

If $P(B_W) = 1$, then B_W is the least proposition of probability 1 by assumption. But that together with P1 entails the Lockean thesis again. ■

It is easy to see that P-stability^r for $\frac{1}{2} < r < 1$ implies P-stability$^{\frac{1}{2}}$ (compare Observation 21, which I am going to prove later). So Theorem 12 combined with Observation 14 yields:

Observation 15 Assume P and Bel satisfy P1, B1–B6, BP1^r (hence there is a least believed proposition B_W), $\frac{1}{2} \leq r < 1$, and assume additionally that if $P(B_W) = 1$ then B_W is the least proposition of probability 1. Then the (unconditional) Lockean thesis follows.

The threshold value in such an instance of the unconditional Lockean thesis is not r but rather $P(B_W)$. Therefore, what the *Lockean* threshold is like is sensitive to the properties of the given probability measure P—one cannot choose the threshold independently of P. That is exactly what was observed in Chapter 3. In the current chapter, I will keep presupposing just the left-to-right direction of the Conditional

247 But note that 'P1' in Chapter 3 denoted a postulate different from the one that it denotes in the present chapter.

Lockean thesis, in which the (as it were, *Humean*) threshold value of 'r' can be chosen *independently* of what P is like.

From the same assumptions as those of Observation 15 one can also derive the Humean thesis from Chapter 2:

Observation 16 Assume P and Bel satisfy P1, B1–B6, BP1^r (hence there is a least believed proposition B_W), $\frac{1}{2} \leq r < 1$, and assume additionally that if $P(B_W) = 1$ then B_W is the least proposition of probability 1. Then the Humean thesis HT^r (= HT$^r_{Poss}$) from Chapter 2 follows:

(HT^r) For all X: $Bel(X)$ iff for all Y, if $Poss(Y)$ and $P(Y) > 0$, then $P(X|Y) > r$ where $Poss(Y)$ if and only if not $Bel(\neg Y)$.

Proof. (I have given that proof already when I proved Theorem 5 in Appendix B. For the sake of self-containment, I will state it again in slightly abbreviated form.)

Let $X \in \mathfrak{A}$. The left-to-right direction of HT$^r_{Poss}$ follows from: assume $Bel(X)$, $Poss(Y)$, $P(Y) > 0$. By $Bel(X)$, it holds that $X \supseteq B_W$. Because of $Poss(Y)$, it is the case that $Y \cap B_W \neq \varnothing$. By B_W being P-stable^r, $P(B_W|Y) > r$. But since $X \supseteq B_W$, it follows that $P(X|Y) \geq P(B_W|Y) > r$. The right-to-left direction of HT$^r_{Poss}$ follows from: assume for all Y, if $Poss(Y)$ and $P(Y) > 0$, then $P(X|Y) > r$. Suppose not $Bel(X)$: then $Poss(\neg X)$, that is, $\neg X \cap B_W \neq \varnothing$. If $P(B_W) = 1$, then B_W is the least proposition of probability 1, which cannot have any non-empty subset of probability 0. If $P(B_W) < 1$, then B_W cannot have any non-empty subset of probability 0 either, by Observation 10. Either way it follows with $\neg X \cap B_W \neq \varnothing$ that $P(\neg X) > 0$. So we have $Poss(\neg X)$, $P(\neg X) > 0$, and thus, by assumption, it has to be the case that $P(X|\neg X) > r$. But of course $P(X|\neg X) = 0$, which is a contradiction. Therefore, $Bel(X)$. ■

Finally, we can now also derive *Robustness Persistence* as formulated in section 4.1.3: the claim that updating Bel by belief expansion on E (where not $Bel(\neg E)$) while simultaneously updating P by conditionalization on E (where $P(E) > 0$) leads to Bel' and P', such that the Humean thesis is preserved: if the prior Bel and P satisfy the Humean thesis, then so do the posterior Bel' and P'.

This will follow from the following Observation 17 together with previous results:

Observation 17 Assume $X \in \mathfrak{A}$ is P-stable^r, and E is a member of $\mathfrak{A}$, such that $E \cap X \neq \varnothing$ and $P(E) > 0$. Then $E \cap X$ is $P(\cdot|E)$-stable^r.

Proof. Consider any $Y, Z \in \mathfrak{A}$, such that $Y \subseteq E \cap X$ and $P(Y|E) = P_E(Y) > 0$, and where $Z \subseteq \neg(E \cap X) = \neg E \cup \neg X = (E \cap \neg X) \cup \neg E$: then by X being P-stable^r, Y being a subset of X, $P(Y) > 0$, $E \cap \neg X$ being a subset of $\neg X$, and Observation 9, it holds that $P(Y) > \frac{r}{1-r} \cdot P(E \cap \neg X)$. Therefore, $\frac{P(Y \cap E)}{P(E)} = \frac{P(Y)}{P(E)} > \frac{r}{1-r} \cdot \frac{P(E \cap \neg X)}{P(E)}$ and thus $P(Y|E) > \frac{r}{1-r} \cdot P(E \cap \neg X|E)$. Moreover, $P(\neg E|E) = 0$. Thus, $P(Y|E) > \frac{r}{1-r} \cdot [P(E \cap \neg X|E) + P(\neg E|E)] = \frac{r}{1-r} \cdot P(\neg(E \cap X)|E) \geq \frac{r}{1-r} \cdot P(Z|E)$. In other words, by Observation 9 again: $E \cap X$ is $P(\cdot|E)$-stable^r. ■

As we already know (recall Theorem 5 from Appendix B), if the Humean thesis HT^r holds, then B_W is P-stable^r, and also if $P(B_W) = 1$ then B_W is the least proposition of probability 1. Because of B_W being P-stable^r, B_W is one of the Xs in the previous Observation 17. Updating by expansion works by intersection: so the agent's new logically strongest proposition after the update is $E \cap X = E \cap B_W$. And updating by conditionalization leads to $P(\cdot|E)$. By Observation 17, $E \cap B_W$ is $P(\cdot|E)$-stable^r again. By another application of Theorem 5, in order to derive that $E \cap B_W$ and $P(\cdot|E)$ satisfy the Humean thesis HT^r again, the only remaining point to observe is: if $P(E \cap B_W|E) = 1$, then $E \cap B_W$ must also be the least proposition of probability 1 with respect to $P(\cdot|E)$, as required for the intended application of Theorem 5. This is so because otherwise $E \cap B_W$ would need to have a subset of zero probability according to $P(\cdot|E)$ and thus also according to P, hence B_W would need to have a zero subset according to P, which cannot be, whether in the case $P(B_W) < 1$ (by Observation 10) or in the case $P(B_W) = 1$ (by the assumption that if $P(B_W) = 1$ then B_W is the least proposition of probability 1). So, as promised at the end of section 4.1.3, update by expansion/conditionalization preserves the Humean thesis.

Theorem 12 has lots of interesting implications and applications. One such application consists in (partially) defining a belief set defined from a P-stable^r set to the effect that the belief set and P taken together satisfy all of the postulates above. But of course that does not mean that a perfectly rational agent's actual belief set would always be definable just from the same agent's degree-of-belief function P: what Theorem 12 tells us is rather that such an agent's belief set (or the proposition B_W that generates it) always corresponds to *some* P-stable^r set. If there were additional means of defining from P the very P-stable^r proposition X that coincides with the agent's least believed proposition B_W, we could indeed define explicitly from P the part of *Bel* that concerns all pairs $\langle Z, Y\rangle$ for which $Y \cap X \neq \varnothing$ holds. Amongst those conditional beliefs, in particular, we would find all of the agent's absolute beliefs, and therefore the set of absolutely believed propositions would be definable explicitly on the basis of P.

So are we in the position to identify the P-stable^r proposition X that gives us the agent's actual beliefs, if we are handed only her subjective probability measure? I will deal with that question in Appendix C. Ultimately, I will argue for a negative answer.

Since P-stable^r propositions play such a distinguished role in all of that, the question arises: how difficult is it to determine whether a proposition is a non-empty P-stable^r set? I will turn to that question now.

COMPUTING P-STABLE^r SETS

At least in the case where W is finite, it turns out not to be difficult at all to determine all and only P-stable^r sets. Let W be finite, let $\mathfrak{A}$ again be the power set algebra on W, and let P be a probability measure on $\mathfrak{A}$. We have seen already that all sets with probability 1 are P-stable^r and that the empty set is trivially P-stable^r. So let us focus just on how to generate all non-empty P-stable^r sets X that are non-trivial, that is, which have a probability of less than 1. As I observed before (Observation 10), such sets do not contain any non-empty subsets of probability 0, which in the present context means that if $w \in X$, $P(\{w\}) > 0$.

For any such non-empty X with $P(X) < 1$ we have by Observation 11:

$$X \text{ is } P\text{-stable}^r \text{ if and only if for all } w \text{ in } X \text{ it holds that } P(\{w\}) > \frac{r}{1-r} \cdot P(W \setminus X).$$

In particular, for $r = \frac{1}{2}$, this yields (where again X is assumed non-empty and $P(X) < 1$):

$$X \text{ is } P\text{-stable}^{\frac{1}{2}} \text{ if and only if for all } w \text{ in } X \text{ it holds that } P(\{w\}) > P(W \setminus X).$$

Thus it turns out to be very simple to decide whether a set X is P-stable^r, and even more so whether it is P-stable$^{\frac{1}{2}}$: one only needs to check for what was called the Outclassing Condition on $X = B_W$ and P in Appendix B, which was reconsidered later in section 3.2 and in Observation 11.

From this it is easy to see that in the present finite context there is also an efficient procedure that computes all non-empty non-trivial P-stable^r subsets X of W. I only give a sketch for the case $r = \frac{1}{2}$:[248] since such sets X do not have singleton subsets of probability 0, let us also disregard all worlds whose singletons are zero sets. Assume that after dropping all worlds of zero probabilistic mass, there are exactly n members of W left, and $P(\{w_1\}), P(\{w_2\}), \ldots, P(\{w_n\})$ are already in (not necessarily strictly) decreasing order. The algorithm is a recursive procedure: if $P(\{w_1\}) > P(\{w_2\}) + \ldots + P(\{w_n\})$ then $\{w_1\}$ is the first P-stable$^{\frac{1}{2}}$ set determined, one keeps w_1 as a member of any set to be produced, and one moves on to the list $P(\{w_2\}), \ldots, P(\{w_n\})$ (now comparing $P(\{w_2\})$ with $P(\{w_3\}) + \ldots + P(\{w_n\})$). If $P(\{w_1\}) \leq P(\{w_2\}) + \ldots + P(\{w_n\})$ then consider $P(\{w_1\}), P(\{w_2\})$: if the latter of them is greater than $P(\{w_3\}) + \ldots + P(\{w_n\})$ then $\{w_1, w_2\}$ is the first P-stable$^{\frac{1}{2}}$ set, one keeps both worlds as members of any set to be produced, and one moves on to the list $P(\{w_3\}), \ldots, P(\{w_n\})$. If $P(\{w_2\})$ is less than or equal to $P(\{w_3\}) + \ldots + P(\{w_n\})$ then consider $P(\{w_1\}), P(\{w_2\}), P(\{w_3\})$: and so forth, until the set $\{w_1, w_2, \ldots, w_n\}$ has been reached which then coincides with the least subset of W of probability 1, that is, the smallest set that is but a trivial instance of P-stability$^{\frac{1}{2}}$. This recursive procedure yields precisely all non-empty non-trivial P-stable$^{\frac{1}{2}}$ sets, and it does so with polynomial time complexity (cf. Krombholz 2013). The same procedure can be applied in cases in which W is countably infinite, but then the procedure will not terminate in finite time.

What Theorem 12 gives us, therefore, is not just a representation result, but even, in the case of a given finite probability space with a measure P, an *efficient construction procedure* for all classes *Bel*, so that *Bel* together with the given P satisfies all of our postulates.

248 That algorithm was sketched already in section 3.2.

Example 11 (Example 7 from section 4.1.3 Reconsidered)
By means of the algorithm from above it is easy to compute all non-empty and non-trivial P-stable$^{\frac{1}{2}}$ sets: $\{w_1\}, \{w_1, w_2\}, \{w_1, \ldots, w_4\}, \{w_1, \ldots, w_5\}, \{w_1, \ldots, w_6\}$.

$\{w_1, \ldots, w_7\}$ is also P-stable$^{\frac{1}{2}}$, but it is trivial in the sense of having probability 1. I left out w_8 from the start, since $P(\{w_8\}) = 0$.

Accordingly, for instance, $P(\{w_1\})$ is greater than the sum of all probabilities of other singletons, $P(\{w_2\})$ is greater than $P(\{w_3\})+\ldots+P(\{w_7\})$, both $P(\{w_3\})$ and $P(\{w_4\})$ are greater than $P(\{w_5\}) + \ldots + P(\{w_7\})$, and so on. But it is neither the case that $P(\{w_3\})$ is greater than $P(\{w_4\}) + \ldots + P(\{w_7\})$, nor is it the case that $P(\{w_4\})$ is greater than $P(\{w_3\})+P(\{w_5\})+\ldots+P(\{w_7\})$, which is why neither $\{w_1, w_2, w_3\}$ nor $\{w_1, w_2, w_4\}$ are P-stable$^{\frac{1}{2}}$.

On the other hand, for $r = \frac{3}{4}$ the corresponding non-empty and non-trivial P-stable$^{\frac{3}{4}}$ sets are: $\{w_1, w_2\}, \{w_1, \ldots, w_4\}, \{w_1, \ldots, w_5\}, \{w_1, \ldots, w_6\}$.

With BP1^r in place, as stated by Theorem 12, it is no longer possible to determine our agent's beliefs by choosing a non-empty $B_W \subseteq W$ *arbitrarily*: for B_W must now be a non-empty *P-stable*^{r} set. While that is the case e.g. for $B_W = \{w_1\}$ (for $r = \frac{1}{2}$, but not for $r = \frac{3}{4}$) and $B_W = \{w_1, w_2\}$ (for both $r = \frac{1}{2}$ and $r = \frac{3}{4}$), it would not be possible to choose e.g. $B_W = \{w_1, w_3\}$ or $B_W = \{w_3, w_7, w_8\}$, whatever the value of 'r' in the interval $[\frac{1}{2}, 1)$, as these sets are not P-stable^r for any such r.

FURTHER PROPERTIES OF P-STABLE^r SETS

In the following I will study P-stable^r sets in more formal detail.

The next theorem summarizes two important properties of (non-empty and non-trivial) P-stable^r sets:[249]

Theorem 18 Let $P : \mathfrak{A} \rightarrow [0,1]$ such that P1 is satisfied. Let $\frac{1}{2} \leq r < 1$. Then the following is the case:

A. For all $X, X' \in \mathfrak{A}$: if X and X' are P-stable^r and at least one of $P(X)$ and $P(X')$ is less than 1, then either $X \subseteq X'$ or $X' \subseteq X$ (or both).
B. There is no infinitely descending chain of sets in $\mathfrak{A}$ that are all subsets of some P-stable^r set X_0 of $\mathfrak{A}$ with probability $P(X_0)$ less than 1. That is, there is no countably infinite sequence

$$X_0 \supsetneq X_1 \supsetneq X_2 \supsetneq \ldots$$

249 These properties correspond to some of the properties of so-called belief cores in Arló-Costa and Parikh (2005) (see also van Fraassen 1995 and Arló-Costa 2001), which are special sets of absolute probability 1 in a setting in which probabilities are determined by a primitive conditional probability measure or a Popper function. In fact, this is not a mere coincidence: once our theory has been generalized in section 4.2.3 to arbitrary conditional belief, one can show that by defining $P^*(Y|X) = P(Y|B_X)$, a Popper function P^* is defined from our P and *Bel* (and given r); and by this definition our P-stable^r sets are being transformed into belief cores as being given by $P*$. One can also show that every Popper function on a finite space can be represented in this way in terms of an absolute probability measure and *Bel*.

of sets in $\mathfrak{A}$ (and hence no infinite sequence of such sets in general), such that X_0 is P-stable^r, $P(X_0) < 1$, and each X_n is a proper superset of X_{n+1} (hence $P(X_n) < 1$ for all $n \geq 0$).
A fortiori, there is then no infinitely descending chain of P-stable^r sets in $\mathfrak{A}$ with probability less than 1 either. And, with $\mathfrak{A}$ being the power set algebra on W, it follows that every P-stable^r set of probability less than 1 must be finite.[250]

Proof.

- Ad A. First of all, let X and X' be P-stable^r, and $P(X) = 1$, $P(X') < 1$: as observed before (Observation 10), there is then no non-empty subset Y of X', such that $P(Y) = 0$. But if $X' \cap \neg X$ were non-empty, then there would have to be such a subset of X'. Therefore, $X' \cap \neg X$ is empty, and thus $X' \subseteq X$. The case for X and X' taken the other way around is analogous.
  So we can concentrate on the remaining logically possible case. Assume for contradiction that there are P-stable^r members X, X' of $\mathfrak{A}$, such that $P(X), P(X') < 1$, and neither $X \subseteq X'$ nor $X' \subseteq X$. Therefore, both $X \cap \neg X'$ and $X' \cap \neg X$ are non-empty, and they must have positive probability since again P-stable^r propositions with probability less than 1 do not have non-empty zero sets as subsets. It holds that $P(X \mid (X \cap \neg X') \cup \neg X)$ is greater than r by: X being P-stable^r, $(X \cap \neg X') \cup \neg X \supseteq (X \cap \neg X') \neq \varnothing$ having non-empty intersection with X, and the probability of $(X \cap \neg X') \cup \neg X \supseteq (X \cap \neg X')$ being positive. The same must hold, mutatis mutandis, for $P(X' \mid (X' \cap \neg X) \cup \neg X')$. Because $\frac{1}{2} \leq r$ by assumption, we have

  $$\text{(i) } P(X \mid (X \cap \neg X') \cup \neg X) > r \geq \frac{1}{2}$$

  and

  $$P(X' \mid (X' \cap \neg X) \cup \neg X') > r \geq \frac{1}{2}.$$

  Next I show that

  $$P(X \cap \neg X') > P(\neg X).$$

  For suppose otherwise, that is (ii) $P(X \cap \neg X') \leq P(\neg X)$: since it must be the case that

  $$P(X \cap \neg X' | (X \cap \neg X') \cup \neg X) + P(\neg X | (X \cap \neg X') \cup \neg X) = 1,$$

  and since we know from (i) that the second summand must be strictly less than $\frac{1}{2}$, the first summand has to strictly exceed $\frac{1}{2}$. On the other hand, it also follows

[250] I am grateful to Martin Krombholz and Laurenz Hudetz for highlighting this last point in discussions. If $\mathfrak{A}$ is a set algebra that differs from the power set algebra on W, then it does not necessarily hold that every P-stable^r member of $\mathfrak{A}$ with a probability less than 1 is a finite set.

that: $\frac{1}{2} > P(\neg X | (X \cap \neg X') \cup \neg X) = \frac{P(\neg X)}{P((X \cap \neg X') \cup \neg X)} \geq$ (by (ii)) $\frac{P(X \cap \neg X')}{P((X \cap \neg X') \cup \neg X)} = P(X \cap \neg X' | (X \cap \neg X') \cup \neg X)$. But this contradicts our conclusion from before that $P(X \cap \neg X' | (X \cap \neg X') \cup \neg X)$ exceeds $\frac{1}{2}$. Therefore, $P(X \cap \neg X') > P(\neg X)$. Analogously, it follows also that

$$P(X' \cap \neg X) > P(\neg X').$$

Finally, from this we can derive: $P(X \cap \neg X') > P(\neg X) \geq P(X' \cap \neg X) > P(\neg X') \geq P(X \cap \neg X')$, which is a contradiction.

- Ad B. Assume for contradiction that there is a sequence $X_0 \supsetneq X_1 \supsetneq X_2 \supsetneq \ldots$ of sets in $\mathfrak{A}$ with probability less 1, with X_0 being P-stable^r as described. None of these sets can be empty, or otherwise the subset relationships holding between them could not be proper. Now let $A_i = X_i \setminus X_{i+1}$ for all $i \geq 0$, and let $B = \bigcup_{i=0}^{\infty} A_i$.[251] Note that every A_i is non-empty and indeed has positive probability, since as observed before P-stable^r sets with probability less than 1 do not contain subsets with probability 0. Furthermore, for $i \neq j$, $A_i \cap A_j = \varnothing$.

  Now we show that the sequence $(P(A_i))$ must converge to 0 as $i \to \infty$. For if not, then there must be an infinite subsequence (A_{i_k}) of (A_i) and a real number $t > 0$, such that $P(A_{i_k}) > t$ for all k. But that would mean that there is an n, such that $P(A_{i_1} \cup \ldots \cup A_{i_n}) = P(A_{i_1}) + \ldots + P(A_{i_n}) > 1$, in contradiction with P1.

  Because, by assumption, X_0 has a probability of less than 1, $P(\neg X_0)$ is a real number greater than 0. It follows that the sequence of real numbers of the form $\frac{P(A_i)}{P(A_i) + P(\neg X_0)} = \frac{P(X_0 \cap (A_i \cup \neg X_0))}{P(A_i \cup \neg X_0)} = P(X_0 | A_i \cup \neg X_0)$ also converges to 0 as $i \to \infty$, where for every i, $(A_i \cup \neg X_0) \cap X_0 \neq \varnothing$ and $P(A_i \cup \neg X_0) > 0$. But this contradicts X_0 being P-stable^r by which every such number $P(X_0 | A_i \cup \neg X_0)$ would have to be greater than r. ■

We may draw three conclusions from this. First of all, in view of part B, P-stable^r sets of probability less than 1 have a certain kind of groundedness property: they do not allow for infinitely descending sequences of subsets. That proves what I claimed to be the case at the end of section 2.3.2 concerning 'quasi-Humean' scepticism.

Secondly, in view of parts A and B taken together, the whole class of P-stable^r propositions X in $\mathfrak{A}$ with $P(X) < 1$ is well-ordered with respect to the subset relation. In particular, if there is a non-empty P-stable^r proposition with probability less than 1 at all, there must also be a *least* non-empty P-stable^r proposition with probability less than 1. A different way of expressing this fact is: if we only look at non-empty P-stable^r propositions with a probability of less than 1, we find that they constitute a so-called *sphere system* that satisfies the so-called Limit Assumption (by well-orderedness) in the sense of Lewis (1973). For every proposition that has non-empty intersection with

[251] Even in the case in which $\mathfrak{A}$ is merely assumed to be a σ-algebra, the set B will in fact be a member of $\mathfrak{A}$.

some 'sphere', that is, with some P-stable^r of probability less than 1, there must be a least 'sphere'—a least P-stable^r of probability less than 1—with which it has non-empty intersection.

Finally, by part A (and P1), we immediately have the following claim, which I put on record for further use:

Observation 19 If $\frac{1}{2} \leq r < 1$, then: all P-stable^r propositions X in $\mathfrak{A}$ with $P(X) < 1$ are subsets of all propositions in $\mathfrak{A}$ of probability 1. (We know already that the latter are all P-stable^r.)

For a given P (and given W), such that P satisfies P1, and for a given $\frac{1}{2} \leq r < 1$, let us denote the class of all non-empty P-stable^r propositions X with $P(X) < 1$ (that is, which are non-trivial) by: $\mathcal{X}_P^r$. What Theorem 18 says is that $\langle \mathcal{X}_P^r, \subseteq \rangle$ is a well-order. So by standard set-theoretic arguments, there is a bijective and order-preserving mapping from $\mathcal{X}_P^r$ into a uniquely determined ordinal β_P^r, where β_P^r is a well-order of ordinals with respect to the subset relation. β_P^r simply measures the length of the well-ordering $\langle \mathcal{X}_P^r, \subseteq \rangle$. Hence, $\mathcal{X}_P^r$ is identical to a strictly increasing sequence of the form $(X_\alpha^r)_{\alpha < \beta_P^r}$. X_0^r is then the least non-empty P-stable^r proposition in $\mathfrak{A}$ with probability less than 1, if there is one at all. If there are none, then β_P^r is simply equal to 0 (that is, the ordinal $\varnothing$). Each world in the union of all X_α^r can be assigned a uniquely determined ordinal rank: the least ordinal α, such that $w \in X_\alpha^r$. So we find that the non-empty P-stable^r propositions X with probability less than 1, if they exist, determine ordinal rankings of those possible worlds that are members of at least one of them. Or equivalently: the resulting assignment of ordinal ranks determines a total pre-order $\preccurlyeq$ for those worlds, such that $w \preccurlyeq w'$ iff the ordinal rank of w is less than or equal to the ordinal rank of w' iff for every sphere X in $\mathcal{X}_P^r$: if $w' \in X$ then $w \in X$.

In the case of a finite probability space, it is useful to extend this assignment of ordinal ranks to all worlds w that are not a member of any P-stable^r set of probability less than 1, but whose singletons $\{w\}$ still have positive probability: naturally, these worlds are then assigned the ordinal rank β_P^r, which in this case is the successor ordinal of the largest ordinal rank that had been assigned previously. If one does so, only worlds whose singletons $\{w\}$ are zero sets will not be assigned any ordinal rank at all. The resulting ordinal ranking of worlds corresponds to the ranking given by the sphere system that consists of all P-stable^r sets of probability less than 1 taken together with the uniquely determined least set of probability 1 (which is also P-stable^r). Later in section 4.2.3 I will turn to sphere systems like that again but without assuming that necessarily *all* P-stable^r sets of probability less than 1 are included in it. The sphere systems in section 4.2.3, the members of which will be P-stable^r sets, will be seen to correspond to conditional belief sets that satisfy variants of the AGM postulates for belief revision.

Furthermore, by P1, Theorem 18, and the fact that no non-empty P-stable^r of probability less than 1 has a non-empty subset of probability zero (Observation 10), each such X in $\mathcal{X}_P^r$ determines a number $P(X) \in (r, 1]$ and no non-empty P-stable^r

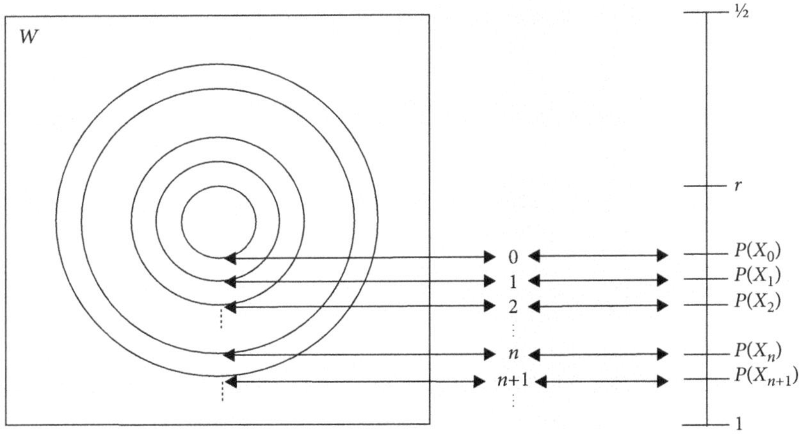

Figure 4.4. P-stable sets for $r \geq \frac{1}{2}$

proposition of probability less than 1 other than X could determine the same number $P(X)$. By P1, the greater the set X with respect to the subset relation, the greater its probability $P(X)$. So we have: for $\alpha < \alpha' < \beta^r_P$ it holds that $r < P(X^r_\alpha) < P(X^r_{\alpha'})$. It follows that there is also a bijective and order-preserving mapping from the set of probabilities of the members of $\mathcal{X}^r_P$ to the set of ordinals below β^r_P (that is, to the set β^r_P). The situation is summarized by Figure 4.4.

From this we can also determine a boundary for β^r_P in case $r \geq \frac{1}{2}$:

Observation 20 Let P satisfy P1 on $\mathfrak{A}$ over W. Let $\frac{1}{2} \leq r < 1$.

The ordinal β^r_P (as defined before) is either finite or equal to ω.

(Hence, the class $\mathcal{X}^r_P$ of all non-empty P-stable^r propositions X with probability less than 1 is countable.)

Proof. Assume for contradiction that $\beta^r_P \geq \omega + 1$: then there certainly exist non-empty P-stable^r propositions X with probability less than 1. Now, for $X^r_\alpha \in \mathcal{X}^r_P$ as defined above, and for all $0 \leq n < \omega$, let $Y_n = X^r_{n+1} \setminus X^r_n$, and let $Z_n = \bigcup_{m \geq n} Y^r_m$.[252] By Theorem 18 and the definition of 'X^r_α' it is the case that for all n, $Z_n \subseteq X^r_\omega$; by assumption we have $P(X^r_\omega) < 1$; and furthermore for all n, $P(Z_n) < 1$, and the sequence (Z_n) is strictly monotonically decreasing. So there is a sequence $X^r_\omega \supsetneq Z_0 \supsetneq Z_1 \supsetneq \ldots$ of sets in $\mathfrak{A}$ with probability less than 1, with X^r_ω being P-stable^r, in contradiction with part B of Theorem 18. ■

We also find that if there are countably infinitely many non-empty P-stable^r propositions X with probability less than 1, then the union of all non-empty P-stable^r propositions X with probability less than 1 is itself P-stable^r, non-empty, and it must

252 Even when $\mathfrak{A}$ were merely assumed to be a σ-algebra, each such set Z_n would be a member of $\mathfrak{A}$.

have probability 1:[253] if $Y \cap \bigcup_{\alpha<\omega} X_\alpha^r \neq \varnothing$ for $Y \in \mathfrak{A}$ with $P(Y) > 0$, then there must be an X_α^r with $\alpha < \omega$, such that $Y \cap X_\alpha^r \neq \varnothing$. Because X_α^r is P-stable^r, it follows that $P(X_\alpha^r|Y) > r$. But $P(\bigcup_{\alpha<\omega} X_\alpha^r|Y) \geq P(X_\alpha^r|Y)$, hence $P(\bigcup_{\alpha<\omega} X_\alpha^r|Y) > r$. So $\bigcup_{\alpha<\omega} X_\alpha^r$ is P-stable^r (and non-empty, of course). If $P(\bigcup_{\alpha<\omega} X_\alpha^r)$ were less than 1, then β_P^r would have to be at least of the order type $\omega + 1$, as $\bigcup_{\alpha<\omega} X_\alpha^r$ is certainly a proper superset of any single set X_α^r; but $\beta_P^r \geq \omega+1$ was ruled out by Observation 20. So $P(\bigcup_{\alpha<\omega} X_\alpha^r) = 1$ (and $\beta_P^r = \omega$).

Since, as noted in Observation 10, no non-empty P-stable^r proposition X with probability less than 1 contains a non-empty zero set as a subset, no union of such sets X could do so either. So in the case in which β_P^r is infinite, the union of all non-empty P-stable^r propositions with probability less than 1 does not just have probability 1, it also does not have any non-empty zero subset. But that means that that union must then be the least set in $\mathfrak{A}$ with probability 1, which thus must exist in that case.

Summing up: with the X_α^r sets for $\alpha < \beta_P^r$ being all and only the non-empty P-stable^r sets of probability less than 1, if $\beta_P^r = \omega$, then it holds that $P(\bigcup_{\alpha<\omega} X_\alpha^r) = 1$ and $\bigcup_{\alpha<\omega} X_\alpha^r$ is the least member of $\mathfrak{A}$ that has this property.

Example 12 (Example 7 from section 4.1.3 Reconsidered)
Clearly, the non-empty and non-trivial P-stable$^{\frac{1}{2}}$ sets are well-ordered with respect to $\subseteq$: $\{w_1\} \subseteq \{w_1, w_2\} \subseteq \{w_1, \ldots, w_4\} \subseteq \{w_1, \ldots, w_5\} \subseteq \{w_1, \ldots, w_6\}$. $\{w_1, \ldots, w_7\}$ and $\{w_1, \ldots, w_8\}$ are also non-empty and P-stable$^{\frac{1}{2}}$, but they are trivial in the sense of having probability 1. The ordinal β_P^r from above is thus $\{0, 1, 2, 3, 4\}$, that is, the ordinal number 5. As mentioned above, one may extend the corresponding assignment of ordinal ranks to worlds also to those worlds that have positive probability but which are not included in any non-trivial P-stable$^{\frac{1}{2}}$ set: in the present case this would apply to w_7, which one might thus assign an ordinal rank of 5 ($= \beta_P^r$). w_8 is not assigned a rank, since it has zero probability. Accordingly, for the non-empty and non-trivial P-stable$^{\frac{3}{4}}$ sets: $\{w_1, w_2\} \subseteq \{w_1, \ldots, w_4\} \subseteq \{w_1, \ldots, w_5\} \subseteq \{w_1, \ldots, w_6\}$.

Figure 4.5 depicts in boldface the ordinal ranks (natural numbers) of all worlds with positive probabilistic mass for the case $r = \frac{1}{2}$ and Example 7 (equivalently, Example 6 from Chapter 3, using the notation from section 3.6). Interpreting ranks in terms of plausibility—the lower the rank is the more plausible is the corresponding world—this would mean: the most plausible world overall is w_1 ($= T \wedge H \wedge \neg E$), which therefore has rank 0. w_2 ($= T \wedge \neg H \wedge \neg E$) has rank 1. Both w_3 and w_4 have rank 2. And so forth.

CHOOSING THE VALUE OF 'r'

What should one choose as the threshold value of 'r' in our bridge postulate BP1^r from section 4.2.2?

253 The countable union $\bigcup_{\alpha<\omega} X_\alpha^r$ is a member of $\mathfrak{A}$ even in the case in which $\mathfrak{A}$ is only assumed to be a σ-algebra.

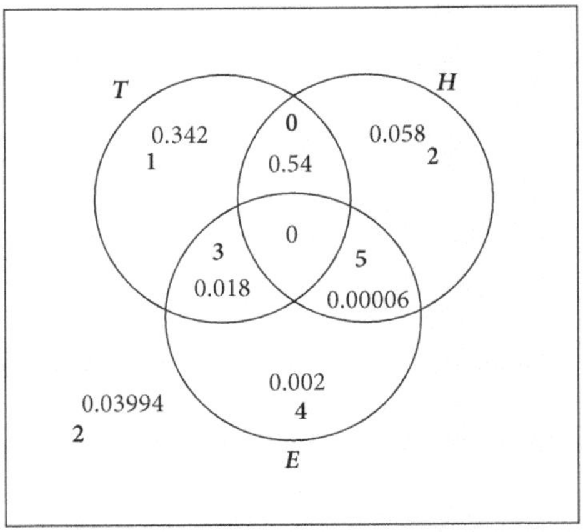

Figure 4.5. Ordinal ranks for the example measure (with $r = \frac{1}{2}$)

As I mentioned before when I introduced BP1^r, if a perfectly rational agent believes a proposition it is hardly acceptable for the same agent to assign to that proposition a probability that is less than or equal to the probability of its negation. That in itself yields a strong argument against $r < \frac{1}{2}$. What I want to add to this now is that Theorem 18 may also be used to support $r \geq \frac{1}{2}$.

In order for the proof of part A in Theorem 18 to go through it was crucial that $r \geq \frac{1}{2}$. Indeed, one can show by means of examples that if $r < \frac{1}{2}$ then A can fail to hold: it is possible then that there are P-stable^r members X, X' of $\mathfrak{A}$, such that neither $X \subseteq X'$ nor $X' \subseteq X$. In fact, it is even possible that there are non-empty P-stable^r members X, X' of $\mathfrak{A}$, such that $X \cap X' = \varnothing$. This means: if our agent's probability measure P is held fixed for the moment, and if $r < \frac{1}{2}$, then depending on what P is like, our postulates P1, B1–B6, and BP1^r may allow for two classes *Bel* and *Bel'* such that all of these postulates are satisfied for each of them (by Theorem 12) and yet some absolute beliefs according to the one class *Bel* contradict some absolute beliefs according to the other class *Bel'*. And that is so although each of them satisfies BP1^r jointly with one and the same subjective probability measure P. Figure 4.6 visualizes a situation like that, where the circles represent different sets B_W (non-empty P-stable^r sets) for different sets *Bel*.

It seems advisable then to demand that $r \geq \frac{1}{2}$ in order to be able to derive as a *law* that a situation like that cannot occur. For if P is fixed, then one might think that our postulates should suffice to rule out systems of qualitative belief that contradict each other. As van Fraassen (1995, p. 350) puts it, the assumed role of full belief is 'to form a single, unequivocally endorsed picture of what things are like'. If $r \geq \frac{1}{2}$, then although Theorem 18 does not pin down such a 'single, unequivocally endorsed picture of what things are like', at least the linearity condition in part A guarantees the following: given P, let X and X' be possible choices of strongest possible believed propositions, such that all of our postulates are satisfied, and where additionally at

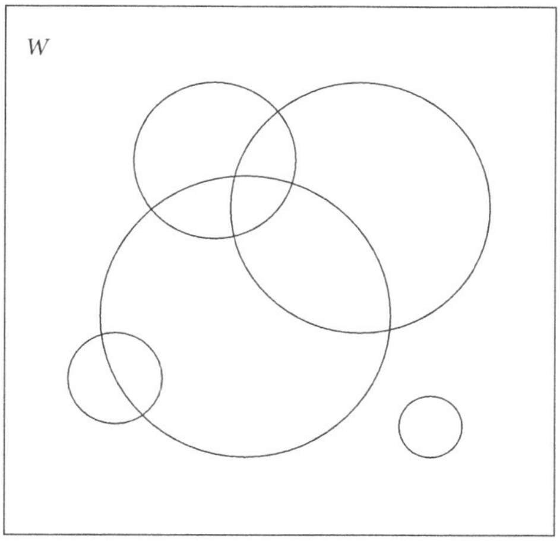

Figure 4.6. P-stable sets for $r < \frac{1}{2}$

least one of them has a probability less than 1. By Theorem 12, X and X' are both non-empty P-stable^r members of $\mathfrak{A}$, and so by the linearity of P-stable^r sets with respect to the subset relation either everything that the agent believes absolutely according to $B_W = X$ would also be believed if it were the case that $B_W = X'$, or vice versa.

Summing up: a strong case can be made against choosing r to be less than $\frac{1}{2}$. Indeed, in Chapter 2 I required from the start that the threshold r was greater than or equal to $\frac{1}{2}$ in all the different versions of the Humean thesis that I considered back then, and in particular this applied to what became my official Humean thesis HT^r. Part A of Theorem 18 yields additional reason to think that assuming $r \geq \frac{1}{2}$ is justified.

But that does not mean that choosing $r < \frac{1}{2}$ would not be an attractive choice if '*Bel*' were taken to express not belief but some weaker epistemic attitude. E.g. let us suppose that '$Bel(Y|X)$' expresses something like: *supposing X, proposition Y is an interesting or salient thesis that is to be investigated further.* We might then be interested in determining *systems* of such interesting hypotheses (given the supposition X) that 'cohere' with each other logically in the way believed propositions do; so B1–B5 would then be plausible again. BP1^r with $r \geq \frac{1}{2}$ would be too much to ask of propositions that are merely considered interesting or salient; but demanding of them that their probabilities are at least above some minimal threshold $r < \frac{1}{2}$ would still be reasonable, for if a proposition is too unlikely, it is probably not worth further investigating either. In this case, all of our postulates so far would be plausible, but the threshold value of 'r' would *not* be assumed to be greater than or equal to $\frac{1}{2}$.[254] I will leave the treatment of this alternative application of the present theory at that.

[254] While part A of Theorem 18 would not apply in that case, it is easy to see from the proof of Theorem 18 that its part B (well-foundedness) would still apply for $r < \frac{1}{2}$.

In the present context, I conclude $r \geq \frac{1}{2}$. I will not exclude any values of 'r' in the present chapter over and above that constraint: none of the options for $r \geq \frac{1}{2}$ seems unnatural or 'irrational' per se, although pragmatic considerations might well favour one number over another one in particular contexts. It all depends on how much stability one demands of all-or-nothing belief.

What I will do, however, is to study formally the general effects that choosing different values of r will have. The following observation tells us more about that:

Observation 21 Let P satisfy P1 on $\mathfrak{A}$ over W. Let $X \in \mathfrak{A}$, and assume that $\frac{1}{2} \leq r < r' < 1$. Then it holds:

If X is P-stable$^{r'}$, X is P-stable^{r}.

Proof. If X is P-stable$^{r'}$, then for all $Y \in \mathfrak{A}$ with $Y \cap X \neq \varnothing$ and $P(Y) > 0$, it holds that $P(X|Y) > r'$. But then it also holds for such Y that $P(X|Y) > r$, since $r' > r$ by assumption. So X is P-stable^{r} as well. ■

Hence, the smaller the threshold value r is, the more inclusive is the class of P-stable^{r} sets that it determines. What this tells us, in conjunction with our previous results, is that if we choose r *minimally* such that $\frac{1}{2} \leq r < 1$, that is, if we choose $r = \frac{1}{2}$, then we do not exclude any of the (in principle) rationally permissible options for B_W. $r = \frac{1}{2}$ maximizes the rationally admissible choices of one's set B_W of doxastically accessible worlds.

Example 13 (Example 7 from section 4.1.3 Reconsidered)
In line with Observation 21, all of the P-stable$^{\frac{3}{4}}$ sets are also P-stable$^{\frac{1}{2}}$ sets. This is clear for the empty set and for the two sets of probability 1, that is, $\{w_1, \ldots, w_7\}$ and W, but it also holds for the non-empty and non-trivial cases of P-stability$^{\frac{3}{4}}$: all of $\{w_1, w_2\}$, $\{w_1, \ldots, w_4\}$, $\{w_1, \ldots, w_5\}$, $\{w_1, \ldots, w_6\}$ are to be found amongst the P-stable$^{\frac{1}{2}}$ sets $\{w_1\}$, $\{w_1, w_2\}$, $\{w_1, \ldots, w_4\}$, $\{w_1, \ldots, w_5\}$, $\{w_1, \ldots, w_6\}$.

Should our agent exclude some P-stable$^{\frac{1}{2}}$ sets by choosing $r > \frac{1}{2}$? So far as the effect of conditional belief on belief revision is concerned, by determining the value of 'r' one lays down how brave a belief can be maximally, or how cautious a belief needs to be minimally, after belief revision (here: belief expansion). Choosing $r = \frac{1}{2}$ is the bravest possible option, and for most purposes this might actually be the right choice. The reason is that the case $r = \frac{1}{2}$ still leaves open whether one demands in addition that one's *absolute* or *un*conditional beliefs should also have a probability above some Lockean threshold $s > r$ (such as: demanding for all and only the believed X that $P(X) \geq 0.9 = s$). So, if one likes, one may still be cautious about one's present *unconditional* beliefs, without being overly cautious about what is going to happen if new evidence comes along: which in many cases might be just the right 'extent' of cautiousness. Some of the results in this book rely on setting $r = \frac{1}{2}$ and thereby add to its plausibility.[255] Finally, many of my examples in this book use a threshold of $r = \frac{1}{2}$ (though maybe a Lockean threshold of greater value), and since the agent's all-or-nothing beliefs in these examples look plausible, this seems to indicate that $r = \frac{1}{2}$ may be a plausible choice in many concrete cases. But then again, maybe for other purposes also one's *conditional* beliefs ought to be especially cautious, in which case a choice of r greater than $\frac{1}{2}$ might be asked for.

FURTHER FORMAL EXAMPLES

I conclude this part on restricted conditional belief by illustrating the theory with the aid of some formal examples of probability measures P and their P-stable^r sets. In all of them, $\mathfrak{A}$ is again the full power set algebra of W.

If W contains exactly two worlds, then the situation is trivial in so far as for given $\frac{1}{2} \leq r < 1$, if there is a world w in W, such that $P(\{w\}) > r$, then the singleton $\{w\} \subseteq W$ is the least non-empty P-stable^r set, and W is the only other P-stable^r set. Otherwise W itself is the only (and thus least) non-empty P-stable^r set.

So let us turn to the first non-trivial case, that is, where W is a set $\{w_1, w_2, w_3\}$ of three elements. Let $r = \frac{1}{2}$. We can view each probability measure on that set W as being represented by a uniquely determined point in a triangle, such that $P(\{w_1\})$, $P(\{w_2\})$, $P(\{w_3\})$ become the scalar factors of the convex combination of three given vectors that we associate with the worlds w_1, w_2, w_3. Then, depending on where P is represented in that triangle, P determines different classes of P-stable^r sets. See Figure 4.7.[256]

The diagram should be read as follows: the vertices of the outer equilateral triangle represent the probability measures that assign 1 to the singleton set of the respective world and 0 to all other singleton sets. Each non-vertex on any of the edges of the outer equilateral triangle represents a probability measure that assigns 0 to exactly one of the three worlds. Each edge of the inner equilateral triangle separates the representatives of probability measures of the following kinds: probability measures that assign to the singleton set of some world a probability that is *greater* than the sum of probabilities that it assigns to the singleton sets of the two other worlds; and probability measures that assign to the singleton set of some world a probability that is *less* than the sum of probabilities that it assigns to the singleton sets of the two other worlds. For instance, to the left-below of the left edge of the inner equilateral triangle we find such probability measures represented that assign to $\{w_1\}$ a greater probability than the sum of what they assign to $\{w_2\}$ and $\{w_3\}$. Each straight line segment that connects a vertex with the mid-point of the opposite edge of the outer equilateral triangle separates the representatives of probability measures of the following kinds: probability measures

[255] Examples are: the results in Chapter 3, those of section 5.2, and Theorem 31 in section 6.1.

[256] I have described this triangle before in section 3.2. But I will include more details about it in the present section.

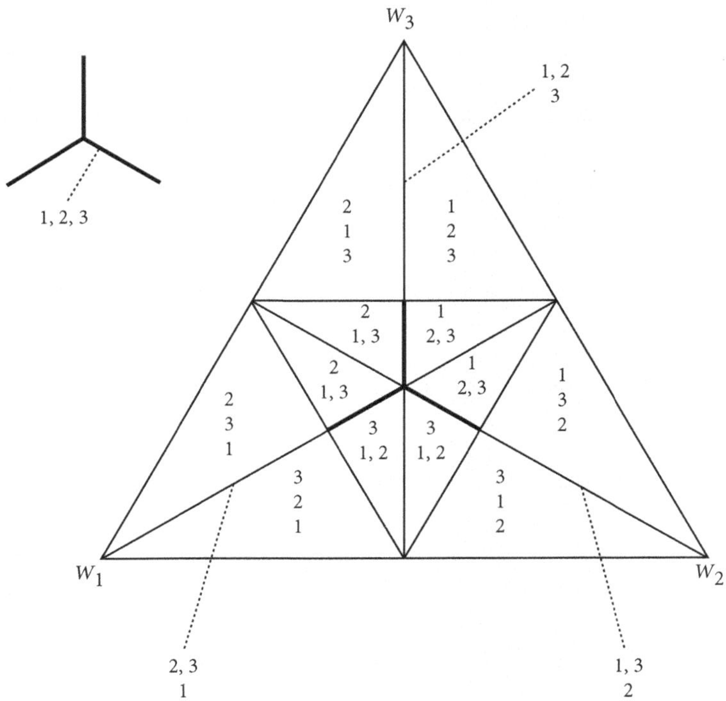

Figure 4.7. Rankings from P-stable^r sets for $r = \frac{1}{2}$

that assign to the singleton set of one world a *greater* probability than to the singleton set of another world; and the probability measures that do so *the other way round.* Accordingly, the straight line segment that connects w_3 and the mid-point of the edge from w_1 to w_2 separates the probability measures that assign more probability to $\{w_1\}$ than to $\{w_2\}$ from those that assign more probability to $\{w_2\}$ than to $\{w_1\}$. The centre point of both equilateral triangles represents the probability measure that is uniform over $W = \{w_1, w_2, w_3\}$, that is, which assigns probability $\frac{1}{3}$ to each (singleton of a) world.

Given all of that, and using the construction procedure for P-stable$^{\frac{1}{2}}$ sets that I have sketched in section 4.2.2, it is easy to read off for each point, and hence for the probability measure that this point represents, all the non-empty P-stable$^{\frac{1}{2}}$ sets that are determined by it. I will turn to this now.

First of all, the points on the outer equilateral triangle are quite special. The probability measure represented by the vertex for w_i has $\{w_i\}$ as its least non-empty P-stable$^{\frac{1}{2}}$ set, all supersets of that set are P-stable$^{\frac{1}{2}}$, too, and all of these P-stable$^{\frac{1}{2}}$ sets have probability 1. Secondly, the probability measures represented by the inner part of the edge between the vertices that belong to two worlds w_i and w_j have either $\{w_i\}$ (which has probability less than 1), or $\{w_j\}$ (which has probability less than 1), or $\{w_i, w_j\}$ (which has probability 1) as their least non-empty P-stable$^{\frac{1}{2}}$ set, depending on whether the representing point is closer to the vertex of w_i than to the vertex of w_j, or vice versa, or equidistant from both of them; all proper supersets of each of them, respectively, are P-stable$^{\frac{1}{2}}$ again, and all of these proper supersets have probability 1 again.

But the really interesting part of the diagram concerns the interior of the outer equilateral triangle. Relative to the probability measures that are represented in that interior region only the set W is assigned probability 1 (and that set is of course P-stable$^{\frac{1}{2}}$). So we can concentrate on non-empty P-stable$^{\frac{1}{2}}$ sets with probability less than 1. As we have learned from Theorem 18, these form a sphere system or a ranked system of sets. In the diagram, I denote these sphere systems by enumerating in different lines the numeral indices of worlds of equal rank in the sphere system, starting with the worlds of rank 0 which I take to correspond to the entries in the bottom line of each numerical inscription. For example: consider the interior of the two smallest right-angled triangles that are adjacent to w_1. Probability measures which are presented by points in the upper one yield a sphere system of three non-empty P-stable$^{\frac{1}{2}}$ sets: $\{w_1\}, \{w_1, w_3\}, \{w_1, w_2, w_3\}$. So w_1 has rank 0, w_3 has rank 1, and w_2 has rank 2. Accordingly, probability measures represented by points in the lower one of the two triangles determine a sphere system of the three non-empty P-stable$^{\frac{1}{2}}$ sets $\{w_1\}, \{w_1, w_2\}, \{w_1, w_2, w_3\}$. In both of these cases, the least P-stable$^{\frac{1}{2}}$ set is $\{w_1\}$. And this makes quite good sense intuitively: for all of (the geometrical representatives of) these measures are pretty close to the vertex for w_1. While, as it were, the measures in the upper right-angled triangle prefer w_2 over w_3 in the ranking that they determine, this is just the other way around for the measures in the lower right-angled triangle. And again this makes good sense: after all, from the viewpoint of the 'dominating' vertex (w_1), the measures in the upper right-angled triangle inhabit the w_3-half of the total equilateral triangle, while the measures in the lower right-angled triangle inhabit the w_2-half of the total equilateral triangle.

The further one moves geometrically towards the centre point of the two equilateral triangles, the more coarse-grained the orderings given by the sphere systems of the represented probability measures become. Probability measures whose points lie on the boldface part in the diagram are treated separately in the little graphic left of the triangle; they all lead to the three worlds being ranked equally. This includes the center point of the outer equilateral triangle which represents the uniform probability measure. The points on the three edges of the inner equilateral triangle—or rather the six halves of those (without their midpoints which fall into the boldfaced lines)—yield sphere systems which coincide with those of the areas to which they are adjacent on the inside. Finally, the three straight (and non-boldfaced) line segments in the interior of the inner equilateral triangle simply inherit their sphere systems from the right-angled triangle areas that they separate.

One might wonder about why in Figure 4.7 sphere systems with one world of rank 0 and two worlds of rank 1 are determined only by points or probability measures in one-dimensional line segments rather than in two-dimensional areas. In one sense, this is really just a consequence of dealing with precisely three worlds. If W had four members, then sphere systems with one world of rank 0, two worlds of rank 1, and hence one world of rank 2 would be represented within proper areas again. However, the following does hold generally: sphere systems with precisely two worlds of maximal rank can only be represented by points or probability measures of areas of dimension $n - 1$, if W has n members. That is because the probabilities of these two worlds of maximal rank must be the same (which follows from $r = \frac{1}{2}$ and the Outclassing Condition in Observation 11), which means that the points of the represented probability measures must lie on one of the distinguished hyperplanes that generalize the distinguished line segments in our diagram to the higher-dimensional case.

For analogous reasons, the following is true: the set of points in the diagram that represent probability measures for which a set of probability 1 is the least P-stable$^{\frac{1}{2}}$ set has Lebesgue measure ('geometrical measure') 0.[257] In words: *Almost all probability measures (in the Lebesgue-measure sense) have non-empty and non-trivial P-stable*$^{\frac{1}{2}}$ *sets!* The reason is that for any such P for which all non-empty P-stable$^{\frac{1}{2}}$ sets are trivial (have probability 1) the following must be the case: if there were a unique world whose singleton had least probability amongst all singletons, then W without that world would be P-stable$^{\frac{1}{2}}$ and non-trivial; so for any such measure there must be at least two worlds whose singleton sets have the same probability. The rest follows from the remark in the previous paragraph: such measures must occupy an area of dimension $n - 1$, and any such measure has Lebesgue measure 0. This is not just so in the case of three possible worlds, such as in our diagram, but for all finite probability spaces. Summing up:

> *For all finite algebras $\mathfrak{A}$ of propositions, almost all probability measures over $\mathfrak{A}$ have a least P-stable*$^{\frac{1}{2}}$ *set with a probability less than 1.*

Taking observations such as that one together with the more concrete examples throughout this monograph suggests: there are sufficiently many ways of determining *Bel* and P so that they may correspond to a perfectly rational agent's belief system that satisfies all of the postulates of the theory for a threshold of $r = \frac{1}{2}$ and where *a proposition of probability less than 1 is believed.* As mentioned before, setting r to $\frac{1}{2}$ also leaves open the absolute degree of belief that an agent assigns to her least *unconditionally* believed proposition B_W: $P(B_W)$ may well be close to 1, such as 0.9 or the like.

If $r > \frac{1}{2}$, then a diagram similar to Figure 4.7 can be drawn, but now with all of the interior straight line segments being pushed towards the three vertices to an extent that is proportional to the magnitude of r: the greater r is, the stronger the constraint that

257 See any textbook in measure theory for the exact definition of the Lebesgue measure.

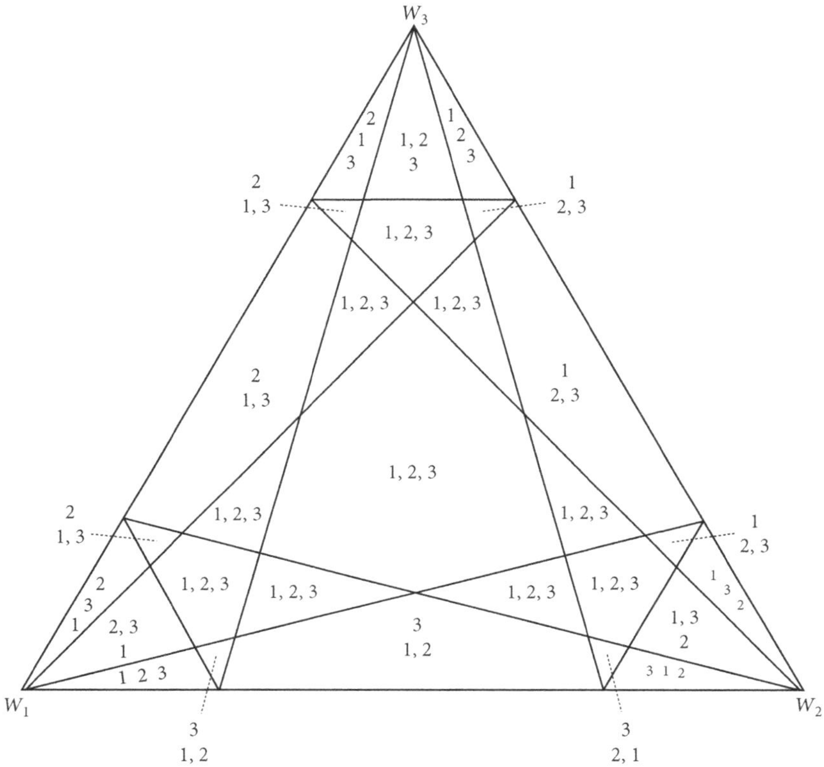

Figure 4.8. Rankings from P-stable^r sets for $r = \frac{3}{4}$

the postulates of the theory impose jointly on P and *Bel*, that is, the smaller the number of P-stable^r sets (as follows from Observation 21). Compare Figure 4.8, which depicts the case in which $r = \frac{3}{4}$. The same conventions about how to read the diagram apply as before. In particular, in its large central region and in some of its neighbouring regions the only P-stable$^{\frac{3}{4}}$ set to exist can be seen to be $W = \{w_1, w_2, w_3\}$ itself. With respect to the agent's absolute beliefs, this means that if the agent's degree-of-belief function P occupies a place within any of these regions, the agent's least believed proposition B_W must be W, which in these cases is also the only proposition of probability 1. In other words: so far as unconditional belief is concerned, the present theory collapses in these cases into the Certainty Proposal again (for all X, $Bel(X)$ iff $P(X) = 1$), as discussed in detail already in section 3.3 of Chapter 3.[258] So far as the agent's *conditional* beliefs are concerned, every consistent proposition X must then be doxastically possible ($B_W \cap X \neq \varnothing$), conditional belief must therefore correspond dynamically to the case of belief expansion, the least believed proposition conditional on any such X must thus be $B_X =$

[258] Makinson (n.d.) raises worries about the theory along these lines.

$B_W \cap X$, which is nothing but $W \cap X = X$, and hence $Bel(Y|X)$ must be the case if and only if $X \subseteq Y$. This might be called the conditional version of the traditional Certainty Proposal about belief. Summing up: the greater the stability threshold r is, the more cautious the agent must be about her beliefs: not just about her absolute beliefs but also about her conditional ones. The closer r is to 1, the larger the region of probability measures is in which the stability theory of belief yields the same results as the conditional variant of the Certainty Proposal for belief; and in the limit of $r \to 1$, that region is simply the full equilateral triangle of *all* (geometrical representatives of) probability measures. A similar phenomenon occurs if r is held fixed but the number n of worlds is increased, as discussed already in section 3.3, in which I also gave reasons to believe that at least in everyday contexts the number of possibilities that an agent needs to distinguish is small: even perfectly rational all-or-nothing belief can afford to be a simple creature in such contexts. And, of course, the same pattern emerges again if *both* r and n are increased: the set of probability measures that force B_W to coincide with the least proposition of probability 1 will be of ever greater Lebesgue measure (which will again be 1 in the limit).

Finally, here is a simple infinite example: let $W = \{w_1, w_2, w_3, \ldots\}$ be countably infinite, let $\mathfrak{A}$ be again the power set algebra on W, and let P be the unique regular countably additive probability measure that is given by: $P(\{w_1\}) = \frac{1}{2} + \frac{1}{4}$, $P(\{w_2\}) = \frac{1}{8} + \frac{1}{16}$, $P(\{w_3\}) = \frac{1}{32} + \frac{1}{64}$, and so on. Then the resulting non-empty P-stable$^{\frac{1}{2}}$ sets are:

$$\{w_1\}, \{w_1, w_2\}, \{w_1, w_2, w_3\}, \ldots, \{w_1, w_2, \ldots, w_n\}, \ldots \text{ and } W.$$

That is a case in which β_P^r from Observation 20 in section 4.2.2 is equal to ω. Or in words: the order type of the set of non-empty P-stable^r propositions of probability less than 1 is just like that of the ordinal number ω. W is the least P-stable^r set of probability 1, which is also equal to the union of all P-stable^r propositions of probability less than 1 in this case.

Once we have covered conditional belief in full in the next section, I will return to some of these formal examples very briefly. I will also add a separate section (section 4.3) with examples that have a more concrete interpretation (some of which will derive from previous chapters).

4.2.3 Conditional Belief in General

As promised, I will now generalize the postulates of the previous section to conditional belief in general: including belief conditional on propositions that may be inconsistent with what our perfectly rational agent believes absolutely.

THE GENERALIZED POSTULATES

The probabilistic postulate P1 remains unchanged. The generalizations of B1–B5 simply result from dropping their antecedent '$\neg Bel(\neg X|W)$' condition:

B1* (Reflexivity) $Bel(X|X)$.

B2* (One Premise Logical Closure)
For all $Y, Z \in \mathfrak{A}$: if $Bel(Y|X)$ and $Y \subseteq Z$, then $Bel(Z|X)$.

B3* (Finite Conjunction)
For all $Y, Z \in \mathfrak{A}$: if $Bel(Y|X)$ and $Bel(Z|X)$, then $Bel(Y \cap Z|X)$.

B4* (General Conjunction)
For $\mathcal{Y} = \{Y \in \mathfrak{A} \mid Bel(Y|X)\}$, $\bigcap \mathcal{Y}$ is a member of $\mathfrak{A}$, and $Bel(\bigcap \mathcal{Y}|X)$.

The Consistency postulate stays the same:

B5* (Consistency) $\neg Bel(\varnothing|W)$.

The same comments and arguments as before apply: in particular, B4* now entails that for every $X \in \mathfrak{A}$ there is a *least set* Y, such that $Bel(Y|X)$, which by B1* must be a subset of X. I denote this proposition again by: B_X. This is consistent with the corresponding notations that I used in section 4.2.2. And once again, we have for all $Y \in \mathfrak{A}$:

$$Bel(Y|X) \text{ if and only if } Y \supseteq B_X.$$ [259]

The following postulate extends the previous Expansion postulate B6 to all cases of conditional belief whatsoever. It corresponds to the standard AGM postulates K*7 and K*8 (Superexpansion and Subexpansion) for belief revision taken together and translated again into the current context:

B6* (General $Bel(\cdot|\cdot)$ / Revision)
For all $X, Y \in \mathfrak{A}$ such that $Y \cap B_X \neq \varnothing$:
for all $Z \in \mathfrak{A}$, $Bel(Z \mid X \cap Y)$ if and only if $Z \supseteq Y \cap B_X$.

So *any* $X \in \mathfrak{A}$ can now take over the role of W in the original B6 postulate on expansion. Equivalently:

B6* (General $Bel(\cdot|\cdot)$ / Revision)
For all $X, Y \in \mathfrak{A}$, such that for all $Z \in \mathfrak{A}$, if $Bel(Z \mid X)$ then $Y \cap Z \neq \varnothing$:
for all $Z \in \mathfrak{A}$, $Bel(Z \mid X \cap Y)$ if and only if $Z \supseteq Y \cap B_X$.

That is: if the proposition Y is consistent with B_X—equivalently: Y is consistent with everything the agent believes conditional on X—then she believes Z conditional on the conjunction of Y and X just in case Z is logically entailed by the conjunction of Y with B_X.

As with the original B6 postulate, this can be justified again based on a sphere semantics (which is formally like David Lewis's semantics for counterfactuals) or total plausibility pre-orders or plausibility rankings (as in belief revision theory

[259] With B6*, it is also going to follow again: $Bel(Y|B_X)$ if and only if $Bel(Y|X)$ if and only if $Y \supseteq B_X$.

and nonmonotonic reasoning)—recall section 4.1.2. What a conditional belief in a consequent given an antecedent expresses according to these semantics is that the most plausible antecedent-worlds are consequent-worlds as given by the total pre-order. For an antecedent proposition X, it holds that B_X is the set of most plausible X-worlds. Now, if some of the most plausible X-worlds are Y-worlds—the set B_X has non-empty intersection with Y—these worlds must be precisely the most plausible $X \cap Y$-worlds, and hence the most plausible $X \cap Y$-worlds are Z-worlds if and only if all the most plausible worlds X-worlds that are also Y-worlds are Z-worlds.

Analogously to the last section, this is yet another equivalent formulation of B6*:

B6* (General $Bel(\cdot|\cdot)$ / Revision) For all $X, Y \in \mathfrak{A}$ such that $Y \cap B_X \neq \varnothing$: $B_{X \cap Y} = Y \cap B_X$.

From the last version of B6* it should become very clear that the postulate corresponds to AGM's K*7 and K*8 combined (compare section 4.1.2). If formulated in the original AGM terminology: in the case in which the formula $\neg B$ is a member $K * A$, it follows that K*8 is not applicable, and K*7 does not impose any constraint on $*$ really, since $(K * A) + B$ gets trivialized and becomes the full language. But if $\neg B$ is *not* a member $K * A$ (in my terminology: $Y \cap B_X \neq \varnothing$), then K*7 and K*8 together correspond to my B6*: so B6* is just the propositional version of K*7 and K*8 taken together.

Like in the case of the AGM postulates for belief revision, B6* together with our other postulates does not pin down a unique conditional belief set. Rather the postulates impose constraints on any conditional belief set of a perfectly rational agent whatsoever. What the present theory (with BP1^{r*} below) adds to the original AGM context is that now also an agent's degree-of-belief function P will need to 'play along': certain AGM-like *Bel*s will exclude certain Ps, and vice versa.

In terms of nonmonotonic reasoning, and with the right additional postulates on nonmonotonic consequence relations $\mid\!\sim$ in the background, B6* corresponds to the rule of Rational Monotonicity (Kraus et al. 1990, Lehmann and Magidor 1992):

$$\frac{X \mid\!\sim Z,\; X \not\mid\!\sim \neg Y}{X \wedge Y \mid\!\sim Z}$$

In my terminology, $X \mid\!\sim Z$ means that $B_X \subseteq Z$. $X \not\mid\!\sim \neg Y$ says that $B_X \nsubseteq \neg Y$, that is, $B_X \cap Y \neq \varnothing$. B6* or Rational Monotonicity demands in this case that $X \wedge Y \mid\!\sim Z$, that is, $B_{X \cap Y} \subseteq Z$. A counterfactual version of Rational Monotonicity is also valid in Lewis's (1973) logic of counterfactuals.

Semantically, it is B6* or rational monotonicity that expresses the totality or linearity of the (pre-)order of worlds (or the corresponding nestedness of spheres) that correspond to conditional *Bel*: for all worlds w, w', $w \leq w'$ or $w' \leq w$. AGM belief revision and rational nonmonotonic consequence relations share this totality presumption with other areas, such as the Lewis–Stalnaker logic of counterfactuals (totality of the closeness order on worlds), decision theory (totality of the preference order over acts), and social choice theory (totality of the preference order over alternatives). In the context of the present theory, totality comes with three benefits: it enforces simplicity (total pre-orders are simpler than arbitrary pre-orders), which goes well with the intended simplicity of all-or-nothing belief in comparison with degrees of belief. It matches the linearity of degrees of belief on the numerical side. And most importantly (continuing the point that I made at the very end of section 4.1.2): if one holds the agent's degree-of-belief function P fixed, the different candidate sets for B_W were found to be P-stable^r as early as in Theorem 12—for which B6* had not played a role as yet. Assume that an agent's fallback positions for cases in which the evidence contradicts one's unconditional beliefs (or contradicts the P-stable^r set B_W) have to be amongst those sets that had been possible candidates for B_W but that did not actualize. That is: assume fallback positions to correspond to P-stable^r sets other than the actual set B_W. Then these fallback positions are ordered linearly by the subset relation at least so far as P-stable^r sets of probability less than 1 are concerned, simply because these sets are always ordered like that according to Theorem 18. This, in turn, induces a total pre-order of all worlds with positive probability, as explained after Theorem 18 in section 4.2.2. Roughly: stability in the sense of the Humean thesis from Chapter 2, which only concerned *un*conditional belief, delivers a total pre-order of worlds all by itself.

The generalized version BP1^{r*} of the previous bridge postulate BP1^r from section 4.2.2 arises from dropping the '$Y \cap B_W \neq \varnothing$' restriction again. So we have:

BP1^{r*} (Likeliness) For all $Y \in \mathfrak{A}$ with $P(Y) > 0$:
for all $Z \in \mathfrak{A}$, if $Bel(Z|Y)$, then $P(Z|Y) > r$.

Finally, I add yet another bridge principle for conditional belief and the degree-of-belief mapping. The principle has mainly an auxiliary role: it will simplify the formulation of the representation theorem below. Without the principle more would have to be said below about P-stable^r sets of probability 1, and how they can figure as spheres or fallback positions for a conditional belief set *Bel*. Additionally, the principle will remove one slight difference between the stability accounts of Chapters 2 and 3 and the one developed so far in the present chapter: back then it was the case that if B_W had probability 1, then B_W would be the least subset of W with probability 1.[260] That was not a consequence of the theory developed so far in this chapter, but it will be once we adopt the following principle.

The downside of the bridge principle will be that it modifies AGM's original Consistency postulate K*5 a bit: as mentioned already in section 4.1.2, in AGM only revision by inconsistent evidence can yield an inconsistent belief set, whereas by the next postulate also (and only) evidence of probability 0 has that effect:

BP2* (Zero Supposition) For all $Y \in \mathfrak{A}$: $P(Y) = 0$ if and only if $B_Y = \varnothing$.

[260] Compare Theorem 5 in Appendix B and Theorem 7 in Chapter 3.

In belief revision terms, the left-to-right direction of this means: if the agent revises her all-or-nothing beliefs based on a piece of evidence of degree of belief 0—a piece of evidence for which standard conditionalization is undefined—her all-or-nothing beliefs will get trivialized. Every proposition will be believed conditional on any such piece of evidence, since every proposition is entailed by (is a superset of) the empty set. The right-to-left direction says: revision leads to trivialization only in cases in which the evidence has probability 0. In a sense, by BP2*, conditional probability and conditional belief are getting more in sync: where the one is undefined the other one is trivialized, and vice versa. If the right-to-left direction of BP2* failed, then there would be cases in which all-or-nothing belief revision given Y would lead to an inconsistent belief set, whereas conditionalization on Y would yield a coherent degree-of-belief function. If the left-to-right direction of BP2* failed, then there would be cases in which conditionalization on Y would not be defined at all, whereas all-or-nothing belief revision given Y would determine a consistent belief set.[261]

Here is a consequence of the left-to-right direction of BP2*: assume $P(Y) = 0$; by that left-to-right direction, $B_Y = \varnothing$. Hence, by B6 (and thus also by the stronger B6*), it must be the case that $Y \cap B_W = \varnothing$. Finally, plugging in '$\neg Y$' for 'Y' gives us: if $P(Y) = 1$, then $B_W \subseteq Y$. In particular, this means that if B_W has probability 1 itself, then B_W must be the least proposition in $\mathfrak{A}$ with probability 1—as promised.

For the same reason, in the case in which $P(B_W) = 1$, BP2* forces our probability space to be such that there exists a least set with probability 1, which is a non-trivial constraint that is not satisfied generally. In our case, in which $\mathfrak{A}$ is the power set of W, the intersection of all sets of probability 1 is a member of $\mathfrak{A}$, but that intersection will not necessarily have probability 1 itself. If $\mathfrak{A}$ had only been assumed to be a σ-algebra, then it would not necessarily have been the case either that there would be a least member of $\mathfrak{A}$ with probability 1: the Lebesgue measure on the (measurable subsets of the) unit interval would be a counterexample. Indeed, one might well diagnose that the theory that is developed in this chapter does not go together well with *continuous* probability distributions. On the other hand, there are of course also lots of probability measures for which there exist least sets of probability 1:

- All probability measures on finite algebras $\mathfrak{A}$, and hence also all probability measures on algebras $\mathfrak{A}$ that are based on a finite set W of worlds.
- All countably additive probability measures on the power set algebra of a set W where W is countably infinite: in that case the conjunction of all sets of probability 1 is a member of the algebra of propositions again, and it is the least set of probability 1.

[261] A different way of handling the case of conditionalizing on zero sets would have been to start on the probabilistic side with primitive conditional probability measures, which indeed allow for conditionalization on zero sets. For an extension of the present theory of stable conditional belief to the case of such primitive conditional probability measures, see Pedersen and Arló-Costa (2012).

- All countably additive probability measures (on a σ-algebra) that are regular (or strictly coherent); that is, where it holds: for all $X \in \mathfrak{A}$, $P(X) = 0$ if and only if $X = \varnothing$. Here W itself happens to be the least set of probability 1. Regularity does not enjoy general support, even though authors such as Carnap, Shimony, Stalnaker, and others have argued for it in the past to be a plausible constraint on subjective probability measures, some of them in view of a special variant of the Dutch book argument that favours Regularity.[262]

BP2* in combination with B6* leads to further conclusions. Even though we know that every proposition of probability 1 is P-stable^r, we have just seen that with BP2* only one such proposition of probability 1 can coincide with an agent's set B_W: the *least* proposition of probability 1. And so, if $P(B_W) = 1$, B_W cannot have non-empty subsets of probability 0 either, just as we found to be the case before when $P(B_W) < 1$ (by B_W being P-stable^r and Observation 10).

More generally, we have:

Observation 22 Our postulates entail that for all $X \in \mathfrak{A}$: B_X does not contain a non-empty subset of probability 0.

Proof. For assume otherwise, that is: there is a $Y \subseteq B_X$ ($\subseteq X$), $Y \neq \varnothing$, and $P(Y) = 0$. By B6*, $B_Y = B_{X \cap Y} = Y \cap B_X = Y$. So B_Y would have to be non-empty, too. But by BP2*, since $P(Y) = 0$, it must be that $B_Y = \varnothing$, which is a contradiction. ■

But let me stress again that BP2* still has mainly an auxiliary role: it makes the theory work 'more smoothly' in a context, such as the present one, in which P is assumed to be a standard probability measure by which conditional probabilities are defined through the ratio formula. In a different formal framework, for instance one in which P would have been assumed to be a primitive conditional probability measure, the theory might be developed just as easily without adopting BP2*.

THE SECOND REPRESENTATION THEOREM

We are now ready to prove the main representation theorem for conditional beliefs in general. Its 'soundness' direction (right-to-left) incorporates the corresponding direction of Grove's (1988) representation theorem for belief revision operators in terms of sphere systems.[263]

[262] See Hájek (2012b) for a recent survey and appraisal of that debate. Hájek himself argues against Regularity as a norm of rationality.

[263] I assumed $\mathfrak{A}$ to be the class of all subsets of W. But at the same time I wanted my results to be prepared for being applied also in cases in which $\mathfrak{A}$ is merely assumed to be a σ-algebra. That is the reason why I did not simply translate the more difficult 'completeness' part of Grove's representation theorem into the present context in order to apply it in the proof of the left-to-right direction of Theorem 23. Grove's construction of spheres involves taking unions of propositions that would not be guaranteed to be members of a given σ-algebra $\mathfrak{A}$. That is why my own proof of that part of the theorem differs quite significantly from Grove's.

Theorem 23 (Representation Theorem for Full Conditional Belief)
Let *Bel* be a class of ordered pairs of members of $\mathfrak{A}$, and let $P : \mathfrak{A} \to [0,1]$. Then the following two statements are equivalent:

I. P and *Bel* satisfy P1, B1*–B6*, BP1^{r*}, BP2*.
II. P satisfies P1, P and $\mathfrak{A}$ are such that $\mathfrak{A}$ contains a least set of probability 1, and there is a class $\mathcal{X}$ of non-empty P-stable^r propositions in $\mathfrak{A}$, such that (i) $\mathcal{X}$ contains the least set of probability 1 in $\mathfrak{A}$, (ii) all other members of $\mathcal{X}$ have probability less than 1, and:
   - For all $Y \in \mathfrak{A}$ with $P(Y) > 0$: if, with respect to the subset relation, X is the least member of $\mathcal{X}$ for which $Y \cap X \neq \varnothing$ holds (which exists), then for all $Z \in \mathfrak{A}$:

$$Bel(Z \mid Y) \text{ if and only if } Z \supseteq Y \cap X.$$

   - Additionally, for all $Y \in \mathfrak{A}$ with $P(Y) = 0$, for all $Z \in \mathfrak{A}$: $Bel(Z|Y)$.

Furthermore, if condition I is the case, then $\mathcal{X}$ in condition II is uniquely determined.

Theorem 23 generalizes Theorem 12 from section 4.2.2 to conditional beliefs in general: accordingly, Theorem 12 simply dealt with the special case of a sphere system of just *one* P-stable^r set. Theorem 23 tells us that general conditional belief is always representable by *some* sphere system of P-stable^r sets, whether of cardinality 1 or of a higher cardinality. In any case, by our postulates, conditional belief is stable again (in the sense of P-stability^r): we have a conditional version of the stability theory of belief from the previous chapters.

Proof. The right-to-left direction (II to I) is like the one in Theorem 12, except that one shows first that the equivalence for *Bel* entails for all $Y \in \mathfrak{A}$ with $P(Y) > 0$ that $B_Y = Y \cap X$, where X is the least member of $\mathcal{X}$ for which $Y \cap X \neq \varnothing$ (and thus also $B_Y \neq \varnothing$). The existence of that least member follows from the following facts: from Theorem 18 (P-stable^r sets of probability less than 1 being well-ordered by $\subseteq$), from the fact that every non-empty P-stable^r proposition with probability less than 1 is a subset of the least set in $\mathfrak{A}$ with probability 1 (by Observation 19), and from the fact that the least set of probability 1 in $\mathfrak{A}$—which is a member of $\mathcal{X}$ by (i), and which is the only member of $\mathcal{X}$ with probability 1 by (ii)—must have non-empty intersection with every proposition of positive probability (by P1). The proof of B6* is straightforward (and analogous to Grove 1988), given this characterization of B_Y. BP2* follows immediately, too, from the 'Additionally, ...' assumption in part II of Theorem 23.

So we can concentrate on the left-to-right direction: P1 is satisfied by assumption. Now we define $\mathcal{X}$ by recursion as the class of all sets X_α of the following kind: for all ordinals $\alpha < \beta_P^r + 1$ (the successor of the ordinal that was defined in section 4.2.2), let

$$X_\alpha = \bigcup_{\gamma<\alpha} (X_\gamma) \cup B_{W \setminus (\bigcup_{\gamma<\alpha} X_\gamma)}.$$

(So, in particular, $X_0 = B_W$.)

At first I make a couple of observations about this class $\mathcal{X}$.

(a) Every member of $\mathcal{X}$ is also a member of $\mathfrak{A}$.[264]

(b) For all $\gamma < \alpha < \beta_P^r+1$: $X_\gamma \subseteq X_\alpha$. This follows directly from the definition of the members of $\mathcal{X}$. From this it also follows that for all $\alpha+1 < \beta_P^r+1$: $X_{\alpha+1} = X_\alpha \cup B_{W\setminus X_\alpha}$.

(c) For all $\alpha < \beta_P^r + 1$: $X_\alpha = \bigcup_{\gamma<\alpha} B_{W\setminus \bigcup_{\delta<\gamma} X_\delta} \cup B_{W\setminus \bigcup_{\gamma<\alpha} X_\gamma}$. By induction. Assume that for all $\gamma < \alpha$: $X_\gamma = \bigcup_{\delta<\gamma} B_{W\setminus \bigcup_{\eta<\delta} X_\eta} \cup B_{W\setminus \bigcup_{\delta<\gamma} X_\delta}$. Substituting this for the first occurrence of 'X_γ' in the original definition of X_α, we conclude: $X_\alpha = \bigcup_{\gamma<\alpha}[\bigcup_{\delta<\gamma} B_{W\setminus \bigcup_{\eta<\delta} X_\eta} \cup B_{W\setminus \bigcup_{\delta<\gamma} X_\delta}] \cup B_{W\setminus \bigcup_{\gamma<\alpha} X_\gamma}$. But this can be simplified, by eliminating double counting of ordinals, to: $X_\alpha = \bigcup_{\gamma<\alpha}[B_{W\setminus \bigcup_{\delta<\gamma} X_\delta}] \cup B_{W\setminus \bigcup_{\gamma<\alpha} X_\gamma}$, which was to be shown.

(d) For all $\alpha < \beta_P^r+1$: for all $Y \in \mathfrak{A}$ with $Y \cap X_\alpha \neq \varnothing$, it holds that $B_Y \subseteq X_\alpha$. This is because: if $Y \cap X_\alpha \neq \varnothing$, then by (c) there is a $\gamma \leq \alpha$, such that $Y \cap B_{W\setminus \bigcup_{\delta<\gamma} X_\delta} \neq \varnothing$, and by the well-orderedness of the ordinals, there must be a least such ordinal γ. Note that for that least ordinal γ it holds that $Y \cap \bigcup_{\delta<\gamma} X_\delta = \varnothing$, by (c) again, and hence $Y \subseteq W\setminus(\bigcup_{\delta<\gamma} X_\delta)$. By B6* and $Y \cap B_{W\setminus \bigcup_{\delta<\gamma} X_\delta} \neq \varnothing$, it holds that $B_{[W\setminus \bigcup_{\delta<\gamma} X_\delta]\cap Y} = Y \cap B_{W\setminus \bigcup_{\delta<\gamma} X_\delta}$, which is equivalent to $B_Y = Y \cap B_{W\setminus \bigcup_{\delta<\gamma} X_\delta}$ by $Y \subseteq W\setminus(\bigcup_{\delta<\gamma} X_\delta)$. Finally, because $Y \cap B_{W\setminus \bigcup_{\delta<\gamma} X_\delta} \subseteq B_{W\setminus \bigcup_{\delta<\gamma} X_\delta} \subseteq X_\alpha$ by (c) again, it follows that $B_Y \subseteq X_\alpha$.

(e) For all $\alpha < \beta_P^r + 1$: X_α is P-stable^r. This can be derived by applying Definition 8 of P-stability^r: for all $Y \in \mathfrak{A}$, if $Y \cap X_\alpha \neq \varnothing$ and $P(Y) > 0$, then by (d), $B_Y \subseteq X_\alpha$, and hence by the properties of B_Y: $Bel(X_\alpha|Y)$. But this implies by BP1^{r*} that $P(X_\alpha|Y) > r$. That is: X_α is P-stable^r.

(f) There exists a least proposition $X \in \mathfrak{A}$ with probability 1, $X \in \mathcal{X}$, and X is the only member of $\mathcal{X}$ with probability 1. Here is the proof of that fact.

First of all, assume for contradiction that all sets X_α with $\alpha < \beta_P^r+1$ have probability less than 1. Since they are all P-stable^r by (e), it follows from (b) that there is a well-ordered sequence of (not necessarily strictly) increasing P-stable^r sets of probability less than 1, where the length of that whole sequence is $\beta_P^r + 1$. That sequence could not be one of *strictly* increasing P-stable^r sets of probability less than 1 throughout the sequence, for by the definition of β_P^r in section 4.2.2, β_P^r itself was already the ordinal type of the sequence of *all* P-stable^r sets of probability less than 1 whatsoever. So there must be $\alpha < \alpha' < \beta_P^r + 1$, such that $X_\alpha = X_{\alpha+1}$. Hence, by (b) again: $X_\alpha = X_\alpha \cup B_{W\setminus X_\alpha}$, and therefore $B_{W\setminus X_\alpha} \subseteq X_\alpha$. Since additionally $B_{W\setminus X_\alpha} \subseteq W\setminus X_\alpha$ by the definition of '$B_{W\setminus X_\alpha}$' and B1*–B4*, it follows that $B_{W\setminus X_\alpha} = \varnothing$. But because $P(X_\alpha) < 1$ by assumption, it also holds that $P(W\setminus X_\alpha) > 0$ by P1, so by the right-to-left

[264] This is trivial, since $\mathfrak{A}$ is the power set algebra on W. But it would also hold if $\mathfrak{A}$ had only been assumed to be a σ-algebra: by induction. For assume that all X_γ are in $\mathfrak{A}$ for $\gamma < \alpha < \beta_P^r+1$: by Observation 20, β_P^r is countable and so are its predecessors, and therefore by $\mathfrak{A}$ being a σ-algebra, $\bigcup_{\gamma<\alpha} X_\gamma \in \mathfrak{A}$; thus $W\setminus \bigcup_{\gamma<\alpha} X_\gamma \in \mathfrak{A}$, and therefore $B_{W\setminus \bigcup_{\gamma<\alpha} X_\gamma} \in \mathfrak{A}$ (by B4*); hence, $X_\alpha \in \mathfrak{A}$.

direction of BP2* it follows that $B_{W\setminus X_\alpha} \neq \varnothing$, which is a contradiction. Hence, we have that there must be at least one set X_α with $\alpha < \beta^r_P + 1$ that has probability 1.

Secondly, consider any such set X_α with $\alpha < \beta^r_P + 1$ and $P(X_\alpha) = 1$: by (c), any such set X_α is a union of sets of the form 'B_X'. By Observation 22, any such set X_α does not have any non-empty zero subsets. But that means, by P1, any such set X_α must be identical to the unique least set $X \in \mathfrak{A}$ with probability 1, which therefore must exist. So we have that there exists a least proposition $X \in \mathfrak{A}$ with probability 1, $X \in \mathcal{X}$, and X is the only member of $\mathcal{X}$ with probability 1.

Now we can conclude the proof of II. We derive first the main equivalence claim for *Bel*. Let $Y \in \mathfrak{A}$ with $P(Y) > 0$: by P1 and (f), there is a member of $\mathcal{X}$ with which Y has non-empty intersection. Let $\alpha < \beta^r_P + 1$ be least, such that $Y \cap X_\alpha \neq \varnothing$: because of (b), X_α is then, with respect to the subset relation, the least member of $\mathcal{X}$ for which this holds.

I will now show that $B_Y = Y \cap X_\alpha$, from which the relevant part of II follows by means of the definition of 'B_Y' and B1*–B4*. Consider $Y \cap X_\alpha$, which by assumption is non-empty: by (c), $X_\alpha = \bigcup_{\gamma<\alpha} B_{W\setminus\bigcup_{\delta<\gamma} X_\delta} \cup B_{W\setminus\bigcup_{\gamma<\alpha} X_\gamma}$. If Y had non-empty intersection with any set of the form $B_{W\setminus\bigcup_{\delta<\gamma} X_\delta}$ for $\gamma < \alpha$, then $Y \cap X_\gamma \neq \varnothing$, by (c) again, in contradiction with the way in which α was defined before. Therefore, $Y \cap X_\alpha = Y \cap B_{W\setminus\bigcup_{\gamma<\alpha} X_\gamma} \neq \varnothing$. The latter implies with B6* that $B_{[W\setminus\bigcup_{\gamma<\alpha} X_\gamma]\cap Y} = Y \cap B_{W\setminus\bigcup_{\gamma<\alpha} X_\gamma}$. By the defining property of α again, $Y \cap \bigcup_{\gamma<\alpha} X_\gamma$ is empty, and thus $[W\setminus\bigcup_{\gamma<\alpha} X_\gamma]\cap Y = Y$. So we have $B_Y = Y \cap B_{W\setminus\bigcup_{\gamma<\alpha} X_\gamma} = Y \cap X_\alpha$, and we are done.

Finally, consider $Y \in \mathfrak{A}$ with $P(Y) = 0$: by BP2*, $B_Y = \varnothing$, from which the last line of part II of Theorem 23 follows by the definition of 'B_Y' and B1*–B4* again.

Uniqueness follows from: if there are two distinct such classes $\mathcal{X}$, $\mathcal{X}'$ with the stated properties, then they must differ with respect to at least one P-stable^r set of probability less than 1. Without loss of generality, let X_α be the first member of $\mathcal{X}$ that is not also a member of $\mathcal{X}'$: since X_α is P-stable^r and has probability less than 1, it follows from Observation 20 that α is finite. If $\alpha = 0$, then B_W could not be the same as being given by $\mathcal{X}$ and $\mathcal{X}'$, which would be a contradiction. If α is a successor ordinal $\gamma + 1$, then, by (b), $B_{W\setminus X_\gamma} = X_\alpha \setminus X_\gamma$ according to $\mathcal{X}$. Because X_γ is a member of $\mathcal{X}'$ while X_α is not, $B'_{W\setminus X_\gamma}$ as given by $\mathcal{X}'$ (which must be of the form $X \setminus X_\gamma$ for some $X \in \mathcal{X}'$) must differ from $B_{W\setminus X_\gamma} = X_\alpha \setminus X_\gamma$. Hence, $B_{W\setminus X_\gamma}$ would not be the same as being given by $\mathcal{X}$ and $\mathcal{X}'$, which would again be a contradiction. ∎

The right-to-left direction of Theorem 23 gives us a recipe of how to build models for the conjunction of all our postulates from any probability measure for which there exists a least set of probability 1: just pick some non-empty P-stable^r sets of probability less than 1 (if there are any), take also the least set of probability 1, and put them together into a set $\mathcal{X}$. Then by defining conditional *Bel* by means of what is said in II above, one ends up with such a model. And the left-to-right direction of the theorem shows that *every* model of all of our postulates taken together can be built in such a manner.

Example 14 (Example 7 from section 4.1.3 Reconsidered)
We know already what the non-empty and non-trivial P-stable$^{\frac{1}{2}}$ sets for the example measure P are: $\{w_1\}, \{w_1, w_2\}, \{w_1, \ldots, w_4\}, \{w_1, \ldots, w_5\}, \{w_1, \ldots, w_6\}$.

By means of Theorem 23, it is now easy to define *Bel* so that all of our postulates on probability and conditional belief in general are satisfied (including all of our bridge axioms for them). For instance, pick $\{w_1, w_2\}, \{w_1, \ldots, w_5\}, \{w_1, \ldots, w_6\}$ from the P-stable$^{\frac{1}{2}}$ sets (say, $r = \frac{1}{2}$), add the least set $\{w_1, \ldots, w_7\}$ of probability 1, and the resulting set

$$\mathcal{X} = \{\{w_1, w_2\}, \{w_1, \ldots, w_5\}, \{w_1, \ldots, w_6\}, \{w_1, \ldots, w_7\}\}$$

will do, amongst many others. Note that for all $Z \subseteq W$, $Bel(Z \mid \{w_8\})$, if we follow the 'Additionally, ...' clause in II of Theorem 23, since $P(\{w_8\}) = 0$. In other words: $B_{\{w_8\}} = \varnothing$.

By Grove's (1988) semantics for AGM, the 'fallback positions' to which a perfectly rational agent must withdraw when encountering evidence that contradicts her current beliefs are the spheres in the very sphere system that corresponds to her conditional belief set $Bel(\cdot|\cdot)$. In the present theory, according to Theorem 23, these spheres are P-stable^r sets. By Theorem 12, P-stable^r sets are precisely the sets that are the permissible candidates for the agent's logically strongest *un*conditionally believed proposition B_W. Therefore, as promised at the end of section 4.1.2, AGM fallback positions are nothing but possible but non-actualized choices for the agent's prior set B_W.

In the case in which W is finite, let us finally combine Theorem 23 with the previous Observation 11 that concerned the Outclassing Condition for P-stable^r sets, where additionally these sets either have probability less than 1 or coincide with the least proposition of probability 1. Let me explain what is going on just for the special case of $r = \frac{1}{2}$: by the Outclassing Condition, the probability of every (singleton of a) member of such a P-stable^r set is greater than the probability of the complement of the set, that is, greater than the sum of probabilities of (the singletons of) the members of such a complement. Now take several such P-stable^r sets to be put together in a sphere system which, in turn, corresponds to a total pre-order $\preccurlyeq$ of worlds: then the probability of every member w of a sphere will be greater than the probability of the complement of the sphere. In particular, if w is a world that is a member of at least some sphere in such a sphere system, that condition will apply to the *least* sphere that includes w as a member: hence, the probability of w will be greater than the probability of the complement of that least sphere, that is, greater than the sum of probabilities of the worlds w' that are 'less plausible' in the order $\preccurlyeq$ than w ('higher up' in the ordering). An analogous conclusion can be drawn, mutatis mutandis, if $r > \frac{1}{2}$.

This yields immediately the following *Sum Condition* for conditional belief or sphere systems as described in Theorem 23:

Observation 24 Assume W is finite, $\mathfrak{A}$ is the power set algebra on W. Then the following two statements are equivalent:

- $\mathcal{X}$ is a sphere system of P-stable^r propositions as described in clause II of Theorem 23 (or equivalently, $\mathcal{X}$ is given by *Bel*, such that P and *Bel* satisfy P1, B1*–B6*, BP1^{r*}, BP2*, as stated in clause I of Theorem 23).
- $\mathcal{X}$ is a sphere system that corresponds to a uniquely determined total pre-order $\preccurlyeq$ on all worlds w in W with positive probability $P(\{w\})$ (given by: $w \preccurlyeq w'$ iff for every X in $\mathcal{X}$, if $w' \in X$ then $w \in X$), such that $\preccurlyeq$ satisfies the following *Sum Condition* relative to P and r: for all $w \in W$ with $P(\{w\}) > 0$,

$$P(\{w\}) > \frac{r}{1-r} \cdot \sum_{w':\, w \prec w'} P(\{w'\})$$

(where $w \prec w'$ iff $w \preccurlyeq w'$ and $w' \not\preccurlyeq w$).

I am going to return to this Sum Condition for $\prec$ relative to P and to the special threshold $r = \frac{1}{2}$ in section 5.2 of Chapter 5.

The theory that was developed in this chapter extends the theory from the previous chapters to the case of rational conditional belief. Therefore, it also inherits its features, whether attractive or controversial. In particular, the context-sensitivity of rational belief that was observed and discussed in Chapters 2 and 3 extends to rational conditional belief as well. But I will not go into this here once again.

FURTHER FORMAL EXAMPLES, ONCE AGAIN

Let us reconsider $W = \{w_1, w_2, w_3\}$, with $r = \frac{1}{2}$. Figure 4.7 from section 4.2.2 already depicted the rankings of worlds as being given by the largest possible sphere system of all P-stable$^{\frac{1}{2}}$ sets of probability less than 1 plus the least set of probability 1, for all the different possible probability measures on the power set algebra of W. For example: consider again the interior of the two smallest right-angled triangles that are adjacent to w_1. Probability measures that are presented by points in the upper one determine a sphere system of three non-empty P-stable$^{\frac{1}{2}}$ sets: $\{w_1\}, \{w_1, w_3\}, \{w_1, w_2, w_3\}$ (only the last one of which has probability 1). So w_1 has rank 0, w_3 has rank 1, and w_2 has rank 2. Accordingly, if *Bel* is determined by the sphere system $\{\{w_1\}, \{w_1, w_3\}, \{w_1, w_2, w_3\}\}$, it holds: $Bel(\{w_1\} \mid W)$, and $Bel(\{w_2\} \mid \{w_2, w_3\})$, that is, $Bel(\{w_2\} \mid \neg\{w_1\})$. In an analogous way, all the examples from the end of section 4.2.2 can be turned into examples of *Bel* and P satisfying all of the postulates from before, including the postulates on *general* or unrestricted conditional belief. Given P, in each of these examples, *Bel* would either have to be determined by the sphere system of *all* P-stable^r of probability less than 1 taken together with the least proposition of probability 1, or alternatively *Bel* would be defined by a sphere system of certain *selected* P-stable^r of probability less than 1 taken together with the least proposition of probability 1.

In our little infinite example from the end of section 4.2.2, the ranking of worlds simply coincided with the predecessors of their natural number indices: it follows e.g.

that $Bel(\{w_2\} \mid \{w_2, w_4, w_6, w_8, \ldots\})$, assuming that Bel is determined from the sphere system of all P-stable$^{\frac{1}{2}}$ sets taken together with W.

4.3 Some Examples with a Concrete Interpretation

With the machinery from this chapter in place, we are in the position to reconsider some of the examples in the previous chapters and to interpret them now in conditional-belief terms; additionally, I will also consider some new examples. In all of the examples, the underlying algebra of propositions is simply the full power set algebra on the respective set of worlds again.

Example 15 (Apple Spritzer Example 2 from Chapter 2 Reconsidered)
Let $W = \{w_1, w_2, w_3\}$. w_1 corresponds to the bottle of apple spritzer being in the fridge in the kitchen, w_2 to the bottle of apple spritzer being in the shopping bag in the kitchen, w_3 to the bottle of apple spritzer not being in either of these places.

Let P assign probabilities as follows: $P(\{w_1\}) = \frac{6}{10}$, $P(\{w_2\}) = \frac{3}{10}$, $P(\{w_3\}) = \frac{1}{10}$. Hence, as intended in the story, it is more likely than not that the bottle of apple spritzer is in the fridge.

Let Bel be determined by the following sphere system: $\{\{w_1, w_2\}, \{w_1, w_2, w_3\}\}$. Equivalently, let Bel be determined by the total pre-order of worlds that is given by: $w_1, w_2 \prec w_3$. So e.g. $Bel(\{w_1, w_2\} | W)$, since the most plausible worlds overall (the least ones in the $\prec$-order) are w_1 and w_2. In words: the agent believes unconditionally that the bottle is in the kitchen. Moreover, the agent does not believe anything more specific than that unconditionally. But it also holds that $Bel(\{w_2\} | \neg\{w_1\})$, since the most plausible world in $\neg\{w_1\} = \{w_2, w_3\}$ is w_2. In suppositional terms: on the supposition that the bottle is not in the fridge, the agent believes it is in the shopping bag, as required by the story from Chapter 2.

P and Bel satisfy all of the postulates in this chapter, and indeed $\{w_1, w_2\}$ and $\{w_1, w_2, w_3\}$ are P-stable^r with a threshold $r = \frac{1}{2}$.

Example 16 (Example of Tracey's Sprinkler from Section 2.3.1 Reconsidered)
If (conditional) Bel is given by the sphere system of all P-stable^r sets in the Tracey's Sprinkler example (where r was $\frac{1}{2}$ again), then Tracey ends up having the following conditional beliefs:

$Bel(T = 1 \land J = 1 \mid R = 1)$	$(R = 1 \Rightarrow J = 1 \land T = 1)$
$Bel(T = 1 \mid S = 1)$	$(S = 1 \Rightarrow T = 1)$
$Bel(R = 1 \land S = 0 \mid T = 1 \land J = 1)$	$(T = 1 \land J = 1 \Rightarrow R = 1 \land S = 0)$
$Bel(S = 1 \mid T = 1 \land R = 0)$	$(T = 1 \land R = 0 \Rightarrow S = 1)$
$Bel(R = 1 \mid R = 1 \lor S = 1)$	$(R = 1 \lor S = 1 \Rightarrow R = 1)$

In the first two of them the conditional beliefs 'proceed' in the direction of the causal edges in the corresponding Bayes net. That is not so for the other three cases of conditional beliefs: e.g. the third line expresses that, given that both Tracey's and Jack's lawns are wet, Tracey believes that it rained but that her sprinkler was not left on. All of that seems to make good sense given the story and the corresponding Bayesian network.

To the right I have also reformulated the conditional belief ascriptions in the form of conditionals; this will be taken up again in section 6.2, and I will also use such reformulations in the examples to follow.

Example 17 (Example 7 from Section 3.6 Reconsidered)
That is the example that I have been discussing throughout the present chapter. Compare Figure 4.5 in section 4.2.2. Let $r = \frac{1}{2}$ again, and let *Bel* be given by the sphere system of all non-empty P-stable^r sets (except for W, which includes the world w_8 that has probability 0 and which I am going to suppress in what follows). Then the ideal physicist at the time would have had the following conditional all-or-nothing beliefs (where T is Newtonian physics, H comprises the auxiliary hypotheses, and E is the evidence):

Based on all P-stable$^{\frac{1}{2}}$ sets:

$Bel(T \wedge H \mid T)$	$(T \Rightarrow T \wedge H)$
$Bel(T \wedge H \mid H)$	$(H \Rightarrow T \wedge H)$
$Bel(T \wedge H \mid T \vee H)$	$(T \vee H \Rightarrow T \wedge H)$
$Bel(T \mid E)$	$(E \Rightarrow T)$
$\neg Bel(H \mid E)$	$(E \not\Rightarrow H)$
$Bel(T \mid E \wedge \neg H)$	$(E \wedge \neg H \Rightarrow T)$

For instance, '$Bel(T \mid E)$' concerns a case of conditional belief in which the given proposition (E) contradicts what is believed unconditionally (since $Bel(\neg E)$, that is, $Bel(\neg E | W)$).

If we switch to the case of $r = \frac{3}{4}$, then determining *Bel* from the sphere system of all P-stable^r sets (except for W) corresponds to determining *Bel* from the ranking in Figure 4.9, which depicts in boldface the ordinal ranks of all worlds with positive probabilistic mass for the case $r = \frac{3}{4}$.

In that case one has:

Based on all P-stable$^{\frac{3}{4}}$ sets
(which are particular P-stable$^{\frac{1}{2}}$ sets):

$\neg Bel(T \wedge H \mid T)$	$(T \not\Rightarrow T \wedge H)$
$Bel(T \wedge H \mid H)$	$(H \Rightarrow T \wedge H)$
$\neg Bel(T \wedge H \mid T \vee H)$	$(T \vee H \not\Rightarrow T \wedge H)$
$Bel(T \mid E)$	$(E \Rightarrow T)$
$\neg Bel(H \mid E)$	$(E \not\Rightarrow H)$
$Bel(T \mid E \wedge \neg H)$	$(E \wedge \neg H \Rightarrow T)$

Alternatively, we can think of the last example as one in which r still equals $\frac{1}{2}$ but where *not* all P-stable$^{\frac{1}{2}}$ sets are used as spheres in the sphere system for *Bel* but

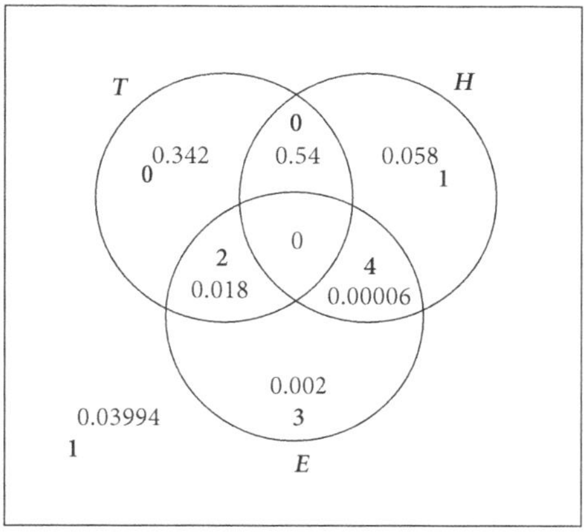

Figure 4.9. Ordinal ranks for the example measure with $r = \frac{3}{4}$

only particular ones (in that case: only those that happen to be P-stable$^{\frac{3}{4}}$ as well). As Theorem 23 told us, a sphere system for a conditional belief set *Bel* does not need to include all P-stable^r sets.

Example 18 (Taking up a Thought from Appendix B: Stability from Simplicity)
In Appendix B I discussed different possibilities of how the stability of belief might come about. One option that I mentioned but which I did not discuss back then was: *prior* inductive judgements of simplicity or regularity.

Here is an example of that kind. Let us assume that our perfectly rational agent is concerned with four individuals (1, 2, 3, 4) and their colours. Although I am going to turn to probabilities only later, let us assume already that the agent is probabilistically certain that each of these four individuals must be red or blue, but one individual cannot be red and blue at the same time. For simplicity, I will identify W with the set of 16 quadruples with entries in $\{R, B\}$ from the start, suppressing all other worlds (which the agent will be able to rule out probabilistically). Furthermore, the agent has certain prior expectations concerning the 'uniformity of nature'. These expectations lead her to order worlds according to the extent of regularity that they satisfy: worlds in which all individuals have the same colour (that is, RRRR and BBBB) maximize simplicity or regularity. They are assigned rank 0, which is the rank of maximal plausibility. The worlds with a distribution of 3 + 1 (e.g. RRRB or BRBB) are already 'less orderly'—they have rank 1. Finally, the worlds with a distribution of 2 + 2 (e.g. RRBB or BRBR) are of the least preferred rank 2: they are 'random'. I do not want to defend this kind of simplicity ordering of worlds in any way: let us simply assume that our agent's conditional belief set *Bel* is determined by it.

If $R(m)$ is the proposition that the individual with number m is red (with $1 \leq m \leq 4$), and if $B(m)$ is the corresponding proposition for the colour blue, then the agent will thus have e.g. the following conditional beliefs:

(1) $Bel(R(2) \mid R(1))$ $(R(1) \Rightarrow R(2))$
(2) $Bel(R(3) \mid R(1))$ $(R(1) \Rightarrow R(3))$
(3) $Bel(R(2) \wedge R(3) \mid R(1))$ $(R(1) \Rightarrow R(2) \wedge R(3))$
(4) $Bel(R(3) \mid R(1) \wedge R(2))$ $(R(1) \wedge R(2) \Rightarrow R(3))$
(5) $Bel(R(4) \mid R(1) \wedge B(2) \wedge R(3))$ $(R(1) \wedge B(2) \wedge R(3) \Rightarrow R(4))$
(6) $\neg Bel(\neg B(4) \mid R(1) \wedge R(2))$ $(R(1) \wedge R(2) \not\Rightarrow \neg B(4))$
(7) $Bel(R(3) \mid R(1) \wedge R(2) \wedge B(4))$ $(R(1) \wedge R(2) \wedge B(4) \Rightarrow R(3))$
(8) $Bel([R(3) \wedge R(4)] \vee [B(3) \wedge B(4)] \mid R(1) \vee B(2))$
$(R(1) \vee B(2) \Rightarrow [R(3) \wedge R(4)] \vee [B(3) \wedge B(4)])$

For instance, line (1) holds because the most plausible world in which $R(1)$ is the case is RRRR, in which $R(2)$ is true. Line (5) holds since the most plausible world in which $R(1) \wedge B(2) \wedge R(3)$ is the case is RBRR, in which $R(4)$ is satisfied. Line (8) is due to the fact that the most plausible worlds in which $R(1) \vee B(2)$ is true are RRRR and BBBB—in both of them $[R(3) \wedge R(4)] \vee [B(3) \wedge B(4)]$ is the case. And the like.

If these conditional beliefs were to trigger corresponding revisions or inferences based on evidence, then clearly these revisions or inferences would instantiate patterns of inductive reasoning, where the patterns in question would be due to a presumption of the 'uniformity of nature'. At the same time, because the agent's set of conditional beliefs satisfy all of the postulates from before, the corresponding revisions or inferences would satisfy all of the standard closure properties of AGM belief revision or rational nonmonotonic inference relations. For instance, in the terms of nonmonotonic reasoning: the transition from lines 1 and 2 to 3 is called Conjunction (in the conclusion), the transition from lines 1 and 2 to 4 is called Cautious Monotonicity, the transition from lines 4 and 6 to 7 is called Rational Monotonicity, and so on. See Lehmann and Magidor (1992) for a statement of these closure conditions (some of which will reappear as rules for conditionals in section 6.2).

Given *Bel* e.g. the following is a probability measure P that—together with *Bel*—satisfies all of our postulates from this chapter: if w is of rank 0, let $P(\{w\}) = \frac{1}{2} \cdot \frac{9}{10} = 0.45$; for w of rank 1, let $P(\{w\}) = \frac{1}{8} \cdot \frac{9}{100} = 0.01125$; and if w has rank 2, let $P(\{w\}) = \frac{1}{6} \cdot \frac{1}{100} = 0.001666\ldots$ Then the previous ranking of worlds satisfies, with respect to that measure P, the condition that is familiar already from the application of the algorithm in section 4.2.2 and from Observation 24 (the Sum Condition): the probability of each world is greater than the sum of the probabilities of all worlds of greater rank (or of all of the less plausible worlds). By the Separation Property of P-stable^r sets from section 4.2.2 (see Observation 9), this means that the following sets are P-stable$^{\frac{1}{2}}$: the set of worlds of rank 0, the set of worlds of rank 0 or 1, and the set of all worlds. That is exactly the sphere system that determines *Bel* as defined before, and so by our Representation Theorem 23 it follows that P and *Bel* satisfy all of the postulates from this chapter. In particular, BP1^{r*} (Likeliness) is the case: for all propositions Y with $P(Y) > 0$, for all propositions Z, if $Bel(Z|Y)$, then $P(Z|Y) > \frac{1}{2}$.

Example 19 (A Little Statistical Example)
Consider a binomial distribution for a sequence of $n = 10$ independent coin tosses each of which yields heads with probability $p = 0.5$. Let W be $\{w_0, \ldots, w_{10}\}$, where w_k represents that there are exactly k heads outcomes amongst the n tosses. The probability of each singleton $\{w_k\}$ is then given by: $\binom{n}{k}p^k(1-p)^{n-k}$.

One can show that the P-stable$^{\frac{1}{2}}$ sets are: $\{w_4, \ldots, w_6\}$, $\{w_3, \ldots, w_7\}$, $\{w_2, \ldots, w_8\}$, $\{w_1, \ldots, w_9\}$, $\{w_0, \ldots, w_{10}\}$.

If $n = 10$ but now $p = 0.8$, the P-stable$^{\frac{1}{2}}$ sets are 'biased' towards a greater number of heads: $\{w_7, \ldots, w_{10}\}$, $\{w_6, \ldots, w_{10}\}$, $\{w_5, \ldots, w_{10}\}$, $\{w_4, \ldots, w_{10}\}$, $\{w_3, \ldots, w_{10}\}$, $\{w_2, \ldots, w_{10}\}$, $\{w_1, \ldots, w_{10}\}$, $\{w_0, \ldots, w_{10}\}$.

All of this can be translated into conditional *Bel* sets again by picking a sphere system of such P-stable$^{\frac{1}{2}}$ sets and defining *Bel* to be determined by that set. P and *Bel* will then satisfy all of the postulates in this chapter again.

For instance, if for each of the two values for 'p' the corresponding sphere system is given by the set of *all* P-stable$^{\frac{1}{2}}$ sets, it follows: with $p = 0.5$, $Bel(\{w_4, \ldots, w_6\})$ (belief in there to be k outcomes of heads where $4 \leq k \leq 6$), while $Bel(\{w_3, w_7\}|\neg\{w_4, \ldots, w_6\})$ (belief in there to be $k = 3$ or $k = 7$ outcomes of heads given that it will not be the case that $4 \leq k \leq 6$).

On the other hand, with $p = 0.8$, it follows that not $Bel(\{w_4, \ldots, w_6\})$ (lack of belief in there to be k outcomes of heads where $4 \leq k \leq 6$), $Bel(\{w_7, \ldots, w_{10}\})$ (belief in there to be k outcomes of heads where $7 \leq k \leq 10$), and $Bel(\{w_6\}|\neg\{w_7, \ldots, w_{10}\})$ (belief in there to be $k = 6$ outcomes of heads given that it will not be the case that $7 \leq k \leq 10$). All of these results seem quite plausible independently.

Appendix C

Does Rational Belief Reduce to Subjective Probability? Does it Supervene?

Does all-or-nothing belief reduce to degrees of belief? I discussed this topic in Chapter 1, where the Reduction Option (ii) from section 1.3.2 answered the question positively, while the Irreducibility Option (iii) from section 1.3.3 answered it in the negative. I did not settle which of these options was the right one.

In the meantime I have developed what is, more or less, one stability theory of rational belief presented in three different ways (Chapters 2–4). With that theory in place, it is time to reconsider the issue.

If all-or-nothing belief reduces to—is nothing but, is constituted by, is grounded in, is explained by, or the like—degrees of belief, then in spite of the ambiguity and the vagueness that affect the term 'reduction', this should entail that all-or-nothing belief *supervenes on* degrees of belief: no difference concerning belief without a difference concerning degrees of belief. And if this general metaphysical supervenience statement about belief states is true, then it should apply also to the belief states of perfectly rational agents:[265] a perfectly rational agent's beliefs would have to supervene on her degrees of belief.

Is that the case? Unfortunately, expressions such as 'supervene on' or 'no difference . . . without a difference' require precisification themselves, and more than one exact concept of supervenience can emerge from this[266]—depending on e.g. whether modal operators are involved, and, if so, where they are positioned vis-à-vis the required quantifiers. Without entering a discussion of the corresponding literature, I will pick one of them and focus on it in the following.

Is a perfectly rational agent's total belief state (*Bel*) at a time a function of the agent's total degree-of-belief state (P) at the same time? More precisely: fix a perfectly rational agent. Is there a function f (which may differ from one rational agent to the next), such that for all times t, if P is the agent's degree-of-belief function at t, then $f(P)$ is the same agent's all-or-nothing belief set at t? So the corresponding supervenience claim has the

[265] Compare the discussion at the end of section 1.3.3 of Chapter 1.

[266] See e.g. Kim (1984).

form: for all perfectly rational agents there is a function f, such that for all times t and for all probability measures P, if P is the agent's degree-of-belief function at t, then $f(P)$ is the agent's belief set at t. For example, this would be the case if '*Bel*' (for an agent at an arbitrary time t) were explicitly definable in terms of 'P' (for the same agent at the same time t).[267] No metaphysical modalities are involved in a supervenience claim like that, but of course every argument against that kind of claim will a fortiori constitute an argument against any modal supervenience claim that is logically stronger than it. The question is now: does that supervenience claim hold?

I will give two reasons to believe that the answer to this supervenience question is *negative*: rational belief does *not* supervene on rational degrees of belief in this sense. Both reasons will, in a way, rely on certain postulates of my stability theory of rational belief. The first argument presupposes as premises the rationality postulates on belief from Chapter 2, or alternatively those of Chapter 3, or those of Chapter 4: it applies to all of the chapters. The second argument requires among its premises specifically some of the postulates from the last chapter: the AGM-like postulates for conditional belief as described in Chapter 4. (The bridge principle for belief and degrees of belief from Chapter 4 will not be required.) That second argument will be based on a formal result by Lin and Kelly (2012b).

If it is right that rational belief does not supervene on rational degrees of belief, then belief cannot reduce to degrees of belief either. However, this finding will still be compatible with a refined reductive view according to which e.g. belief does not reduce to degrees of belief alone though it does reduce to degrees of belief *plus x*, where x might comprise certain 'practical' features of either the belief ascriber or of the agent of belief (or both) within a context, such as attention, interest, and the like. In short: belief might still reduce to degrees of belief *plus context*, in either of the two possible senses of 'context' (semantic or non-semantic) that were explained in section 3.3. Whether or not belief supervenes on them, and if so, how to fill in the blank 'x' exactly, I will not be able to settle.

Assume that all-or-nothing belief does reduce to graded belief plus context: for certain purposes, it might not be too misleading even then to speak of belief as reducible just to graded belief, without mentioning the contextual component explicitly. For instance: the Lockean thesis is sometimes given a reductive interpretation,[268] in which case it may be said to *reduce belief to high enough subjective probability*. That is so even when, strictly speaking, the choice of threshold in the Lockean thesis is normally not taken to be given by the subjective probability measure itself. Expressing oneself in this way might be excusable for the purpose of highlighting the priority of subjective

[267] With sufficient linguistic resources around, the existence of such a function f would even be *equivalent* to the definability of '*Bel*' in terms of 'P'. Note also that in this formulation of the supervenience thesis, f may depend on the agent—or maybe on a 'time period' of an agent within a sufficiently extended interval of time—but not on the exact point of time (within such an interval of time): in particular, in the case of belief change, f is supposed to remain invariant.

[268] Although we saw in Chapter 3 that it need not be interpreted reductively.

probability over categorical belief in any such reductive account, but it should not make one forget that belief would really be reduced to probability taken together with whatever determines the Lockean threshold. In this appendix I will take things more strictly: when I consider the question whether an agent's belief set supervenes on her degree-of-belief function, then I wonder whether *Bel* supervenes on *P taken by itself*, without suppressing further relevant parameters by pushing them into the background.

C.1 The First Argument Against Supervenience

We have seen in each of Chapters 2–4 that fixing a perfectly rational agent's degree-of-belief function *P*, while assuming the stability theory of belief, does not always determine the same agent's all-or-nothing belief set *Bel* uniquely.

For instance: fixing the probability measure *P* as in the Tracey's Sprinkler example from section 2.3.1, and assuming at the same time the Humean thesis for *P* and *Bel* (and also that not $Bel(\varnothing)$), did not uniquely determine *Bel*. Each of six possible options for *Bel* could be combined with the given measure *P* so that the Humean thesis was satisfied. These six ways of defining *Bel* corresponded to the six *P*-stable$^{\frac{1}{2}}$ sets in that example.

Similarly in Example 6 from section 3.1 (which ultimately turned into the Dorling example of section 3.6): given the probability measure *P* in the example, there were six ways of determining *Bel*, based on six *P*-stable$^{\frac{1}{2}}$ sets, such that *Bel* was consistent and closed under logic, and *P* and *Bel* jointly satisfied an instance of the right-to-left direction of the Lockean thesis ('if high enough probability, then belief'), which taken together entailed an instance of the full Lockean thesis.

Finally, we encountered various examples of the same sort also in section 4.3 of Chapter 4: even with a probability measure *P* specified, there was generally more than one conditional *Bel*, such that *Bel* satisfied the AGM postulates, and *P* and *Bel* together satisfied an instance of the left-to-right direction of the conditional Lockean thesis ('if conditional belief, then high enough conditional probability'). Each of these conditional belief sets *Bel* corresponded to a sphere system of *P*-stable^r sets, and given *P* and *r* there was in general more than just one such system.

In other words: the stability theory of belief from Chapters 2–4 does not by itself *entail* that rational belief is a function of rational degrees of belief. There are cases in which it is possible to change one belief set *Bel* into a distinct belief set *Bel*′ such that all of the axioms of our stability theory hold true and yet *P* is held fixed.[269] So there can be change concerning the agent's total state of all-or-nothing belief without changing the agent's total state of graded belief in any way. For the same reason, our theory so far does not entail the explicit definability of '*Bel*' on the basis of '*P*' either.

[269] Actually, as we know from the previous chapters, with a Humean threshold of $r = \frac{1}{2}$ that will be so for *almost all* probability measures on a finite space of worlds; compare e.g. n. 134 from Chapter 2.

Obviously, this observation alone does not settle the question whether rational belief is a function of rational degrees of belief or not: after all, our theory so far might simply be *incomplete*. Once strengthened by further correct rationality postulates, rational belief might indeed follow to be a function of rational degrees of belief.

So how would one, conceivably, attempt to strengthen the theory, such that this would be the case? Given an agent's subjective probability measure P at a time, and given r, there would have to be a uniquely determined manner of picking the 'right' P-stable^r set B_W that would generate the agent's unconditional belief set. More generally: there would have to be a uniquely determined manner of picking the 'right' sphere system of P-stable^r sets that would generate the agent's conditional belief set. Unfortunately, it is unclear what that unique method of 'picking' these sets should be like, and hence which postulates one would have to add such that the unique existence of such a method would become derivable.

In Leitgeb (2013a) I did make a suggestion to that effect: the method in question might be one that *maximizes the agent's set of beliefs*. Given P and r, let unconditional *Bel* be given by choosing B_W to be the *least* P-stable^r set with respect to the subset relation (which must exist, by the results from Chapter 4, at least in those cases in which there is a least set of probability 1). By B_W being P-stable^r, the postulates of the stability theory of belief will be satisfied. By B_W being least amongst the P-stable^r sets, the set of believed propositions (the set of supersets of B_W) will be maximized. In the case of conditional belief, the analogous result can be obtained by postulating *Bel* to be given by the sphere system of *all* P-stable^r sets (except for the sets of probability 1 that do not coincide with the least set of probability 1): a choice that can be shown again to satisfy all of the postulates from Chapter 4, and, at the same time, to maximize the agent's conditional belief set. By adding postulates to precisely this effect, '*Bel*' becomes definable explicitly in terms of 'P' and 'r'. Finally, one might require r to be $\frac{1}{2}$—the least admissible value of 'r', assuming that a believed proposition should always have a greater probability than its negation. In this way, the class of P-stable^r sets is also maximized, as follows from Observation 21 in section 4.2.2. By taking this last step of maximization (which I did not take in Leitgeb 2013a), '*Bel*' would become definable in terms of 'P' alone: B_W would be the least P-stable$^{\frac{1}{2}}$ set, and conditional *Bel* would be given by the sphere system of all P-stable$^{\frac{1}{2}}$ sets (except for the sets of probability 1 that differ from the least set of probability 1).

But that kind of maximizing all-or-nothing belief by first minimizing r and then maximizing *Bel* for the minimal r does not seem plausible at all. With P being fixed and r being set to the minimal permissible value $\frac{1}{2}$,[270] why would rationality require an agent to be maximally brave concerning her beliefs? For instance: in the continuation of the example of Tracey's Sprinkler in section 2.3.2, why and in what sense would Bold Tracey be 'more rational' than Cautious Tracey? Is this meant to be in the Popperian spirit of making risky hypotheses? But it is one thing to test one's bold scientific

[270] Compare the discussion on 'Choosing the Value of "r"' in section 4.2.2.

conjectures and another to act upon one's risky beliefs: in the former case one's theory might perish, while in the latter case one might perish oneself. A general norm to the effect that one's categorical beliefs should always be maximally brave (given one's degree-of-belief function) would also mean that one should always maximize the risk of believing a proposition that is false (which will become important in section 5.1). To the extent to which assertability is given by belief (an issue that will be taken up in section 6.2), it would mean that one should always be maximally willing to assert a proposition independently of what one believes one's audience is like and how they will act upon it;[271] and so on. All of this seems wrong: a general norm like that seems invalid.

At best, perhaps, r might be fixed to be *any* number in the interval $\frac{1}{2} \leq r < 1$, such that r determines a lower 'cautiousness' boundary for the degrees of belief of the agent's believed propositions: their probabilities are not supposed to fall short of that threshold. And then the agent's set of belief is maximized given that value of 'r' *because matters of cautiousness have already been taken care of by means of* r. That is a conceivable strategy that I cannot rule out. But then '*Bel*' would be definable only in terms of 'P' and the threshold numeral 'r', where the value of 'r' might itself be determined by practical features of either the belief ascriber or the subject of belief: their attention, interest, perceived stakes, and so on. If one adds to this the other contextual feature of rational belief that we encountered in Chapter 2—the sensitivity of belief to the partition of possibilities—then it becomes quite clear that rational belief would not be a function of rational degrees of belief but rather a function of rational degrees of belief plus another substantial component: the context (in one of the two senses explained in section 3.3) or components thereof. Since such contextual parameters vary in time, it would *not* be the case that there is a function f, such that for all times t and for all probability measures P, if P is the agent's degree-of-belief function at t, then $f(P)$ is the agent's belief set at t. Or, perhaps, rational belief is not even a function of P and the context.

I will not be able to decide these matters here. In any case, rational belief does not seem to supervene on rational degrees of belief alone.

C.2 The Second Argument Against Supervenience

There is a second argument for the same conclusion that relies specifically on the postulates from Chapter 4. The argument uses a very nice 'no-go' theorem by Hanti Lin and Kevin Kelly (Lin and Kelly 2012b, theorem 1 and corollary 1) which I will state in my own terminology and without giving every formal detail.

Combine the following assumptions:

(a) The set W of worlds has at least three members.
(b) At any point in time, our agent's degree-of-belief function P is a probability measure.

[271] I am grateful to Chris Gauker for suggesting that maximizing belief might be implausible because of such social consequences.

(c) At any point in time, the agent's conditional belief set *Bel* satisfies the AGM-like postulates from Chapter 4. In fact, it is sufficient to assume just the postulates on restricted conditional belief from section 4.2.2, including B6, which was my conditional-belief version of AGM's Preservation postulate (K*4).[272]

(d) At any point in time, updating P and *Bel* proceeds in line with the respective diachronic norms from sections 4.1.1 and 4.1.2: by conditionalizing P and by belief revision as determined by $Bel(\cdot|\cdot)$.

(e) Belief is a function of degrees of belief in the sense explained at the beginning of this appendix: there a function f, such that for all times t, if P is the agent's degree-of-belief function at t, then $f(P)$ (or Bel_P) is the agent's conditional belief set at t. (Lin and Kelly call such an f an 'acceptance rule', which in Lin and Kelly 2012b is assumed to be determined uniquely and which is meant to remain invariant under belief change.)[273]

(f) Some auxiliary quasi-geometrical assumptions on the (uniquely determined) function f from (e) are satisfied (f needs to be 'sensible': see Lin and Kelly 2012b, p. 965).

(g) We have seen in section 4.1.2 that update by conditionalization and update by AGM belief revision cohere with each other: if P and *Bel* satisfy the Humean thesis initially, then their respective updates will do so, too. But Lin and Kelly require more than that as another assumption for their result: they assume that P and *Bel* commute *modulo the function f from before*.

Here is what they demand: at any point in time, conditionalization and belief revision commute modulo f in the sense that if P is the agent's degree-of-belief function at a time, and Bel_P $(= f(P))$ is the agent's conditional belief set at the same time, then for all propositions E, X: $Bel_{P(\cdot|E)}(X|W)$ if and only if $Bel_P(X|E)$. In words: assume that one updates the prior P on a piece of evidence E, which results in the posterior $P(\cdot|E)$; one determines then the conditional belief set $Bel_{P(\cdot|E)}$ by applying f to $P(\cdot|E)$; and in the resulting belief state $Bel_{P(\cdot|E)}$ the agent believes absolutely or unconditionally that X: $Bel_{P(\cdot|E)}(X|W)$. Then the agent should also believe X conditional on E in the belief state that results from applying f to the prior P: $Bel_P(X|E)$. And vice versa. Or reformulated again: it does not matter whether one updates first probabilistically and then determines conditional belief (and hence absolute belief) by f; or whether one determines first conditional belief (and hence absolute belief) by f and then updates absolute belief qualitatively by belief revision. The two pathways should always yield the same absolute or unconditional beliefs.

[272] See section 4.1.2.

[273] Lin and Kelly (2013) weaken this assumption from their 2012b article a bit by replacing their functional acceptance rule f by a relation R that is meant to hold between probability measures and conditional belief sets. But then they need an additional assumption called 'Diachronic Admissibility of Case Reasoning' in order for another version of their no-go theorem to go through again. In Leitgeb (2013b) I criticize this additional assumption on grounds of stability.

This seems to be a natural requirement in a context in which *Bel* has been assumed to supervene on P (as in (e) above) modulo a function f: $Bel = f(P)$. The existence of such a function f (as postulated in (e)) is a presupposition of the present assumption (g).

As Lin and Kelly (2012b) show, a contradiction can be derived from these assumptions (a)–(g) taken together. So something needs to go. Lin and Kelly themselves take (e) for granted, that is, belief being a function of subjective probability, and they also accept (g) which presupposes (e). Their proposal is to reject (c): the AGM postulates (especially Preservation).

I have already committed myself to the AGM postulates when I stated the 'Logic' Assumption 8 in Chapter 1, and when I assumed my postulates on conditional belief in Chapter 4. AGM is the standard theory of belief revision in the relevant literature. I also gave various bits and pieces in defence of AGM in the course of Chapter 4: in particular, I mentioned at the end of my discussion of B6 in section 4.2.2 that Preservation may be viewed as a stability principle for all-or-nothing belief and that the stability that it supplies may be just as plausible to assume as that conveyed by the Humean thesis in Chapter 2. Semantically, Preservation follows from the assumption that a perfectly rational agent's doxastic order of worlds is total or linear—a linearity requirement that is similar to the one made in other areas (the Lewis–Stalnaker semantics of counterfactuals, decision theory, social choice) and that is also satisfied by probabilities (qua real numbers). In addition, we will encounter two further arguments for Preservation in sections 6.2 and 6.3 of Chapter 6. So my proposal is instead to reject (e)—belief being a function of subjective probability—and therefore also (g), which presupposes (e). That is exactly the same conclusion as in the first argument of this appendix.

So my diagnosis is that rational belief does not supervene on rational degrees of belief (alone). That is so at least in the sense of supervenience that was used in this appendix and that was also presupposed by Lin and Kelly (2012b).[274]

274 For the special case in which $f(P)$ is defined to be the conditional belief set that is given by the sphere system of *all* P-stable^r sets whatsoever (except for those of probability 1 that are not least with that property), there is a much shorter proof that conditionalization and belief revision cannot commute modulo f in each and every case; see Leitgeb (2013a, pp. 1387–8) for the argument.

5

Stability and Epistemic Decision Theory

Chapter 2 formulated the stability theory of belief for perfectly rational agents in terms of what I called the Humean thesis on belief. Chapter 3 reformulated that theory (for a Humean threshold of $\frac{1}{2}$) by means of postulates that made the synchronic coherence aspects of the Humean thesis more explicit: synchronic coherence amongst beliefs (logical closure and consistency of belief), synchronic coherence amongst degrees of belief (the axioms of probability), and synchronic coherence amongst beliefs and degrees of belief (the Lockean thesis). Chapter 4 extended the stability account to *conditional* belief or, equivalently, to the doxastic *rankings* of worlds that play such a salient role in belief revision theory and nonmonotonic reasoning. The chapter also added corresponding norms on *diachronic* coherence among beliefs and among degrees of belief, and it derived diachronic norms of coherence between beliefs and degrees of belief from them.

In this chapter I will present two further reformulations of the theory: this time they are going to derive from the 'aiming at the truth' aspect of belief that I highlighted early on in Chapter 1 when I stated Assumption 2 on belief in section 1.1: belief aims at the truth. I will demonstrate that the stability conception of belief that I defend in this book makes rational belief not just aim at the truth but even, in a sense, makes it 'get sufficiently close to the truth'. And the other way around: the stability conception of belief can be recovered from belief 'getting sufficiently close to the truth' taken together with some additional assumptions.

In section 5.1, rational all-or-nothing belief will be determined to 'get sufficiently close to the truth' in a sense that will be made precise by means of a perfectly rational agent's belief set *Bel* having *positive expected epistemic utility*. The epistemic utility of a belief set at a world will be given solely by the truth values of her beliefs in that world: positive utility will correspond to truth, negative utility to falsity. That is precisely what will make the utility in question epistemic. The *expected* epistemic utility of her beliefs will be given by taking the expected value of such epistemic utilities. Taking such expected epistemic utilities will amount to an internalist conception of 'getting sufficiently close to the truth'. For instance, demanding of (inferentially) perfectly rational agents to believe all and only truths would exceed their epistemic capabilities. In line with an epistemic Ought-Can principle, it would therefore not make sense to demand of such an agent to 'get perfectly close to the truth' in that sense. That is why I will instead require a conception of 'the agent ought to get sufficiently close to the truth' that the believing agent is in principle epistemically capable of realizing, and that will be: positive expected epistemic utility. Assessing doxastic attitudes in terms of notions of expected epistemic utility (or accuracy) is not a new idea at all—what will be new is just that I will also add a stability component to this. I will prove another representation theorem (Theorem 27) according to which the Humean thesis (part 2 of the theorem) is equivalent to the combination of the logic of belief with the thesis that rational all-or-nothing belief is *stably* getting sufficiently close to the truth (the expected epistemic utility of belief being *stably* positive—part 1 of the theorem). Stable positivity of the expected epistemic utility of an agent's beliefs will amount to yet another bridge principle concerning an agent's belief set *Bel* and her degree-of-belief function *P*, since the degree-of-belief function symbol '*P*' will figure in the underlying definition of expected epistemic utility. As the bridge principle will express a constraint on degrees of belief and *unconditional* belief, the section will continue the presentation of the theory from Chapters 2 and 3.[275] At the end of the section I will also argue that a perfectly rational agent is not required to *maximize* expected (stable) epistemic utility in this sense, for the same reason for which a perfectly rational agent's unconditional belief set does not have to be maximally cautious: while e.g. aiming at the truth by only believing tautologies is certainly not irrational on purely epistemic grounds, it is not mandated on these grounds either. In contrast, requiring positive (but not necessarily maximal) expected epistemic utility seems to be a norm of just about the right normative strength, and the Humean thesis will be shown to realize that norm.

Section 5.2 will characterize belief with respect to a different 'aiming property': an agent's belief set *Bel* aiming at the same agent's *degree-of-belief function P*. Given that *P* itself aims at the truth, *Bel* will still aim at the truth, but now only indirectly so: *Bel* aims at *P*, and *P* aims at the truth. The section will continue Chapter 4 in so far as *Bel* will be assumed to be conditional or corresponding to a doxastic ranking or ordering of worlds (and thereby propositions). It will make *Bel*'s 'getting sufficiently close to' *P* precise by defining what it means to say that such a doxastic ranking or ordering is an *error-free approximation* of *P*. Essentially, it will turn out that *Bel* approximates *P* without errors if and only if *Bel* and *P* satisfy the postulates from Chapter 4 (with a threshold $r = \frac{1}{2}$), that is, the postulates that generalized the stability theory from Chapters 2 and 3 to conditional belief. That will be the content of yet another representation theorem (Theorem 30).[276] The thesis that *Bel* corresponds to

[275] It would be less clear to say what it would mean for *conditional* belief to approximate the truth: conditional belief in the sense of Chapter 4 was not belief in a single proposition but rather belief in one proposition given another, and one would first have to determine in what sense two propositions with their two truth values approximate 'the' truth. I will actually give a hint at that at the end of section 5.2, but I will avoid the issue in section 5.1.

[276] Even if r is chosen to be greater than $\frac{1}{2}$, it will still follow that *Bel* approximates *P* without errors.

a doxastic ranking of worlds (and thus propositions), such that the ranking amounts to an error-free approximation of *P*, is yet another bridge principle for rational belief and degrees of belief that is equivalent to a version of the stability theory of belief that I am defending in this book. I will end the section with an argument for why a perfectly rational agent is not required to *maximize* the accuracy by which her conditional belief set approximates her degrees-of-belief function: this will be for the same reason for which a perfectly rational agent's conditional belief set does not have to be maximally brave. Rationality does not require one to minimize 'gaps' in one's doxastic ordering, that is, cases of indifference on the all-or-nothing side of belief as to whether a certain proposition is ranked above another. Requiring error-free (but not necessarily gap-free) accuracy of *Bel* with respect to *P* is a norm that seems to have just about the right normative strength, and once again the Humean thesis will be shown to realize that norm.

Both of the two sections 5.1 and 5.2 will exemplify a certain asymmetry between *P* and *Bel* so far as their truth-aiming properties are concerned. With respect to *P*, I will simply build on existing work in epistemic decision theory in which *P*'s aiming-at-the-truth is made precise and in which the axioms of probability for *P* are derived from this. None of this involves the agent's all-or-nothing belief set *Bel*. However, in both sections of the present chapter I will clarify *Bel*'s aiming properties in ways that do involve the agent's degree-of-belief function *P*. In a nutshell: *P*'s aiming-at-the-truth does not involve *Bel*, while *Bel*'s aiming-at-the-truth does involve *P*.

However, not too much ought to be read into this asymmetry: in particular, none of this will show that *P* is conceptually or metaphysically prior to *Bel* (even for perfectly rational agents); further argument and additional premises would be required for any such conclusion. Instead, the asymmetry in question might only demonstrate that *P* is a more complex 'creature' than *Bel*: as we have seen already in Chapters 3 and 4, the Humean thesis entails *Bel* to be something like a coarse-grained or simplified version of *P*: different probability measures may give rise to one and the same belief set, and the belief set captures, to some extent and in certain respects, what these probability measures have in common.[277] But that did not entail that *Bel* was conceptually or ontologically *nothing but* a coarse-graining or simplification of *P*. In certain contexts, *Bel* might actually be determined prior to *P*, and *P* might then be determined such that *Bel* and *P* jointly satisfy the Humean thesis: this would entail ultimately that *Bel* happens to be a coarse-grained version of *P*, but only *because P was determined such that this would be the case*. The fact that it is possible to formulate substantive epistemic norms for *Bel* solely on the basis of *P*, while it does not seem possible to formulate substantive epistemic norms for *P* solely on the basis of *Bel*, might simply be due

277 See e.g. the discussion of the equilateral triangle of (representatives of) probability measures in section 4.2.2.

to P containing 'more information' than *Bel*. But that additional information is not necessarily given prior to *Bel*.

5.1 Belief's Aiming at the Truth

By Assumption 2 from Chapter 1, beliefs aim at the truth. The way in which I presented this in Chapter 1 was: belief aiming at the truth is a normative claim about belief, and it is partially constitutive of belief to realize this norm at least to a great extent and in normal circumstances. I will turn to this constitutive feature of belief in more detail now.

Wedgwood's (2002, 2013) helpful normative analysis of belief's aiming at the truth is a representative instance of the literature on this topic.[278] According to Wedgwood, a belief is *correct* if and only if the proposition believed is true.[279] He regards this as the fundamental epistemic norm for belief: one that explains all other epistemic norms for belief, such as norms concerning the rationality of belief or a knowledge norm of belief. And he maintains that the norm applies to all types of belief, including all-or-nothing belief and degrees of belief. For categorical belief this means that if a proposition X is true, then believing X is perfectly correct, which is better—for purely epistemic purposes—than suspending judgement about X, while suspending judgement is better than disbelieving X (that is, believing $\neg X$), which is maximally incorrect in this case. Things are just the other way around if X is false. For degrees of belief, if a proposition X is true, then the closer one's degree of belief in X is to the maximal degree of 1 the better it is—the more correct the degree of belief is (cf. Wedgwood 2002, p. 272). Accordingly, the closer one's degree of belief in X is to the minimal degree of 0, the worse it is. The 'distance' between the agent's degree of belief in X and the truth value of X can be measured by what is called an inaccuracy measure or a scoring function in epistemic (or cognitive) decision theory:[280] a measure of incorrectness for degrees of belief.

While this goes some way towards clarifying belief's aiming at the truth, Wedgwood (2002, p. 274) also argues that 'The fundamental epistemic norm of correct belief... does not determine any unique way of balancing the value of having a correct belief about p against the disvalue of having an incorrect belief about p.' Indeed, different ways of making such an epistemic norm formally exact have been proposed

[278] Additionally, see Chan (2013, introduction) for a survey of different accounts of belief's aiming at the truth.

[279] Compare also Stalnaker (1987, p. 80): 'a correct belief is a true belief'. Gibbard (2005) also uses the notion of correctness in order to analyse belief's aiming at the truth. Fantl and McGrath (2010, p. 141) convey a similar thought when they speak of the 'truth standard' of belief: 'If you believe p and p is false, then you are mistaken about whether p, and if you believe p and p is true, then you are right about whether p'. In my discussion of Assumption 2 in Chapter 1 I have given the term 'correct' in the corresponding norm a wide scope interpretation—it is correct to: *believe a proposition if and only if it is true*. But other interpretations would be conceivable as well. Wedgwood (2013) also qualifies the notion of correctness as an ex post or retrospective one that applies to beliefs *had* by an agent, rather than an ex ante or prospective one by which one would say that it *would* be correct for an agent to have a certain belief.

[280] See e.g. Joyce (1998).

and studied in epistemic decision theory: e.g. Hempel (1962), Levi (1967), Easwaran and Fitelson (2014), Easwaran (2016), Fitelson (n.d.), Dorst (n.d.), and Leitgeb (n.d.) have done so for all-or-nothing belief (or acceptance), while e.g. Joyce (1998, 2009), Greaves and Wallace (2006), Gibbard (2007), and Leitgeb and Pettigrew (2010a, 2010b) have done the same for degrees of belief. Different epistemic utility measures (accuracy measures) or epistemic *dis*utility measures (*in*accuracy measures) are invoked by some of these authors in order to determine whether one doxastic state is better than another one by being 'closer to the truth', and the outcomes of such assessments of closeness to the truth are often interpreted differently, too.

In the following, I will opt for an approach that is closest to those of Leitgeb and Pettigrew (2010a, 2010b), Hempel (1962, section 12), Easwaran (2016), and Leitgeb (n.d.) among the references stated before. With Leitgeb and Pettigrew, I accept the norm that a perfectly rational agent's degree-of-belief function minimizes expected inaccuracy or, equivalently, maximizes expected epistemic utility, and in this sense such an agent's degrees of belief get as close as possible to their aim: truth. Maximizing *expected* epistemic utility amounts to an internalist understanding of 'getting as close to the truth as possible': an agent's degrees of belief ought to maximally approximate the truth *within her epistemic boundaries*, that is, given the evidence that is presently available to her and given her present degree-of-belief function. Leitgeb and Pettigrew manage to derive the axioms of probability for such an agent's degree-of-belief function P from this norm, if taken together with some background assumptions concerning the relevant inaccuracy or distance-to-the-truth functions. So if the premises of their argument are right, then the standard coherence norms for an agent's degrees of belief actually derive from a truth norm to the effect that an agent's degrees of belief ought to approximate the truth to the best possible extent. I will take that for granted now as far as the numerical side of belief is concerned. Fortunately, the exact details will not matter for the rest of this section[281] in which I will turn to the expected epistemic utility of an agent's *all-or-nothing* beliefs instead.

More importantly for this section, with Hempel, I will determine a norm on what is to be categorically believed, disbelieved, or left in suspense by calculating expected epistemic utilities again. As I am going to argue, rational all-or-nothing belief need not maximize expected epistemic utility—'get as close to the truth as possible'—since in the case of categorical belief, matters of closeness to the truth need to be balanced also against matters of informativeness: while e.g. believing all logical truths and suspending judgement on all other propositions can be shown to maximize expected epistemic utility in the sense that is to be developed below, there is no valid norm of rationality that would require an agent always to have beliefs as uninformative as that. Instead, all-or-nothing belief merely ought to 'get *sufficiently* close to the truth', which I am going to explain in terms of *Bel* having *positive* expected epistemic utility. How informative an agent's belief set should be will be up to the agent, as long as its expected

281 See Leitgeb and Pettigrew (2010a, 2010b) for the details.

epistemic utility is high enough, that is, above 0. Note that it is not clear whether issues like that arise also in the case of numerical belief, as subjective probability measures do not allow for numerical suspension of judgement (at least not in the sense that propositions would be allowed to lack a degree of belief).[282]

In what follows, let W be a finite and non-empty set of possible worlds, which one may think of again as the set of logically possible worlds for a simple propositional language with finitely many propositional letters. Propositions are subsets of W again, and the set-theoretic complement of a proposition X with respect to W is denoted by: $\neg X$ $(= W \setminus X)$. Any perfectly rational agent's degree-of-belief function P satisfies the axioms of probability by the truth norm that I have already accepted (and the results in Leitgeb and Pettigrew 2010b). Assume that the same agent's belief set *Bel* is closed under logical consequence, and it is not the case that the agent believes the contradictory proposition, in which case we already know from previous chapters[283] that there must be a least believed proposition B_W (a subset of W) such that: for all propositions X, $Bel(X)$ if and only if $B_W \subseteq X$. In other words: B_W generates *Bel*. The set B_W determines *Bel* uniquely, which is why instead of focusing on *Bel* we may just as well focus on B_W instead; that is what I am going to do in the following. In the case in which P and *Bel* also jointly satisfy the Humean thesis HT^r from Chapter 2 (and where it is not the case that the agent believes the contradictory proposition), it follows from Theorem 2 in Chapter 2 that *Bel* is closed under logical consequence, and hence there must be such a least believed proposition B_W by the Humean thesis. But the epistemic decision theory framework that I am going to develop will also work in a more general setting in which only the logical closure and consistency of *Bel* are assumed (and thus the existence of B_W, by W being finite), but not the Humean thesis.

Now hold a (non-empty) set X of worlds fixed. Suppose w is a possible world in W: what is the epistemic utility of the agent's logically strongest believed proposition being identical to X if assessed from the viewpoint of that world w? If w is a member of X, then X is true at w, as will be all believed propositions if *Bel* is generated from X. Clearly, this is better than w not being a member of X, in which case X is false at w (and hence there will be false believed propositions). The corresponding notion of betterness is an epistemic one in the sense that only considerations of truth will matter now: whatever practical or prudential merits believing a certain falsehood may have, for purely epistemic purposes it holds that B_W being false at w is always worse than B_W being true at w. (X or B_W, and thus the belief set *Bel* as a whole, have been fixed already, which is why considerations of the informativeness of an agent's beliefs do not matter here any more.)[284]

282 As David Makinson notes in personal communications, *credibility functions*, understood as functions on a subclass of propositions of a given algebra, such that any such function is assumed extendable in at least one way to a probability measure on the entire algebra, would allow for numerical suspension of judgement.

283 See e.g. section 3.2.

284 Both Hempel (1962) and Levi (1967) invoke also measures of the content of believed propositions in order to determine epistemic utilities, which I am not going to do in what follows.

In line with Easwaran (2016) and Leitgeb (n.d.), I will measure the epistemic difference between truth and falsity in numerical terms. More concretely, I will measure the epistemic difference between *the logically strongest believed proposition* B_W *being true at a world* and *the same proposition* B_W *being false at a world* in the following quantitative terms: let a and b be real numbers, such that $a > 0$ and $b < 0$. Let $u_w(X)$ be the epistemic utility at w of belief being generated by the set X: I will assume that $u_w(X)$ is a if $w \in X$, and $u_w(X)$ is b otherwise. That is: a is the reward of being right, while b is the penalty of being wrong. u_w is an *epistemic* utility measure, because truth is valued positively ($a > 0$) while falsity is valued negatively ($b < 0$). The numbers themselves will only matter up to a positive factor c: the numbers $c \cdot a$ and $c \cdot b$ will be indistinguishable from a and b, respectively, for all of this section's purposes. 0 is used as the 'neutral' epistemic utility in this case and hence is salient and has a meaning.[285] Indeed, both Easwaran (2016) and Leitgeb (n.d.) reserve a particular numerical value for indifference or suspension of judgement, and in the present account that value is the number 0; but this will not play much of a role in the following, since I will only determine the epistemic utility of a proposition that is supposed to be believed already (B_W).

The *magnitude* of the values of truth and falsity can be determined by taking their corresponding absolute values. For instance, if $-b = |b| \geq |a| = a$, that is, $\frac{|b|}{|a|+|b|} \geq \frac{1}{2}$, then the magnitude of the penalty either exceeds the reward or is equal to it. One interpretation of such circumstances would be: the strength of one's desire for it *not to be the case that X is believed and false* is greater than or equal to the strength of one's desire for it *to be the case that X is believed and true*. The ratio $\frac{|b|}{|a|+|b|}$ might be interpreted as measuring cautiousness with respect to the agent's strongest believed proposition being false: the greater $\frac{|b|}{|a|+|b|}$ is, the more the worry about the falsity of belief subdues the aspiration for truth, and hence the greater is one's cautiousness. The smaller $\frac{|b|}{|a|+|b|}$ is (or the greater $\frac{|a|}{|a|+|b|}$ is), the more the longing for truth overcomes the wish to avoid falsity, and thus the greater is one's boldness. But I will not deal with such 'epistemic desires' (desires concerning truth and falsity) in any more detail here.[286]

Since normally even an inferentially perfect agent will not be certain which of the worlds in W happens to be the actual world @, one cannot simply define the epistemic utility of generating belief from X in terms of $u_@(X)$. At least one cannot do so from

[285] That is a difference compared to standard decision theory for actions in which utility measures are normally assumed to be measured on the interval scale, whereas according to the epistemic decision theory that is developed here utility measures u are measured on the ratio scale.

[286] One might perhaps argue that every perfectly rational agent's epistemic desires must satisfy the condition '$|b| \geq |a|$ ($= a$)', since every *case in which X is true and the agent believes X* is, by necessity, *not a case in which X is false and the agent believes X*. (Note that the other direction does not hold generally: if it is *not the case that X is false and the agent believes X*, then this might be a case in which the agent suspends judgement on X, which would not be a *case in which X is true and the agent believes X*.) But this would require a further discussion of such 'epistemic desires', which I will not go into. Of course, all of this relates to the famous Clifford–James debate concerning the ethics of belief: cf. James (1896).

the internalist perspective on rationality that I am subscribing to[287] (but which I will not defend here): the notion of epistemic utility of belief that I am interested in is one according to which epistemic utility only depends on *what is internally accessible to the believing agent*—epistemic utility that is within her epistemic reach at the time. Therefore, one rather needs to turn to the *expected* epistemic utility of the least believed proposition being the set X: *summing up* the various utilities $u_w(X)$ over all worlds (not just the actual world), where additionally each summand that corresponds to a world w is weighted by the agent's degree of belief in that world being the actual one, that is: by $P(\{w\})$.

So my first shot at defining the expected epistemic utility for belief to be generated by X is:

$$\sum_{w \in W} (P(\{w\}) \cdot u_w(X)).$$

(Soon I am going to improve this proposal by taking into account not just the agent's degrees of belief at a time but also pieces of evidence that the agent might acquire.)

Roughly: the preferences that are represented by such expected epistemic utilities are preferences for propositions being true 'more often', where the exact meaning of 'more often' is modified by the utilities a and b and the subjective probability measure P. If such an expected epistemic utility for X is positive, then belief being generated by X is epistemically useful (as judged according to the agent's standards and from within the agent's epistemic boundaries); if it is negative, it is epistemically adverse; and if it is zero, it is epistemically neutral. Since for each w, $u_w(X)$ reflects the truth value of X at w, and since also the agent's degrees of belief (such as $P(\{w\})$) have been assumed already to get as close to the truth as possible (again in an epistemic utility sense), generating belief from a proposition X of positive expected epistemic utility may be said to formally capture the agent's categorical beliefs getting sufficiently close to the truth, at least in one salient internalist sense of the word: the expected epistemic utility of X being positive, or it being epistemically useful to generate belief from the set X.

It is easy to simplify this formula for expected epistemic utility: clearly, $\sum_{w \in W}(P(\{w\}) \cdot u_w(X))$ equals $\sum_{w \in X}(P(\{w\}) \cdot a) + \sum_{w \notin X}(P(\{w\}) \cdot b)$, since $u_w(X) = a$ for $w \in X$, and $u_w(X) = b$ for $w \notin X$. In turn (by the axioms of probability), this last sum is nothing but: $a\, P(X) + b\, (1 - P(X))$. This means that, holding a and b fixed, the expected epistemic utility for belief to be generated by X goes with the same agent's subjective probability of X, which is just as intended since a perfectly rational agent's degree of belief in X is supposed to be the agent's best possible estimate of the truth value of X.

For instance, if $b = -a$, and if B_W is again an agent's strongest believed proposition at a time, then the expected epistemic utility for B_W is $a\, P(B_W) - a\, (1 - P(B_W)) = a\, (2P(B_W) - 1)$, which is positive if $P(B_W) > \frac{1}{2}$, negative if $P(B_W) < \frac{1}{2}$, and neutral

[287] See Leitgeb and Pettigrew (2010a, s 3.3) for more on this.

if $P(B_W) = \frac{1}{2}$. For the same reason, the expected epistemic utility for B_W is positive in this case if and only if the expected epistemic utility for $\neg B_W$ is negative, and vice versa. All of this is just like in Hempel's (1962) account of expected epistemic utility from which he derives that one ought to accept a hypothesis if its probability exceeds $\frac{1}{2}$, one ought to reject it if its probability is below $\frac{1}{2}$, and one ought to suspend judgement concerning the hypothesis if its probability equals $\frac{1}{2}$.[288] But of course, in my present account, the values of 'a' and 'b' may also be chosen such that $|b| \neq |a|$, in which case the conclusions would be different. (Compare also Easwaran's 2016 generalization and improvement of Hempel's account, which also involves such numerical values for truth and falsity, but which does not assume the logical closure of belief.)

Now I would like to relate this simple epistemic decision theory for all-or-nothing belief (under the assumption of the logical closure and consistency of belief) to the Humean thesis HT^r from Chapter 2.[289] The Humean thesis involved a Humean threshold r, with $\frac{1}{2} \leq r < 1$, to the effect that r functions as a lower boundary for the probabilities of believed propositions X conditional on possible propositions Y (in the sense that both not $Bel(\neg Y)$ and $P(Y) > 0$ are satisfied). In this sense, r encodes how cautious an agent is about her beliefs. One of these believed propositions is of course the logically strongest believed proposition B_W, which by the Humean thesis must exist and which must be non-empty by the assumption that belief is consistent.

Let us assume now that the Humean threshold r does not just satisfy $\frac{1}{2} \leq r < 1$ but that it also satisfies $\frac{|b|}{|a|+|b|} \leq r$. We have already seen before that $\frac{|b|}{|a|+|b|}$ may be regarded as measuring cautiousness with respect to the agent's strongest believed proposition being false, and '$\frac{|b|}{|a|+|b|} \leq r$' expresses now that this sort of cautiousness in the present sense of our little epistemic decision theory is 'in sync' with cautiousness as regards the Humean thesis from Chapter 2. $\frac{|b|}{|a|+|b|}$ is then the definitive lower boundary for what will be an admissible threshold in the Humean thesis. In fact, with the values of 'a' and 'b' in place—where $a > 0$, $b < 0$, and hence $\frac{|b|}{|a|+|b|}$ satisfying the strict inequalities $0 < \frac{|b|}{|a|+|b|} < 1$—one might even think of r to be determined such that $max(\frac{1}{2}, \frac{|b|}{|a|+|b|}) \leq r < 1$. Or r is taken to be prior to the choice of a and b, and the values of 'a' and 'b' are determined, such that $\frac{|b|}{|a|+|b|} \leq r < 1$ (and $a > 0$ and $b < 0$) for given r. Or neither is prior to the other, and they are chosen jointly in order to satisfy this requirement.

In any case, $r \geq \frac{|b|}{|a|+|b|}$ means: the lower boundary for the probabilities of believed propositions conditional on possible propositions is assumed to be greater than or

[288] In contrast to the present approach, Hempel's result does not presuppose that belief or acceptance is generated by a least believed or accepted proposition. There are also two further differences: in Hempel's case the probabilities in question are meant to be probabilities conditional on the total body of scientific knowledge at the given time (see Hempel 1962, p. 149), and the epistemic utility of a hypothesis is assumed to depend on the measure of its logical content (see Hempel 1962, pp. 154–5).

[289] I am grateful to Seamus Bradley, Catrin Campbell-Moore, and David Makinson for very helpful discussions on this matter.

equal to the ratio of the magnitude of falsity's value over the sum of the magnitudes of truth's value and falsity's value. Equivalently: $\frac{1-r}{r}|b| \leq |a| = a$. Or in another equivalent formulation: $r|a| \geq (1-r)|b|$. Assuming the Humean thesis for P, Bel, and r, and noting with section 2.3.2 that the Humean thesis entails the Lockean thesis with a Lockean threshold $s > r$, this last inequality being the case tells us that for an arbitrary belief, the probability of it being right multiplied by the reward for being right is at least as great as the probability of it being wrong multiplied by the penalty for being wrong.[290] And this remains so even if P is conditionalized on a proposition that is doxastically possible, by the Humean thesis.

It follows from the Humean thesis (and not $Bel(\varnothing)$) that $P(B_W) > \frac{r}{1-r} \cdot (1-P(B_W))$:[291] thus, $a\,P(B_W)+b\,(1-P(B_W)) > a\,\frac{r}{1-r}\cdot(1-P(B_W))+b\,(1-P(B_W)) = (a\,\frac{r}{1-r}+b)(1-P(B_W)) \geq (\frac{1-r}{r}|b|\,\frac{r}{1-r}+b)(1-P(B_W)) = 0$. (I have used that $\frac{1-r}{r}|b| \leq a$ and $b = -|b|$.) In other words: the Humean thesis entails that the expected utility for belief to be generated by B_W is always strictly greater than 0.[292]

Let me put this on record:

Theorem 25 Let P be a probability measure, assume that Bel and P satisfy the Humean thesis HT^r (with $\frac{1}{2} \leq r < 1$), and suppose not $Bel(\varnothing)$ (the contradictory proposition is not believed). This implies that there is a strongest believed proposition B_W, which is non-empty.

Let a and b be real numbers, such that $a > 0$, $b < 0$, and assume additionally that a and b relate to the threshold r in the Humean thesis, such that $\frac{|b|}{|a|+|b|} \leq r$ (or $r|a| \geq (1-r)|b|$).

Then the following is the case: the expected epistemic utility of the belief set being generated by B_W is *positive*:

$$\sum_{w \in W} (P(\{w\}) \cdot u_w(B_W)) > 0,$$

where

- the epistemic utility $u_w(X)$, at world w, for belief to be generated by a proposition X is defined by

$$u_w(X) = \begin{cases} a, \text{ in case } w \in X \\ b, \text{ in case } w \notin X. \end{cases}$$

This means: if the Humean thesis holds (and also all other assumptions from above are satisfied), then the logically strongest believed proposition is guaranteed to be

[290] I owe this formulation to David Makinson.

[291] This is a consequence of B_W satisfying the Outclassing Condition: recall Theorem 5 in Appendix B.

[292] I did not have to use here that $\frac{1}{2} \leq r$; however, in order to derive the logical closure of belief from the Humean thesis one needs that assumption; see section 4.2.2.

epistemically useful—its expected epistemic utility is positive. Belief succeeds in 'approaching its target', that is, truth, at least to the extent that the epistemic utility of belief is positive; in this sense of expected epistemic utility exceeding a threshold of 0, belief 'gets sufficiently close to the truth'.[293]

But one can actually show much more than that: by the Humean thesis, belief will keep approaching its epistemic target even if new evidence Y comes along that is possible in the all-or-nothing sense of belief ($Poss(Y)$, or $Y \cap B_W \neq \varnothing$), and where the expected epistemic utility of B_W is now taken by summing up the epistemic utilities of B_W only at those worlds w that are still live options, that is, where $w \in Y$. Or in other words: belief aims at the truth, and the Humean thesis makes it *stably* hit its target in this sense of the word. The resulting notion of stable positive expected utility is closely related to the notion of stably high probability that was underlying the Humean thesis from Chapter 2. Formally, the main difference will be that a sum of the form '$\sum_{w \in Y} \ldots$' is well-defined even in cases in which $P(Y) = 0$, whereas P can only be conditionalized on Y—as required by stability in the sense of the Humean thesis—if additionally $P(Y) > 0$. However, this difference will not matter once Y is assumed possible and the Humean thesis is in place, since we already know from previous chapters that propositions that are possible (the negations of which are not believed) have positive probability by the Humean thesis.[294] The proposal of incorporating propositional evidence by summing up epistemic quantities (whether utilities or inaccuracies) only over those worlds that are compatible with the evidence in question is well-known from epistemic decision theory, e.g. Leitgeb and Pettigrew (2010a, 2010b) do so when they epistemically justify update by conditionalization, as do others. I am going to follow this literature here.

B_W has *stable* positive expected utility in the required sense for the following reason: if the expected epistemic utility of B_W is calculated by restricting attention just to the worlds in W that are compatible with a piece of evidence Y, then it holds that $\sum_{w \in Y}(P(\{w\}) \cdot u_w(B_W)) = \sum_{w \in (Y \cap B_W)}(P(\{w\}) \cdot a) + \sum_{w \in (Y \cap \neg B_W)}(P(\{w\}) \cdot b) = a\, P(Y \cap B_W) + b\, P(Y \cap \neg B_W)$. We know already from Theorem 5 in Appendix B that if the Humean thesis is satisfied (and not $Bel(\varnothing)$), then for all w in B_W it holds that $P(\{w\}) > \frac{r}{1-r} \cdot P(W \setminus B_W)$. If additionally $Poss(Y)$ (and hence $Y \cap B_W \neq \varnothing$), then it follows from this with the axioms of probability that $P(Y \cap B_W) > \frac{r}{1-r} \cdot P(Y \cap \neg B_W)$, which implies as before: $a\, P(Y \cap B_W) + b\, (P(Y \cap \neg B_W) > a \frac{r}{1-r} \cdot P(Y \cap \neg B_W) + b\, P(Y \cap \neg B_W) = (a \frac{r}{1-r} + b) P(Y \cap \neg B_W) \geq (\frac{1-r}{r}|b| \frac{r}{1-r} + b) P(Y \cap \neg B_W) = 0$.

[293] It is not just B_W that approaches the target in this sense. Obviously, if X is a superset of B_W, then for all w, $u_w(X) \geq u_w(B_W)$: so the expected epistemic utility of *any* believed proposition is always greater than or equal to the expected epistemic utility of the strongest believed proposition B_W. Consequently, if the latter is positive, the same must be true of the former.

[294] This follows again e.g. from the Outclassing Condition in Theorem 5 of Appendix B.

Hence we get the following strengthening of Theorem 25:

Theorem 26 (Stable Positive Expected Epistemic Utility)
With the same assumptions in place as in Theorem 25, the following holds: for all pieces Y of evidence with *Poss*(Y), the expected epistemic utility of the belief set being generated by B_W, restricted to worlds in Y, is positive:

$$\sum_{w \in Y} (P(\{w\}) \cdot u_w(B_W)) > 0$$

where the epistemic utility $u_w(B_W)$ is defined as in Theorem 25.

What is more, assuming the axioms of probability for P and the consistency of *Bel* in the background, this consequence of the Humean thesis can actually be turned into another recovery result again (much as the Humean thesis for $r = \frac{1}{2}$ could be recovered from some of its consequences in Chapter 3). If one combines the assumption that belief has stable positive epistemic utility with the logical closure of belief (another consequence of the Humean thesis), and if one additionally postulates stability in this sense for *all* epistemic utilities a and b, for which $a > 0$, $b < 0$, and $\frac{|b|}{|a|+|b|} \leq r$ are the case, then the Humean thesis can be derived from this. This leads us finally to the following representation theorem according to which any strongest believed proposition B_W that has stable positive expected epistemic utility can be represented in terms that are very familiar already from previous chapters (e.g. Appendix B):

Theorem 27 (Representation Theorem for Stable Positive Expected Epistemic Utility)
Let W be finite and non-empty. Let P be a probability measure on W. Let r be a threshold, such that $\frac{1}{2} \leq r < 1$. Assume that not *Bel*($\varnothing$). Then the following four conditions are equivalent:

1. (Logical Closure and Consistency)
   There is a (non-empty) logically strongest believed proposition B_W.
   (Stable Positive Expected Epistemic Utility)
   For all pieces Y of evidence with *Poss*(Y) (or $Y \cap B_W \neq \varnothing$), for all numbers $a > 0$ and $b < 0$, such that $\frac{|b|}{|a|+|b|} \leq r$ (or $r|a| \geq (1-r)|b|$), the expected epistemic utility of the belief set being generated by B_W, given evidence Y, is positive:

   $$\sum_{w \in Y} (P(\{w\}) \cdot u_w(B_W)) > 0$$

   where the epistemic utility $u_w(X)$, at world w, for belief to be generated by a proposition X is defined by

   $$u_w(X) = \begin{cases} a, \text{ in case } w \in X \\ b, \text{ in case } w \notin X. \end{cases}$$

2. P and *Bel* satisfy the Humean thesis HT^r.

3. There is a (non-empty) logically strongest believed proposition B_W, that proposition B_W is P-stable^r, and if $P(B_W) = 1$ then B_W is the least proposition of probability 1.
4. There is a (non-empty) logically strongest believed proposition B_W, and that proposition B_W satisfies the Outclassing Condition with respect to P and r: for all w in B_W it holds that $P(\{w\}) > \frac{r}{1-r} \cdot P(W \setminus B_W)$.[295]

The Humean thesis does not just entail that belief gets sufficiently close to the truth *stably* (given the background assumptions, which I now take for granted). Additionally, if this consequence is combined with another consequence of the Humean thesis—the logical closure of belief—then the Humean thesis can even be recovered from this combination. That is what the (1) → (2) direction of Theorem 27 tells us.

One final thought on this: suppose one required the expected epistemic utility of the least believed proposition not just to be stably positive but to be *maximal* given that property. Theorem 27 would then entail that B_W would have to be the least set of probability 1 (which must exist by W being finite): because it is easy to see that this choice of B_W would maximize, for each X with $Poss(X)$, the expected epistemic utility $\sum_{w \in X}(P(\{w\}) \cdot u_w(B_W))$. For the same reason, that choice of B_W would also minimize to zero the agent's risk of having a false belief among all belief sets that satisfy the Humean thesis (with P). But it would be a mistake to conclude from this that a perfectly rational agent's least believed proposition B_W would *have* to coincide with the least set of probability 1. As usual—and as discussed by Hempel (1962), Levi (1967), and many others—matters of truth also need to be balanced against matters of informativeness or content, which is why the most cautious (and least informative or least contentful) choice is merely permissible but not mandatory. Each of the 'braver' or 'bolder' P-stable^r propositions of probability less than 1 (if there are such) would be options for a perfectly rational agent's least believed proposition B_W, too.

The epistemic requirement that I suggest not to regard as negotiable is the *positivity* of the expected epistemic utility of the strongest believed proposition B_W. If the expected epistemic utility of a perfectly rational agent's set B_W is negative, then she should expect it to be epistemically detrimental to generate her beliefs from it: she should expect her beliefs to approach falsity more than she prefers. Belief's aiming at the truth should then make her generate *Bel* from a different proposition that does have positive expected epistemic utility. What is more, the stability of belief should make her generate *Bel* from a proposition that has positive expected epistemic utility *stably*. For perfectly rational agents whose beliefs and degrees of belief satisfy the Humean thesis, this norm is always realized.

[295] Proof: We already know (2), (3), and (4) to be equivalent to each other from previous chapters (see e.g. Theorem 5 in Appendix B). It was shown in the main text that (2)/(3)/(4) imply (1). The converse can be shown by deriving (4) from (1): for $w \in B_W$, choose Y to be $\{w\} \cup (W \setminus B_W)$, and set $a = 1-r$, $b = -r$, which indeed satisfies $\frac{|b|}{|a|+|b|} = \frac{r}{1-r+r} \leq r$. Then $Poss(Y)$, and (1) entails that $(1-r)\,P(\{w\}) - r\,P(W \setminus B_W) > 0$, which means: $P(\{w\}) > \frac{r}{1-r} \cdot P(W \setminus B_W)$. Thus, (4) holds.

5.2 Belief's Aiming at Subjective Probability

Now I will change gears. Let us take the following for granted: the Bayesian reconstruction of a rational agent's degrees of belief in terms of a subjective probability measure; and the reconstruction of a rational agent's *conditional* all-or-nothing beliefs in terms of a total doxastic (pre-)ordering of possible worlds, as standard AGM belief revision theory or the theory of rational consequence relations in nonmonotonic reasoning has it.[296] In other words: I will take postulates P1 (the axioms of probability) and B1*–B6* (the axioms for conditional all-or-nothing belief) from Chapter 4 as given; but I will not presuppose any bridge principles for probability and conditional belief as yet. If one additionally assumes again that a rational agent has beliefs and degrees of belief simultaneously, then in order for them to cohere, it is plausible that the agent's doxastic order must *approximate*, in some sense, his subjective probability measure. If the agent is perfectly rational, it must do so without errors: the approximation will 'get sufficiently close to the agent's degrees of belief' in this sense. Finally, if we assume again with epistemic decision theory about numerical belief that a perfectly rational agent's degrees of belief 'get as close as possible to the truth'—as assumed already at the beginning of section 5.1—then by approximating these degrees of belief also an agent's conditional belief will 'get sufficiently close to the truth', if only indirectly and in a certain sense.

These are the questions that I am going to answer in this section: what could it mean to say that one doxastic ordering over worlds approximates a probability measure more accurately than another? When is such an approximation error-free and, in this sense, sufficiently close to a probability measure? What are such error-free approximations like?

As I am going to show, ultimately, the answers to these questions will determine and justify a particular theory of absolute and conditional belief for perfectly rational agents, and they will do so in a way that is similar to arguments for Bayesianism in epistemic decision theory. The theory that is to be justified in that manner will be the one of Chapters 2–4 again. The difference to standard epistemic decision theory will be that where normally either subjective probability theory or a theory of all-or-nothing belief is justified by considerations on closeness to the truth, I am going to justify a theory of all-or-nothing belief by considerations on closeness to *subjective probability*.

5.2.1 Probabilistic Order vs Doxastic Order over Worlds

I presuppose again a finite (and non-empty) set W of possible worlds. Propositions will be identified with subsets of W, and all the probability measures that I will consider will assign degrees of belief to propositions in this sense.

The first step in the argument will be to determine a doxastic order $\preccurlyeq_P$ over propositions from any subjective probability measure P. It is obvious how to do so: for given P, define

296 See section 4.1.2 for an overview of rational conditional belief and how it relates formally to so-called sphere systems of propositions, or equivalently, to total 'plausibility' pre-orders of worlds.

$$A \preccurlyeq_P B$$

to be the case if, and only if,

$$P(A) \leq P(B).$$

It follows that $\preccurlyeq_P$ is a total pre-order, that is, it is reflexive, transitive, and total (linear):[297] the usual formal requirements on (preference) orders. Suppressing the reference to P, I will read '$A \preccurlyeq_P B$' as 'the agent doxastically ranks B over A, or she is doxastically indifferent about the two of them', I will say '$A \approx_P B$' ('the agent is doxastically indifferent with respect to A and B') when $P(A) = P(B)$, and '$A \prec_P B$' ('the agent doxastically ranks B over A') for $P(A) < P(B)$. One should keep in mind that the rankings here are purely epistemic: they are truth-centred rankings in the sense of epistemic decision theory, not pragmatic preferences in the sense of standard Bayesian decision theory for actions. An agent does not have to make any conscious decisions either when she doxastically ranks B over A.

The relation $\preccurlyeq_P$ is of great importance to subjective probability theory and its applications. In fact, there is a tradition in probability theory—the theory of qualitative probability, which goes back to Bruno de Finetti[298]—that reverses direction and aims to determine a probability measure P from a given ordering $\preccurlyeq$ (such that $\preccurlyeq = \preccurlyeq_P$).[299] Accordingly, when I am going to speak of approximating an agent's degree-of-belief function P, I will understand this so that really the corresponding doxastic order relation $\preccurlyeq_P$ is being approximated, without caring much about the difference between P and $\preccurlyeq_P$.

$\preccurlyeq_P$ is a relation between sets of worlds that is determined by a probability measure on a finite set of worlds, which in turn is determined by an assignment of *numbers* to (singletons of) *worlds* (that is, numbers of the form $P(\{w\})$). Now here is a much simpler kind of object: let $\preccurlyeq$ be a total pre-order on W, in other words, a reflexive, transitive, and linear *relation* on *worlds*. If

$$u \preccurlyeq v$$

I am going to say again that either v is doxastically ranked over u or that u and v stand in the relationship of doxastic indifference; analogously for the corresponding relations $\approx$ and $\prec$ on worlds. We know such doxastic orderings of worlds already from Chapter 4, except that back then we interpreted them in the reverse fashion—the further down in the ordering, the more plausible the world—whereas in the present section things will just be the other way around: the higher up in the ordering, the

297 A binary relation $\preccurlyeq$ is reflexive iff for all x, $x \preccurlyeq x$. It is transitive iff for all x, y, z, if $x \preccurlyeq y$ and $y \preccurlyeq z$, then $x \preccurlyeq z$. It is total or linear iff for all x, y, it holds that $x \preccurlyeq y$ or $y \preccurlyeq x$.

298 See Suppes (1994) for an overview of the theory of qualitative probability.

299 For a recent view along these lines that highlights the importance of qualitative probability orderings, see Fitelson (2013). See Leitgeb (2014f) for a comparison between qualitative probability in the de Finetti tradition and an interpretation of conditional all-or-nothing belief in the sense of Chapter 4 as a different kind of 'qualitative probability'.

more plausible the world. The reason for the reversal is that in Chapter 4 I followed the tradition in belief revision theory and nonmonotonic reasoning, whereas in the present section it will be more important for me to compare '$u \preccurlyeq v$' to '$A \preccurlyeq_P B$' from above, such that in both cases I will be able to say that the *second* relatum is at least as plausible as the first one is.

In any case, if one were capable of 'boiling down', in some sense, $\preccurlyeq_P$ to any such relation $\preccurlyeq$, then this would correspond to a reduction of complexity by one level of abstraction in the hierarchy of sets: from sets of worlds, to worlds. Or, if one prefers: from assigning real numbers to worlds, to merely ordering worlds.[300] That is exactly where I aim to go: I will 'compress' probabilistic orderings of propositions to doxastic (all-or-nothing) orderings of worlds. In order to do so, I will need to construct an ordering for propositions from any such doxastic ordering $\preccurlyeq$ over worlds. If what is constructed in this way is identical to $\preccurlyeq_P$, the compression process will have been lossless; if what is constructed is at least sufficiently close or similar to $\preccurlyeq_P$, the process will have been accurate enough.

Fortunately, there is a well-known formal method of determining orderings for sets of individuals from orderings for individuals: given $\preccurlyeq$, first define by

$$max_{\preccurlyeq}(A) = \{w \in A \mid \forall w' \in A, w' \preccurlyeq w\}$$

the set of $\preccurlyeq$-maximal members of A, which, by W being finite, must always exist and must always be non-empty for non-empty A. Such sets play a crucial role in areas such as the theory of choice, belief revision theory/nonmonotonic reasoning, or the semantics for counterfactuals, where $max_{\preccurlyeq}(A)$ is called the *choice set of* A, the *set of most plausible A-worlds*, or the *set of closest A-worlds* (to the actual world), respectively, depending on the intended interpretation of $\preccurlyeq$.[301] In each of these theories, propositions are being judged in terms of their highest ranked or most preferred members, or equivalently, in terms of their sets of maximal elements. In our case, only the doxastic 'plausibility' interpretation of such sets will be relevant.[302]

This being in place, one can determine $\preccurlyeq$ for non-empty (or consistent) *propositions* A and B from the relation $\preccurlyeq$ on *worlds*,[303] by

$$A \preccurlyeq B \text{ iff } \forall w \in max_{\preccurlyeq}(A), \exists w' \in max_{\preccurlyeq}(B)\colon w \preccurlyeq w'.$$

300 Compare the corresponding discussion at the end of section 1.5 about belief being simpler than probability.

301 As mentioned before, in Chapter 4 I read doxastic orderings of worlds in the reverse manner; accordingly, in Chapter 4 I would have considered the set of *minimal* members of A with respect to $\preccurlyeq$ where I am now considering the set $max_{\preccurlyeq}(A)$ of maximal members of A with respect to $\preccurlyeq$.

302 For this type of formal construction, as far as belief revision is concerned, see Grove (1988). See Lehmann and Magidor (1992) for a similar type of construction in nonmonotonic reasoning.

303 I will use the same symbol for both kinds of relations; it should always be clear from the context which one is meant.

So $A \preccurlyeq B$ holds if none of the most plausible A-worlds is more plausible than all of the most plausible B-worlds.[304] Because $\preccurlyeq$ is total as an order on worlds, one might just as well use '$\forall w \forall w'$' or '$\exists w \forall w'$' or '$\exists w \exists w'$' instead in the definition above without changing the extension of '$\preccurlyeq$' as an order for propositions.

It follows that $\preccurlyeq$ is a total pre-order on propositions, '$A \approx B$' holds when the agent is indifferent about choosing between the $\preccurlyeq$-maximal members of A and of B qua candidates for what the actual world is like, and '$A \prec B$' is the case when the agent doxastically ranks all $\preccurlyeq$-maximal worlds of B over all $\preccurlyeq$-maximal worlds in A (and hence over all worlds in A whatsoever). Additionally, one could postulate that $\varnothing \preccurlyeq B$ for all propositions B, but I will not be interested in doxastic rankings involving the empty (or inconsistent) proposition in any of the following considerations.

This is certainly a natural way of reducing doxastic rankings over propositions to doxastic rankings over worlds which I will take for granted now. It will allow us to compare $\preccurlyeq_P$ over propositions with $\preccurlyeq$ over worlds through the intermediate step of first generating $\preccurlyeq$ over propositions by the method from above.

But why would we want to compare $\preccurlyeq_P$ and $\preccurlyeq$ in the first place? Because they are in the same line of work. In both belief revision and nonmonotonic reasoning, $\preccurlyeq$ on worlds is interpreted as a kind of plausibility ordering, where plausibility is often specified in terms of doxastic entrenchment or assessments of normality or the like. This order is then taken to induce a plausibility ordering $\preccurlyeq$ on propositions as explained before. On the side of subjective probability theory, the corresponding 'plausibility' ordering is $\preccurlyeq_P$, where plausibility is given by ordering propositions according to their probabilities.[305] The higher up a proposition according to $\preccurlyeq$, the more plausible it is as determined from the worlds that are its most plausible members; and the higher up a proposition according to $\preccurlyeq_P$, the more plausible it is as determined from summing up the probabilities of all the worlds that figure as its members. Therefore, at least prima facie, it should be natural enough to view $\preccurlyeq$ and $\preccurlyeq_P$ as orderings of propositions that have similar doxastic functions, even when they exercise these functions on different levels of complexity: $\preccurlyeq$ is determined from an order of worlds, whereas $\preccurlyeq_P$ is determined from an assignment of numbers to (singletons of) worlds. And since the ordinal scale is less complex than the numerical scale, $\preccurlyeq$ on propositions is likely to be a coarse-grained version of $\preccurlyeq_P$, at best.

But what does 'best' mean here exactly? If both $\preccurlyeq$'s and $\preccurlyeq_P$'s role in one and the same rational agent's cognitive system is to order propositions according to subjective plausibility, then $\preccurlyeq$ ought to approximate $\preccurlyeq_P$ at least in some way, or otherwise, when the agent reasons on the basis of $\preccurlyeq$, the results might diverge too much from her conclusions as being based on $\preccurlyeq_P$. (I am assuming as usual that the agent cannot help

[304] The definition is equivalent to Halpern's (2003, p. 46) 'for all-there is' definition of his relative-likelihood order in any context such as ours in which linearity of $\preccurlyeq$ is assumed over worlds.

[305] Accordingly, Halpern (2003, s. 2.8) discusses all of these formal approaches in terms of so-called 'plausibility measures'.

reasoning both in terms of $\preccurlyeq$ and $\preccurlyeq_P$—so there is no radically Bayesian way out by simply disregarding $\preccurlyeq$.) One such kind of approximation would be: $\preccurlyeq$ approximates $\preccurlyeq_P$ perfectly. That is: for given P (and hence $\preccurlyeq_P$), $\preccurlyeq$ is an order for worlds, and thus also an order for propositions as being given by the method above, such that

$$A \preccurlyeq_P B \text{ iff } A \preccurlyeq B$$

for all propositions A and B.

It is easy to see that such a perfect 'compression' of $\preccurlyeq_P$ is only possible if P is trivial, in the sense that P assigns to one (singleton of a) world probability 1 and hence probability 0 to all other worlds.[306] It follows that truth value assignments yield the only possible cases of probability measures that can be approximated perfectly by a doxastic order over worlds. This should not be too surprising as a binary truth value mapping is the only case of a probability measure that carries precisely the same amount of information as a belief set: probability 1 corresponds to belief and probability 0 corresponds to disbelief. But that should not cast a damper on our goal of approximating $\preccurlyeq_P$ by $\preccurlyeq$ so far as some weaker sense of approximation is concerned. Here is an analogy: if we want to approximate numbers in the closed real interval $[0, 1]$ by the two 'crisp' integers 0 and 1, then the fact that this can only be done without loss of information for the numbers 0 and 1 themselves does not mean that there would not be any reasonable approximation procedure whatsoever.

Since perfect approximation is not to be had in our context, the more relevant questions should therefore be: is there a plausible comparative notion of accuracy by which we can say that one ordering $\preccurlyeq$ over worlds approximates $\preccurlyeq_P$ *more accurately than* some other $\preccurlyeq'$ on worlds does? And is there, for every P, an ordering $\preccurlyeq$ on worlds that approximates $\preccurlyeq_P$ *without errors*? (Where I always assume that $\preccurlyeq$ on propositions is determined from $\preccurlyeq$ on worlds as explained above, and where the exact meanings of 'accurate' and 'error' are yet to be determined.)

We know already from Chapter 4 why these questions are relevant to *belief*: a doxastic order relation over possible worlds corresponds in well-known and natural ways to a set of conditional beliefs (including a set of absolute beliefs), as determined by AGM belief revision theory and nonmonotonic reasoning. Given a doxastic order $\preccurlyeq$ on worlds, consider the set

$$max_{\preccurlyeq}(W)$$

of worlds that are ranked highest overall to be the agent's set of doxastically possible candidates for what the actual world is like. In other words: $max_{\preccurlyeq}(W)$ is the agent's

[306] For assume there to exist at least two worlds w and w' of positive probability: then $P(\{w\}) < P(\{w, w'\})$ and $P(\{w'\}) < P(\{w, w'\})$, and therefore also $\{w\} \prec_P \{w, w'\}$ and $\{w'\} \prec_P \{w, w'\}$. But in whatever way a doxastic order $\preccurlyeq$ is being defined on worlds, it will always be the case that $max_{\preccurlyeq}(\{w\}) = \{w\}$, $max_{\preccurlyeq}(\{w'\}) = \{w'\}$, and $max_{\preccurlyeq}(\{w, w'\}) \subseteq \{w, w'\}$, which rules out that $\{w\} \prec \{w, w'\}$ and $\{w'\} \prec \{w, w'\}$ are the case simultaneously, for it cannot be that all maximal members of $\{w, w'\}$ are ranked over both all maximal members of $\{w\}$, that is, w, and all maximal members of $\{w'\}$, that is, w'. Contradiction.

strongest believed proposition B_W: the set of most plausible worlds. Absolute or unconditional belief corresponds to $\preccurlyeq$ in the way

$$Bel(B) \text{ iff } B_W = max_{\preccurlyeq}(W) \subseteq B,$$

that is, B is believed just in case every most plausible candidate for the actual world is a member of B. Furthermore, conditional belief—the all-or-nothing counterpart of conditional probability—corresponds to $\preccurlyeq$ as given by

$$Bel(B \mid A) \text{ iff } max_{\preccurlyeq}(A) \subseteq B,$$

which means that the agent believes B given A precisely when all of the most plausible A-worlds are B-worlds. All of these notions of belief correspond naturally to the notion of doxastic order over possible worlds, as discussed e.g. in section 4.1.2 in Chapter 4.[307]

For this reason, as soon as we know how to determine an 'accurate approximation' of an agent's $\preccurlyeq_P$ by a doxastic ordering $\preccurlyeq$ over worlds, we can determine from $\preccurlyeq$ an agent's belief set that will also, in a sense, 'accurately approximate' $\preccurlyeq_P$ or P: the agent's beliefs will be something like accurate (by $\preccurlyeq$ being accurate) simplifications (by determining a doxastic order over propositions from a doxastic order over worlds) of $\preccurlyeq_P$. And indeed, just as it had been the case for $\preccurlyeq$ vs $\preccurlyeq_P$ before, it seems to be in the nature of belief to be formally simpler than subjective probability, and it seems to be normatively required of a perfectly rational agent that her beliefs be, in some sense, close or similar to her degrees of belief: for otherwise, just as mentioned before, her beliefs might commit her to acts that might conflict with her commitments from subjective probability, and this would be so even beyond the inevitable differences that follow from belief being more coarse-grained than probability.

The upshot is: by assuming that $\preccurlyeq$ approximates the same agent's $\preccurlyeq_P$, we should be guaranteed to hit upon a normatively compelling theory of belief in which conditional all-or-nothing belief approximates the same agent's degrees of belief (in a certain sense, and to some extent).

In section 5.2.2 I will propose and defend a definition of comparative accuracy for doxastic orders over worlds, which will be based on a precise notion of *error with respect to* $\prec_P$. Section 5.2.3 will determine those approximations of a given probability measure by means of doxastic order relations over worlds that are not subject to error. Section 5.2.4 will summarize what has been achieved, and it will draw some general conclusions from this on belief. The stability theory of conditional belief from Chapter 4 will emerge from this (for the special threshold $r = \frac{1}{2}$).

307 I am glossing over a couple of issues here. Instead of '$Bel(B \mid A)$', belief revision theorists would say '$B \in K * A$' where $*$ is a belief revision operator and K is the agent's present belief set. In nonmonotonic reasoning, the same state of affairs is described by means of '$A \mid\!\sim B$' where $\mid\!\sim$ is a nonmonotonic inference relation. See section 4.1.2 for more on this. Furthermore, if $P(A) = 0$, postulate BP2* (Zero Supposition) from section 4.2.3 entailed that $Bel(B|A)$ for all B. So in this special case (and only there) it does *not* hold in my theory that for all B, $Bel(B \mid A)$ iff $max_{\preccurlyeq}(A) \subseteq B$.

5.2.2 *Accuracy for Orders over Worlds*

When aiming to approximate the truth by means of belief, and assuming that a rational agent's set of beliefs must be consistent, an agent can encounter just two kinds of deficiencies:[308]

- The (Soundness) Error: believing A when A is false.
- The (Completeness) Gap: neither believing A nor $\neg A$ when A is true.

By substituting '$\neg A$' for 'A' in the soundness clause, also the case of believing $\neg A$ when A is true is covered. Substituting '$\neg A$' for 'A' in the completeness clause covers also the case of neither believing A nor $\neg A$ when $\neg A$ is true.

The soundness case is a proper error: the agent is doing something wrong, at least from a purely epistemic, that is, truth-focused, point of view. In contrast, the completeness case does not really involve an error: by suspending judgement on a proposition, no mistake is made, although an opportunity may have been lost.[309]

In the realm of doxastic ranking, suspension of belief corresponds to indifference between worlds. Indeed, the correspondence is quite tight: in particular, by the definition of $Bel_{\preccurlyeq}$ from the last section, an agent will suspend judgement on A precisely when $max_{\preccurlyeq}(W)$ contains both A-worlds and $\neg A$-worlds, that is, when the agent's ranking of worlds yields a tie between the most plausible A-worlds and the most plausible $\neg A$-worlds.

Replacing suspension by indifference leads us to the corresponding two kinds of deficiencies when aiming to approximate $\preccurlyeq_P$ over propositions by $\preccurlyeq$ over worlds (and derivatively over propositions, as explained in the last section):

- The (Soundness) Error: $A \prec B$ when it is not the case that $A \prec_P B$.
- The (Completeness) Gap: $A \approx B$ when either $A \prec_P B$ or $B \prec_P A$.

Neither of these kinds of deficiency concerns truth; they are rather defects with respect to $\prec_P$ (or P). In the soundness case, $\prec$ either reverses the ordering as given by $\prec_P$ or misrepresents '$P(A) = P(B)$' as a strict ordering. In the completeness case, an instance of $\prec_P$ is merely changed into indifference, that is, into an instance of $\approx$.

In order to define a notion of comparative accuracy for relations $\preccurlyeq$ that are meant to approximate some given $\preccurlyeq_P$, we need to somehow trade deficiencies of the two kinds against each other. There is not necessarily a unique and *sacrosanct* way of doing so, which is why we will have to make a decision here.

In the case of beliefs again, the following would be a natural suggestion to hold for perfectly rational agents: it is of primary importance to avoid false beliefs; and it is only of (far) subordinate importance to avoid incomplete beliefs. In a sense, this type of proposal is very much on the cautious side, since even the empty and

308 James (1896) famously discusses a similar distinction.

309 I am grateful to David Makinson for highlighting this in precisely these terms in personal communication.

hence maximally incomplete set of beliefs will be counted as superior to any belief set that includes at least one false proposition. But then again, *epistemically*, suspension of judgement does not constitute an error, and it is only epistemic issues that I am interested in here. Furthermore, braveness may still be accounted for at least so far as those belief sets are concerned that are affected only by completeness gaps (but not by soundness errors): because for them, the more informative a belief set is, the more accurate it might be taken to be.

I am going to opt for a translation of this proposal into the context of doxastic ordering or ranking. That is not because I think that there would not be other reasonable choices: on the contrary, there surely are, and their properties and consequences ought to be explored and compared. But one needs to start somewhere, and—over and above any intrinsic epistemic plausibility of valuing soundness more highly than completeness—the resulting definition of comparative accuracy will be seen to have some nice properties and consequences:

Definition 28 For all total pre-orders $\preccurlyeq, \preccurlyeq'$ over W (from which total pre-orders $\preccurlyeq, \preccurlyeq'$ over subsets of W can be defined as carried out in section 5.2.1), and for all probability measures P that are defined on arbitrary subsets of W:

$\preccurlyeq$ is more accurate than $\preccurlyeq'$ (when approximating $\preccurlyeq_P$) iff

1. the set of soundness errors of $\preccurlyeq$ is a proper subset of the set of soundness errors of $\preccurlyeq'$:
   - for all A, B: if $A \prec B$ but $A \nprec_P B$, then also $A \prec' B$ (but $A \nprec_P B$),
   - there exist A, B, such that $A \prec' B$ and $A \nprec_P B$, while $A \nprec B$ (and $A \nprec_P B$);
2. or: their sets of soundness errors are equal, but the set of completeness gaps of $\preccurlyeq$ is a proper subset of the set of completeness gaps of $\preccurlyeq'$:
   - for all A, B: if $A \approx B$ but either $A \prec_P B$ or $B \prec_P A$, then also $A \approx' B$ (but either $A \prec_P B$ or $B \prec_P A$),
   - there exist A, B, such that $A \approx' B$ but either $A \prec_P B$ or $B \prec_P A$, but $A \not\approx B$.

This defines an irreflexive[310] and transitive relation of comparative accuracy which, as one might say, orders doxastic ordering relations over worlds lexicographically: by ranking soundness errors over completeness gaps much as the first letters of words are ranked over their second letters in a lexicon. All of the deficiencies in question, and hence the comparative relation of accuracy itself, are determined relative to P (or $\preccurlyeq_P$).

Let us have a look at two examples:

Example 20
Let $W = \{w_1, w_2, w_3\}$, and assume P to be determined by $P(\{w_1\}) = \frac{5}{8}$, $P(\{w_2\}) = \frac{2}{8}$, $P(\{w_3\}) = \frac{1}{8}$. Table 5.1 summarizes what the total preorder $\preccurlyeq_P$ of propositions is like given that measure P, and it lists three further orderings over propositions that can be

[310] That is: no $\preccurlyeq$ is more accurate than itself.

Table 5.1. Table for Example 20

$\preccurlyeq_P$ **on propositions (determined from** P**)**	$\preccurlyeq_1$ **on propositions (determined from** $w_3 \prec_1 w_2 \prec_1 w_1$**)**	$\preccurlyeq_2$ **on propositions (determined from** $w_3 \prec_2 w_1, w_2$**)**	$\preccurlyeq_3$ **on propositions (determined from** $w_3 \prec_3 w_1 \prec_3 w_2$**)**
$\{w_1, w_2, w_3\}$	$\{w_1, w_2, w_3\}$ $\approx_1 \{w_1, w_2\}$ $\approx_1 \{w_1, w_3\}$ $\approx_1 \{w_1\}$	$\{w_1, w_2, w_3\}$ $\approx_2 \{w_1, w_2\}$ $\approx_2 \{w_1, w_3\}$ $\approx_2 \{w_1\}$ $\approx_2 \{w_2, w_3\}$ $\approx_2 \{w_2\}$	$\{w_1, w_2, w_3\}$ $\approx_3 \{w_1, w_2\}$ $\approx_3 \{w_2, w_3\}$ $\approx_3 \{w_2\}$
$\{w_1, w_2\}$	$\{w_2, w_3\}$ $\approx_1 \{w_2\}$	$\{w_3\}$	$\{w_1, w_3\}$ $\approx_3 \{w_1\}$
$\{w_1, w_3\}$	$\{w_3\}$		$\{w_3\}$
$\{w_1\}$			
$\{w_2, w_3\}$			
$\{w_2\}$			
$\{w_3\}$			

determined (by the method described in section 5.2.1) from the order relations over worlds that are stated in the third line of the text in the table. Once again, I ignore the case of the empty proposition.

The higher up in the table, the higher ranked doxastically or 'plausible' the corresponding proposition. E.g. $\{w_1, w_3\} \prec_P \{w_1, w_2\}$ and $\{w_1, w_2\} \prec_P \{w_1, w_2, w_3\}$, as follows from $P(\{w_1, w_3\}) < P(\{w_1, w_2\})$ and $P(\{w_1, w_2\}) < P(\{w_1, w_2, w_3\})$.

Similarly, e.g. $\{w_2, w_3\} \prec_1 \{w_1, w_2\}$, because $max_{\preccurlyeq_1}(\{w_2, w_3\}) = \{w_2\}$, $max_{\preccurlyeq_1}(\{w_1, w_2\}) = \{w_1\}$, and $w_2 \prec_1 w_1$ by the definition of $\prec_1$. But e.g. $\{w_1, w_2\} \approx_1 \{w_1, w_3\}$, because $max_{\preccurlyeq_1}(\{w_1, w_2\}) = \{w_1\}$, $max_{\preccurlyeq_1}(\{w_1, w_3\}) = \{w_1\}$, and $w_1 \approx_1 w_1$.

All the empty entries in the table can be ignored, and no horizontal comparisons between single entries of different columns ought to be made.

An application of our definition of comparative accuracy to $\preccurlyeq_1$, $\preccurlyeq_2$, and $\preccurlyeq_3$, when approximating $\preccurlyeq_P$, delivers the following results: $\preccurlyeq_1$ is more accurate than $\preccurlyeq_2$, and $\preccurlyeq_2$ is more accurate than $\preccurlyeq_3$. That is for the following reason: first of all, $\preccurlyeq_1$ and $\preccurlyeq_2$ do not generate any soundness errors—whenever one of them says that a proposition is ranked below another, this is so also according to $\preccurlyeq_P$. On the other hand, $\preccurlyeq_3$ is affected by soundness errors: e.g. $\{w_1\} \prec_3 \{w_2\}$, although it does not hold that $\{w_1\} \prec_P \{w_2\}$. Therefore, $\preccurlyeq_1$ and $\preccurlyeq_2$ are more accurate than $\preccurlyeq_3$. Secondly, $\preccurlyeq_1$ is more accurate than $\preccurlyeq_2$, since every completeness gap that affects $\preccurlyeq_1$ also affects $\preccurlyeq_2$, however, there are completeness gaps which only concern $\preccurlyeq_2$ but not $\preccurlyeq_1$: an example would be $\{w_1\} \approx_2 \{w_2\}$ and $\{w_2\} \prec_P \{w_1\}$, while indeed $\{w_2\} \prec_1 \{w_1\}$. So amongst the doxastic

Table 5.2. Table for Example 21

$\preccurlyeq'_P$ on propositions (determined from P')	$\preccurlyeq_1$ on propositions (determined from $w_3 \prec_1 w_2 \prec_1 w_1$)	$\preccurlyeq_2$ on propositions (determined from $w_3 \prec_2 w_1, w_2$)	$\preccurlyeq_3$ on propositions (determined from $w_3 \prec_3 w_1 \prec_3 w_2$)
$\{w_1, w_2, w_3\}$	$\{w_1, w_2, w_3\}$ $\approx_1 \{w_1, w_2\}$ $\approx_1 \{w_1, w_3\}$ $\approx_1 \{w_1\}$	$\{w_1, w_2, w_3\}$ $\approx_2 \{w_1, w_2\}$ $\approx_2 \{w_1, w_3\}$ $\approx_2 \{w_1\}$ $\approx_2 \{w_2, w_3\}$ $\approx_2 \{w_2\}$	$\{w_1, w_2, w_3\}$ $\approx_3 \{w_1, w_2\}$ $\approx_3 \{w_2, w_3\}$ $\approx_3 \{w_2\}$
$\{w_1, w_2\}$	$\{w_2, w_3\}$ $\approx_1 \{w_2\}$	$\{w_3\}$	$\{w_1, w_3\}$ $\approx_3 \{w_1\}$
$\{w_1, w_3\}$	$\{w_3\}$		$\{w_3\}$
$\{w_2, w_3\}$			
$\{w_1\}$			
$\{w_2\}$			
$\{w_3\}$			

orders $\preccurlyeq_1$, $\preccurlyeq_2$, $\preccurlyeq_3$ on worlds, it is $\preccurlyeq_1$, or graphically (when reading the order from bottom to top)

$$\begin{array}{c} w_1 \\ w_2 \\ w_3 \end{array}$$

that is the most accurate approximation of $\preccurlyeq_P$. This does look plausible also on intuitive grounds, as the probability of each world according to P is of an order of magnitude greater than the probabilities of all worlds with higher index.[311]

Example 21

Let W be as before, but now let $P'(\{w_1\}) = \frac{4}{9}$, $P'(\{w_2\}) = \frac{3}{9}$, $P'(\{w_3\}) = \frac{2}{9}$. Consider Table 5.2 in which $\preccurlyeq_{P'}$ is determined from the new probability measure P', but in which $\preccurlyeq_1$, $\preccurlyeq_2$, $\preccurlyeq_3$ are just as before. In comparison to the previous table, $\{w_1\}$ and $\{w_2, w_3\}$ have switched ranks in the leftmost column, because now $P'(\{w_2, w_3\}) > P'(\{w_1\})$. Relative to $\preccurlyeq_{P'}$, it follows: $\preccurlyeq_2$ is more accurate than both $\preccurlyeq_1$ and $\preccurlyeq_3$, and $\preccurlyeq_1$ is incomparable to $\preccurlyeq_3$ in terms of accuracy. And the reason is: both $\preccurlyeq_1$ and $\preccurlyeq_3$ are subject to soundness errors—just consider $\{w_2, w_3\} \preccurlyeq_1 \{w_1\}$ and $\{w_1, w_3\} \preccurlyeq_3 \{w_2\}$—while $\preccurlyeq_2$ is not. The set of soundness errors for $\preccurlyeq_1$ is neither a superset, nor a subset, of the set of soundness errors for $\preccurlyeq_3$. Summing up: amongst the doxastic orders $\preccurlyeq_1$, $\preccurlyeq_2$, $\preccurlyeq_3$ on worlds, it is now $\preccurlyeq_2$, or graphically

311 See Halpern (2003, p. 44) and Goldszmidt and Pearl (1992) for more on probabilistic order-of-magnitude reasoning.

$$\begin{array}{c} w_1, w_2 \\ w_3 \end{array}$$

that is the most accurate approximation of $\preccurlyeq_{P'}$. This does not seem counterintuitive either, as the probability of w_1 does not dominate this probability space in the same way as had been the case in Example 20. (Although it has to be admitted that ordering w_2 over w_3 does not seem to follow from one's mere intuitions about the 'orders of magnitude' of their probabilities alone.)

We have determined what seems to be a plausible proposal for a comparative notion of accuracy.[312] In the next section I will determine for every probability measure P those approximations of $\preccurlyeq_P$ that are free of soundness errors, and I will show how these approximations can be ordered easily according to accuracy.

5.2.3 *Error-Free Doxastic Orders of Worlds*

When $\preccurlyeq, \preccurlyeq'$ are two total pre-orders on W, call $\preccurlyeq$ more fine-grained than $\preccurlyeq'$ iff (a) for all u, v in W, if $u \prec' v$ then $u \prec v$, and (b) additionally there exist u, v in W, such that $u \prec v$ while not $u \prec' v$ (that is, using condition (a), where $u \approx' v$). The following result answers the main open questions from before:

Theorem 29 Let W be non-empty and finite. For every probability measure P defined on all subsets of W the following is the case:

- For all total pre-orders $\preccurlyeq$ on W:
  $\preccurlyeq$ is error-free, that is, not subject to any soundness errors (relative to $\preccurlyeq_P$) iff $\preccurlyeq$ satisfies the following Sum Condition with respect to P:[313] for all $w \in W$ with $P(\{w\}) > 0$,

  $$P(\{w\}) > \sum_{w':\, w' \prec w} P(\{w'\}),$$

  and for all $w \in W$ with $P(\{w\}) = 0$: for all $w' \in W$, $w \preccurlyeq w'$.
- For all total pre-orders $\preccurlyeq, \preccurlyeq'$ on W that are error-free (relative to $\preccurlyeq_P$):
  $\preccurlyeq$ is more accurate than $\preccurlyeq'$ (when approximating $\preccurlyeq_P$) iff

  $$\preccurlyeq \text{ is more fine-grained than } \preccurlyeq'.$$[314]

[312] Of course, various refinements of this proposal would be conceivable, such as taking into account the *cardinality* of error sets, or the like.

[313] This Sum Condition is a special case of the Sum Condition that we encountered at the end of section 4.2.3: it concerns the special case in which $r = \frac{1}{2}$. The only difference is that I have added a clause for worlds $w \in W$ with $P(\{w\}) = 0$ that was not included back in Chapter 4. The Sum Condition follows from iterated application of the Outclassing Condition; see the end of section 4.2.3 for the details.

[314] Here is the proof (where I leave out the easy case concerning worlds of probability 0): for the first part, whenever there is a world w, such that $P(\{w\}) \le \sum_{w':\, w' \prec w} P(\{w'\})$, then $P(\{w\}) \le P(\{w' \mid w' \prec w\})$ and thus $\{w' \mid w' \prec w\} \not\prec_P \{w\}$, even though the $\preccurlyeq$-maximal members of $\{w\}$, that is, w itself, are ranked over the $\preccurlyeq$-maximal members of $\{w' \mid w' \prec w\}$, and hence $\{w' \mid w' \prec w\} \prec \{w\}$. In other words: if the Sum Condition is not satisfied, then $\preccurlyeq$ is affected by a soundness error. And vice versa it follows in the same

In words: $\preccurlyeq$ does not lead to any soundness errors precisely when worlds of zero probabilistic weight are assigned least plausibility in $\preccurlyeq$ (which in this section means that they are at the bottom of the doxastic ordering $\preccurlyeq$) and when the probability of each other world is greater than the sum of all probabilities of worlds of lesser plausibility in the doxastic order $\preccurlyeq$. And amongst the orders on worlds that satisfy this Sum Condition, the more fine-grained an order is, the more accurate it happens to be. The Sum Condition should be considered as a constraint on a total pre-order $\preccurlyeq$ of worlds given an arbitrary probability measure P; the strict part $\prec$ of $\preccurlyeq$ figures in the sums in question.

As mentioned already in section 4.1.4 of Chapter 4, there is literature from theoretical computer science[315] in which a condition such as the Sum Condition was studied, though in application to total partial orders rather than total *pre*-orders. However, if ties between distinct worlds are being excluded, this has consequences: the set of probability measures P, such that some total partial order satisfies the Sum Condition with respect to such a P, is quite restrictive. In contrast, for every P there is a total pre-order that satisfies the Sum Condition with respect to P.

It follows immediately from the theorem that for all P (on a finite non-empty W) there is a total pre-order $\preccurlyeq_{acc,P}$, such that:

For all total pre-orders $\preccurlyeq'$ on W that are distinct from $\preccurlyeq_{acc,P}$,

$$\preccurlyeq_{acc,P} \text{ is more accurate than } \preccurlyeq' \text{ (when approximating } \preccurlyeq_P).$$

By Theorem 29, this $\preccurlyeq_{acc,P}$ is simply *the most fine-grained* $\preccurlyeq$ *on* W *that satisfies the Sum Condition*, which, as one can show, must always exist for finite W.[316]

Thus, for every probability measure there is indeed a most accurate approximation to it by a doxastic order relation over worlds. When P is trivial, such that there is a world w for which $P(\{w\}) = 1$, the Sum Condition implies that $\preccurlyeq_{acc,P}$ ranks w over all other worlds and treats these other worlds as being tied amongst each other; in that case, $\preccurlyeq_{acc,P}$ indeed turns out to be the perfect approximation of $\preccurlyeq_P$. But, as we have

immediate manner that if the Sum Condition is satisfied, then $\preccurlyeq$ cannot be affected by a soundness error. (Note that a doxastic order $\preccurlyeq$ according to which there is a tie between every two worlds satisfies the Sum Condition trivially.)

For the second part of the theorem, one only needs to observe: the more fine-grained a total pre-order is, the smaller the set of indifferences must be that it generates; and since only order relations without soundness errors are considered in this second part, the greater must also be the set of true instances of the form '$A \prec B$'.

315 Compare the *big-stepped probabilities* of Benferhat, Dubois, and Prade (1999), and the *atomic bound systems* of Snow (1998), as discussed in section 4.1.4.

316 Theorem 29 from the present section and Observation 24 from Chapter 4 taken together imply that this doxastic order $\preccurlyeq_{acc,P}$ of worlds is simply the one that is given by the (inverse of the) ordering of worlds that corresponds to the sphere system that consists of: all P-stable$^{\frac{1}{2}}$ sets of probability less than 1 (if there are such) and the uniquely determined least proposition of probability 1 (which is P-stable$^{\frac{1}{2}}$, too). The only minor addition in the present section is that all worlds of probability 0 need to be added in that ordering $\preccurlyeq_{acc,P}$ as the least (and hence maximally implausible) layer of worlds.

seen already in section 5.2.1, such probability measures are the only ones that allow for lossless approximation in terms of doxastic orderings over worlds. For all other P, the doxastic order $\preccurlyeq_{acc,P}$ 'merely' approximates $\preccurlyeq_P$ to the best possible extent.

Let us illustrate the findings above by means of the examples from the last section: in Example 20, it follows that $\preccurlyeq_1$, that is,

$$\begin{array}{c} w_1 \\ w_2 \\ w_3 \end{array}$$

is actually the most accurate approximation of $\preccurlyeq_P$ overall: for $P(\{w_1\}) > P(\{w_2\}) + P(\{w_3\})$, $P(\{w_2\}) > P(\{w_3\})$, and clearly there cannot be a more fine-grained total pre-order on worlds than $\preccurlyeq_1$ which also satisfies the Sum Condition, since $\preccurlyeq_1$ is already maximally fine-grained.

Accordingly, in Example 21 it holds that $\preccurlyeq_2$, that is,

$$\begin{array}{c} w_1 \approx_2 w_2 \\ w_3 \end{array}$$

is the most accurate approximation of $\preccurlyeq_{P'}$ overall: for $P'(\{w_1\}) > P'(\{w_3\})$, $P'(\{w_2\}) > P'(\{w_3\})$, but no fine-graining of $\preccurlyeq_2$ would satisfy the Sum Condition from above; in particular, neither w_1 nor w_2 would do as a single top element, because $P'(\{w_1\}) \not> P'(\{w_2\}) + P'(\{w_3\})$ and $P'(\{w_2\}) \not> P'(\{w_1\}) + P'(\{w_3\})$.

In the final section I am going to evaluate the consequences of these results for belief.

5.2.4 Conclusions on Rational Belief

I intended to approximate probabilistic orderings $\preccurlyeq_P$ over propositions by doxastic orderings $\preccurlyeq$ over worlds. In section 5.2.1 I introduced a method of determining a doxastic order on propositions from $\preccurlyeq$ on worlds; in this way, $\preccurlyeq_P$ over propositions became comparable to $\preccurlyeq$ on worlds. In section 5.2.2 I suggested a concept of comparative accuracy by which we were able to say when some $\preccurlyeq$ on worlds approximated $\preccurlyeq_P$ more accurately than some other $\preccurlyeq'$ on worlds. And in section 5.2.3 I characterized those approximations of $\preccurlyeq_P$ by a doxastic order relation that were error-free.

It is about time to return to belief. What are a perfectly rational agent's beliefs like when they correspond to an error-free approximation of (the ordering determined by) her degrees of belief, where this approximation is carried out by means of a doxastic order over worlds?

The answer is given by the following theorem. In order to be able to state the theorem succinctly, I introduce the following abbreviation: let us say that a conditional belief set *Bel corresponds to* a total pre-order $\preccurlyeq$ and a probability measure P (all defined on the same non-empty set W of worlds) iff

- for all $A \subseteq W$ with $P(A) > 0$, for all $B \subseteq W$: $Bel(B \mid A)$ iff $max_{\preccurlyeq}(A) \subseteq B$,
- for all $A \subseteq W$ with $P(A) = 0$, for all $B \subseteq W$: $Bel(B \mid A)$.

Then one has:

Theorem 30 (Representation Theorem for Error-Free Doxastic Orders)
Let W be non-empty and finite. For every probability measure P defined on all subsets of W, and for all sets *Bel* of pairs of subsets of W, the following three conditions (on P and *Bel*) are equivalent:

1. There is a total pre-order $\preccurlyeq$ on W, such that *Bel* corresponds to $\preccurlyeq$ and P, and $\preccurlyeq$ is error-free, that is, not subject to any soundness errors (relative to $\preccurlyeq_P$).
2. There is a total pre-order $\preccurlyeq$ on W, such that *Bel* corresponds to $\preccurlyeq$ and P, and $\preccurlyeq$ satisfies the Sum Condition with respect to P: for all $w \in W$ with $P(\{w\}) > 0$,
$$P(\{w\}) > \sum_{w':\, w' \prec w} P(\{w'\}),$$
and for all $w \in W$ with $P(\{w\}) = 0$: for all $w' \in W$, $w \preccurlyeq w'$.
3. P and *Bel* satisfy the postulates from Chapter 4 for the threshold $r = \frac{1}{2}$.[317]

With the axioms of probability for a perfectly rational agent's degrees of belief in place, we find that the same agent's conditional beliefs amount to an error-free approximation of her degrees of belief—and hence, in the sense of this section, 'get sufficiently close to' their aim—if, and only if, they satisfy the principles of the stability theory of conditional belief from Chapter 4 for the least possible numerical threshold $r = \frac{1}{2}$.[318] Note that if (3) in Theorem 30 had been formulated for a threshold $r > \frac{1}{2}$, then both (1) and (2) would still be derivable from (3), just not vice versa.

If additionally such an agent's degrees of belief get as close to the truth as possible (as epistemic decision theory for degrees of belief has it), then belief still aims at, and gets sufficiently close to, the truth: though only indirectly. Or at least this should be so for absolute or unconditional belief in a proposition B, which the theory in Chapter 4 identified with belief in B conditional on the trivial or tautological information W, and for which 'aiming at the truth' has the straightforward paraphrase: *aiming for the believed proposition B to be true*. (On the other hand, the meaning of 'aiming at the truth' for a *conditional* belief in B given A, such that $A \neq W$, would be much less clear-cut. But see (ii) below for what might go some way towards a proposal.)

[317] The relevant postulates from Chapter 4 are: P1, B1*–B6*, BP1^{r*}, BP2* from section 4.2.3. The clause in the definition of 'corresponds to' that concerns 'all A with $P(A) = 0\ldots$' matches the auxiliary principle BP2* from section 4.2.3. Proof: The equivalence of (1) and (2) follows immediately from the first part of Theorem 29. The equivalence of (2) and (3) follows from Observation 24 from the end of section 4.2.3. In the '(3) $\rightarrow$ (2)' direction one needs to extend the total pre-order $\preccurlyeq$ on worlds with positive probability from Observation 24 to one that is defined on all worlds in W by means of: for all $w \in W$ with $P(\{w\}) = 0$, for all $w' \in W$, $w \preccurlyeq w'$.

[318] We know from Observation 21 from section 4.2.2 that every P-stable^r set is also a P-stable$^{\frac{1}{2}}$ set (as long as $\frac{1}{2} \leq r < 1$). So $r = \frac{1}{2}$ is a salient choice in so far as it allows for the greatest class of permissible candidates for spheres in any sphere system that corresponds to a rational conditional belief set. (Compare Theorem 23 from Chapter 4.) More generally, I have collected some arguments for what makes the choice $r = \frac{1}{2}$ especially salient in subsection 'Choosing the Value of "r"' of section 4.2.2.

Let me conclude this chapter with a remark that complements the one from the end of section 5.1. At the end of that section we saw that *maximizing* the stably positive expected epistemic utility of an agent's least believed proposition was not epistemically mandatory: the outcome of this kind of maximization would have been the maximally cautious option of choosing B_W to be the least proposition of probability 1. But matters of truth also needed to be balanced against matters of content, which is why minimizing one's risk of having a false belief was not more than just permissible. The proper norm on belief was to get *sufficiently* close to the truth.

So far as belief's aiming at subjective probability is concerned, one can draw a similar conclusion, but now for inverse reasons: this time, given P, *maximizing* the accuracy of $\preccurlyeq$ with respect to P would lead to the maximally fine-grained error-free ordering $\preccurlyeq_{acc,P}$ of worlds that I considered briefly at the end of section 5.2.3. By being maximally fine-grained, $\preccurlyeq_{acc,P}$ can be seen to maximize (given P) the agent's corresponding conditional belief set that corresponds to $\preccurlyeq$ and P, such that all of the postulates from Chapter 4 are satisfied. That is: the corresponding conditional belief set would be maximally *brave*. Let me explain this in more detail: the strongest (unconditionally) believed proposition B_W would be as small as possible, and hence there would be as many (unconditionally) believed propositions as possible (as many supersets of B_W as possible). Accordingly, for every proposition A, the strongest believed proposition conditional on A would be as small as possible,[319] and hence there would be as many propositions B believed conditional on A as possible. While this would get *Bel* as close as possible to P—in the sense developed before (and always based on a total pre-ordering of *worlds*)—by maximizing *Bel* also the agent's 'falsity risks' would be maximized: the risks of (i) having a false unconditional belief, and also of (ii) having a conditional belief in B given A, when A is true but B is false. Being error-free with respect to P is no guarantee for being error-free with respect to truth, and minimizing completeness gaps with respect to P maximizes the risk of falsity (amongst the orders that are error-free with respect to P). It is not epistemically mandatory for a perfectly rational agent to take this kind of risk, which is why $\preccurlyeq_{acc,P}$ is merely permissible again.

What I suggest not to be negotiable is a perfectly rational agent's doxastic ordering of worlds being *error-free*: this captures belief's approximating degrees of belief in a way that does not *distort* the probabilistic ordering of propositions. Conditional all-or-nothing belief may be more or less cautious, but it should cohere with the same agent's degree-of-belief function at least in so far as it should not rank the plausibility of B over that of A when the degree-of-belief function does not. For perfectly rational agents whose beliefs and degrees of belief satisfy the Humean thesis, this norm is always satisfied.[320]

319 In the terminology of section 4.2.3, that set would be the set B_A.

320 See Leitgeb (2014e) for yet another, and substantially different, justification of the stability theory of belief along the lines of epistemic decision theory.

6

Action, Assertability, Acceptance

While the previous chapters were mostly (though not exclusively) concerned with the role that stability plays for the theoretical rationality of belief, this chapter will deal with the practical rationality of belief: consequences that stable rational belief has for decision-making (recall the 'Action' Assumption 3 from section 1.1), for assertion (compare the 'Assertion' Assumption 5 from Chapter 1), and for the mental act of accepting a proposition. The chapter will conclude with an analysis of the Preface Paradox: a story according to which, as it were, all the statements in a book get asserted. The analysis of the paradox, which will continue my first shot at it from section 3.5, will be based on insights from the present chapter as well as from previous ones.

Section 6.1 will return to the simple decision theory for all-or-nothing belief that was presented in section 2.3.3, in which utilities were assumed to be as binary as categorical belief. Back then a little theorem[321] showed that if a perfectly rational agent's degrees of belief and her categorical beliefs jointly satisfy the Humean thesis from Chapter 2 (and if the contradictory proposition is not believed by the agent), then the simple decision theory from Chapter 2 commands the agent to decide about actions in a way that coheres with Bayesian decision theory. One part of that 'decision-theoretic coherence' was: actions that are permissible in the all-or-nothing sense are always superior in expected value to actions that are impermissible in the all-or-nothing sense. A second part was: what the decision theory for all-or-nothing belief deems permissible (those actions that the agent *believes to be useful*) cannot differ too much in expected utility from what Bayesian decision theory judges permissible (those actions that *maximize expected utility*). Section 6.1 will extend this account in three different respects. The first one concerns another representation theorem (Theorem 31): the Humean thesis can be recovered from two of its consequences, that is, from the logical closure of belief taken together with (the first part of) decision-theoretic coherence as explained before. In other words: given the logical closure of belief, it is hard to avoid the Humean thesis even on practical grounds. The second extension will relieve our simple decision theory for categorical belief from the assumption that *every* function from worlds to outcomes counts as an action. Finally, I will state an interesting variation of my original simple decision theory for categorical belief: according to that variant, an action is permissible for an agent just in case the agent regards it as *possible that it*

[321] See Theorem 4 in section 2.3.3.

is useful. This weaker notion of permissibility may apply to actions even when no action is permissible in the stronger *belief* sense from before. Another theorem will demonstrate that this variant of our qualitative decision theory still somehow coheres with Bayesian decision theory, though in a sense of coherence that has been weakened correspondingly.[322]

Section 6.2 will consider how the stability of belief shapes subjective assertability (that is, assertability internal to, and solely from the perspective of, the asserting agent): subjective assertability for non-conditional propositions and for indicative conditionals. Subjective assertability will be expressed both on a quantitative and a categorical scale again, where the theory of subjective numerical assertability for (indicative) conditionals will correspond to Adams's and Edgington's suppositional theory of conditionals. The theory of subjective *all-or-nothing* assertability for conditionals will be the all-or-nothing counterpart thereof: it will include closure conditions for assertability that correspond simultaneously to well-known rules from conditional logic and to some of the postulates for conditional belief from Chapter 4. Finally, the two kinds of assertability of conditionals will be related by a bridge principle of the form: if a conditional is subjectively assertable in the all-or-nothing sense, then its degree of assertability is high enough. (This corresponds to the bridge principle $BP1^r$ from Chapter 4.) Another representation theorem (Theorem 34) will represent coherence conditions on rational subjective assertability: the coherence conditions will correspond to a perfectly rational agent's degrees of belief and conditional beliefs satisfying a certain subset of the postulates from Chapter 4.[323] The section will also include an example (Example 25) in which joint constraints on degrees of belief and conditional belief are derived from a given set of (conditional) assertions. The corresponding derivations will result from an application of the theory from this chapter and from previous ones to the given assertions in the example. This will exemplify the 'super-additivity' value that a *joint* theory of belief and degrees of belief has to offer. Additionally, I will demonstrate in the same section that my account of assertability relates nicely to some of Jackson's (1979, 1981) independent findings on 'robust assertability'.

Section 6.3 argues that the stability of rational belief that is guaranteed by the Humean thesis might not be enough for certain practical long-term purposes: such purposes might ask for a mental state with an extreme form of stability that results, on the numerical side, from assigning propositions the maximal probability of 1. Since contingent propositions are not often believed by an agent to that maximal degree,

[322] I am grateful to Alexandru Baltag here, who suggested to me to include in the present chapter the second and third extension of this joint decision theory for graded and binary belief.

[323] The postulates from Chapter 4 will be: the axioms of subjective probability joined with those for restricted conditional belief (or, in dynamic terms, belief expansion) as explained in section 4.2.2. In particular, these postulates for restricted conditional belief include the Preservation principle: this is postulate K^*4 in the AGM theory of belief revision, as discussed in section 4.1.2, which corresponds to my postulate B6 for conditional belief from section 4.2.2.

the probabilities in question cannot actually be degrees of *belief*. Instead I suggest that the mental state in question coincides with what various authors in different fields call 'acceptance': the corresponding probabilities are degrees of *acceptance* and the corresponding all-or-nothing state is acceptance *simpliciter*. The section discusses how acceptance in this sense relates to belief—what they have in common and how they differ. The main difference between them will be that aiming-at-the-truth is constitutive of belief[324] but not of acceptance. The section will also highlight a salient special case of acceptance: *accepting one's beliefs* in which an agent accepts propositions that she also believes (categorically). It will follow from the definitions in that section and from the postulates in previous chapters that, on the categorical scale, accepted belief coincides extensionally with belief, whereas degrees of accepted belief may well differ from degrees of belief. This will also lead to another argument for the Preservation principle B6 for conditional belief from Chapter 4.

Section 6.4 will continue the discussion of the Preface Paradox (cf. Makinson 1965) from section 3.5 in Chapter 3. Let $A_1, \ldots, A_n$ be the statements in the main part of the book (after the preface): I will argue that by asserting them *taken as a whole* an author normally does not assert each single statement A_i, nor does she express her belief in each single statement A_i. Instead she only asserts, and expresses her belief, that the *great majority of the statements* $A_1, \ldots, A_n$ are the case. This is perfectly consistent with asserting in the preface, and thereby expressing her belief, that not all of $A_1, \ldots, A_n$ are true. While normally an author does not *believe* each and every statement in the book that she publishes—and *could not* do so rationally—she may well *accept* each and every such statement. But acceptance in that sense differs from belief proper (as discussed in section 6.3). All of this will be consistent with the Humean thesis on belief.

6.1 Action

In section 2.3.3 I derived a compatibility result for Bayesian and categorical decision-making from the Humean thesis (and the consistency of belief).

The simple framework was this: let W be a finite non-empty set of worlds; as usual, the standard choice would be the set of logically possible worlds for a simple propositional language with finitely many propositional variables. I fix a perfectly rational agent for which I am going to formulate a simple all-or-nothing decision theory based on all-or-nothing beliefs and all-or-nothing utilities. O is the set of potential outcomes of the agent's actions: I assume that O has at least two members. $u : O \to \{u_{max}, u_{min}\}$ is the agent's 'all-or-nothing' utility function that is assumed to be onto: it has precisely the two real numbers $u_{max} > u_{min}$ as its values, such that at least one outcome has utility u_{max} and at least another outcome has utility u_{min}. An outcome of utility u_{max} is a 'good' or useful outcome, one of utility u_{min} is an outcome that is 'bad' or that does not serve the agent's desires. Actions A are *all* the

[324] Compare the 'Truth-Aiming' Assumption 2 from Chapter 1 and the discussion of it in Chapter 5.

functions from W to O—a very tolerant conception of actions that will be modified later in this section. When A is an action and w is a world, $A(w)$ is the outcome in O of carrying out A in w, and $u(A(w))$ is the utility of that outcome. So I assume that the utility of an outcome does not depend on worlds (or, in the decision-theoretic literature, 'states'): there is just one utility function u applied to outcomes of actions at arbitrary worlds in W. $Use(A)$ is defined to be set $\{w \in W \mid u(A(w)) = u_{max}\}$ ($= \{w \in W \mid A$ is successful with regard to u in $w\}$). $Use(A)$ may be regarded as the proposition that is expressed by the sentence 'Action A is useful', which is true precisely in those worlds that are members of $Use(A)$. Since $Use(A)$ is a proposition, that is, a subset of W, the agent will believe it ($Bel(Use(A))$), or disbelieve it ($Bel(\neg Use(A))$, that is, $Bel(W \setminus Use(A))$), or suspend judgement on it (neither $Bel(Use(A))$ nor $Bel(\neg Use(A))$). Finally, according to the decision theory from section 2.3.3, an action A is practically permissible in the all-or-nothing sense (relative to Bel and u) if and only if $Bel(Use(A))$: so an action is permissible from the viewpoint of the agent just in case the agent believes it to be useful. I will also consider a different conception of permissibility shortly, but for the moment let us stick to this belief version of permissibility from section 2.3.3.

The expected utility of an action A with respect to the agent's degree-of-belief function P and the 'binary' utility measure u was defined as: $E_P(u(A)) = \sum_{w \in W}(P(\{w\}) \cdot u(A(w)))$. It is easy to see that our assumptions taken together yield: $E_P(u(A)) = P(Use(A)) \cdot u_{max} + [1 - P(Use(A))] \cdot u_{min}$.[325]

The first extension of section 2.3.3 consists in another recovery result for the Humean thesis: given that our perfectly rational agent does not believe the contradictory proposition (not $Bel(\varnothing)$), we know already that the Humean thesis with a Humean threshold r, such that $\frac{1}{2} \leq r < 1$, entails the logical closure of belief (Theorem 2 in section 2.3.1). Furthermore, it also entails that $E_P(u(A)) > E_P(u(B))$ for all actions A and B for which $Bel(Use(A))$ (A is permissible) but not $Bel(Use(B))$ (B is not permissible): see part (i) of Theorem 4 in section 2.3.3.

It turns out that the Humean thesis with the special threshold $r = \frac{1}{2}$ can also be *derived* from these two consequences taken together (and background assumptions):

Theorem 31 (Representation Theorem for the Logic of Belief and Decision-Theoretic Coherence)
Let W be finite and non-empty.

Let O be a set with at least two members. Let $u_{max} > u_{min}$ be two real numbers. Let the set of actions be the set of all functions A from W to O.

Let Bel be a set of subsets of W, and let P assign to each subset of W a number in the interval $[0, 1]$.

Then the following two conditions are equivalent:

[325] See the proof of Theorem 4 for the details.

- P is a probability measure on W, Bel is closed under logical consequence, not $Bel(\varnothing)$, and Bel and P are such that for all onto utility measures $u : O \to \{u_{max}, u_{min}\}$ it holds that:
    - for all actions A, B: if $Bel(Use(A))$ and not $Bel(Use(B))$ then
    $$E_P(u(A)) > E_P(u(B)).$$
- P is a probability measure on W, Bel and P satisfy the Humean thesis $HT^{\frac{1}{2}}$ (with Humean threshold $\frac{1}{2}$), and not $Bel(\varnothing)$.[326]
  In turn, we know already that the Humean thesis $HT^{\frac{1}{2}}$ can be represented in purely probabilistic terms: in terms of P-stability$^{\frac{1}{2}}$ or the Outclassing Condition, as stated in Theorem 5 in Appendix B.

For illustration, reconsider the first Apple Spritzer Example 1 from Chapter 2. (I have already reconsidered the second Apple Spritzer Example 2 from Chapter 2 in section 4.3.)

Example 22 (The First Apple Spritzer Example 1 from Chapter 2 Reconsidered)
Let $W = \{w_1, w_2, w_3\}$. w_1 corresponds to the bottle of apple spritzer being in the fridge in the kitchen, w_2 to the bottle of apple spritzer being in the shopping bag in the kitchen, w_3 to the bottle of apple spritzer not being in either of these places.

Let P assign probabilities as follows: $P(\{w_1\}) = \frac{6}{10}$, $P(\{w_2\}) = \frac{3}{10}$, $P(\{w_3\}) = \frac{1}{10}$. Hence, as intended in the story, it is more likely than not that the bottle of apple spritzer is in the fridge.

Let (unconditional) Bel be given by defining the least believed proposition to be $B_W = \{w_1, w_2\}$; a fortiori, Bel is consistent and logically closed. In words: the agent believes unconditionally that the bottle is in the fridge or the shopping bag, she does

[326] Proof: the proof of the '$\leftarrow$' (right-to-left) direction is contained in the proof of Theorem 4 from section 2.3.3 (for $r = \frac{1}{2}$). So let us focus on the '$\rightarrow$' direction now: by assumption, Bel is closed under logical consequence, and it is not the case that $Bel(\varnothing)$, which entails that there is a least believed proposition $B_W \neq \varnothing$. Assume for reductio that there is a w in B_W, such that $P(\{w\}) \leq P(W \setminus B_W)$. Consider an onto utility measure $u : O \to \{u_{max}, u_{min}\}$ and actions A and B, such that the following is the case: for all w', $u(A(w')) = u_{max}$ iff $w' \in B_W$, and for all w', $u(B(w')) = u_{max}$ iff $w' \neq w$. Such u, A, and B exist in view of O having at least two members and because of the tolerant conception of actions that is presupposed in the theorem. It follows that $Use(A) = B_W$ and $Use(B) = W \setminus \{w\}$, which is why $Bel(Use(A))$ but not $Bel(Use(B))$ (since $w \in B_W \not\subseteq Use(B)$).

As in the proof of Theorem 4 it holds that $E_P(u(A)) = P(Use(A)) \cdot u_{max} + [1 - P(Use(A))] \cdot u_{min} = P(B_W) \cdot u_{max} + [1 - P(B_W)] \cdot u_{min}$, and $E_P(u(B)) = P(Use(B)) \cdot u_{max} + [1 - P(Use(B))] \cdot u_{min} = P(W \setminus \{w\}) \cdot u_{max} + [1 - P(W \setminus \{w\})] \cdot u_{min}$. Furthermore, we have (by the axioms of probability): $P(W \setminus \{w\}) = P((W \setminus B_W) \cup (B_W \setminus \{w\})) = P(W \setminus B_W) + P(B_W \setminus \{w\}) \geq P(\{w\}) + P(B_W \setminus \{w\}) = P(\{w\} \cup (B_W \setminus \{w\})) = P(B_W)$.

So $P(W \setminus \{w\}) \geq P(B_W)$ and $u_{max} > u_{min}$, from which it follows that the convex combination $P(B_W) \cdot u_{max} + [1 - P(B_W)] \cdot u_{min}$ (that is, $E_P(u(A))$) must be less than or equal to the convex combination $P(W \setminus \{w\}) \cdot u_{max} + [1 - P(W \setminus \{w\})] \cdot u_{min}$ (that is, $E_P(u(B))$). But that contradicts the assumption of (the left-hand side of) the theorem. Therefore, for all w in B_W, $P(\{w\}) > P(W \setminus B_W)$. Thus, B_W satisfies the Outclassing Condition with respect to P and $r = \frac{1}{2}$, which with Theorem 5 from Appendix B implies the Humean thesis $HT^{\frac{1}{2}}$.

not believe anything more specific than that, but she also believes its consequences (here: W).

We know already from Example 15 in section 4.3 and the results from Chapter 4 that *Bel* and P jointly satisfy the Humean thesis $\mathrm{HT}^{\frac{1}{2}}$: the agent's all-or-nothing beliefs are stable with respect to P (and $r = \frac{1}{2}$), since $\{w_1, w_2\}$ is P-stable$^{\frac{1}{2}}$. As long as evidence comes along that is possible from the viewpoint of the agent—equivalently: that is consistent with B_W, or which, in that sense, is not 'utterly surprising'—the probability that she assigns to a believed proposition will always remain high enough after conditionalizing on the evidence (by the Humean thesis from Chapter 2). Similarly, none of her categorical beliefs will have to be given up after revising or expanding her beliefs on the basis of such evidence (by the Preservation postulate B6 from section 4.2.2 of Chapter 4).

Now let $O = \{\text{thirst quenched, still thirsty}\}$, let $u_{max} = 1$, $u_{min}, = 0$, and let $u(\text{thirstquenched}) = u_{max}$, $u(\text{still thirsty}) = u_{min}$.

Finally, let A be the action of walking to the kitchen, checking the fridge and the shopping bag, getting the bottle (if it is there), and emptying it: so $A(w_1) = A(w_2) =$ thirst quenched (since the bottle is in the kitchen in these worlds). On the other hand, $A(w_3) =$ still thirsty (because in that world the bottle is in neither of the two places that get checked by doing A). Consequently, $Use(A) = \{w_1, w_2\}$, $Bel(Use(A))$, and therefore A is permissible for the agent in question.

Contrast this with the action B of walking to the kitchen, checking only the fridge, getting the bottle (if it is there), and emptying it: in that case, $B(w_1) =$ thirst quenched (since in that world the bottle is in the fridge), while $B(w_2) = B(w_3) =$ still thirsty (because in these worlds the bottle is not in the fridge). Therefore, $Use(B) = \{w_1\}$, but not $Bel(Use(B))$, which is why B is not permissible for the agent in question (given her beliefs and her utility measure). The agent regards it as a serious possibility that the bottle is in the shopping bag, which is why merely checking the fridge does not yield enough of a guarantee to find the bottle and to quench her thirst, from the viewpoint of the agent. That is why the action is not rationally permissible for her.

Accordingly, it holds that $E_P(u(A)) = P(Use(A)) \cdot u_{max} + [1 - P(Use(A))] \cdot u_{min} = P(\{w_1, w_2\}) = \frac{9}{10}$, while $E_P(u(B)) = P(Use(B)) \cdot u_{max} + [1 - P(Use(B))] \cdot u_{min} = P(\{w_1\}) = \frac{6}{10}$, and so $E_P(u(A)) > E_P(u(B))$. The permissible action is superior in expected value to the impermissible one.

Similarly, let C be the action of doing nothing (not even trying to find the bottle anywhere). It follows that $C(w_1) = C(w_2) = C(w_3) =$ still thirsty, $Use(C) = \varnothing$, not $Bel(Use(C))$, C is therefore impermissible, and indeed $E_P(u(A)) = \frac{9}{10} > E_P(u(C)) = P(Use(C)) \cdot u_{max} + [1 - P(Use(C))] \cdot u_{min} = 0$.

In the simple decision-theoretic framework so far (and also back in Chapter 2), I presupposed unrealistically that every function A from W to O counted as an action.[327] Indeed, that assumption was required for the proofs of Theorems 4 and 31. So far as

[327] An assumption like that is also included in Savage's decision-theoretic framework: see Joyce (1999, ch. 3).

Theorem 4 is concerned, the assumption was needed in order to conclude from the maximality of $E_P(u(A))$ that $Bel(Use(A))$ was the case: every action that is permissible in the Bayesian sense is also permissible in the all-or-nothing sense. Another—less problematic—background assumption that was needed for the same purpose was that $u : O \to \{u_{max}, u_{min}\}$ was *onto*: in particular, at least some outcomes are 'good' or as desired by the agent.[328] However, it turns out that the rest of Theorem 4, which includes the '$\leftarrow$' (right-to-left) direction of Theorem 31 as a proper part, does not depend on either of these two assumptions.

Let me make this explicit now. Let us change the framework in the way that a non-empty *set of actions, Act*, is presupposed now: where before *Act* was assumed to coincide with the set of *all* functions A from W to O, in the new and more general framework it might well coincide with but a restricted range of actions $A \in Act$ that are available to the agent. Furthermore, I also omit the restriction that the given utility measure u is onto (and that O has at least two members).

In that case, one can still show:

Theorem 32 Let W be finite and non-empty. Let P be a probability measure on W. Let Bel be a set of subsets of W.

Let O be a non-empty set. Let $u_{max} > u_{min}$ be two real numbers, and $u : O \to \{u_{max}, u_{min}\}$. Let *Act* be a non-empty set of actions, such that every A in *Act* is a function from W to O.

If *Bel* and P satisfy the Humean thesis HT^r (with $\frac{1}{2} \leq r < 1$), and not $Bel(\varnothing)$, then it holds that

(i) for all A, B in *Act*: if $Bel(Use(A))$ and not $Bel(Use(B))$ then

$$E_P(u(A)) > E_P(u(B)),$$

(ii) for all A in *Act*: if

$$E_P(u(A)) \text{ is maximal (among actions in } Act),$$

then for all B in *Act*, such that $Bel(Use(B))$, it is the case that

$$E_P(u(A)) - E_P(u(B)) < (1 - P(B_W))\,(u_{max} - u_{min}) < (1 - r)\,(u_{max} - u_{min}).$$[329]

The only difference to Theorem 4 from section 2.3.3 is that it is no longer the case that if $E_P(u(A))$ is maximal then $Bel(Use(A))$: the best Bayesian options do not

[328] In order to satisfy the requirement that u was onto it was also necessary to presuppose that O had at least two members. I will be able to drop that requirement in what follows.

[329] Proof: the proof is the same as that of Theorem 4, except for: one needs to take out the part 'if $E_P(u(A))$ is maximal, then by our liberal definition of an action and u being onto, it must be the case that $Use(A) = W$. So $Bel(Use(A))$, because $Bel(W)$.' And one needs to observe that if $E_P(u(A))$ is maximal within *Act* and $Bel(Use(B))$, then $E_P(u(A)) - E_P(u(B)) \leq u_{max} - (P(Use(B)) \cdot u_{max} + [1 - P(Use(B))] \cdot u_{min})$ which follows to be less than or equal to $(1 - P(B_W))(u_{max} - u_{min}) < (1 - r)(u_{max} - u_{min})$ as derived in the proof of Theorem 4.

necessarily belong to the best all-or-nothing options any more. Other than that the decision theory from section 2.3.3 goes through as before even for a given set *Act* that does not include all functions from W to O whatsoever (and without assuming u to be surjective). It is easy to see that, in view of the Humean thesis, one could also formulate part (ii) above like this: if $E_P(u(A))$ is maximal, then either (ii.i) $Bel(Use(A))$ and for all B in *Act*, such that $Bel(Use(B))$, the inequalities in (ii) above are the case, or (ii.ii) not $Bel(Use(A))$ and indeed there is no B in *Act* at all, such that $Bel(Use(B))$ holds (in which case the 'for all B' quantifier in (ii) is vacuous). I have only chosen the formulation of (ii) in Theorem 32 because it is more continuous with that of Theorem 4.

This leads me to the third and final extension of section 2.3.3, which concerns the following question: let us assume again that only a certain non-empty set *Act* of actions is available to the agent. We defined an action to be permissible for the agent just in case the agent believes it to be useful. But what if none of the actions in *Act* is permissible in that sense? There would not be any action that the agent believes to do the job, that is, realizing her desires. The next best thing to do, from the viewpoint of categorical belief, would then be for the agent to turn to those actions that will *possibly* do the job: those actions for which the agent holds it possible that they are useful. Let us call the decision-theoretic conception of practical permissibility that corresponds to this new proposal *weak permissibility*: an action A in *Act* is weakly permissible for the agent if and only if $Poss(Use(A))$, that is, not $Bel(\neg Use(A))$.[330]

This 'weak permissibility' or *Poss*-variant of our simple decision theory for categorical belief can still be linked with Bayesian decision theory for a 'binary' utility measure, though in a much weaker sense if compared to Theorem 32:

Theorem 33 Let W be finite and non-empty. Let P be a probability measure on W. Let *Bel* be a set of subsets of W.

Let O be a non-empty set. Let $u_{max} > u_{min}$ be two real numbers, and $u : O \to \{u_{max}, u_{min}\}$. Let *Act* be a non-empty set of actions, such that every A in *Act* is a function from W to O.

If *Bel* and P satisfy the Humean thesis HT^r (with $\frac{1}{2} \leq r < 1$), and not $Bel(\varnothing)$, then it holds that

(i) for all A, B in *Act*: if $Poss(Use(A))$ and not $Poss(Use(B))$ then

$$E_P(u(A)) > E_P(u(B)),$$

(ii) for all A in *Act*: if

$$E_P(u(A)) \text{ is maximal (among actions in } Act),$$

[330] I am grateful to Alexandru Baltag for suggesting this to me. In personal communication he related the proposal to game-theoretic conceptions of permissibility as *lack of knowledge of being dominated*.

then for all B in Act, such that $Poss(Use(B))$, it is the case that

$$E_P(u(A)) - E_P(u(B)) < \left(1 - \frac{r}{1-r} \cdot P(W \setminus B_W)\right) (u_{max} - u_{min}).$$[331]

Condition (i) is as in Theorem 32: weakly permissible actions are better in terms of epistemic utility than weakly impermissible actions. However, condition (ii) is significantly weakened now: the difference in expected utility between a permissible action A in the Bayesian sense and a weakly permissible action B in the all-or-nothing sense is only bounded by $(1 - \frac{r}{1-r} \cdot P(W \setminus B_W))\,(u_{max} - u_{min})$. For instance: with $r = \frac{1}{2}$ and $P(B_W) = 0.9$, that is, $P(W \setminus B_W) = 0.1$, this means that $E_P(u(A)) - E_P(u(B)) < 0.9\,(u_{max} - u_{min})$. This allows for weakly permissible actions to significantly lag behind the optimal Bayesian solutions: the difference between their expected utilities can be large. But at least weakly permissible actions come with *non-negligible* expected utility, which might be the best lower bound that one can aim for categorically in a case in which no action is available that one believes to be useful. Finally, even in such a case, a weakly permissible action may have an expected utility that is still reasonably high. Reconsider Example 22: if the restricted set of available actions had been the set $Act = \{B, C\}$, then neither of the two options would have been permissible in the stronger belief sense. Yet, B would be the rational thing to do on Bayesian grounds, and indeed B is weakly permissible while C is not, and $E_P(u(B)) = \frac{6}{10} = \frac{6}{10}\,(u_{max} - u_{min})$ is reasonably high.

My extended proposal of a decision theory for all-or-nothing beliefs is therefore: if there are actions in Act that the agent believes to be useful, she is permitted to carry them out, and she is not permitted to carry out any of the others. If there is no such action, but there are actions in Act such that the agent regards it as possible for them that they are useful, then she is (weakly) permitted to carry out those actions but none of the others. In both cases, the Humean thesis guarantees a form of compatibility with Bayesian decision theory, as explained before.[332] If not even actions of the latter 'weak permissibility' kind are available, then the agent is in trouble so far as decisions based on her categorical beliefs are concerned: she might well pick an action in Act at random in that case. But not even that would be particularly problematic on Bayesian terms since in such a situation (for high enough $P(B_W)$) all members of Act will be bound

[331] Proof: the proof is analogous to that of Theorem 4: for (i) one derives from the assumptions and from the Outclassing Condition (see Theorem 5 in Appendix B) that there is a world w in $B_W \cap Use(A)$, such that $P(\{w\}) > \frac{r}{1-r} \cdot P(W \setminus B_W) \geq P(W \setminus B_W)$. From this one derives that $P(Use(A)) \geq P(\{w\}) > P(W \setminus B_W) \geq P(Use(B))$ and concludes from this that $E_P(u(A)) > E_P(u(B))$. Similarly, in (ii) there must be a world w in $B_W \cap Use(B)$, such that $P(\{w\}) > \frac{r}{1-r} \cdot P(W \setminus B_W)$. From this one derives: $E_P(u(A)) - E_P(u(B)) = P(Use(A)) \cdot u_{max} + [1 - P(Use(A))] \cdot u_{min} - (P(Use(B)) \cdot u_{max} + [1 - P(Use(B))] \cdot u_{min}) \leq$ [by $u_{min} < u_{max}$] $u_{max} - (P(Use(B)) \cdot u_{max} + [1 - P(Use(B))] \cdot u_{min}) \leq$ [by $w \in Use(B)$ and reasoning about convex combinations] $\leq u_{max} - (P(\{w\}) \cdot u_{max} + [1 - P(\{w\})] \cdot u_{min}) = u_{max} - u_{min} - P(\{w\})(u_{max} - u_{min}) < u_{max} - u_{min} - \frac{r}{1-r} \cdot P(W \setminus B_W)(u_{max} - u_{min}) = (1 - \frac{r}{1-r} \cdot P(W \setminus B_W))\,(u_{max} - u_{min})$.

[332] This overall extension of the decision theory from section 2.3.3 was suggested to me by Alexandru Baltag in personal communication.

to have a low expected utility (on a scale between u_{min} and u_{max}): their differences in expected utility will be small enough to be negligible.

It would be easy to extend this joint decision theory based on all-or-nothing belief and subjective probability by expanding the simple decision theory for all-or-nothing belief that is part of it. On the one hand, u might be taken to be more refined than (binary) *Bel*: for instance, u might have an intermediate neutral value additional to u_{max} and u_{min}, or u might even take a greater range of real numbers. Accordingly, rather than believing that an action is useful ($Bel(Use(A))$) the agent might believe an action to be *sufficiently useful* ($Bel(u(A) > t)$) or an action to be *more useful* than another one ($Bel(u(A) > u(B))$) or the like. On the other hand, the theory might be extended to full *conditional* belief: an additional dimension of complexity that e.g. Lin (2013) exploits for his decision theory.[333] Any such decision theory based on conditional all-or-nothing belief will need to steer clear of running into counterparts of Lewis's (1988b) triviality result for the so-called Desire-as-Belief thesis in quantitative decision theory (see Collins 1988 for more on this), but all of that is doable. For my own purposes, the decision theory based on unconditional belief from before is sufficiently illustrative: it it simple, plausible, consistent (as follows from the existence of a wide variety of models), and it coheres with Bayesian decision theory (as follows from the Humean thesis).

6.2 Assertability

There is a great variety of theories of assertion and assertability.[334] I will not be able to discuss them here. Instead I will proceed quickly to a 'belief view' of assertion and assertability. My focus will rather be on what the stability conception of belief that was developed in the previous chapters has to say about assertion and assertability *once that 'belief view' is in place.*

I will take the following for granted. Assertions are speech acts; if they are sincere (no jokes, no lies, or the like), they express beliefs. My first step will be to turn to a specific normative way of making this more precise. In what follows, for the sake of

333 Lin (2013) also contains further references to 'qualitative decision theories' that are based on all-or-nothing notions of belief; see also Dubois et al. (2009).

334 For a survey of different accounts of assertion, see MacFarlane (2011a). MacFarlane distinguishes four different types of answers to the question 'what is an assertion?' that one can find in the literature: an assertion is (i) an expression of an attitude (belief is the standard option), (ii) a move that is defined by its constitutive rules (i.e. by certain norms on assertion), (iii) a proposal to add information to the conversational common ground, (iv) the undertaking of a commitment. See MacFarlane (2011a) for details and references. (One might add another Gricean version of assertion to this.) In what follows I will opt for a combination of (i) and (ii). For an overview of different kinds of norms of assertion, see Williamson (2000, ch. 11). Amongst the norms that he considers, one finds a truth norm (one must: *assert X only if X is true*), a warrant norm (one must: *assert X only if one has warrant to assert X*), a knowledge norm (one must: *assert X only if one knows X*), and more. Williamson himself argues for the knowledge norm, while my starting point will be the weaker belief norm: one must: *assert X only if one believes X*.

simplicity, let us understand the term 'assertion' to be restricted to sincere and serious assertions from the start.

In the 'Assertion' Assumption 5 from Chapter 1, I have already committed myself to the following simple, traditional, and quite plausible belief norm for assertion: an agent ought *to assert (sincerely) X only if she believes X* (where 'ought' has wide scope); or, an agent must: *assert X only if she believes X*. I will not distinguish between the two formulations.[335] Furthermore, I regard a proposition as assert*able* for an agent just in case the agent is *permitted* to assert it. I will not offer any definitions or postulates for the terms 'assertion', 'speech act', 'sincere', or 'express belief', and I will not try to unfold the exact interpretation of the deontic modalities 'ought' and 'permitted' either, except that I regard the norms in question at least partially as constitutive of assertion and assertability.

With respect to assertability, the account thus far is deductively weak. I am going to strengthen it now. One way of achieving this would be to add a truth norm (one must: *assert X only if X is true*) or a knowledge norm (one must: *assert X only if one knows X*). In fact, the truth norm already follows from the belief norm on assertion (one must: *assert X only if X is believed*) together with one version of an 'aiming at the truth' or '(partially) reaching the truth' norm for belief (one must: *believe X only if X is true*).[336] The knowledge norm entails the belief norm, assuming that knowledge entails belief; but dealing with knowledge would take me too far afield. Instead I will extend the account in a different direction.

First of all, reconsider the norm 'one ought: *to assert X only if one believes X*', which I regard as constitutive of assertion and hence as having the status of a quasi-logical normative law. One can show that this law is derivable in a standard system of deontic logic that includes the similarly plausible, and closely related, normative law 'one is permitted *to assert* X only if one believes X' (where 'permitted' has narrow scope now), as long as that narrow-scope permission norm is assumed to have the status of a (quasi-)logical law, too.[337] Indeed, that will be my first step towards strengthening

[335] Lewis (1976, p. 297) maintains a closely related probabilistic version of this: 'The truthful speaker wants not to assert falsehoods, wherefore he is willing to assert only what he takes to be very probably true'. Jackson (1979, p. 574) does the same but adds a robustness or stability requirement to it. (I will return to this later in this section.) Milne (2012, p. 332) comments on Lewis's view as follows: 'Running Locke [i.e. the Lockean thesis] and Lewis together, one asserts only what one believes. This gives us an appealing, and appealingly simple, if incomplete, picture of assertion.' Douven (2006, p. 449) argues for the again closely related view that 'One should assert only what is rationally credible to one'. For a criticism of such belief-related norms on assertability, see again Milne (2012).

[336] Compare the brief discussion of the 'Truth-Aiming' Assumption 2 in section 1.1 and the discussion in section 5.1.

[337] Assume a normal (Kripke-style) system of deontic logic (cf. Chellas 1980) that includes the plausible logical axiom scheme $O(O(X) \rightarrow X)$ and the quasi-logical axiom scheme $P(Asst(X)) \rightarrow Bel(X)$: it follows that all instances of $O(Asst(X) \rightarrow Bel(X))$ are provable in the system and therefore are (quasi-)logical theorems. More briefly: if $\vdash P(Asst(X)) \rightarrow Bel(X)$, then $\vdash O(Asst(X) \rightarrow Bel(X))$. ('$O$' is short for 'ought', 'P' is short for 'permitted'.) Here is a sketch of the proof: apply necessitation to $P(Asst(X)) \rightarrow Bel(X)$ to derive $O(P(Asst(X)) \rightarrow Bel(X))$; this is admissible since $P(Asst(X)) \rightarrow Bel(X)$ is understood to be a quasi-logical axiom. Use normality and the scheme $O(O(X) \rightarrow X)$ to derive $O(\neg X \rightarrow \neg O(X))$

the account from before: I assume that 'one is permitted *to assert* X only if one believes X' is also constitutive of assertion. This yields: if X is assertable for an agent, then X is believed by the agent.

Secondly, I would like to extend this conditional statement to the following biconditional one: X is assertable for an agent *just in case* X is believed by the agent. Or equivalently: an agent is permitted to assert X if and only if the agent believes X. Some authors have defended a thesis like that: e.g. Kripke's (1979) so-called biconditional disquotational principle is a version of it. However, even if restricted to perfectly rational agents, that biconditional would still be quite controversial: e.g. typical defenders of a knowledge norm on assertion would dispute it. At least this will be so if assertability is understood in the way they understand it, that is, where the assertability of a proposition is meant to depend also on factors that are external to the asserting agent—when also an agent's physical and social environment need to play along in order for X to be assertable for the agent. So I will have to move more cautiously here.

Let me introduce the term '*subjectively* assertable' as a remedy for worries of that sort. Subjective assertability is supposed to stand to assertability *simpliciter* as belief stands to knowledge: whereas assertability may depend on factors that are beyond a speaker's control or awareness, subjective assertability will only depend on factors that are internal to the agent. Or in other words: subjective assertability is something like an internalist version of assertability. A different way of making the same point is: it will still be fine to call a proposition (subjectively) assertable for an agent just in case the agent is permitted to assert it, it is just that the permission operator in question will have to be a subjective one, too.[338] Furthermore, I will understand subjective assertability as not being governed as yet by Gricean (cf. Grice 1989) requirements of efficient communication: subjective assertability is supposed to be the kind of assertability that only takes into account the *speaker's* perspective, and not what might be especially relevant or useful to the listener; it is the kind of assertability that is unaffected by the cancellation of a conversational implicature; and the like. Of course, it is yet to be seen how fruitful such a subjective notion of assertability will be, but I hope the rest of this section will demonstrate enough of its utility.

This being said, it should be sufficiently plausible that (i) assertability (*simpliciter*) entails subjective assertability, (ii) subjective assertability entails belief (which is also why 'if X is assertable for an agent, then X is believed by the agent' did sound plausible before), and while belief does not entail assertability, it does entail subjective assertability. This will finally allow me to take the following biconditional for granted, which I am going to formulate just for perfectly rational agents:

and thus $O(\neg\neg Asst(X) \rightarrow \neg O(\neg Asst(X)))$, that is, $O(Asst(X) \rightarrow P(Asst(X)))$. Finally, derive from these conclusions and from normality that $O(Asst(X) \rightarrow Bel(X))$. I am grateful to Edgar Morscher for a discussion on this.

[338] For more on the distinction between objective and subjective normative operators, see e.g. Hansson (2010), Carr (2015), and Wedgwood (2016).

(1) X is subjectively assertable for a perfectly rational agent (at a time) if and only if the agent believes X (at that time).

Or more briefly: $Asst(X)$ iff $Bel(X)$.

In fact, this will be so almost 'by design': the present notion of subjective assertability is supposed to track those aspects of assertability that are most naturally captured by the agent's beliefs. If an agent asserts something, then this expresses her subjective assertability in this sense: her belief in what is asserted. But, of course, if a proposition X is subjectively assert*able* for an agent, this does not mean that the agent will necessarily assert it: she might simply not desire to assert X, for whatever reasons.

In the following, '*Asst*' (whether in the main text or used as a subindex) will always express *subjective* assertability of propositions in that sense. '*Asst*' itself is tacitly indexed by a name for the corresponding agent relative to whom assertability is determined (just as '*Bel*' carries a tacit reference to an agent). The agent in question will always be assumed perfectly rational.[339] In addition, subjective assertability and belief are also relativized to time, which I will normally suppress as well.

Next I will turn to a quantitative version of categorical subjective assertability: assignments Deg_{Asst} of *numerical degrees* of assertability. So '$Deg_{Asst}(X)$' will denote a perfectly rational agent's subjective degree of assertability assigned to the proposition X.

While it seems to be much more common to speak of assertability in categorical terms, numerical assertability is not unheard of either: in particular, in his early work before *The Logic of Conditionals* (Adams 1975), Ernest Adams does speak of degrees of (justified) assertability in such a manner (see e.g. Adams 1965, 1966; similarly, Jackson 1979, 1981). Adams is mostly interested in assigning such degrees to conditionals (about which more below), but his theory also accounts for the degrees of assertability of 'factual' descriptive sentences which in turn may be regarded as deriving from the degrees of assertability of the propositions that are expressed by these sentences. Adams identifies such degrees of assertability with an agent's subjective probabilities for these sentences or propositions. I am going to do the same, without arguing for it.[340]

339 If we turned from the propositional to the linguistic level, principle (1) from before would have the following counterpart for declarative *sentences* A: A is subjectively assertable for an agent just in case the proposition that is expressed by A is believed by the agent. An account like that would ignore all additional questions concerning *how* that proposition is expressed by A, for instance, Gricean questions such as: was A brief enough, or would the agent have been able to convey the same proposition in a more efficient manner? In the present section, I will only deal with assertability on the propositional level. In section 6.4 about the Preface Paradox, I will partially address also the linguistic level when I will deal with the question what exactly a 'mass assertion' of statements in a book expresses.

340 One way of approaching such a kind of argument would be: (i) to argue that permissibility may come in numerical degrees, (ii) to identify an agent's degree of assertability of X with the degree of permissibility for the agent to assert X, and (iii) to argue that such degrees-of-permissibility-to-assert coincide extensionally with the agent's corresponding subjective probabilities. But I will have to leave this to one side.

In his later work (from Adams 1975), Adams avoids speaking of degrees of assertability in that way—or indeed of assertability more generally—and only talks about *probabilities* of sentences directly. To the best of my knowledge, he never explains in writing why he changed his way of expressing himself, but presumably his reasons were twofold:[341] first, the notion of assertability is employed by too many philosophers in too many different ways, which is why using it might not be particularly conducive to the understanding of Adams's own theory. Secondly, it is questionable whether the pre-theoretic concept of assertability comes in degrees at all. So far as the second worry is concerned, I am happy to understand *degree of assertability* as a technical term. And I will also have an all-or-nothing concept of assertability around which will be ready to be applied whenever required. Furthermore, although I do share the first kind of worry, I hope that qualifying the concept of assertability that I am interested in as *subjective*, and taking this together with what I am going to say about it in the rest of this section, will go at least some way towards disambiguating the concept in question.

Which leads me to the following numerical counterpart of (1) above:

(2) The degree of subjective assertability of X for a perfectly rational agent (at a time) equals the agent's degree of belief in X (at the time).

Or more briefly: $Deg_{Asst}(X) = P(X)$.

Adams (1966, p. 266) proposes 'to replace the vague and unquantified notion of "justified assertability" by that of "high probability" (i.e. probability very close to 1)'. As should be clear by now, I am not going to follow him in that respect and instead keep both *categorical* assertability and *numerical* assertability around without eliminating or reducing either of the two concepts. It will be the job of yet another bridge principle to tell us how the two relate to each other. (This said, categorical assertability for perfectly rational agents will indeed entail high enough numerical assertability, in line with the analogous 'Likeliness' bridge principle $BP1^r$ from Chapter 4.)

(1) and (2) together may be regarded as a normative way of making the claim '(sincere) assertions express beliefs' more precise. The beliefs in question are absolute or unconditional ones. However, there is also *conditional* belief, as we have seen in Chapter 4. My next step will be to extend the present account of assertion, assertability, and belief to conditionals and conditional belief: assertions of (indicative) conditionals express conditional beliefs.

Let me do this first on the numerical side. In fact, this is familiar territory: according to the suppositional theory of conditionals, as developed by Adams (1965, 1966, 1975), Edgington (1995), Bennett (2003), and others, a person's degree of assertability or acceptability for a conditional is given by the person's corresponding conditional probability in the consequent given the antecedent.

[341] I am grateful to Dorothy Edgington for her leads on this matter and for sharing her remembrances of corresponding discussions with Adams.

In order to express this in more formal terms, it is useful to abuse notation a bit: when X and Y are propositions, let me speak of the ordered pair $\langle X, Y \rangle$—which I am going to denote by '$X \Rightarrow Y$'—as the 'conditional' with antecedent X and consequent Y. Literally, a conditional should be a linguistic item (whether token or type) rather than a set-theoretic construction on propositions, but adhering to the level of propositions will simplify matters significantly and will keep the following considerations continuous with the assertability of single propositions (such as X) as developed before. Since 'X' and 'Y' in '$\langle X, Y \rangle$' are meant to denote sets of worlds, conditionals in my sense will be 'first-degree' or 'flat': X and Y do not involve any constructions involving $\Rightarrow$ again.

That being in place, I am ready to formulate a version of the central thesis of the suppositional theory:

(3) If a perfectly rational agent's degree of belief in X (at a time) is greater than 0, then the agent's degree of subjective assertability of $X \Rightarrow Y$ (at the time) equals the agent's degree of belief in Y on the supposition of X (at the time).

Or more briefly: if $P(X) > 0$, then $Deg_{Asst}(X \Rightarrow Y) = P(Y|X)$.

If A and B are descriptive sentences in natural language, such that A expresses the proposition X, and B expresses the proposition Y, then the degree of subjective assertability of the linguistic item $A \Rightarrow B$ ('if A then B') for an agent at a time may also be identified with $Deg_{Asst}(X \Rightarrow Y)$. But in what follows I will rather focus on X, Y, and $X \Rightarrow Y$ directly.

Suppositionalists take (3) to be an explication of Ramsey's (1931) famous footnote—now called the 'Ramsey test for conditionals'—on a numerical scale:

> If two people are arguing 'If p will q?' and are both in doubt as to p, they are adding p hypothetically to their stock of knowledge and arguing on that basis about q . . . We can say that they are fixing their degrees of belief in q given p. (Ramsey 1931)

The idea is this: in order to determine one's degree of assertability of $X \Rightarrow Y$, one enters a kind of thought experiment or simulation.[342] One supposes first the antecedent X ('adding p hypothetically to their stock of knowledge'): this will not affect one's *actual* degree-of-belief function P, as might have been the case if X had been *learned*, but it will determine a new 'hypothetical' or 'offline' degree-of-belief function P_X that results from P by supposing X. That degree-of-(hypothetical)-belief function P_X as employed in that suppositional context may well differ from one's actual degree-of-belief function P outside of that context. Then, still within the same suppositional context, one determines one's degree of (hypothetical) belief in the consequent Y as given by P_X: $P_X(Y)$, that is, the degree of belief in Y on the supposition of X. Afterwards, one ends the thought experiment and takes the number $P_X(Y)$ to be one's

[342] More on this can be found in Leitgeb (2012c).

degree of assertability for the conditional $X \Rightarrow Y$: this final step takes place outside of the suppositional context.

Suppositionalists regard the operation that maps P to P_X to be conditionalization: P_X is the result of conditionalizing P on X, that is, $P_X(Y) = P(Y|X)$. This is plausible at least as long as the conditional in question is understood to be in the indicative mood (rather than the subjunctive mood) and hence the corresponding act of supposition is indicative supposition or supposition *as a matter-of-fact* (rather than subjunctive or *counterfactual* supposition).[343] *Supposing that X is the case* plausibly rules out all $\neg X$-worlds as candidates for the actual world—if only hypothetically—by setting their probability to 0. Afterwards, the probabilities of X-worlds need to be 'renormalized', such that they sum up to 1, and hence probabilistic coherence will be restored again. None of this should affect the ratios of probabilities of X-worlds. That is exactly what conditionalizing P on X achieves, assuming the underlying set W of worlds to be finite. (This corresponds to Ramsey's 'fixing their degrees of belief in q given p'.) The 'renormalization' step consists in dividing all of the original probabilities by $P(X)$, which is well-defined in standard probability theory only if $P(X) > 0$.[344] But if P is the asserting agent's degree-of-belief function, this condition will normally (though maybe not always) be satisfied anyway when the *indicative* conditional $X \Rightarrow Y$ is asserted, and normally it is even pragmatically implied by that. (Ramsey's 'in doubt as to p' may also be interpreted as entailing that condition.) In contrast, asserting a subjunctive conditional normally (though not always) implies one's *disbelief* in the antecedent, which in the extreme case might correspond to $P(X)$ being 0. In the following I will only deal with indicative conditionals: so $X \Rightarrow Y$ is meant to be the indicative conditional with antecedent X and consequent Y.

It should be emphasized that suppositionalists such as Adams do not conceive of '$Deg_{Asst}(X \Rightarrow Y)$' as denoting the unconditional probability of a proposition or a set of worlds; indeed, they do not think of indicative conditionals as expressing propositions at all. In my notation: $X \Rightarrow Y$ is *not* the result of applying a propositional operation $\Rightarrow$ that takes two propositions X and Y as its input and that maps them to an output proposition $X \Rightarrow Y$ (another subset of W). While, by (2) above, it is indeed the case that $Deg_{Asst}(X)$ is $P(X)$ and $Deg_{Asst}(Y)$ is $P(Y)$, (3) merely says that $Deg_{Asst}(X \Rightarrow Y)$ is $P(Y|X)$, that is, $\frac{P(X \cap Y)}{P(X)}$, where '$P(Y|X)$' is not of the form '$P(Z)$' with 'Z' denoting a subset of W. Adams himself never understood his 'probabilities of conditionals' other than as conditional probabilities. Since Lewis's (1976) famous triviality results it is well-known that he *could not* have done otherwise, at least as long as (3) above and some plausible background assumptions are satisfied: as Lewis demonstrated, given these assumptions, there are no non-trivial probability measures according to which the

[343] For more on this distinction, see Joyce (1999, ch. 6) and Leitgeb (2012a).

[344] As mentioned in previous chapters, if P were a Popper function or primitive conditional probability measure, then conditionalization on a proposition of absolute probability 0 would be well-defined. See McGee (1994) for a corresponding improvement of Adams's explication of the Ramsey test on the basis of Popper functions.

conditional probability of Y given X would always equal the unconditional probability of a proposition $X \Rightarrow Y$. But none of this will be particularly important for anything that follows.

Although the suppositional theory is generally regarded as one of the prime contenders for a successful theory of (indicative) conditionals, it is not without problems, of course: in particular, it is not clear how it should handle the application of propositional connectives to conditionals or the nesting of conditionals (assuming these make general sense in natural language). Consequently, I will not consider negations, conjunctions, or disjunctions of conditionals in the following, and the 'X' and 'Y' in '$X \Rightarrow Y$' will always denote propositions, not conditionals.[345]

On the brighter side, it would be truly surprising if the 'Suppose X: . . . ' of suppositional reasoning were not closely related to the assertion of 'if X then . . . ' statements in natural language. Assuming that conditionalization is a reasonable enough numerical explication of supposition, the suppositional theory nicely captures and explicates this affinity between 'suppose' and 'if-then' on a numerical scale. In any case: I will take (3) above for granted now, at least for the sake of the argument.

But doing so still leaves open the corresponding question about *all-or-nothing* assertability: when is an indicative conditional $X \Rightarrow Y$ subjectively assertable *simpliciter* for a person (at a time t)? The standard Bayesian suppositional theory is silent on this matter, which constitutes a gap in their theory.

Fortunately, there is an obvious way of addressing that problem: we can simply fill in the all-or-nothing analogue to (3) from above. That is:

(4) If a perfectly rational agent regards X as possible (at a time), then $X \Rightarrow Y$ is subjectively assertable for the agent (at a time) if and only if the agent believes Y conditional on X (at the time).

Or more briefly, if $Poss(X)$, then: $Asst(X \Rightarrow Y)$ iff $Bel(Y|X)$.

The corresponding notion of all-or-nothing conditional belief is the one from Chapter 4: one that corresponds formally to the postulates of AGM belief revision operators, or, in case $Poss(X)$ holds, the belief *expansion* operator. Absolute or unconditional belief was regarded as a special case of conditional belief back then: $Bel(X)$ iff $Bel(X|W)$ (just as $P(X)$ equals $P(X|W)$). '$Poss(X)$' is short for 'not $Bel(\neg X)$', which in turn can be identified with 'not $Bel(\neg X|W)$'. (Recall section 4.1.2.) Restricting (4) to the '$Poss(X)$' case will be sufficient for my purposes and covers the more standard instances of assertions of indicative conditionals in which the antecedent is a live option from the viewpoint of the asserting agent. With (1) from before, we can also reformulate the assumption '$Poss(X)$' in terms of: not $Asst(\neg X)$.

Just as conditional probabilities may be interpreted either in terms of update (learning) or in terms of supposition, the same applies to conditional belief. On the one hand,

[345] See Hájek (2012a) for a recent criticism of the suppositional theory of conditionals.

'$P(Y|X)$' may be regarded as coinciding with the agent's posterior degree of belief in Y given a new piece of evidence X, just as $Bel(Y|X)$ may be taken to entail that the agent is disposed to believe Y (unconditionally) given that a new piece of evidence X comes along: these are the interpretations that were discussed and exploited in Chapter 4. On the other hand, '$P(Y|X)$' may also be regarded to coincide with the agent's *hypothetical* degree of belief in Y on the indicative *supposition* of X, and similarly $Bel(Y|X)$ may be taken to determine that the agent is disposed to *hypothetically* believe Y on the indicative *supposition* of X: which is the interpretation that will be salient right now. I have also mentioned before that these two interpretations in terms of learning and supposition do not always perfectly run in parallel. But the differences are negligible for most purposes, at least so long as introspective statements and introspective beliefs are disregarded (see n. 209 from Chapter 4).

In the present section the focus is on the suppositional manifestations of conditional belief, and it is these manifestations that turn (4) into a plausible summary of what is going on in the Ramsey test if applied on a qualitative scale. The agent aims to determine whether $X \Rightarrow Y$ is assertable for her or not; she supposes X; she determines whether Y is believable in that hypothetical context; and just in case this is so also the conditional $X \Rightarrow Y$ will be assertable for her outside of that thought experiment. Since X is assumed to be consistent with everything that she believes unconditionally (by $Poss(X)$), the hypothetical belief set that results from supposing X will be given simply by hypothetically 'throwing' X into her actual unconditional belief set and closing deductively—the operation that is called belief expansion in the theory of AGM belief revision and that I dealt with before in Chapter 4 (see sections 4.1.2 and 4.2.2).

(4) is not a new proposal either: just as there is a suppositional theory of conditionals on a numerical scale, there is also one on a classificatory scale—Levi (1988, 1996) and Gärdenfors (1988) are typical references. Analogous considerations apply: (4) may be seen as an explication of the Ramsey test on a qualitative scale; and at least Levi is very clear on *not* regarding (in my terminology) '$Asst(X \Rightarrow Y)$' as expressing the belief or acceptance of a *proposition*.[346] Finally, Gärdenfors's (1986b, Chapter 7 of 1988) proved some triviality results that may be interpreted in analogy to what we found before when discussing Lewis's results: indicative conditionals *could not* express propositions, at least as long as the following strengthening (4*) of (4) is assumed,

(4*) (Whether or not $Poss(X)$:) $Asst(X \Rightarrow Y)$ iff $Bel(Y|X)$,

and some plausible background assumptions are satisfied. As Gärdenfors's results demonstrate, given these assumptions, there are no non-trivial belief sets K according to which for all X, Y, it holds that $Y \in K^*X$ just in case $X \Rightarrow Y \in K$, where '$X \Rightarrow Y$' denotes a descriptive sentence now and where the belief revision operator * satisfies the AGM axioms (see section 4.1.2) even if applied to such conditional statements. Or in my terminology: there are no non-trivial conditional belief sets $Bel(\cdot|\cdot)$, such

[346] Gärdenfors's case is less clear on this matter.

that for all X, Y, it holds that $Bel(Y|X)$ iff $Bel(X \Rightarrow Y)$, where '$X \Rightarrow Y$' denotes a set of possible worlds now, and where the conditional belief set satisfies my postulates for general conditional belief from section 4.2.3 of Chapter 4. But these triviality results will not be important for the rest of this section.[347]

For me it is only important to observe that all-or-nothing conditional belief allows for a suppositional treatment of indicative conditionals, too, and that the corresponding theory can be developed along similar lines (and with similar merits and shortcomings) as its quantitative sibling. In short: conditional belief as in Chapter 4 yields a reasonable enough categorical explication of supposition, and the corresponding suppositional theory of conditionals nicely captures and explicates the affinity between 'suppose' and 'if-then' on a categorical scale. In the following I am going to presuppose suppositional theories of indicative conditionals on both the quantitative and the qualitative sides.

This said, the following—perhaps surprising—corollary to (4) can be derived with the help of the postulates B1–B6 for restricted conditional belief and the corresponding definitions from section 4.2.2.

Assume $Poss(X)$. Then:

$Asst(X \Rightarrow Y)$	iff	(by (4) above)
$Bel(Y|X)$	iff	(by B1–B4 and the definition of 'B_X')
$B_X \subseteq Y$	iff	(since $B_X = X \cap B_W$ by $Poss(X)$ and B6)
$X \cap B_W \subseteq Y$	iff	(by plain set theory)
$B_W \subseteq \neg X \cup Y$	iff	(by B1–B4 and the definition of 'B_W')
$Bel(\neg X \cup Y|W)$	iff	(by the definition of unconditional belief)
$Bel(\neg X \cup Y)$	iff	(by (4) above)
$Asst(X \supset Y)$,		

where in the last line $X \supset Y$ is the *material* conditional (proposition).

In other words: (4) with the postulates from Chapter 4 entails that an indicative conditional is subjectively assertable for an agent just in case the corresponding material conditional is.[348] This does *not* mean the *degree* of acceptability of $X \Rightarrow Y$ would have to coincide with the *degree* of acceptability of $X \supset Y$: indeed $Deg_{Asst}(X \Rightarrow Y) = P(Y|X)$ *differs* from $Deg_{Asst}(X \supset Y) = P(\neg X \cup Y)$ except for special cases of extreme probability. But, as it happens, that numerical difference is 'washed out' by taking the step from the numerical to the qualitative scale. It does not follow from this either that indicative conditionals would have to express propositions after all: the derivation only tells us that in the case in which the antecedent of an indicative

347 Levi does not regard conditionals as expressing propositions, which is why he does not include them as members in K (or in Levi's terminology, a corpus) either. Hence, his account is not threatened by Gärdenfors's results. A different way of responding to Gärdenfors's results would be to give up some of the AGM postulates that are used in Gärdenfors's results, such as the Preservation postulate K*4; see Leitgeb (2010) for more about the available options. Gärdenfors himself leaves open how to interpret his triviality theorems.

348 This has been observed before: see e.g. Stalnaker (1987, pp. 106–7).

conditional is open, such that *Poss*(X), the subjective assertability conditions for the indicative conditional happen to coincide with those of the corresponding material conditional. As mentioned by Leitgeb (2010, p. 30), this might go some way towards explaining why material conditionals are not *so* bad as logical representatives of indicative conditionals after all, at least in the following circumstances: when the antecedent is open; when neither the antecedent nor the consequent includes the conditional operator again; and when propositional connectives do not get applied to conditionals. It is also worth pointing out that the Preservation principle B6 from Chapter 4 was required for the derivation to go through.

Now that the (subjective) assertability conditions for propositions and conditionals have been clarified on both a numerical and a classificatory scale, I will turn to what might be called rationality postulates, or logical closure conditions, or logical rules, concerning all-or-nothing assertability. As Milne (2012, p. 332) puts it: 'In making sincere and serious assertions, we take on commitments to the consistency of what we assert and commitments to the logical consequences of what we assert: challenged on a consequence of what one has said, one stands by the consequence or withdraws one of the assertions.' In fact, we are already able to *derive* such closure conditions from (1)–(4) and the postulates from Chapter 4, but for some authors (including Milne), the validity of logical constraints on qualitative assertability may even be *more* plausible than that of their counterparts for qualitative belief. So let me first state and discuss such closure conditions on subjective assertability independently of (1)–(4). Ultimately, this will allow us to recover our postulates for restricted conditional belief from Chapter 4 (see section 4.2.2) from logical constraints on the assertability of indicative conditionals with 'open' antecedents.

Here are the requirements on subjective all-or-nothing assertability for perfectly rational agents that I am going to presuppose. Once again I will abuse notation a bit by using logical symbols for propositions and operations on propositions: $\top = W$, $\bot = \varnothing$, $\neg$ will be set-theoretic complement with respect to W, $\wedge = \cap$, $\vee = \cup$. As in previous chapters, the subset relation serves as logical implication relation for propositions. (This is because '$X \subseteq Y$' means that: every world w that makes X true, that is, where $w \in X$, also makes Y true, that is, $w \in Y$.) Additionally, I will state in the form of 'logical rules' what are actually *postulates* on rational assertability.

These are my logical postulates or 'rules' on subjective assertability for propositions:

$$Asst(\top) \text{ (Taut)}$$

$$\text{not } Asst(\bot) \text{ (Cons)}$$

$$\frac{Asst(X),\ X \subseteq Y}{Asst(Y)} \text{ (Weak1)}$$

$$\frac{Asst(X),\ Asst(Y)}{Asst(X \wedge Y)} \text{ (And1)}$$

Premise-free 'rules' are meant to express unconditional constraints on a perfectly rational agent's set *Asst* of assertable propositions (at an arbitrary time). In particular,

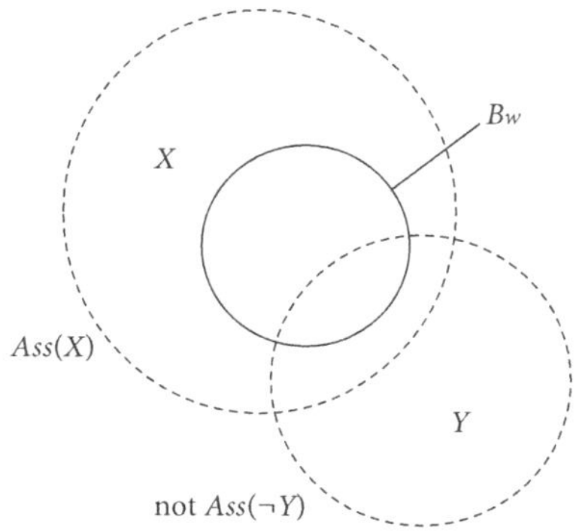

Figure 6.1. Logical postulates for assertability of propositions

Taut expresses that the tautological proposition is assertable, while Cons expresses that the contradictory proposition is not. Rules with premises postulate certain closure conditions for *Asst*: e.g. in the case of Weak(ening), if X is assertable, and X logically implies Y, then Y must also be assertable. My idealized perfectly rational agents will satisfy all of these requirements.

From the rules we get that the set of assertable propositions is closed under logical consequence, and it is a consistent set that does not include a proposition X and its negation $\neg X = (W \setminus X)$ at the same time. Furthermore, assuming the set W of worlds to be finite again, it follows that there must be a non-empty proposition B_W for which it holds that: for all X, $Asst(X)$ iff $B_W \subseteq X$. In other words: there is a uniquely determined least or strongest assertable proposition B_W which must be consistent, by the postulates from before.[349] It also follows that for all Y: not $Asst(\neg Y)$ iff $Y \cap B_W \neq \varnothing$. Any such Y, such that not $Asst(\neg Y)$, is a proposition for which assertability is not ruled out (which is consistent with B_W) and which is a 'live option' in that sense of the word. All of this follows in the same way as it followed for belief (see e.g. section 3.2 in Chapter 3). Figure 6.1 illustrates the situation; note that the dashed lines are for X and Y, whereas e.g. '$Ass(X)$' does not correspond to any region at all, in the diagram.

Next I consider the first set of logical postulates of assertability for indicative conditionals (viewed as pairs of propositions again). In all of them I will only consider premises of the form '$Asst(X \Rightarrow Y)$' for which 'not $Asst(\neg X)$' is assumed as well. That is: I will only deal with the assertability of indicative conditionals that have live antecedents from the viewpoint of the asserting agent. This is analogous to the case of 'restricted conditional belief' from section 4.2.2 in Chapter 4 in which conditional belief was restricted to cases in which the 'given' proposition X was consistent with

349 Here I use the same notation 'B_W' for the strongest *assertable* proposition as I do normally for the strongest *believed* proposition. The context should make it clear what is meant, and postulate (1) from above entails that the strongest assertable proposition must coincide with the strongest believed proposition anyway. Analogously for 'B_X' below.

everything that the agent believed unconditionally. On the side of belief dynamics, this corresponded to the special case of belief revision by expansion, which was governed, amongst others, by the so-called Preservation postulate (see section 4.1.2). All of these analogies will become more precise by Theorem 34.

Here are my postulates for the assertability of indicative conditionals:

$$\frac{\text{not } Asst(\neg X)}{Asst(X \Rightarrow X)} \text{ (Ref)}$$

$$\frac{Asst(X \Rightarrow Y),\ Y \subseteq Z,\ \text{not } Asst(\neg X)}{Asst(X \Rightarrow Z)} \text{ (Weak2)}$$

$$\frac{Asst(X \Rightarrow Y),\ Asst(X \Rightarrow Z),\ \text{not } Asst(\neg X)}{Asst(X \Rightarrow Y \wedge Z)} \text{ (And2)}$$

Clearly, these postulates simply extend the previous ones to conditionals with 'possible' antecedents. Still assuming W to be finite, it follows that for every X, such that not $Asst(\neg X)$, there must be a least proposition B_X for which it holds: for all Y, $Asst(X \Rightarrow Y)$ iff $B_X \subseteq Y$. See Figure 6.2. We will see soon that the 'B_X' notation is consistent with the previous 'B_W' notation.

While the rules would be equally plausible without the 'not $Asst(\neg X)$' restriction, the next rule is designed especially for indicative conditionals the antecedents of which are live possibilities (in line with Ramsey's 'in doubt as to *p*'):

$$\frac{Asst(Y),\ \text{not } Asst(\neg X)}{Asst(X \Rightarrow Y)} \text{ (Pres)}$$

Pres(ervation) guarantees that there is substantial *logical* interaction between the assertability of propositions and the assertability of indicatives with live antecedents: this seems plausible, at least so long as we allow for some of the conclusions of Pres to be read in terms of 'even if' or 'still'. As in (omitting occurrences of '*Asst*'): *We won't catch the train. It might well be that we leave right now.* Therefore: *Even if we leave right*

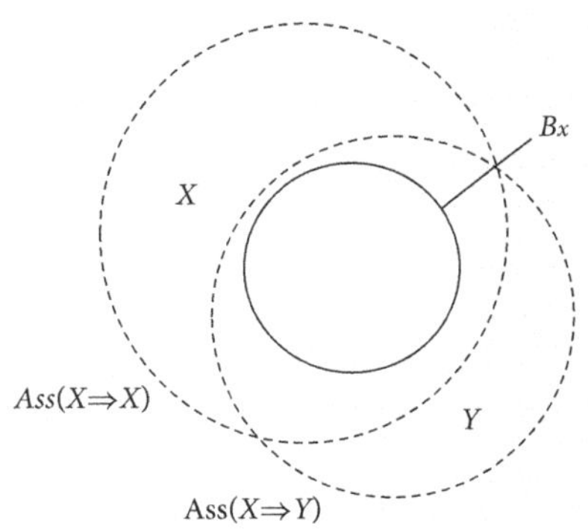

Figure 6.2. Logical postulates for assertability of conditionals

now, we still won't catch the train. Unlike counterfactual conditionals in the subjunctive mood—the antecedents of which are understood not to be true in the actual world—the antecedents of indicative conditionals are indeed supposed to apply to the actual world. The Pres rule postulates that if Y is subjectively assertable—an agent takes Y to hold in the actual world—then this remains so even if the actual world is additionally assumed to be such that X is the case, where X is a live option. Hence, '(even) if X is the case, Y is (still) the case' is subjectively assertable, too.

Since we know already that not $Asst(\neg\top)$ is the case, because of Cons from above (and $\neg\top = \bot = \varnothing$), this is one especially important instance of Pres:

$$\frac{Asst(Y)}{Asst(\top \Rightarrow Y)} \quad (\top 1\text{, derivable from Pres})$$

More generally, Pres (together with the other rules from before) can be shown to imply that $B_X \subseteq B_W \cap X$, for all X with not $Asst(\neg X)$. Compare Figure 6.3.[350] Later, Pres will be seen to correspond to AGM's Preservation postulate K*4 (see section 4.1.2) or (given also the rules below) to the Preservation postulate B6 for restricted conditional belief from section 4.2.2.

Next, I add the converse of the previous rule as another postulate for subjective assertability:

$$\frac{Asst(\top \Rightarrow X)}{Asst(X)} \quad (\top 2)$$

So $\top 2$ is not derived but taken as given. $\top 1$ and $\top 2$ together immediately guarantee the consistency of the 'B_W' and the 'B_X' notations that were introduced before. $\top 1$ and $\top 2$ are the all-or-nothing counterparts of Adams's and other suppositionalists' numerical assumption that $Deg_{Asst}(X)$ $(= P(X))$ equals $Deg_{Asst}(\top \Rightarrow X)$ $(= P(X|W))$.

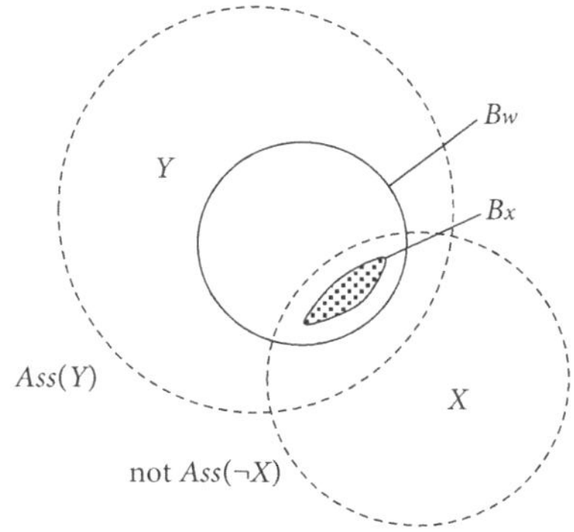

Figure 6.3. The consequences of Pres

[350] The proof of this is contained in the proof of Theorem 34.

One final note on Pres above: Lewis's (1973) logic of counterfactual conditionals of the form $X \,\Box\!\!\rightarrow Y$ also contains special axioms for the interaction between the truth of non-counterfactual statements and truth for counterfactuals (the so-called Centering Axioms). The rule

$$\frac{Y,\ \neg(\neg X)}{X \,\Box\!\!\rightarrow Y}$$

which vaguely resembles Pres in some ways, can easily be shown to derive from them. But it might be too misleading to compare that rule with Pres from above, since the truth of Y is not the same as its assertability, and especially $\neg(\neg X)$ does not mirror our original premise 'not $Asst(\neg X)$' of Pres very well.

In order to relate our logic of assertability with Lewis's more properly, we can take a cue from T1 and T2 before: if we represent our '$Asst(Y)$' within Lewis's language by means of his '$\top \,\Box\!\!\rightarrow Y$' (and similarly 'not $Asst(\neg X)$' by '$\neg(\top \,\Box\!\!\rightarrow \neg X)$'), and if we represent our '$Asst(X \Rightarrow Y)$' by means of his '$X \,\Box\!\!\rightarrow Y$', then the resulting counterpart of Pres is

$$\frac{\top \,\Box\!\!\rightarrow Y,\ \neg(\top \,\Box\!\!\rightarrow \neg X)}{X \,\Box\!\!\rightarrow Y}$$

which is indeed derivable in Lewis's system even independently of the Centering Axioms. The same is true of Stalnaker's logic of conditionals. While Pres is certainly not sacrosanct,[351] this shows that it has valid counterparts in standard systems of conditional logic. This said, I should also point out that Pres is not valid in Adams's (1975) own logic of conditionals, even though it is valid in strengthenings of Adams's logic that still have a probabilistic semantics (see e.g. Schurz 1998 and the appendix to Lehmann and Magidor 1992).

Finally, I add the following three closure conditions, which (if '*Asst*' is dropped) are well-known from conditional logic and nonmonotonic reasoning:[352]

$$\frac{Asst(X \Rightarrow Y),\ Asst(X \Rightarrow Z),\ \text{not } Asst(\neg X)}{Asst(X \wedge Y \Rightarrow Z)} \quad \text{(CM)}$$

$$\frac{Asst(X \Rightarrow Y),\ Asst(X \wedge Y \Rightarrow Z),\ \text{not } Asst(\neg(X \wedge Y))}{Asst(X \Rightarrow Z)} \quad \text{(CC)}$$

$$\frac{Asst(X \Rightarrow Z),\ Asst(Y \Rightarrow Z),\ \text{not } Asst(\neg X),\ \text{not } Asst(\neg Y)}{Asst(X \vee Y \Rightarrow Z)} \quad \text{(Or)}$$

Note that in CC, 'not $Asst(\neg X)$' is entailed by the premise 'not $Asst(\neg(X \wedge Y))$' and Weak2, which is why I did not have to state it explicitly as another premise.

C(autious) M(onotonicity) expresses that importing X's consequents (Y) into the antecedent (so that X becomes $X \wedge Y$) does not *subtract* from the original antecedent's

351 Compare the related discussion of the Preservation principle B6 in section 4.2.2.

352 See e.g. Kraus et al. (1990) for a detailed discussion of these closure conditions.

(X's) 'inferential power'. C(autious) C(ut) expresses that importing consequents in this way does not *add* to the antecedent's 'inferential power' either: consider the denial of CC's conclusion, but assume $X \Rightarrow Y$ to be assertable, such that Y is a 'consequence' of X (and assume all relevant antecedents to be live options). Then CC maintains that $X \wedge Y \Rightarrow Z$ cannot be assertable either, which means that $X \wedge Y$ does not have 'more consequences' (Z) than X does. CM is a restricted from of monotonicity or strengthening of the antecedent, while CC may be viewed as a restricted form of transitivity or Cut. CM and CC taken together are often summarized by the term 'cumulativity': cumulativity was suggested first by Gabbay (1984) and has become a principal feature of logical systems of nonmonotonic reasoning.[353]

The Or-rule is simply the standard rule for the introduction of disjunctions into the antecedent of conditionals.

One can show that if these three rules are combined with the previous ones, then the previous '$B_X \subseteq B_W \cap X$', where not $Asst(\neg X)$, is strengthened to: for all X, such that not $Asst(\neg X)$, it holds that $B_X = B_W \cap X$. See Figure 6.4.[354]

This concludes my list of closure conditions on subjective assertability for perfectly rational agents.

I will not discuss rationality constraints on degrees of assertability separately. (They simply coincide with the axioms of probability.) But I do add one bridge principle for categorical and numerical assertability—the assertability variant of postulate BP1^r from section 4.2.2 in Chapter 4. This is really a statement concerning $Asst$ and Deg_{Asst}, but for the sake of continuity with the 'rules' for assertability from above, I will give it the form of a rule of inference again:

$$\frac{Asst(X \Rightarrow Y) \ \ (\text{and not } Asst(\neg X), Deg_{Asst}(X) > 0)}{Deg_{Asst}(X \Rightarrow Y) > r} \ (\text{HP}^r)$$

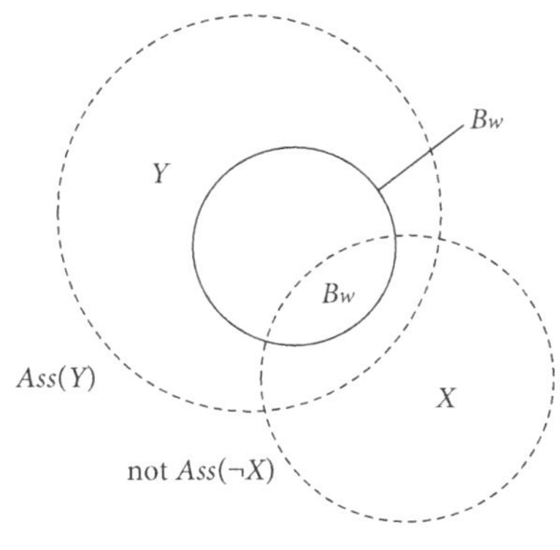

Figure 6.4. More logical postulates for assertability of conditionals

353 I already discussed cumulativity after introducing my Preservation principle B6 in section 4.2.2.

354 The proof of this claim is also contained in the proof of Theorem 34.

H(igh) (P)robability with a contextually determined threshold r demands: if $X \Rightarrow Y$ is assertable, where X is a live possibility both in terms of *Asst* and P, then the degree of assertability of $X \Rightarrow Y$ is greater than r (where $\frac{1}{2} \leq r < 1$). In a nutshell: the all-or-nothing assertability of indicative conditionals is probabilistically reliable.

Taking all of the postulates so far together, I am ready to formulate another representation theorem that relates quantitative and qualitative assertability to quantitative and qualitative belief:

Theorem 34 (Representation Theorem for Subjective Assertability)
Let W be a finite and non-empty set of worlds. Let $Asst \subseteq \wp(W) \cup (\wp(W) \times \wp(W))$, where $\wp(W)$ is the set of all subsets of W. (So every member of *Asst* is either a subset of W or a pair of subsets of W, where the first captures propositions and the second captures conditionals.) Let *Bel* be a set of pairs of subsets of W (so $Bel \subseteq (\wp(W) \times \wp(W))$). Finally, let Deg_{Asst} and P be mappings from the set of subsets of W to the unit interval $[0, 1]$.

Assume (1)–(4) from above, that is, for all $X, Y \subseteq W$:

(1) $Asst(X)$ iff $Bel(X)$ (which is short for: $Bel(X|W)$).
(2) $Deg_{Asst}(X) = P(X)$.
(3) If $P(X) > 0$, then $Deg_{Asst}(X \Rightarrow Y) = P(Y|X) = \frac{P(X \cap Y)}{P(X)}$.
(4) If $Poss(X)$ (equivalently: not $Bel(\neg X)$, or not $Bel(\neg X|W)$, or with (1), not $Asst(\neg X)$), then: $Asst(X \Rightarrow Y)$ iff $Bel(Y|X)$.

Then the following three statements are equivalent:

I. *Asst* satisfies Taut, Cons, Weak1, And1, Ref, Weak2, And2, Pres, $\top$2, CM, CC, Or. Deg_{Asst} is a probability measure. *Asst* and Deg_{Asst} jointly satisfy HP^r.
II. P and *Bel* satisfy P1, B1–B6, and $BP1^r$ from section 4.2 of Chapter 4.
III. Deg_{Asst} is a probability measure, and *Asst* corresponds to stably high numerical degree of assertability in the following sense: there is a (uniquely determined) proposition X, such that X is a non-empty Deg_{Asst}-stable^r proposition (for the definition of P-stability^r see Appendix B), and:
   - For all propositions Z:

$$Asst(Z) \text{ if and only if } X \subseteq Z$$

   (and hence, $B_W = X$).
   - For all propositions Y, such that $Y \cap X$ is non-empty, for all propositions Z:

$$Asst(Y \Rightarrow Z) \text{ if and only if } Y \cap X \subseteq Z.$$ [355]

[355] Proof: we already know the equivalence of II and III: this is Theorem 12 from section 4.2.2 (while using assumptions (1)–(4) above). So I concentrate on proving I and II to be equivalent.

'I $\rightarrow$ II': using (1)–(4), all postulates in II follow immediately from the postulates in I, except for B6: for all X such that $X \cap B_W \neq \emptyset$, $B_X = X \cap B_W$. I will turn to the proof of B6 in a moment. Here, B_W is both the least assertable and the least believed proposition (by (1)): for all X, $Asst(X)$ iff $B_W \subseteq X$. Such a

If '*Asst*' is replaced by '*Bel*', and 'Deg_{Asst}' is replaced by '*P*', then the equivalence of II and III corresponds to Theorem 12 in section 4.2.2—the representation theorem for restricted conditional belief. The equivalence of I and II reformulates the conditions on belief from Chapter 4 in terms of the conditions on assertability from above, and vice versa. In this way, all of the results of the previous chapters become applicable to subjective assertability, too, including all of the findings on conditional belief from Chapter 4, which thus become findings on the subjective assertability of conditionals (such as the Conditional Lockean thesis, which I derived immediately after Theorem 12 in section 4.2.2).

proposition exists and is non-empty by Taut, Cons, Weak1, And1. Similarly, if not $Asst(\neg X)$, B_X will be the least proposition for which it holds: for all Y, $Asst(X \Rightarrow Y)$ iff $B_X \subseteq Y$. Such a proposition exists by Ref, Weak2, And2. If B_X were empty, then $B_X \subseteq \neg X$, thus $Asst(X \Rightarrow \neg X)$; by Ref, $Asst(\neg X \Rightarrow \neg X)$; therefore, by Or, $Asst((X \vee \neg X) \Rightarrow \neg X)$, that is, $Asst(\top \Rightarrow \neg X)$ (since $\top = W = X \vee \neg X$). So with $\top 2$ it would follow that $Asst(\neg X)$, which would contradict not $Asst(\neg X)$. Therefore, B_X must be non-empty, too, whenever not $Asst(\neg X)$.

Now assume that $X \cap B_W \neq \varnothing$. First I show $B_X \subseteq B_W \cap X = X \cap B_W$: we have $Asst(B_W)$, and not $Asst(\neg X)$ (since otherwise it would hold that $B_W \subseteq \neg X$, which would contradict $X \cap B_W \neq \varnothing$). Hence, by Pres: $Asst(X \Rightarrow B_W)$. Thus: $B_X \subseteq B_W$. Additionally, because $Asst(X \Rightarrow X)$ holds by Ref, it is also the case that $B_X \subseteq X$. And2 entails that $B_X \subseteq B_W \cap X$.

Next I will strengthen this to: $B_X = B_W \cap X$. We have to consider two cases (in both cases it is still assumed that $X \cap B_W \neq \varnothing$, that is, as before, not $Asst(\neg X)$).

Case 1: $\neg X \cap B_W \neq \varnothing$ (that is, analogously as before, not $Asst(X)$). By Weak2 (and not $Asst(X)$), it holds that $Asst(\neg X \Rightarrow \neg X \vee B_X)$. By the defining property of B_X, it also holds that $Asst(X \Rightarrow B_X)$; applying Weak2 (and not $Asst(X)$) yields that $Asst(X \Rightarrow \neg X \vee B_X)$ is the case. Hence, with Or: $Asst((\neg X \vee X) \Rightarrow \neg X \vee B_X)$. So by the defining feature of B_W, it follows that B_W must be a subset of $\neg X \vee B_X$. If B_X were a proper subset of $B_W \cap X$, then there would be a world in $B_W \cap X$ which would not be in B_X (but in X). Which means B_W would not be a subset of $\neg X \vee B_X$, which would contradict what we have shown before. So $B_X = B_W \cap X$, as required.

Case 2: $\neg X \cap B_W = \varnothing$, that is, $B_W \subseteq X$. Because we know already that $B_X \subseteq B_W \cap X \subseteq B_W$, by the defining property of B_X it follows that: $Asst(X \Rightarrow B_W)$. By the properties of B_X again, also $Asst(X \Rightarrow B_X)$. With CM we can derive: $Asst(X \wedge B_W \Rightarrow B_X)$. In our special Case 2 this means: $Asst(B_W \Rightarrow B_X)$, or equivalently, $Asst(W \wedge B_W \Rightarrow B_X)$. By the defining properties of B_W we also have: $Asst(W \Rightarrow B_W)$. Thus, by CC: $Asst(W \Rightarrow B_X)$. Which means again: $B_W \subseteq B_X$, that is, with $B_X \subseteq B_W \cap X \subseteq B_W$ from before, $B_W = B_X$. So we are done: $B_X = B_W = B_W \cap X$, as required.

Hence, B6 is the case.

'II $\to$ I': presupposing again (1)–(4) from above, the only non-obvious closure conditions to derive from the postulates in II are CM, CC, and Or. B_W and B_X are defined in their usual manner (in terms of '*Bel*'), and I will apply the defining features of B_W and B_X without further comments now (see section 4.2.2). The same holds for (1) and (4).

About CM: if $Asst(X \Rightarrow Y)$, $Asst(X \Rightarrow Z)$, not $Asst(\neg X)$, then $B_X \subseteq Y$, $B_X \subseteq Z$, and $X \cap B_W \neq \varnothing$. By B6, $B_X = X \cap B_W \neq \varnothing$. It follows that $X \cap Y \cap B_W = B_X \neq \varnothing$, thus by B6: $B_{X \cap Y} = X \cap Y \cap B_W = B_X$. Since $B_X \subseteq Z$, also $B_{X \cap Y} \subseteq Z$, which means that $Asst(X \wedge Y \Rightarrow Z)$.

About CC: if $Asst(X \Rightarrow Y)$, $Asst(X \wedge Y \Rightarrow Z)$, not $Asst(\neg(X \wedge Y))$, then $B_X \subseteq Y$, $B_{X \cap Y} \subseteq Z$, and $X \cap Y \cap B_W \neq \varnothing$ (and thus $X \cap B_W \neq \varnothing$). B6 implies that $B_{X \cap Y} = X \cap Y \cap B_W$ and $B_X = X \cap B_W$. Therefore, $B_{X \cap Y} = B_X \cap Y$, which with $B_X \subseteq Y$ yields: $B_{X \cap Y} = B_X$. Since $B_{X \cap Y} \subseteq Z$, also $B_X \subseteq Z$, which means that $Asst(X \Rightarrow Z)$.

About Or: if $Asst(X \Rightarrow Z)$, $Asst(Y \Rightarrow Z)$, not $Asst(\neg X)$, not $Asst(\neg Y)$, then $B_X \subseteq Z$, $B_Y \subseteq Z$, $X \cap B_W \neq \varnothing$, and $Y \cap B_W \neq \varnothing$. Hence, $(X \cup Y) \cap B_W \neq \varnothing$, so with B6 $B_{X \cup Y} = (X \cup Y) \cap B_W$, which by distributivity is equal to $(X \cap B_W) \cup (Y \cap B_W)$. B6 also gives us that $B_X = X \cap B_W$, $B_Y = Y \cap B_W$, from which we can derive $B_{X \cup Y} = B_X \cup B_Y$. From $B_X \subseteq Z$, $B_Y \subseteq Z$ we have that $B_{X \cup Y} \subseteq Z$, which means that $Asst(X \vee Y \Rightarrow Z)$.

It would be possible to extend this assertability account for indicative conditionals with 'live' antecedents to all indicative conditionals whatsoever, which, on the belief side, would correspond to extending restricted conditional belief to general conditional belief (the topic of section 4.2.3). Formally, it should be clear by now how this would go and what the result would be: the sphere systems of P-stable^r sets from Theorem 23 in Chapter 4 that corresponded to conditional belief sets would become sphere systems of Deg_{Asst}-stable^r sets that would correspond to assertability sets of conditionals. But I will not deal with this in any more detail here.

Let me turn to three examples instead. In the first two of them belief will determine assertability. In the third one it will be the other way around.

Example 23

Consider the following story, which is an abbreviated version of an example by Bradley (2006, p. 65) (which in turn goes back to an earlier example by Stalnaker):

Lord Russell has been murdered. There are three suspects: the butler, the cook and the gardener. The gardener does not seem a likely candidate, since he was seen pruning the roses at the time of the murder. The cook could easily have done it, but she had no apparent motive. But the butler was known to have recently discovered that his lordship had been taking liberties with the butler's wife. Moreover, he had had every opportunity to do the deed. So it was probably the butler, but if it wasn't the butler, then it was most likely the cook. (Bradley 2006)

The detective in that murder case believes all of the above. Let me reconstruct this in more formal terms now. Let us assume that the detective distinguishes between four possibilities: $g \wedge \neg c \wedge \neg b$ (it was only the gardener), $\neg g \wedge c \wedge \neg b$ (it was only the cook), $\neg g \wedge \neg c \wedge b$ (it was only the butler), and the negation of the disjunction of the previous three cases (so this fourth 'fat' world captures all remaining logical options). Therefore, W has four members. I will abuse notation a bit by using 'g', 'c', 'b' as denoting both propositional letters and the propositions in which the respective propositional letters are true.

I determine the detective's degrees of belief as follows: $P(g \wedge \neg c \wedge \neg b) = 0.1$, $P(\neg g \wedge c \wedge \neg b) = 0.3$, $P(\neg g \wedge \neg c \wedge b) = 0.6$, and the probability that the detective assigns to the remaining 'catch-all hypothesis' is 0. Intuitively, this matches the story from above, although other numbers would of course do so as well.

Now assume that the detective's least believed proposition is determined as follows:

- $B_W = \{\neg g \wedge \neg c \wedge b,\ \neg g \wedge c \wedge \neg b\}$.

Let $r = \frac{1}{2}$: B_W follows to be P-stable^r (as can be seen e.g. by checking for the Outclassing Condition in Appendix B).

Assuming our postulates (1) and (4) concerning *Bel* vs *Asst* from above, this means:

- Assertable: $\neg g$, $\top \Rightarrow \neg g$, $\neg b \Rightarrow c$, $\neg b \Rightarrow \neg g$, $\neg b \Rightarrow c \wedge \neg g, \ldots$

For instance, $Bel(\neg g)$, since all worlds in B_W are $\neg g$-worlds, which is why also $Asst(\neg g)$: it is subjectively assertable for the detective that it was not the gardener.

Similarly: $Poss(\neg b)$, as there is a $\neg b$-world in B_W. $Bel(c|\neg b)$, because $B_{\neg b} = \neg b \cap B_W = \{\neg g \wedge c \wedge \neg b\}$, which is a subset of the set of all c-worlds; therefore $Asst(\neg b \Rightarrow c)$: it is subjectively assertable for the detective that if it was not the butler then it was the cook. This is just as stated in the story above.

The corresponding degrees of assertability are given by the corresponding absolute or conditional probabilities as determined by P: e.g. $Deg_{Asst}(\neg g) = P(\neg g) = 0.9$, and $Deg_{Asst}(\neg b \Rightarrow c) = P(c|\neg b) = 0.75$.

If the detective's all-or-nothing beliefs are more specific, for instance,

- $B_W = \{\neg g \wedge \neg c \wedge b\}$,

which is P-stable^r again, then this means:

- Assertable: b, $b \Rightarrow \neg g \wedge \neg c, \ldots$

In that case, it is subjectively assertable for the detective that it was the butler, again as in the story above.[356] On the numerical side it holds: $Deg_{Asst}(b) = P(b) = 0.6$.

Example 24 (Tracey's Sprinkler from Section 2.3.1 Reconsidered Again)
In Example 16 of section 4.3 I have already used conditionals as a means of making Tracey's conditional beliefs more transparent. (I should add that not all of them will follow to be assertable on the basis of the postulates in the present section, since not all of them have antecedents that are possible from Tracey's viewpoint.)

Let us assume now that $B_W = \{w_1, w_2, w_3\}$, where the worlds are as follows: w_1: $T = 0, J = 0, R = 0, S = 0$. w_2: $T = 1, J = 1, R = 1, S = 0$. w_3: $T = 0, J = 1, R = 0, S = 0$. From this and postulates (1) and (4) the following can be derived:

$$Asst(R = 1 \Rightarrow J = 1 \wedge T = 1)$$
$$Asst(T = 1 \wedge J = 1 \Rightarrow R = 1 \wedge S = 0)$$
$$Asst(R = 1 \vee S = 1 \Rightarrow R = 1)$$

Similarly, (non-)assertability claims about propositions follow as well, such as:

$$Asst(S = 0)$$
$$\text{not } Asst(\neg(R = 1))$$

Of course, assertability is closed under identities between propositions and under the logical rules from above. Using this allows one to derive further assertability claims, for instance:

$$Asst((R = 1 \vee S = 1) \wedge R = 1 \Rightarrow J = 1 \wedge T = 1)$$

356 But one could not derive any more that '$Asst(\neg b \Rightarrow c)$' holds, using only the theory of assertability as developed before, since $\neg b$ would no longer be regarded as doxastically possible by the agent. Instead one would have to turn to a proper belief *revision* version of subjective assertability along the lines of section 4.2.3.

follows from the very first conditional assertability claim from before and the fact that $(R = 1 \vee S = 1) \wedge R = 1$ is the same proposition as $R = 1$. For the same reason, it follows from not $Asst(\neg(R = 1))$ that:

$$\text{not } Asst(\neg[(R = 1 \vee S = 1) \wedge R = 1])$$

Applying CC (Cautious Cut) to the claims from above (and using not $Asst(\neg[(R = 1 \vee S = 1) \wedge R = 1]))$, that is, applying

$$\frac{Asst(R = 1 \vee S = 1 \Rightarrow R = 1),\ Asst((R = 1 \vee S = 1) \wedge R = 1 \Rightarrow J = 1 \wedge T = 1)}{Asst(R = 1 \vee S = 1 \Rightarrow J = 1 \wedge T = 1)}$$

yields the conclusion $Asst(R = 1 \vee S = 1 \Rightarrow J = 1 \wedge T = 1)$: it is assertable for Tracey that if it rained or her sprinkler was on, then both her neighbour Jack's and her own lawns are wet. (That is because she believes that if it rained or her sprinkler was on, then it rained.)

Example 25 (From Assertion to Mixed Constraints on Belief)
The Secretary-General of the United Nations (UNSG) prepares a meeting in which—if all goes well—a new treaty will be signed. The three countries left to sign are P, Q, R, but the UNSG does not know whether they will sign the treaty or not. She sends off an assistant to each of two NGOs in order to find out about their experts' assessments of what the countries P, Q, R will do: one to LOG and another one to PROB.

Her first assistant reports back and tells the UNSG the following about Expert 1 from LOG:

- Expert 1 asserted that P will sign the treaty or that Q will do so or even both.
- Expert 1 asserted that if P signs the treaty, then either Q will sign it, or R will do, or both.
- Expert 1 did not want to commit herself to: if P signs the treaty, then Q will sign it.
- Expert 1 did not want to commit herself to: P will not sign the treaty.

The second assistant reports back and tells the UNSG the following about Expert 2 from PROB:

- Expert 2 found it equally assertable as not that Q will sign the treaty.
- Expert 2 found it less assertable that R won't sign the treaty than that both P and Q won't sign it.

Of course the UNSG is well-versed in techniques from logic and formal epistemology. She formalizes the situation like this: propositions p, q, r represent 'P signs the treaty', 'Q signs the treaty', 'R signs the treaty', respectively. The following are her formal representations of what is subjectively assertable for Expert 1, and also of what Expert 2's degrees of assertability must be like, given what the two experts actually asserted in conversation.

In the case of Expert 1 ($Asst = Asst_1$):

(a) $Asst(p \vee q)$.
(b) $Asst(p \Rightarrow q \vee r)$.
(c) not $Asst(p \Rightarrow q)$.
(d) not $Asst(\neg p)$.

For Expert 2 ($Deg_{Asst} = Deg_{Asst_2}$):

(e) $Deg_{Asst}(q) = \frac{1}{2}$ $(= Deg_{Asst}(\neg q))$.
(f) $Deg_{Asst}(\neg r) < Deg_{Asst}(\neg p \wedge \neg q)$.

Let (1)–(4) be as in Theorem 34. Applying (1) and (4), this is what Expert 1's doxastic state seems to be like (exploiting in the case of (4) that not $Asst(\neg p)$ from (d) above):

(a) $Bel(p \vee q)$.
(b) $Bel(q \vee r \mid p)$.
(c) not $Bel(q \mid p)$.
(d) not $Bel(\neg p)$.

And this is what (2) tells the UNSG about the formal representation of Expert 2's doxastic state (she does not need to apply (3) here):

(e) $P(q) = \frac{1}{2}$.
(f) $P(\neg r) < P(\neg p \wedge \neg q)$.

The UNSG trusts the two experts, she takes seriously what they asserted in front of her assistants, and she wants to take on board what both of them said. *So what should she believe now?*

The first step that she takes in order to answer that question is to conduct the following thought experiment: suppose Expert 1's belief set *Bel* from above were 'aggregated' with Expert 2's degree-of-belief function *P as if they belonged to one and the same subject*. What would this amalgamated state of mind be like? She is going to use the stability theory of belief to answer that question.

Once an answer has been determined, her next step will be to try to put herself into the shoes of both experts at the same time by learning from that answer (to the extent to which this is possible for her at all). Of course, she also worries whether the two sets of information that were conveyed by the two experts are compatible at all—this is not clear, as the two experts might themselves possess conflicting information; or the reasons that might support the one expert's beliefs might undermine the reasons for the other expert's beliefs; or the like. Let her find out if this is so.

As a working hypothesis, the UNSG takes the two experts to be rational, she assumes the expert from LOG to be coherent with that from PROB in the sense of the Humean thesis, and hence she freely applies all of the postulates from the previous chapters: postulates for *Bel*, postulates for *P*, and bridge postulates for *Bel* and *P* jointly. All of the following could also be carried out by applying the postulates and rules above for *Asst* and Deg_{Asst}, but I will work now with *Bel* and *P* directly.

(d) means that $p \cap B_W \neq \emptyset$, which entails with B6 from section 4.2.2 that $B_p = p \cap B_W$. (c) tells the UNSG that $\neg q \cap B_p \neq \emptyset$, that is (by what was just shown), $\neg q \cap p \cap B_W \neq \emptyset$.

From this she can conclude with B6 that $B_{p \wedge \neg q} = p \cap \neg q \cap B_W$. (b) says that $B_p \subseteq q \vee r$, therefore (applying what was shown before) $p \cap B_W \subseteq q \vee r$, thus (by set theory) also $p \cap \neg q \cap B_W \subseteq q \vee r$, and hence (by what was also shown before) $B_{p \wedge \neg q} \subseteq q \vee r$. By B1–B6 and the definition of '$B_{p \wedge \neg q}$', it also holds that $B_{p \wedge \neg q} \subseteq p \wedge \neg q$. Taking the two together, $B_{p \wedge \neg q} \subseteq p \wedge \neg q \wedge r \subseteq r$. That is: $Bel(r \mid p \wedge \neg q)$.

So the UNSG may conclude that B_W needs to include $p \wedge \neg q$-worlds, and that all of the $p \wedge \neg q$-worlds in B_W must be r-worlds.

(f) with the axioms of probability yields: $P(r) > P(p \vee q)$. On the one hand, this implies with the axioms of probability that $P(p \vee q) < 1$, and on the other hand it entails with (a), that is, $Bel(p \vee q)$, and the Lockean thesis (which follows from the Humean thesis on belief, as shown in Chapter 2, for any Humean threshold $r \geq \frac{1}{2}$):

$$Bel(r).$$

This last conclusion required taking the constraints on *Bel* and the constraints on *P* together. By (a) and the closure of belief under conjunction: $Bel((p \vee q) \wedge r)$. Which implies with the Humean thesis (or the Lockean thesis): $P((p \vee q) \wedge r) > \frac{1}{2}$.

Now the UNSG builds some models for this set of given and derived claims. The existence of such models will prove to her that what the two experts conveyed to her assistants is at least coherent in the Humean thesis sense. She chooses the underlying set of worlds in the obvious manner, that is, only taking into account distinctions that were expressed in the two experts' assertions (or in what her assistants regarded to be their take-home message). So W is the set of eight state ascriptions for p, q, r, or equivalently the set of all eight truth-value assignments to the three propositional letters p, q, r.[357] Ultimately the UNSG is going to find out that the constraints underdetermine what the amalgamated doxastic state of the two experts must be like: there is indeed an infinite set of pairs $\langle Bel, P \rangle$ on W that satisfy all of the constraints.

So far as *Bel* is concerned, there is already more than one option. A salient one is for the doxastic possibilities in B_W to be given by: $B_W = \{p \wedge q \wedge r, p \wedge \neg q \wedge r, \neg p \wedge q \wedge r\}$. Or more briefly:

- $B_W = \{pqr, p\bar{q}r, \bar{p}qr\}$.

As the UNSG knows, this determines unconditional *Bel* completely, and the same holds for conditional *Bel* with possible antecedents (the belief expansion case), which will be good enough for her purposes.

[357] I am again abusing notation here, as e.g. 'p' represents both a propositional letter and a proposition (the set of worlds in which the propositional letter p is assigned the truth value 1). But the context should always make clear what is meant.

With W and Bel determined in this way, the considerations concerning P from before together with the Humean compatibility requirements for Bel and P (and a Humean threshold of $r = \frac{1}{2}$) yield the following formal constraints on P:

- From what B_W is like, and exploiting the P-stability$^{\frac{1}{2}}$ of B_W (as implied by the Humean thesis, and expressible in terms of inequalities as determined by the Outclassing Condition—see Theorem 5 in Appendix B):
  (each of) $P(pqr), P(p\overline{q}r), P(\overline{p}qr) > P(pq\overline{r}) + P(p\overline{qr}) + P(\overline{p}q\overline{r}) + P(\overline{pq}r) + P(\overline{pqr})$.
  And also: $P(pqr) + P(p\overline{q}r) + P(\overline{p}qr) > \frac{1}{2}$.
- From $P(q) = \frac{1}{2}$ as determined before (see (e)):
  $P(pqr) + P(pq\overline{r}) + P(\overline{p}qr) + P(\overline{p}q\overline{r}) = \frac{1}{2}$.
- From $P(r) > P(p \vee q)$ as determined before:
  $P(\overline{pq}r) > P(pq\overline{r}) + P(p\overline{qr}) + P(\overline{p}q\overline{r})$.

One can show that this set of equalities and inequalities is satisfied by an infinite set of probability measures P.

Observing this, the UNSG restricts herself to construct some illustrative examples of what such probability measures can look like. By $P(q) = \frac{1}{2}$, it holds that $P(pqr)$, $P(\overline{p}qr) \leq \frac{1}{4}$. In her examples she assumes additionally, for convenience or by some Laplacean 'equal probability' assumption, that $P(pqr) = P(\overline{p}qr)$ (which is not itself entailed by the constraints above).

The first class of probability measures that the UNSG constructs is given by the following probabilities (relative to some parameter $0 < \epsilon < \frac{1}{16}$):

- Members of $W \setminus B_W$: $pq\overline{r}$: 0, $\overline{p}q\overline{r}$: 0, $\overline{pq}r$: 2ϵ, $p\overline{qr}$: ϵ, $\overline{pqr}$: ϵ.
- Members of B_W: pqr: $\frac{1}{4}$, $p\overline{q}r$: $\frac{1}{2} - 4\epsilon$, $\overline{p}qr$: $\frac{1}{4}$.

For instance, for $\epsilon = \frac{1}{20}$, this yields:

- Members of $W \setminus B_W$: $pq\overline{r}$: 0, $\overline{p}q\overline{r}$: 0, $\overline{pq}r$: $\frac{1}{10}$, $p\overline{qr}$: $\frac{1}{20}$, $\overline{pqr}$: $\frac{1}{20}$.
- Members of B_W: pqr: $\frac{1}{4}$, $p\overline{q}r$: $\frac{3}{10}$, $\overline{p}qr$: $\frac{1}{4}$.

The second class of probability measures that the UNSG determines is given by the following probabilities (relative to some parameter $0 < \epsilon < \frac{1}{36}$):

- Members of $W \setminus B_W$: $pq\overline{r}$: ϵ, $\overline{p}q\overline{r}$: ϵ, $\overline{pq}r$: 4ϵ, $p\overline{qr}$: ϵ, $\overline{pqr}$: ϵ.
- Members of B_W: pqr: $\frac{1}{4} - \epsilon$, $p\overline{q}r$: $\frac{1}{2} - 6\epsilon$, $\overline{p}qr$: $\frac{1}{4} - \epsilon$.

For instance, for $\epsilon = \frac{1}{40}$:

- Members of $W \setminus B_W$: $pq\overline{r}$: $\frac{1}{40}$, $\overline{p}q\overline{r}$: $\frac{1}{40}$, $\overline{pq}r$: $\frac{1}{10}$, $p\overline{qr}$: $\frac{1}{40}$, $\overline{pqr}$: $\frac{1}{40}$.
- Members of B_W: pqr: $\frac{9}{40}$, $p\overline{q}r$: $\frac{14}{40}$, $\overline{p}qr$: $\frac{9}{40}$.

The UNSG is now equipped with the belief that R will be signing the contract ($Bel(r)$ from above), and with a clearer understanding of what the constraints on the serious all-or-nothing possibilities (the members of B_W) and on her degrees of belief in the various logically possible circumstances (P defined on W) are like. She is ready to enter

the forthcoming UN meeting: the rest will be due to receiving further evidence and her updating B_W and P, accordingly, in line with the diachronic norms on belief revision and on conditionalization from Chapter 4. (The UNSG knows that this will preserve the Humean compatibility between the two sides of belief—see the end of section 4.1.3 in Chapter 4.)

This last example illustrates the benefits of a joint theory of belief and degrees of belief and should go some way towards meeting the Bayesian challenge (Kaplan 1996), as mentioned at the very end of Chapter 2.

Assume that, ultimately, a radical Bayesian epistemologist decided to use from the derivations above only what concerned P, and assume that she regarded '*Bel*' as merely an auxiliary term by which the transition from the experts' assertions to the constraints on P could be expressed succinctly. Such a radical Bayesian would be like an antirealist scientist who regards theoretical concepts as instruments by which 'purely empirical' claims can be derived conveniently from hypotheses that involve these theoretical concepts. At least in a similarly weak and 'quasi-antirealist' sense of acceptance, even such a radical Bayesian might be said to 'accept' the theory above, if only instrumentally.

Of course, over and above the present theory, there might well be alternative systematic ways for the UNSG to make sense of the two experts' verdicts—their plausibility and success will depend on the details. Maybe the report on Expert 1's assertions could be interpreted probabilistically more directly, e.g. in terms of an expert's high enough absolute or conditional probability, or in terms of some constraints on the expert's expected utility assessments? But what if Expert 1 understands her assertions such that these assertions commit herself to all of their logical consequences, as given by the logic of propositions and conditionals? Probabilistically, this would then have to be taken into account somehow. Perhaps the two reports could be exploited probabilistically by conditionalizing the UNSG's degree-of-belief function on them, or by describing the effects of such conditionalizations in theoretical terms. But how exactly will this work? For instance, conditionalizing the UNSG's degree-of-belief function on the *conditionals* that are said to be assertable in (b) and (c) would be impossible, if these conditionals do not express propositions (as the Bayesian suppositional theory of conditionals has it).[358] Or maybe the UNSG's degree-of-belief function is to be conditionalized on the *assertability statements* (a)–(f) themselves and perhaps also on statements concerning the logical laws of *Asst* and Deg_{Asst}. But that would mean that the UNSG's degree-of-belief function would have to be *really* complex and higher-order: the theoretical description of that kind of conditionalization would involve statements of the form 'the UNSG's degree of belief in Expert 2's degree of belief in ... *being so-and-so* is so-and-so'. How complicated would the corresponding

358 This relates to van Fraassen's Judy-Benjamin problem: see van Fraassen (1981).

epistemological theory become? After all, already the application of the relatively simple theory from above did not seem completely straightforward.

Fortunately, I do not have to settle these questions here. In any case, the theory that has been developed so far is on offer.

So much for examples. Let me conclude this section by returning to the topic of stability. Clearly, the emerging account of subjective assertability is a stability account: by Theorem 34, a proposition Z is assertable for a perfectly rational agent if and only if the agent assigns a stably high (enough) degree of assertability to that proposition; and similarly for conditionals. Stability is again explicated as resilience under salient cases of conditionalization, as developed in Chapter 2 for belief. Other than for our theory of belief, the stability in question may be regarded to derive from logical rules on all-or-nothing assertability, the axioms of probability for degrees of assertability, and the high-enough-probability constraint HP^r: that is what the I-to-III direction of Theorem 34 tells us.

But this is not the first time that probabilistic stability or robustness has been highlighted to be an important feature of assertability: in fact, the theory above may explain some of the independent findings of Jackson (1979, 1981) and Lewis (1987) about assertability and probabilistic robustness.

Compare:

High probability is an important ingredient in assertability. Everyone accepts that. But so is robustness. (Jackson 1979, p. 576)

where

'P is robust with respect to I' will be true just when both $Pr(P)$ and $Pr(P|I)$ are close and high. (Jackson 1979, p. 569)

Jackson argues for, and applies, this thought in two cases: indicative conditionals and disjunctions.[359] So far as indicative conditionals are concerned, he defends the view that

it is proper to assert $P \rightarrow Q$ when $P \supset Q$ is highly probable and robust with respect to P, that is, when $Pr(P \supset Q|P)$ is also high. (Jackson 1979, p. 576)

Jackson thinks that the truth conditions of indicative conditionals $P \rightarrow Q$ (in my terminology: $X \Rightarrow Y$) are those of the corresponding material conditionals $P \supset Q$. That will not be important in anything that follows, and I do not take up that part of Jackson's theory. But Jackson also argues (using my terminology again) that the degree of assertability of $X \Rightarrow Y$ equals the conditional probability $P(X \supset Y|X)$, which, by the axioms of probability, is equal to $P(Y|X)$: Adams's degree of assertability of the

[359] Jackson regards the truth conditions of indicative conditionals to be those of the corresponding material conditionals $P \supset Q$, which in turn may be viewed as disjunctions of the form $\neg P \vee Q$. That is why the cases of indicative conditionals and disjunctions are closely related in Jackson's eyes.

indicative conditional $X \Rightarrow Y$.[360] So Jackson argues for Adams's numerical account of assertability, which I took for granted from the start, by stability considerations. What is more, as expressed in the quotation, Jackson also takes the all-or-nothing assertability of $X \Rightarrow Y$ to entail both $P(X \supset Y)$ and $P(X \supset Y|X)$ to be high. His reason for this is the following one:

What is the point of signalling the robustness of $(P \supset Q)$ with respect to P? The answer lies in the importance of being able to use *Modus Ponens*. (Jackson 1979, p. 577)

Indicative conditionals are like inference tickets that permit the application of Modus Ponens. But it will only be pragmatically useful to apply Modus Ponens when learning or supposing X does not at the same time leave $X \supset Y$ (and thus Y) with a small probability: hence the pragmatic requirement to assert indicative conditionals only when the consequent retains a high enough probability given the antecedent—when $P(Y|X)$ is high enough.

All of this holds also according to the present theory. The assertability of $X \Rightarrow Y$ (given $Poss(X)$) entails with postulate (4) from above the asserting agent's conditional belief in Y given X, which, by postulate BP1^r from Chapter 4, entails $P(Y|X)$ to be greater than a threshold r. What the theory in this section adds to Jackson's account is the thesis that the all-or-nothing assertability of an indicative conditional $X \Rightarrow Y$ (where $Poss(X)$) yields even more robustness. By the theory of conditional belief in Chapter 4, $Bel(Y|X)$ also entails that $P(Y|X)$ is robust or stably high with respect to further strengthenings of the antecedent: it even holds that $P(Y|X \wedge Z) > r$ for all Z with $Poss(Z)$. So if this combined theory of belief and assertability is correct, then the assertion of an indicative conditional expresses an even stronger form of probabilistic robustness than Jackson thought. Given the stability of theory of belief, the all-or-nothing version of the Ramsey test for indicative conditionals from above has it that the degree of assertability for such conditionals must be stably high with respect to all propositions that are live options for the agent in the all-or-nothing sense.

With respect to the assertability of disjunctions, Jackson claims that

putting it [the disjunction] in explicitly disjunctive form you signal robustness with respect to the negation of each disjunct taken separately. (Jackson 1979, pp. 574–5)

Jackson argues that for that reason it can even make good pragmatic sense—contra Grice—to assert a disjunction when one would be in a position to assert one of the disjuncts and hence something that is logically stronger than the disjunction itself:

Consider 'Either Oswald killed Kennedy or the Warren Commission was incompetent.' This is highly assertable even for someone convinced that the Warren Commission was not incompetent [...] The disjunction is... highly assertable for them, because it would still be probable were information to come to hand that refuted one or the other disjunct. The

[360] Jackson (1981) distinguishes between two kinds of assertability—assert*a*bility and assert*i*bility—but that will not be important in the following.

disjunction is robust with respect to the negation of either of its disjuncts taken separately–and just this may make it pointful to assert it. (Jackson 1979, p. 571)

While I will not be able to say more about the part of Jackson's claim that goes beyond Grice, I am able to derive a properly *Gricean* version of Jackson's robustness thought from the stability theory of belief.

Let us assume that (i) asserting a proposition $X \vee Y$—in the disjunctive syntactic form $A \vee B$, where A expresses X and Y expresses B—expresses one's belief that $X \vee Y$ is the case, in line with postulate (1) from above. That is: $Bel(X \cup Y)$ (or $Bel(X \vee Y)$). Additionally, let us assume that by Gricean conversational implicature (cf. Grice 1989), (ii) asserting the disjunction also signals pragmatically that one does not believe either of the two disjuncts: not $Bel(X)$, not $Bel(Y)$. Or otherwise asserting the disjunction would be misleading, as one could have (and should have by the Gricean Cooperative Principle) asserted the stronger information. In other words: one signals that not $Bel(\neg\neg X)$ and not $Bel(\neg\neg Y)$, that is, $Poss(\neg X)$ and $Poss(\neg Y)$. Note that (ii) goes *beyond* the general account of assertability that was developed on the basis of (1)–(4): it is a matter of conversational implicature.

By the Humean thesis from Chapter 2, it follows now from (i) and (ii) that $P(X \cup Y|\neg X) > r$ and $P(X \cup Y|\neg Y) > r$. The asserted disjunction is robust with respect to the negation of either of its disjuncts taken separately: which is precisely what the disjunction was meant to signal, if Jackson is right.[361]

For instance: reconsider Apple Spritzer Example 3 from Chapter 2. (I have already reconsidered Apple Spritzer Example 2 in section 4.3 of Chapter 4, and Apple Spritzer Example 1 in section 6.1 above.)

Example 26 (The Third Apple Spritzer Example 3 from Chapter 2 Reconsidered) Let again $W = \{w_1, w_2, w_3\}$: in w_1 the bottle of apple spritzer is in the fridge, in w_2 it is in the shopping bag, in w_3 it is not in either of these places. Let $P(\{w_1\}) = \frac{6}{10}$,

[361] The present account of subjective assertability could be made to approach Gricean assertability even more closely by replacing postulate (1) from above by means of (1'): X is (subjectively) Grice-assertable iff $X = B_W$, that is, when X coincides with the strongest believed proposition in the relevant context. In this sense, a proposition would be assertable for an agent in a context just in case it is the maximally specific information that is available to the agent in that context. The relativization to the context—and the partition of possibilities that it determines—should make sure the information in question is relevant and does not become overly specific. The Humean thesis HT^r would thus entail: if

- $B_W = \{w_1\} \cup \{w_2\} \cup \ldots \cup \{w_n\}$ (for $n > 1$), that is,

  the disjunction $\{w_1\} \cup \{w_2\} \cup \ldots \cup \{w_n\}$ *is (subjectively) Grice-assertable*

  (in line with the Gricean maxims of belief, relevance, and informativeness, or Quality, Relation, and Quantity),

then

- $P(B_W \mid \neg\{w_i\}) > r$ (for all $1 \leq i \leq n$), that is,

  the disjunction is robust with respect to the negation of any disjunct

  as intended by Jackson.

But I will not explore this option any further here.

$P(\{w_2\}) = \frac{3}{10}$, $P(\{w_3\}) = \frac{1}{10}$. Finally, let $B_W = \{w_1, w_2\}$. *Bel* and *P* satisfy the Humean thesis $HT^{\frac{1}{2}}$ ($r = \frac{1}{2}$).

My wife asks me about the bottle of apple spritzer. In line with my beliefs about the situation, I assert: 'The bottle of apple spritzer is either in the fridge or in my shopping bag'. In fact, this disjunction corresponds to my total belief in the relevant context. When $X = \{w_1\}$ and $Y = \{w_2\}$, it holds that (i) $Bel(X \cup Y)$, and (ii) $Poss(\neg X)$ and $Poss(\neg Y)$. By the Humean thesis, $P(X \cup Y|\neg X) = 0.75 > r$, and $P(X \cup Y|\neg Y) = 0.857\ldots > r$. From my point of view, it will therefore be useful to convey that disjunction to my communication partner: even if turns out that $\neg X$, the disjunction will remain likely enough, and the same is the case if it turns out that $\neg Y$. Finally, because of $Poss(\neg X)$ and $Poss(\neg Y)$, I do take these possibilities seriously.

Lewis (1986) supports the same Jacksonian view on the assertability of disjunction:

> I speak to you (or to my future self, via memory) in the expectation that our belief systems will be much alike, but not exactly alike [...] Maybe you (or I in future) know something that now seems to me improbable. I would like to say something that will be useful even so. [...] Let me say something... that will not need to be given up... even if a certain hypothesis that I now take to be improbable should turn out to be the case. (Lewis 1986)

It is interesting that Lewis suggests in this passage that what holds for assertion might also hold for belief itself, since belief may be viewed as a kind of 'assertion to one's future self (via memory)'.[362] Indeed, according to the present theory, rational belief satisfies the same stability constraints that also apply to rational assertion and (subjective) assertability.

6.3 Acceptance

By the Humean thesis from Chapter 2, rational belief is stable under certain instances of conditionalization. In section 3.3 of Chapter 3, we found rational belief and its stability to be restricted to a context, where the context in question involves an (implicit or explicit) choice of the underlying partitioning of the space of logical possibilities.

For instance, reconsider the three 'apple spritzer examples' from section 2.1 in Chapter 2, all of which got formally reconstructed in the meantime. The respective contexts concerned time spans that extended over

- an episode of decision-making (from the decision to get something to drink to the completion of the resulting action),
- an episode of suppositional reasoning (from supposing that the bottle is not in the fridge to the conclusion that it must be in the shopping bag),

[362] Recently, Douven (2010) defended, and worked out in detail, a similar view of belief as assertion-to-oneself, by which belief becomes subject to constraints similar to those that apply to assertion. But he does not combine this with considerations on stability.

- an episode of communication (from the first assertion in the dialogue to the last one).

Maybe it is even the case that the relevant contexts would govern periods of reasoning that go beyond that. But if, for some reason, much longer periods of time were to be based on one and the same partition of possibilities, then presumably that partition would have to be significantly larger or more fine-grained than those in my three examples: large enough for the agent to be able to respond rationally to the large range of pieces of evidence that she might encounter in such a longish episode of reasoning, and large enough for the great variety of suppositions that she might need to make in any such episode. By the findings from section 3.3, this would mean that the Lockean threshold for belief would have to be very close to 1 or otherwise the stability that is required of belief could not be maintained. So either a perfectly rational agent would have to be *very* cautious or the partitioning aspects of the relevant belief contexts must be sufficiently short-term in order to keep things simple enough. In short: according to the present theory, rational belief is stable in a context, but the respective contexts and the partitions of possibilities they determine will normally not remain invariant over an agent's long-term activities.[363]

But what if there is practical pressure that requires precisely that: a mental state that is capable of grounding an agent's actions much like belief does but which at the same time affords *long-term* stability? If the present theory is right, that mental state cannot be *belief*, at least in normal circumstances. On the probabilistic side, the required stability under conditionalization could only be achieved by assigning a proposition probability 1 or something very close to that. That is because only a proposition of (almost) probability 1 is (almost) perfectly stable even on a large space or partition of possibilities: only such a proposition will retain that high probability under (almost) all conditionalizations on propositions of positive probability. (Compare the Certainty Proposal (e) in section 2.2.) But other than logical laws, analytic propositions, and evidence that is firm enough, believed propositions normally do not have extreme subjective probabilities like that.

However, there might be a different, though closely related, mental state that might fill the functional role that all-or-nothing belief cannot fill. I am going to argue that the corresponding mental state has been discussed before in the relevant literature, and it is usually called: *acceptance*. Before I reconstruct acceptance in a way that will indeed supply accepted propositions with a perfectly stable probability of 1, let me briefly review the existing literature on acceptance.[364]

The distinction between belief and acceptance has been drawn and discussed by many different authors in many different areas: in the philosophy of action

[363] I am grateful to Philip Ebert for a discussion on this point.

[364] In parts of formal epistemology, 'acceptance' is also used as a technical term that can stand for virtually any kind of 'belief-like' state, whereas the term 'acceptance' as understood in the present context has much more specific functional properties.

(Bratman 1999), philosophy of mind and epistemology (Lehrer 1983, Stalnaker 1987, Cohen 1989, 1992, Engel 1998, 2000, Cresto 2010), philosophy of science (van Fraassen 1980, Maher 1990), and many more.[365]

For instance:

> The three of us need jointly to decide whether to build a house together. We agree to base our desiderations on the assumption that the total cost of the project will include the top of the estimated range offered by each of the subcontractors. We facilitate our group deliberations and decisions by agreeing on a common framework of assumptions. We each accept these assumptions in this context, the context of our group's deliberations, even though it may well be that none of us believes these assumptions or accepts them in other, more individualistic contexts. (Bratman 1999, pp. 24–5)[366]

The same act of accepting a proposition or of 'assuming it' or of 'taking it for granted' is also described in the following passage:

> In planning my day—a June day in Palo Alto—I simply take it for granted that it will not rain even though I am not certain about this. If I were instead figuring out at what odds I would accept a monetary bet from you on the weather I would not simply take it for granted that it will not rain. But in my present circumstances taking this for granted simplifies my planning in a way that is useful, given my limited resources for reasoning. (Bratman 1999, p. 22)

Stalnaker (1987) regards belief as a special case of the broader group of acceptance attitudes, where:

> To accept a proposition is to treat it as a true proposition in one way or the other—to ignore, for the moment at least, the possibility that it is false. (Stalnaker 1987, p. 79)

Cohen (1992) maintains a similar distinction, though without subsuming belief under acceptance:

> To accept that *p* is to have or to adopt a policy of deeming or postulating that *p*—i.e. of including that proposition or rule among one's premisses for deciding what to do or think in a particular context, whether or not one feels it true that *p*. (Cohen 1992, p. 4)

Finally, while van Fraassen (1980) holds that a scientist's acceptance of a scientific theory also involves some beliefs (in the theory being empirically adequate), in other respects acceptance goes beyond these beliefs:

> To accept a theory is to make a commitment, a commitment to the further confrontation of new phenomena within the framework of that theory, a commitment to a research programme, and a wager that all relevant phenomena can be accounted for without giving up that theory. (Van Fraassen 1980, p. 88)

365 For surveys of this kind of literature, see Engel's introduction to Engel (2000), Paglieri (2009), de Vries and Meijers (2013), and Frankish (2004, s. 3.4). Engel (2000) collects various very helpful essays on different accounts of belief vs acceptance.

366 Bratman (1987) still speaks of belief at places at which Bratman (1999) suggests speaking of acceptance instead.

In spite of all the subtle differences between these accounts of acceptance vs belief, what all of them have in common is this: *the act of accepting a proposition consists in taking that proposition as a premise for certain practical purposes.* That acceptance variant of 'taking-as-a-premise' is much like assuming a proposition in suppositional reasoning, except that acceptance is not carried out in the same hypothetical 'offline' manner. Rather, the agent accepts a proposition 'online': once the act of accepting a proposition has led to the corresponding state of acceptance, she acts in that state upon that premise, she uses the premise in reasoning and decision-making, and she is committed to keep doing so for the purpose that made her accept the proposition in the first place. These purposes might be: to facilitate social coordination, as in Bratman's house-building case; to simplify reasoning in the face of limited resources, as in Bratman's day-planning example; to exploit a framework for scientific theorizing and puzzle-solving, as in van Fraassen's case of theory acceptance; and so on. In fact, the relevant purposes could be *anything* really: in particular, they do not necessarily involve any kind of truth-aiming, which is in contrast with the aiming-at-the-truth feature that is constitutive of belief. (See Assumption 2 in section 1.1 and its discussion in Chapter 5.) Accordingly, it is perfectly possible to accept a proposition that one does not believe to be true, as maintained in some of the examples above.

On the other hand, what acceptance and belief do have in common are their action-determining roles and their stability: Bratman (1987, 1999) emphasizes the stability that is required for intention, planning, and action, and Bratman (1999) highlights the role that acceptance plays for this; van Fraassen's (1980) talk of 'commitment' may be taken to entail similar properties. The stability in question may need to extend beyond short-term projects—the aim might be to accept a proposition throughout the processes of: building a house; planning for, and living through, a day in Palo Alto; and carrying out a research programme. In all of this, the acceptance of a proposition is context-relative, where the context is given partially by the purpose that the acceptance is meant to serve. In the examples above these contexts were: the building-of-a-house context, the day-planning context, and the scientific-research-programme context. The accepted proposition is only taken to be a premise inside of its context of acceptance.

Clearly, such (states of) acceptance and belief are very similar to each other. Most of the authors above still drive a wedge between acceptance and belief, and they do so for two reasons: belief aims at the truth while acceptance does not (necessarily); and belief is often thought to be context-*in*sensitive while acceptance is context-sensitive. As we have seen in earlier chapters, the theory of rational belief that was developed in this essay must ascribe an even greater degree of similarity to them: though *degrees* of belief may be independent of context, categorical belief turns out to be context-sensitive, too (as argued in particular in section 3.3 of Chapter 3).

In view of these similarities, it is not surprising that some authors defend theories of all-or-nothing belief that might just as well—and, as I think, maybe even more appropriately—be called theories of *acceptance*. For example, according to Levi (see e.g. Levi 1980, p. 76, and Levi 1996, p. 167), one's all-or-nothing corpus of knowledge or full belief imposes a constraint on one's credal state (one's set of credal probability measures): the constraint being that every sentence in the corpus must receive probability 1 in all of one's credal probability measures. I suggest to think of this in the way that all of the sentences in the corpus are thereby *accepted*.[367] Or consider Frankish (2004, 2009), who takes belief to be given by acceptance-like premising policies. While Frankish distinguishes belief from acceptance (Frankish 2004, section 3.4), he only does so for two reasons: he thinks that we can accept something at will for prudential reasons, while we cannot believe at will; and acceptance is context-relative, whereas belief is not. I have already mentioned that the last point does not stand, at least according to the present theory. The status of the first point is unclear, at best: on the one hand, perhaps one cannot simply accept a proposition at will either. Perhaps one can only do so in non-voluntary response to prior evidence, intentions, and deliberation; and much the same might be true of belief. On the other hand, Frankish himself acknowledges that 'while it is true that we cannot choose to believe anything we like, it is arguable that there is a sense in which belief can be responsive to pragmatic reasons' (Frankish 2009, p. 86); so there is not much of a difference between belief and acceptance after all. Ultimately, Frankish (2009, p. 86) concludes that 'flat-out beliefs form a subset of acceptances'—they are acceptances that are also 'truth-critical with respect to premises' (Frankish 2009, p. 87). So Frankish's theory of belief may indeed be viewed as a theory of a special kind of acceptance. Since what is accepted (and hence gets probability 1) in Levi's theory is already believed (is part of the corpus), and since Frankish presupposes that belief-acceptance must still be 'truth-critical', my best shot at their accounts within the boundaries of my own framework might be the case of *accepted belief* that I am going to deal with below: cases in which believed propositions are also accepted.[368]

It is time to make things formally more precise now. I suggest that the act of *accepting a proposition X* in a context consists in taking X as a premise and acting upon that premise in that context. Within the context, the agent's actual belief state is modified by accepting X as if X had been supposed or learned. However, unlike suppositions, the premise has the same 'online' action-determining consequences that learning X would have; unlike learning, the act of accepting a proposition is not necessarily aiming at the truth (though it may be). In line with my theory of belief, I will analyse acceptance on both the categorical and the numerical scales.[369]

367 It is worth noting that Levi does *not* regard corpora to be context-insensitive either.

368 Similar considerations apply to Weatherson's (2005) preferential account of belief as treating as true for the purposes of practical reasoning, Fantl and McGrath's (2009) pragmatic view of belief, and Ross and Schroeder's view of belief as a defeasible disposition to treat a proposition as true in reasoning. In the terminology of my theory, they might best be viewed as theories of acceptance or accepted belief.

369 Not everyone allows for acceptance to come in degrees—but e.g. Engel (1998) does.

I am not going to characterize acceptability *conditions* now, since the conditions under which a proposition is acceptable for an agent in a context can be quite arbitrary. Instead I will turn to the mental *act* of accepting a proposition more directly. My characterization of such acts will be broad enough to apply also to truth-conducive acts, such as *learning* a proposition, but it will not be restricted to them. But, as usual, I will restrict myself to perfectly rational agents.[370]

This leads me to the following proposal. Let W be a non-empty set of logically possible worlds, or partition cells of such worlds, as given by the context (compare section 3.3). Let *Bel* be a perfectly rational agent's conditional belief set (as described in Chapter 4) at the time and given relative to the context; the members of *Bel* are pairs of propositions over W, that is, subsets of W. Let P be the same agent's degree-of-belief function at the same time; P assigns numerical degrees of belief to the subsets of W. Finally, let X be a set of worlds in W, such that $P(X) > 0$ (hence conditionalization on X will be well-defined). Then my proposal is:

- A perfectly rational agent's act of accepting X on a numerical scale (in the given context) consists in:
  - Determining a degree-of-acceptance measure P_X^{Acc} that is given by $P(\cdot|X)$. So P_X^{Acc} is the probability measure that is defined by, for all Y:

$$P_X^{Acc}(Y) = P(Y|X).$$

  - Acting upon the so-determined degree-of-acceptance measure P_X^{Acc} (within that context).
- A perfectly rational agent's act of accepting X on a categorical scale (in the given context) consists in:
  - Determining the (unconditional) acceptance set Acc_X that is given by $Bel(\cdot|X)$. So Acc_X is meant to be the set of propositions that the agent accepts as a consequence of accepting X. The set is defined by, for all Y:

$$Acc_X(Y) \text{ iff } Bel(Y|X).$$

  - Acting upon the so-determined acceptance set (within that context).

Finally, an agent's act of accepting X (in a context) consists in that agent's act of accepting X on both a numerical and a categorical scale (in that context).

Accepting X in this sense has the following consequences, which I am going to spell out using the terminology and the postulates of Chapter 4. First of all, the degree-of-acceptance measure P_X^{Acc} assigns probability 0 to $\neg X$ and probability 1 to X. In this way, all of the $\neg X$-possibilities are ruled out, or perhaps simply 'ignored' (as

[370] I will leave open whether such perfectly rational agents will ever be *required* to accept propositions in order to simplify their reasoning or to facilitate it in some other way. I will describe what acceptance amounts to *if* they accept a proposition, which should be interesting enough to be studied at least for the sake of us real-world agents who strive to approximate such perfectly rational beings.

in the quote by Stalnaker above), once X is accepted.[371] But acceptance does not merely put probabilities aside by raising the probability of X to 1: by the same move other propositions will simply change their probabilities without these probabilities becoming 1. Even in a state in which one accepts X for whatever reasons, one might still want to draw inductive inferences on the basis of such acceptances, and one can do so rationally by means of P_X^{Acc}.

On the all-or-nothing side, the set of doxastically possible worlds (the least believed set), B_W, is transformed into the set B_X of worlds that are possible given X: B_X is at the same time the least believed set conditional on X and the least accepted proposition in Acc_X. (All of that follows from the postulates in Chapter 4.) Postulate BP2* from section 4.2.3 in Chapter 4 yields that B_X is non-empty (since $P(X) > 0$). Therefore, the set Acc_X of accepted propositions, which is is generated by its least member B_X, is both closed under logical consequence and consistent. As follows from postulate B1* in section 4.2.3, B_X is a subset of X: so X is of course a member of Acc_X itself. But accepting X might make some propositions sufficiently plausible that are logically stronger than X: if so, B_X will be a *proper* subset of X. In any case, by postulate BP1^{r*} from section 4.2.3, $P(B_X|X) = P_X^{Acc}(B_X) > r$. Hence, the agent gives a high enough degree-of-acceptance to the strongest proposition B_X that she categorically accepts as a consequence of accepting X. Furthermore, in the case in which B_X is a large finite set or even infinite, the Humean thesis entails that the probability of B_X must be close or even identical to 1 (compare section 3.3 and Theorem 18 in section 4.2.2): in that case B_X and X will be almost the same proposition up to a set of very small probability, or they will be the same proposition up to a set of probability 0. (In set-theoretic terms: their symmetric difference $(B_X \setminus X) \cup (X \setminus B_X)$ will have low probability or even probability 0.)

In any case, the results about robustness persistence from Chapter 4 (see Observation 17) entail that if P and Bel satisfy the Humean thesis HT^r, and if additionally $Poss(X)$, then also P_X^{Acc} and Acc_X will satisfy the Humean thesis HT^r. Even when the functional role of P_X^{Acc} is not that of a degree-of-belief function, it will still hold that B_X—the least accepted proposition—is P_X^{Acc}-stable^r. In this sense, the resulting theory of acceptance is again a stability theory. Furthermore, since $P_X^{Acc}(X) = 1$, any additional conditionalizations of P_X^{Acc} that might be required for reasoning in the respective context will preserve X's maximal probability. The accepted proposition keeps the same perfectly stable probability of 1 as long as the agent is willing to reason within that context of acceptance: until the house is built or the building project is cancelled, until the end of that sunny day at Palo Alto, or until the underlying research programme is abandoned in favour of a new one. Up to that point in time, the agent

371 The case of accepting X by ignoring $\neg X$ is much as what is emphasized in the debate on knowledge and contextualism: 'S *knows* that P iff S's evidence eliminates every possibility in which not-P—Psst!—except for those possibilities that we are properly ignoring' (Lewis 1996, p. 554).

may (explicitly or implicitly) 'stick' P_X^{Acc} and Acc_X into her quantitative and qualitative decision theories (see section 6.1) and thus translate her acceptance of X into action. In this respect, acceptance behaves like belief again.

But one should not lose sight of the fact that P_X^{Acc} may differ from the agent's degree-of-*belief* function P, just as Acc_X may differ from the agent's unconditional *belief* set, and the transition from P to P_X^{Acc} and from Bel to Acc_X may well result from the agent's decision to merely ignore $\neg X$ for some non-epistemic or not truth-related reasons. On the numerical side, the 'mistake' that is made by $P_X^{Acc}(X)$, if compared to the agent's actual estimate $P(X)$ of X's being true, is not less than $P_X^{Acc}(X) - P(X) = 1 - P(X) = P(\neg X)$, which may well be close to 1 (though short of 1, assuming $P(X) > 0$). Acceptance may completely distort the truth, while belief must not (by its truth-aiming nature). As the examples are meant to show, an agent may still be willing to buy into the epistemic error of accepting X as long as the practical rewards for doing so look promising enough.[372]

In some cases, ignoring $\neg X$ may also be in the service of a large-scale project that is truth-aiming in the long run, such as scientific inquiry. But still the result of accepting X may differ from truth-aiming belief proper. As Stalnaker (1987) formulates this:

> When is it reasonable to accept something one believes is false? When one believes that it is *essentially* true, or close to the truth—as close as one can get, or as close as one needs to get for the purpose at hand. [...] Accepting a certain false proposition may greatly simplify an inquiry, or even make possible an inquiry not otherwise possible, while at the same time it is known that the difference between what is accepted and the truth will have no significant effect on the answer to the particular question being asked. When a scientist makes idealizing assumptions, he is accepting something for this kind of reason. [...] The scientist does not, of course, *believe* the propositions he accepts, but he acts, in a limited context, as if he believed them in order to further his inquiry. (Stalnaker 1987, p. 93)

Finally, in yet a different kind of case, one may even accept a proposition that one believes already, in which case acceptance of X inherits some of the truth-aiming properties of belief in X: I will return to this case in a moment.

I do not claim that this stability theory of acceptance can do justice to all of the accounts of acceptance vs belief that I mentioned before.[373] But I do think it does get pretty close to them while still remaining compatible with the stability theory of belief that was developed before. It should also be quite obvious now how this theory of binary and numerical acceptance could be extended into further directions: for instance, just as the previous section 6.2 built an account of subjective assertability

372 William James's position (James 1896) in the famous Clifford–James debate might perhaps be reconstructed as a proposal about acceptance, while William Kingdon Clifford would be talking about belief *proper*. I am grateful to Johannes Brandl for bringing that connection to my attention.

373 For instance, Levi's complex theoretical framework differs too much from mine in order to be able to do justice to the former within the latter.

on top of our theory of belief, a similar account of 'acceptance' assertability could be added to the explication of acceptance above.[374] And so on. I will not expand on this.

I should also emphasize that the context-sensitivity of acceptance as explicated here differs from that of belief as highlighted previously in section 3.3. First, it is perfectly consistent with my theory of belief that a perfectly rational agent's degree-of-belief function P is independent of context; and secondly, an agent's all-or-nothing belief set *Bel* was found to be context-dependent only in so far as belief is sensitive to partitionings of the underlying space of logical possibilities and to the choice of (Lockean or Humean) thresholds—recall the discussion in section 3.3. Since an agent's all-or-nothing acceptance set Acc_X is defined in terms of *Bel*, it inherits that kind of context-sensitivity. But the context-relativity of acceptance is not exhausted by this: both the agent's degree-of-acceptance function P_X^{Acc} and her all-or-nothing acceptance set Acc_X are also sensitive to *the choice of the proposition being accepted*. Therefore, contexts of acceptance need to supply one parameter more than contexts of belief do—the proposition X that is to be accepted—and degrees of acceptance are context-dependent, by depending on the parameter X, while degrees of belief are not (or it is at least consistent with the present theory to view them as such). Compare the end of section 3.3 for more discussion on this point.

Let me now take one final step of bringing acceptance even closer to belief. Let us assume that the proposition X that is to be accepted by an agent in a context is already *categorically believed* by the same agent in the same context. Such cases of *accepted belief* are indeed discussed in the literature on acceptance (e.g. by Engel 1998), and they are perfectly compatible with the account from above: one only needs to add the assumption that $Bel(X)$.

Here is what is going to happen: on the side of categorical acceptance, *nothing*. If X is believed, then $B_W \subseteq X$, in which case the Preservation postulate B6 entails that $B_X = X \cap B_W = B_W$. In other words: $Acc_X = Bel$. Accepting a believed proposition yields the agent's original belief set again on which the agent is prepared to act anyway. *On the categorical scale, accepted belief coincides extensionally with belief.* On the other hand, accepting a believed proposition may well have effects with respect to numerical acceptance: if X is believed but $P(X) < 1$ (which we know to be compatible with the Humean thesis), $P_X^{Acc}(X)$ will be 1 and thus differ from $P(X)$. However, because $P(X) > r$ by the Humean thesis HT^r (where $\frac{1}{2} \leq r < 1$), the 'mistake' made by $P_X^{Acc}(X)$ will at least be reasonably bounded: $P_X^{Acc}(X) - P(X) = P(\neg X) < 1 - r \leq \frac{1}{2}$. Although still diverging from belief proper, accepted belief inherits from belief at least a weakened form of 'aiming-at-the-truth'. Such a simultaneously believed and accepted proposition X might even coincide with the agent's strongest believed proposition B_W itself, in which case the degree-of-acceptance measure P_X^{Acc} $(= P_{B_W}^{Acc})$ would end up

[374] This might also involve a Stalnakerian view of assertion as an act that makes the participants of a conversation accept (or take for granted) a proposition, in which case the proposition in question will be added to the 'common ground' among the participants in the conversation (cf. Stalnaker 1979, 2002).

assigning probability 1 to *all* believed propositions in the context. In that sense, an agent's accepting her strongest believed proposition B_W may be said to coincide with the agent's accepting *all of her beliefs at the same time* (in the context in question).

There might even be historical precursors to this. Locke's account of belief is usually interpreted in terms of the Lockean thesis for belief (as defended by Foley 1993), but with a little bit of charity one might just as well read an account of *accepted belief* into the following passage—an account along the lines sketched before:

> most of the Propositions we think, reason, discourse, nay act upon, are such, as we cannot have undoubted Knowledge of their Truth: yet some of them border so near upon Certainty, that we make no doubt at all about them; but *assent* to them firmly, and act, according to that Assent, as resolutely, as if they were infallibly demonstrated . . . (Locke, *Essay*, book IV)

The probability of a believed proposition X might 'border near upon Certainty'; e.g. $P(X)$ might be 0.9. By accepting X, its degree of belief is pushed to a degree of acceptance of 1, upon which the agent will act as if X 'were infallibly demonstrated'.

I will conclude this section with an interesting corollary to what seems to be another plausible assumption on accepted belief. Let B_W again be an agent's strongest believed proposition (in a context). Here is one possible course of events:

- The agent accepts her strongest believed proposition B_W (or equivalently, all of her beliefs) in the same context. Afterwards, in that state of acceptance, she receives a new piece of evidence X (where I assume $P(X|B_W) > 0$).
  On the numerical side, this means: by accepting B_W, the agent determines her degree-of-acceptance measure $P^{Acc}_{B_W}$. Afterwards, she updates $P^{Acc}_{B_W}$ on X, which yields the probability measure $[P^{Acc}_{B_W}]_X$, for which it holds: $[P^{Acc}_{B_W}]_X(Y) = P^{Acc}_{B_W}(Y|X) = \frac{P^{Acc}_{B_W}(X \cap Y)}{P^{Acc}_{B_W}(X)} = \frac{P(X \cap Y|B_W)}{P(X|B_W)}$, which is equal to

$$P(Y|X \cap B_W).$$

Here is another possible course of events:

- The agent receives first the new piece of evidence X (where I assume $P(X) > 0$). Afterwards, in that updated state of belief, she accepts her new logically strongest believed proposition, that is, B_X (in the same context); or in other terms: she accepts all of her beliefs at the time (where I assume that $P(B_X|X) > 0$).
  Formally: by learning X, the agent's degree-of-belief function becomes P_X (that is, P conditionalized on X), and her logically strongest believed proposition becomes B_X. Afterwards, she accepts B_X, which on the numerical side yields the probability measure $[P_X]^{Acc}_{B_X}$, for which it holds: $[P_X]^{Acc}_{B_X}(Y) = P_X(Y|B_X) = \frac{P_X(B_X \cap Y)}{P_X(B_X)} = \frac{P(B_X \cap Y|X)}{P(B_X|X)} = P(Y|B_X \cap X)$, which, because of $B_X \subseteq X$, is equal to

$$P(Y|B_X).$$

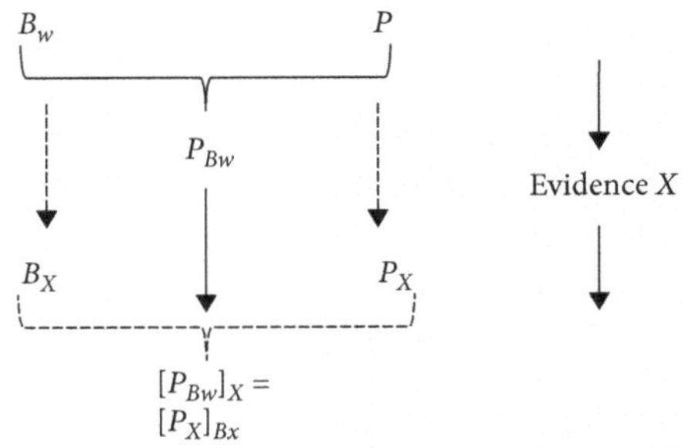

Figure 6.5. Accepted-belief and update commuting

Finally, it seems plausible to assume that both courses of events should rationally lead to one and the same outcome: it should not matter whether one accepts all of one's beliefs first and then learns X, or vice versa. *Accepting belief* and *update on evidence* ought to commute. See Figure 6.5 for an illustration.

If that is granted, it follows that for all propositions Y,

$$P(Y|X \cap B_W) = P(Y|B_X),$$

which, by the axioms of probability, entails that $X \cap B_W$ is identical to B_X up to a set of probability 0. In other words: their symmetric difference $((X \cap B_W) \setminus B_X) \cup (B_X \setminus (X \cap B_W))$ has probability 0.[375] Or again put differently: up to a zero set, the Preservation principle B6, that is, if $X \cap B_W \neq \varnothing$ then

$$B_X = X \cap B_W,$$

from section 4.2.2 in Chapter 4 must be the case.

Note that I did not have to assume the Humean thesis for this conclusion nor the Preservation principle itself: it was only required that there was a strongest believed proposition B_W before the update on X, a strongest believed proposition B_X after the update on X, and that numerically both acceptance and update proceed by conditionalization. This yields another argument for the Preservation principle from Chapter 4.[376]

[375] For contradiction, assume that $P(((X \cap B_W) \setminus B_X) \cup (B_X \setminus (X \cap B_W))) > 0$: then either $P((X \cap B_W) \setminus B_X) > 0$ or $P(B_X \setminus (X \cap B_W)) > 0$ (or both). In the first case, let $Y = ((X \cap B_W) \setminus B_X)$: then $P(Y|X \cap B_W) > 0$, while $P(Y|B_X) = 0$, which contradicts $P(Y|X \cap B_W) = P(Y|B_X)$. Analogously in the other case. (I am making the same assumptions as before about the relevant conditional probabilities being well-defined.)

[376] One can show further niceties about acceptance: as mentioned already in n. 249, given the postulates from Chapter 4, the measure P^* that is defined by $P^*(Y|X) = P(Y|B_X)$ is a so-called Popper function or primitive conditional probability measure. By this definition our P-stable^r sets are being transformed into so-called belief cores as being given by P^* and as described by van Fraassen (1995), Arló-Costa (2001), and Arló-Costa and Parikh (2005). What these authors regard as degrees of belief given X may therefore be taken to coincide with my degrees of belief given the result of *accepting the set of most plausible worlds in X* (or to coincide with the result of conditionalizing P on B_X). Such 'degrees of belief given acceptance' might

6.4 The Preface Paradox Reconsidered

An author advances a great number of sentences $A_1, \ldots, A_n$ in the main chapters (or the main part) of, say, an academic book, but then she also admits in the preface that some of these sentences will be false: it is not the case that $A_1 \wedge \ldots \wedge A_n$. The author seems to be holding logically incompatible beliefs, and yet she seems to be rational in doing so. That is the gist of David Makinson's well-known Paradox of the Preface (cf. Makinson 1965).[377]

In section 3.5 I analysed the Preface Paradox to the extent to which it resembled the Lottery Paradox. The agent may regard each statement A_i in the main part of the book to be likely, and if the relevant context determines the coarse-grained partition $\{A_i, \neg A_i\}$ of possibilities, she may well believe A_i to be true in that context. On the other hand, the context of the preface may determine a different partition of possibilities in which she may believe the negation of the conjunction of all statements in the main part, which she may also regard as likely. All of that is compatible with the Humean thesis on belief from Chapter 2, the exact details depending on the author's degree-of-belief function.

But there are also differences between the Lottery and the Preface cases. In particular, publishing a book seems to come with some sort of commitment to the truth of the statements that are contained in it, whereas there does not seem to be a corresponding commitment to all tickets losing in the lottery case. Which led to the following open questions: what kind of commitment does the author express by asserting all of the statements in the book as a whole? What kind of mental state corresponds to the author's presentation of her theory in its entirety? It is about time to address these questions.

Compare the situation in the Preface case with a scientific lab publishing a database of a great many experimental results $A_1, \ldots, A_n$. Do the members of the lab believe the conjunction of all of the published data to be true? Of course not; they know that there will always be statistical outliers, to say the least, and accordingly they are happy to admit that it is not the case that $A_1 \wedge \ldots \wedge A_n$. And no one would have expected them to believe the conjunction of all of the data in the first place.

What, then, is asserted by the lab's act of publishing a database of a great many data $A_1, \ldots, A_n$? I submit it is the proposition that *the vast majority of the data are correct*,[378] where the exact meaning of 'vast majority' is determined partially by the context. And in terms of mental attitude, in a context in which the issue is the publication of the database as a whole, it is the *belief that the vast majority of the data are correct* that is expressed.

also be closely related to what Wedgwood (2012) calls 'practical credences', as opposed to his 'theoretical credences' which might coincide with *P*.

[377] For recent references on the Preface Paradox, see e.g. Christensen (2004); and for a discussion of the differences between the Preface and the Lottery Paradoxes, see Foley (1993).

[378] More precisely, I should say: *approximately* correct; but I will have to leave this to one side.

Let me make this more precise: abbreviating $A_1, \ldots, A_n$ by 'T', and given a contextual parameter m that is some natural number greater than or equal to 1, and less than or equal to n (but sufficiently close to n), let $S_m(T)$ be the following 'statistical weakening' of T:

$$\bigvee_{I \subseteq \{1,\ldots,n\}, |I|=m} \quad \bigwedge_{i \in I} A_i.$$

Thus, $S_m(T)$ is the disjunction ('$\bigvee$') of all conjunctions ('$\bigwedge$') of length m of sentences from $A_1, \ldots, A_n$. For instance, in the—unrealistic—case in which $n = 3$ and $m = 2$, the sentence $S_m(T)$ would be

$$(A_1 \wedge A_2) \vee (A_1 \wedge A_3) \vee (A_2 \wedge A_3).$$

That sentence says that at least $m = 2$ out of the $n = 3$ statements A_1, A_2, A_3 are true; or: most of A_1, A_2, A_3 are true.

When n is large, and the proposition to be expressed is that at least about 90 per cent of all sentences in the book are true, then m would be such that $\frac{m}{n}$ is about 0.9 (or m is about $0.9n$). And so on. My claim is now that one ought to distinguish between what is uttered or published in terms of a great many declarative sentences and what is thereby asserted: the proposal is that by publishing T, the lab asserts that $S_m(T)$ for some contextually determined m. And the same holds, mutatis mutandis, for the author in the case of the Preface Paradox. Accordingly, just as no one expects the lab personnel to believe the conjunction of all published data, no one should expect the author to believe the conjunction $\bigwedge T$ of all of the sentences in her book. Instead, one should take her to believe just a 'statistical weakening' $S_m(T)$ of T.

A different way of stating the same proposal is this: by publishing her book, the author asserts, as it were, that the frequency or statistical probability of a claim in the main part of the book being true is high (such as at least 90 per cent or 0.9); $S_m(T)$ is just a way of conveying that statistical probability. However, when putting things in these terms, one needs to be careful: first of all, the author is not literally speaking about *frequency* in her book (unless that is amongst her actual topics): $S_m(T)$ still talks about whatever $A_1, \ldots, A_n$ are talking about. Indeed, 'the frequency of a claim in the main part of the book being true is high' is just a handy paraphrase of $S_m(T)$, which is a construction on $A_1, \ldots, A_n$. Secondly, one should not mix up statistical probabilities with the author's degrees of belief or *subjective* probabilities: while the statistical probability in question is a probability of an ensemble of sentences (the set of true sentences in the book) relative to a larger reference class of sentences (the set of all sentences in the main part of the book), the author's subjective probabilities are assigned to (the propositions expressed by) single sentences, such as to A_1 or, indeed, to $S_m(T)$ itself.

The proposal is not meant to affect our everyday oral practice of assertions: if one utters a single declarative sentence A, then $S_m(A)$ will just be A again: one asserts that A. Even for a couple of consecutive utterances, one may still take the speaker perfectly seriously in expressing their conjunction. However, as illustrated very nicely by the Paradox of the Preface, one should no longer do so once an author makes *a great many* consecutive utterances, as it were, *in one fell swoop*, and when the context is such that the attention is directed towards this 'mass utterance' taken as a whole. The linguistic acts of publishing a bulk of data or a book come with different conversational implicatures than the linguistic act of saying a single sentence (or saying one sentence after the other, but with the focus just on one at a time). When one utters a handful of declarative sentences, the default is to mean their conjunction; not so in the case of 'mass utterance', where the sheer amount of information displayed signals the author's fallibility.

Accordingly, if we take assertion as a guide to belief—as it seems plausible to do, in line with section 6.2—then analogous claims apply to belief: by uttering the single sentence A in a sincere assertoric mode, the speaker expresses her belief that A, and, normally, *in that context* she does believe that A. That is just as required by principle (1) in section 6.2. However, by publishing a great number of sentences $A_1, \ldots, A_n$ *jointly*, the author only expresses her belief that $S_m(T)$, and, normally, *in that context* she does believe only that $S_m(T)$. Just as uttering A_i may contribute differently to what is asserted, depending on whether A_i is taken by itself or whether it is part of a larger ensemble, also the respective beliefs differ in the corresponding contexts. It is one consequence of this view, therefore, that belief is relative to context. But that should not come as a huge surprise any more given what we found to be the case in previous chapters (especially in section 3.3 of Chapter 3).

Now I want to show that the proposal seems to tick the relevant boxes as regards the Paradox of the Preface.

First of all, it explains the paradoxical impression that is left by the story: the tension is between an utterance of a single A_i *as taken by itself*—by which the author would assert and believe that A_i—and an utterance of A_i *as part of the joint assertion* of $A_1, \ldots, A_n$—where by means of $A_1, \ldots, A_n$ the author asserts and believes that $S_m(T)$. In the case in which a database or a book is published, the whole is normally more salient than any of its proper parts, which is why what is asserted is $S_m(T)$.[379] However, previously, prior to publishing, when the author might have focused just on a single A_i, when she might have asked herself 'Shall I believe A_i?', and when she ultimately answered that question affirmatively, then: by that very linguistic act she did in fact assert that A_i, and she did express her belief in the truth of A_i. As Makinson (1965, p. 205) formulates it: '*Given each one of these* [my emphasis]', the author 'believes that it is true'. The usual benefits of a contextualist proposal apply: what seemed to

[379] This is not far from Makinson's (1965) own diagnosis according to which one ought to distinguish between the rationality of the belief in a single A_i and the rationality of the belief set $\{A_1, \ldots, A_n\}$ as a whole.

be a conflict of intuitions gets disentangled once the contextual boundaries are made transparent.

Secondly, clearly, the author's belief in $S_m(T)$ is logically compatible with the belief in the negation of $\bigwedge T$ as expressed in the preface, as long as the realistically plausible condition $m < n$ is satisfied.

Thirdly, also a high degree of belief in $S_m(T)$ may be expected to be logically compatible with the author's high degree of belief in the negation of $\bigwedge T$. For simplicity (though unrealistically), assume $A_1, \ldots, A_n$ to be probabilistically independent of each other, and suppose that the author believes each of $A_1, \ldots, A_n$ with a subjective probability of 0.9. Then by the law of large numbers, for sufficiently large n, the author is justified in assigning a high degree of belief to the proposition that about 90 per cent of $A_1, \ldots, A_n$ are true. When m is about $0.9n$, this means that the author is rationally permitted to assign a high degree of belief to $S_m(T)$. At the same time, she will also believe the negation of $\bigwedge T$ to the very high degree of $1-0.9^n$. No contradiction arises.

Fourthly, while the proposal in this section does not *entail* that the author's rational beliefs—as expressed in the book—are closed under logical consequence, it is *compatible* with logical closure within a context: say, by jointly saying $A_1, \ldots, A_n$ in the main part of her book, the author does assert that $S_m(T)$ and she does express her belief that $S_m(T)$ in that context; then, if what she rationally believes is closed deductively, she must also rationally believe in the same context everything that follows logically from $S_m(T)$. But that does not mean that the author would be rationally committed to believing anything contradictory or anything that has a low probability. In particular, the proposal does not entail that by saying $A_1, \ldots, A_n$, the author would be rationally committed to believing $A_1 \wedge \ldots \wedge A_n$: for, according to the proposal, writing $A_1, \ldots, A_n$ jointly does not entail believing each of $A_1, \ldots, A_n$ in that context (it only entails believing $S_m(T)$), and hence, even if rational belief is closed under logical consequence, advancing $A_1, \ldots, A_n$ does not entail rationally believing $A_1 \wedge \ldots \wedge A_n$. Of course, this is good news for the theory in *this* book, since my stability account of belief does entail the logical closure (and consistency) of belief. Indeed, all of these findings are consistent with the Humean thesis on belief.[380]

Fifthly, it still makes good sense for the author—as it does for the lab in the example from before—to publish $A_1 \wedge \ldots \wedge A_n$ rather than $S_m(T)$ itself. This is for pragmatic reasons: as mentioned above, the parameter m is determined by the context, which is why it might be neither possible nor advisable for the author to state $S_m(T)$ directly. The right value of 'm' might not be accessible to the author herself (how many exactly are 'the vast majority' from the viewpoint of the author?), and readers might want

380 For instance, consider a partition of possibilities into three coarse-grained possibilities: $\bigwedge T$; $S_m(T) \wedge \neg \bigwedge T$; and $\neg S_m(T)$. ($\bigwedge T$ logically entails $S_m(T)$.) Let $P(\bigwedge T) = 0.01$, $P(S_m(T) \wedge \neg \bigwedge T) = 0.89$, $P(\neg S_m(T)) = 0.1$, and let the strongest believed proposition B_W be given by $S_m(T) \wedge \neg \bigwedge T$: then the resulting logically closed belief set *Bel* satisfies, together with P, the Humean thesis $HT^{\frac{1}{2}}$ (by the Outclassing Condition in Appendix B).

to assign the value differently (how many exactly are 'the vast majority' from the viewpoint of a reader?). In line with the Gricean Cooperative Principle, the most helpful and efficient way of conveying $S_m(T)$ for the appropriate m might well be to publish T.

Sixthly, as it is the case with other conversational implicatures, also the act of implicating $S_m(T)$—and hence, in this case, of meaning *less* than what had been said (that is, T)—can be cancelled. For instance, there might be precisely one claim, A_1, about which the author cares the most, she might think that she has especially good reasons to believe it to be true, and she might say all of that somewhere in the book. Then, by publishing the book, the author might assert and express her belief that

$$A_1 \wedge \bigvee_{I \subseteq \{2,\ldots,n\}, |I|=m-1} \; \bigwedge_{i \in I} A_i.$$

In the extreme case, she might even be saying explicitly in the preface: 'I honestly and very firmly believe the conjunction of all the claims in this book.' In that case, one might well understand the author in the way that she asserts and believes that $A_1 \wedge \ldots \wedge A_n$. (Of course, in that case she will not say in the preface that there is a mistake in the book, and thus no Preface Paradox case will arise.) It is just in the absence of such additional information that the author is 'merely' asserting, and conveying her belief, that the vast majority of the sentences in the book are true.[381]

Seventhly, given the default character of the proposal, an author of a book is not actually required to express her fallibility in the preface; and accordingly, many authors do not do so. But of course an author may still be polite enough to emphasize that the existing errors are not due to suggestions by other people. As Makinson (1965, p. 205) quotes from a preface: 'the errors and shortcomings to be found herein are not their fault.'

Finally: so far I have only dealt with belief in this section. By publishing her book, the author is able to assert that she believes the majority of the claims in its main part to be true. And the author can still assert consistently in the preface that there is a mistake in the main part of her book and thereby express her belief that this is so. But there is also something else that the author can do. At the same time she can *accept* the conjunction of *all* the statements in that main part for the purpose, say, of simplifying her present and subsequent research. And with them she can *accept* all of their logical consequences. Since acceptance is not belief, as discussed in section 6.3, even that is

[381] I am putting further complications to one side here: for instance, some of the sentences in the book might be presented as following deductively from other such sentences. In that case, maybe, the author should be taken to assert that most of the *non-derivative* sentences in the book are true. Or assume that an author is writing a book that solely consists of sentences that are obviously true: then the contextual reference of 'vast majority' might be pushed to the extreme, that is, to $m = n$, and the author might well assert that all sentences in the book are true. And the like. $S_m(T)$ is really a placeholder for more complicated constructions in which the logical strength of each of $A_1, \ldots, A_n$ is taken into account.

consistent with my proposal.[382] But note that the author in the Preface story does not *merely* intend to convey that she *accepts* all the statements in the main part of the book. Acceptance comes much too cheap: anything can be accepted for whatever crazy purposes. The author wants also, and primarily, to be taken seriously as a scholar: she wants to convey some of her beliefs, too, and in contrast to acceptance, one cannot rationally believe anything for whatever kind of reason. She also wants to convey that she regards $S_m(T)$ to be *true*.

I accept the conjunction of all statements made in *this* book, but I do not believe their conjunction to be true. I only believe most of them to be true; or at least that is what I hope for. It is yet to be seen how stable this belief will turn out to be.

382 So my proposal is close to Stalnaker's (1987, p. 93) who says about his version of the Paradox of the Preface (a preface to some historical narrative): 'The historian in the example believes that his narrative is *mostly* right, and the doubts he does have about it are based on general considerations of fallibility. What more effective way does he have to say just what he is *sure* of than to tell the story as best as he can, and then add, in the preface, that it is probably only roughly true. Here his motive for accepting what he does not believe is that doing so is an efficient means of telling what he does believe.'

Appendix D

On Counterfactuals and Chance

Chapter 4 and parts of Chapters 5 and 6 concerned a perfectly rational agent's conditional belief set $Bel(\cdot|\cdot)$ and her degree-of-belief function P at a time. In order to study systematically how the two relate to each other, I proved several general theorems of the form: if Bel is an arbitrary set of pairs of propositions—'pairs', because Bel is a *conditional* belief set—and if P is an arbitrary probability measure on propositions, such that the two of them satisfy a certain set of assumptions, then certain conclusions follow from this. Since these theorems and their proofs were purely mathematical in nature, the conclusions did not depend in any way on the intended interpretation of '*Bel*' as a set of conditional *beliefs* and of '*P*' as a degree-of-*belief* function. The same conclusions would follow if '*Bel*' and '*P*' were interpreted differently.

In this appendix I will indeed change their interpretation: instead of talking about *conditional beliefs* I will deal with *counterfactuals being true*, and instead of speaking about (conditional) *degrees of belief* I will turn to (conditional) *objective chance*. By reformulating some (though not all) of the postulates from Chapter 4 as assumptions on counterfactuals and chance, some of the conclusions from previous chapters will become applicable even with this revised interpretation.[383]

Here is the plan of the appendix. I will start by presenting a new lottery-style paradox on counterfactuals and chance. The upshot will be: combining natural assumptions on (i) the truth values of ordinary counterfactuals, (ii) the conditional chances of possible but non-actual events, (iii) the manner in which (i) and (ii) relate to each other, and (iv) a fragment of the logic of counterfactuals leads to disaster. In contrast with the usual lottery-style paradoxes, logical closure under conjunction—that is, in this case, the rule of Agglomeration of (consequents of) counterfactuals—will not play a role in the derivation and will not be entailed by the premises either. I will sketch four obvious

383 In more detail: premise P3 in the argument below will correspond to a fragment of the AGM-like postulates B1*–B6* on conditional belief from Chapter 4, and P4 below will correspond to the bridge principle $BP1^{r*}$ (for a threshold of $r = \frac{1}{2}$) from Chapter 4, which concerned conditional belief and conditional subjective probability. (See section 4.2.3 for the details about these postulates.) The semantic condition COMP in section D.5 below will correspond to (a version of) what I called the 'Sum Condition' in sections 4.2.3 and 5.2.3. In the models that I am going to construct in section D.5 below, a counterfactual $\varphi \mathrel{\Box\!\!\rightarrow} \psi$ will be true if, and only if, its corresponding conditional chance of ψ given φ is stably high. This will follow immediately by applying the formal results from Chapter 4 in the present context. But I will not deal with these stability issues in this appendix; instead I will concentrate merely on suggesting and arguing for a certain kind of context-sensitivity of counterfactuals.

but problematic ways out of the dilemma, and I will end up with a new resolution strategy that is non-obvious but (I hope) less problematic: contextualism about what counts as a proposition. This proposal will not just save us from the paradox, it will also save each of its premises in at least some context, and it will be motivated by independent considerations from measure theory and probability theory. The context-sensitivity of belief that I observed in previous chapters (especially in Chapter 3) will thus translate into a new kind of contextualism for counterfactuals. Where my primary interpretation of 'context' in Chapter 3 was an epistemic one, contexts will be semantic in this chapter.

In turn, the new lottery-style paradox on counterfactuals and chance in this appendix may also be translated into the original '*Bel*' and '*P*' terminology with their original intended interpretation: the resulting lottery-style paradox on conditional belief and subjective probability would then constitute an additional and (hopefully) interesting argument for the context-sensitivity of conditional all-or-nothing belief in which the contribution of the context is to provide a salient partition of possibilities again (thus adding to the corresponding discussion in Chapter 4).

Ultimately, my proposal is: why not approach the topic of counterfactuals vs chance with a similar kind of methodology as the topic of belief vs degrees of belief?[384]

D.1 A New Paradox

Once a week, a TV lottery takes place which is hosted by a famous entertainer. One day the host has a serious car accident on his way to the studio; out of respect for his condition, the lottery show is cancelled. At the end of the day, the situation is fairly summarized by the first premise P1:

P1 If A had been the case, B would have been the case.
('If the host had made it to the studio, there would have been the TV lottery that day.')

It happens to be the case that the TV lottery is a lottery with 1,000,000 tickets; let us assume that it would not be the TV lottery any more if this were not so. And for at least one of the tickets we cannot exclude it to have won if the host had made it to the studio. Taking these together, we have:

P2 Necessarily: B if and only if $C_1 \vee \ldots \vee C_{1000000}$; and there is an i,[385] such that the following is not the case: if A had been the case, then C_i would not have been the case.

[384] There are further pairs of concepts that might be treated with a similar kind of methodology: e.g. 'normic' qualitative laws (in the sense of Schurz 2001) vs statistical probability as discussed in the philosophy of science. And more.

[385] Since there are only finitely many tickets, here and elsewhere any quantification over i could always be replaced in terms of a long but finite statement of purely propositional form.

('Necessarily: the TV lottery would have taken place that day if and only if ticket 1 or ticket 2 or . . . or ticket 1,000,000 would have won in the TV lottery that day; and there is a ticket *i*, such that the following is not the case: if the host had made it to the studio, then ticket *i* would not have won.')

The set of true counterfactuals is of course closed under all logical rules and includes all logical laws. I suppose just a couple of rules to be valid (which are all contained e.g. in Lewis's 1973 logic of conditionals):

P3 All of the following rules are valid:[386]

$$\text{Left Equivalence: } \frac{\Box(\varphi \leftrightarrow \psi), \varphi \;\Box\!\!\rightarrow \rho}{\psi \;\Box\!\!\rightarrow \rho},$$

$$\text{Right Weakening: } \frac{\varphi \;\Box\!\!\rightarrow \psi, \Box(\psi \rightarrow \rho)}{\varphi \;\Box\!\!\rightarrow \rho},$$

Intersubstitutivity of Necessary Equivalents—of which Left Equivalence is a special case—and finally,

$$\text{Rational Monotonicity: } \frac{\varphi \;\Box\!\!\rightarrow \rho, \; \neg(\varphi \;\Box\!\!\rightarrow \neg\psi)}{\varphi \wedge \psi \;\Box\!\!\rightarrow \rho}.$$ [387]

What P3 says, is: any of these rules may be applied freely, whether to any of the other premises or in suppositional contexts. Indeed, one may think of the relevant applications as delivering material conditionals that belong to our overall set of premises. In this sense, P3 really constitutes an infinite set of premises. As far as $\Box$ ('necessity') is concerned, I will not need to assume more than what is contained in any so-called normal system of modal logic; but I will not state this explicitly in terms of a premise.

Note that the following rule has *not* been assumed:

$$\text{Agglomeration: } \frac{\varphi \;\Box\!\!\rightarrow \psi, \; \varphi \;\Box\!\!\rightarrow \rho}{\varphi \;\Box\!\!\rightarrow \psi \wedge \rho}$$

Agglomeration is a logical rule that is included in all standard logical systems for counterfactuals (such as Lewis's and Stalnaker's). On the other hand, considerations concerning chance might be thought to cast doubts about it just as considerations about subjective probability do about the logical closure of belief. In any case, the validity of Agglomeration is not assumed as a premise. And it is not hard to show

[386] If $\Box\varphi$ is defined in terms of $\neg\varphi \;\Box\!\!\rightarrow \bot$, then all these rules follow from David Lewis's axioms and rules. I use this notation in this appendix: $\rightarrow$ and $\leftrightarrow$ are the material conditional and the material equivalence connectives, respectively. $\Box\!\!\rightarrow$ is the counterfactual conditional connective. Later on I will also use $\Diamond\!\!\rightarrow$ for the conditional-might connective.

[387] The analogy between these rules for counterfactuals and the closure conditions for conditional belief in Chapter 4 is this: Left Equivalence was not an issue back then in view of the propositional framework of Chapter 4. Right Weakening corresponds to B2* (One Premise Logical Closure) in section 4.2.3. Rational Monotonicity corresponds to B6* (Revision) in section 4.2.3, which in turn is a version of postulates K*7 and K*8 (Superexpansion and Subexpansion) of the standard postulates for AGM belief revision. See sections 4.1.2 and 4.2.3 for the details.

that Agglomeration does not follow either from Rational Monotonicity together with the very weak rules that had been stated before.[388]

If a counterfactual is true—if φ had been the case, ψ would have been the case—then it is plausible to assume that its consequent ψ should have had a greater chance to have been the case than its negation $\neg\psi$, conditional on the antecedent φ. That is:

P4 If a counterfactual of the form $\varphi \mathrel{\Box\!\!\rightarrow} \psi$ is true, then the conditional chance of ψ given φ is greater than $\frac{1}{2}$.[389]

In fact, in many cases, it should be possible to strengthen P4 by replacing '$\frac{1}{2}$' by some threshold closer to 1 that would be given contextually in some way. If so, P4 is really not more than just a minimal requirement. P4 is entailed by theories of counterfactuals such as Kvart (1986) and Leitgeb (2012a), and something close to it is also contained in Loewer (2007).

By 'chance' I mean objective, non-epistemic, single-case probability; and, of course, chances as referred to by P4 are to be determined in the actual world, not in some merely possible world. I will speak interchangeably of the chance of a sentence and of the chance of the proposition that is expressed by that sentence. Obviously, 'wordly' chance (and even more so 'worldly' conditional chance) is a big topic in metaphysics and philosophy of science that has caused a lot of concern throughout the years; however, I will have to put all of that to one side here.[390]

Since chances are usually taken to be time-relative, too, let us presuppose that the conditional chances in question are taken always at some time shortly before the event that is described by the antecedent φ (assuming that φ does in fact specify an event clearly bounded in time).[391] This has the advantage that at least for all ordinary 'common sense' statements φ, if φ is possible at all, then the chance of φ to take place will be greater than 0; hence the conditional chance of ψ given φ will be well-defined by means of the usual ratio formula for conditional probabilities. In this way I can avoid using anything like Popper functions (on which see Makinson 2011 for

[388] One way of seeing this is to give '$\Box\!\!\rightarrow$' a non-standard semantic interpretation that makes Rational Monotonicity and the weak rules from before come out valid but Agglomeration invalid. For instance, assume $\varphi \mathrel{\Box\!\!\rightarrow} \psi$ to express a might-counterfactual with the following semantic rule: $\varphi \mathrel{\Box\!\!\rightarrow} \psi$ is true (at a world w) if and only if there is a closest φ-world (to w) that is a ψ-world. It is easy to see that this does the job. Or choose a semantics for '$\Box\!\!\rightarrow$' that involves a *set* of standard Lewis–Stalnaker models on one and the same set of worlds, and define $\varphi \mathrel{\Box\!\!\rightarrow} \psi$ to be true (at a world w) if and only if there is a model in that set, such that in that model all closest φ-worlds (to w) are ψ-worlds. This said, one can also show that Agglomeration is e.g. entailed by Rational Monotonicity together with the stronger rules that are contained in Hawthorne and Makinson (2007). I am grateful to David Makinson for this additional observation.

[389] This premise corresponds to postulate $BP1^{r*}$ (Likeliness) in section 4.2.3, where $r = \frac{1}{2}$.

[390] For more on this, see s. 4.2 in Leitgeb (2012a).

[391] Actually, both conditional chances and counterfactuals can be assessed relative to different points of time, and determining the time of assessment of a conditional chance statement or a counterfactual to be close to their antecedent time is not generally right; in some contexts, other points of assessment are more appropriate. See s. 4.3 of Leitgeb (2012a) for a discussion on this.

an overview): primitive conditional probability measures that would be well-defined and non-trivial even in cases where the antecedent chance is 0. The theory of such extended probability functions is still not accepted widely, and our considerations will be simplified by not having to rely on them. In terms of properties of conditional chance, I will not have to presuppose more than if an ordinary statement φ is possible, then the chances of statements taken conditional on φ can be determined by the usual ratio formula for conditional probabilities, and if φ and ψ are necessarily equivalent, then the conditional chance of a statement given φ is identical to the conditional chance of that statement given ψ. However, I will not state any of these assumptions on chance as extra premises.

Finally, I add one further supposition concerning our TV lottery and host story: assume that the host had made it to the studio. Even then there would have been a small chance of the lottery being cancelled: maybe the lottery machine would have been broken; maybe a lunatic would have abducted the TV host from the studio; maybe the lottery show would have been dropped by the boss of the TV channel who had found out that the host had an affair with his wife; or the like. Depending on the empirical circumstances, the chance of cancellation might have been bigger or smaller. Let us assume that the chance of the cancellation happening was small but not tiny; indeed, I suppose that the chance of the lottery not taking place given the host had made it to the studio is bounded from below by the (presumably, tiny) chance of any particular ticket i winning in this lottery of 1,000,000 tickets given the host had made it to the studio.

Which leads me to premise P5. Let Case 1 and Case 2 be the following two counterfactual circumstances:

Case 1: $A \wedge C_i$
('The host made it to the studio, and ticket i won.')

Case 2: $A \wedge \neg B$
('The host made it to the studio, but the lottery still did not take place.')

I presuppose that the disjunction of Case 1 and Case 2 describes a possible state of affairs; and I assume that given that state of affairs the proposition $A \wedge C_i$ does not have a chance greater than that of $A \wedge \neg B$:

P5 For all i: $(A \wedge C_i) \vee (A \wedge \neg B)$ is possible; and the conditional chance of $A \wedge C_i$ given $(A \wedge C_i) \vee (A \wedge \neg B)$ is less than, or equal to, $\frac{1}{2}$.

('For all i: The chance of the host making it to the studio and ticket i winning given that either the host had made it to the studio and ticket i had won or the host had made it to the studio and the lottery had not taken place, is less than, or equal to, one-half. The "given" condition describes a possible state of affairs.')

In case one still worries about this, one might additionally assume the lottery to be fair and, maybe, reformulate the story so that it involves an even greater number of tickets—1,000,000,000 or so. Then P5 should be perfectly harmless.

As things stand, I take it that each of these premises is plausible if considered just by itself. However, one can show that, if all of the premises P1–P5 are taken together, they logically imply a contradiction. It is in this sense that the argument from P1–P5 to $\bot$ may be called a paradox.

In section D.2 I will demonstrate that the five premises entail a contradiction. Section D.3 is devoted to a comparison of the paradox to related ones; as we are going to see, the new paradox differs in structure from all of them. Section D.4 deals with the diagnosis of what has gone wrong in the paradox. In particular, I will discuss in detail the options of dismissing one of: P1; the second conjunct of P2; P3; and P4. None of these options will turn out to be particularly attractive. Section D.5 presents a new way out of the paradox: a version of contextualism about what counts as a proposition in a context. This proposal will allow us to save each of the five premises in at least some context, and it will be motivated by independent considerations from measure theory and probability theory. Section D.6 concludes with an evaluation of the new proposal and its prospects.

D.2 The Derivation

Let us now turn to the corresponding derivation. First of all, I consider the last conjunct of P2:

> C1 There is an i, such that the following is not the case: if A had been the case, then $\neg C_i$ would have been the case.

In what follows, keep any such i that exists by C1 as fixed—then we have as another intermediate conclusion:

> C1i The following is not the case: if A had been the case, then $\neg C_i$ would have been the case.

So, in the counterfactual situation in question, the winning of that very ticket i would not have been excluded.

With this being in place, using P1, P2, and P3, one can derive a further intermediate conclusion. I will suppress P3 as an explicit premise, instead I simply apply the rules as being permitted by P3 (and, of course, standard propositional logic):

1. $A \mathrel{\Box\!\!\rightarrow} B$ (P1)
2. $\Box(B \leftrightarrow C_1 \vee \ldots \vee C_{1000000})$ (P2)
3. $\underline{\neg(A \mathrel{\Box\!\!\rightarrow} \neg C_i)\ \ (\mathrm{C1}i)}$
4. $A \mathrel{\Box\!\!\rightarrow} (C_1 \vee \ldots \vee C_{1000000})$ 1, 2 (Right Weakening)
5. $\|$ $\neg\neg(A \mathrel{\Box\!\!\rightarrow} \neg(C_i \vee \neg(C_1 \vee \ldots \vee C_{1000000})))$ (Assumption for Reductio)
6. $\|$ $A \mathrel{\Box\!\!\rightarrow} \neg(C_i \vee \neg(C_1 \vee \ldots \vee C_{1000000}))$ 5 (Elimination of Double Negation)
7. $\|$ $A \mathrel{\Box\!\!\rightarrow} \neg C_i \wedge (C_1 \vee \ldots \vee C_{1000000})$ 6 (Right Weakening)
8. $\|$ $A \mathrel{\Box\!\!\rightarrow} \neg C_i$ 7 (Right Weakening)
9. $\|$ $(A \mathrel{\Box\!\!\rightarrow} \neg C_i) \wedge \neg(A \mathrel{\Box\!\!\rightarrow} \neg C_i)$ 8, 3 (Conjunction)

10. $\neg(A \,\Box\!\!\rightarrow \neg(C_i \vee \neg(C_1 \vee \ldots \vee C_{1000000})))$ 5–9 (Reductio)
11. $A \wedge (C_i \vee \neg(C_1 \vee \ldots \vee C_{1000000})) \,\Box\!\!\rightarrow C_1 \vee \ldots \vee C_{1000000}$ 4, 10 (Rational Monotonicity)
12. $(A \wedge C_i) \vee (A \wedge \neg(C_1 \vee \ldots \vee C_{1000000})) \,\Box\!\!\rightarrow C_1 \vee \ldots \vee C_{1000000}$ 11 (Left Equivalence)

But 12 implies with premise P4:

C2 The conditional chance of $C_1 \vee \ldots \vee C_{1000000}$ given $(A \wedge C_i) \vee (A \wedge \neg(C_1 \vee \ldots \vee C_{1000000}))$ is greater than $\frac{1}{2}$.

By P5, P2, and standard modal logic, $(A \wedge C_i) \vee (A \wedge \neg(C_1 \vee \ldots \vee C_{1000000}))$ is possible (and 'ordinary'), so we can apply the usual ratio formula for conditional probabilities here. But according to this formula, the conditional chance of $C_1 \vee \ldots \vee C_{1000000}$ given $(A \wedge C_i) \vee (A \wedge \neg(C_1 \vee \ldots \vee C_{1000000}))$ is identical to the conditional chance of $A \wedge C_i$ given $(A \wedge C_i) \vee (A \wedge \neg(C_1 \vee \ldots \vee C_{1000000}))$. And since $(A \wedge C_i) \vee (A \wedge \neg(C_1 \vee \ldots \vee C_{1000000}))$ is necessarily equivalent to $(A \wedge C_i) \vee (A \wedge \neg B)$ by P2 and standard modal logic again, the conditional chance of $A \wedge C_i$ given $(A \wedge C_i) \vee (A \wedge \neg(C_1 \vee \ldots \vee C_{1000000}))$ is in turn equal to the conditional chance of $A \wedge C_i$ given $(A \wedge C_i) \vee (A \wedge \neg B)$. Using this we can derive from C2:

The conditional chance of $A \wedge C_i$ given $(A \wedge C_i) \vee (A \wedge \neg B)$ is greater than $\frac{1}{2}$.

However, if put together with P5, this leads to a contradiction.

D.3 Related Arguments

Before I turn to the diagnosis of what has gone wrong here, it is illuminating to compare the new paradox with more familiar ones in order to put it in context and to see where exactly it differs from the others. Our paradox involves a lottery-type situation. Let us therefore contrast it first with a version of Kyburg's (1961) classical Lottery Paradox on belief and degrees of belief (as discussed throughout this book) that is formulated so that it proceeds from the following five premises:

Q1 I am certain that B is the case.
('I am certain that there is one and only one lottery at time t.')

Q2 I am certain that: B if and only if $C_1 \vee \ldots \vee C_{1000000}$.
('I am certain that: there is the lottery at t if and only if ticket 1 wins or ticket 2 wins or . . . or ticket 1,000,000 wins at t.')

Q3 All standard axioms and rules of doxastic logic are valid.

Q4 If my subjective probability of ψ is greater than 0.9, then I believe that ψ is the case.

Q5 For all i: My subjective probability of C_i is $\frac{1}{1000000}$.
('For all i: My subjective probability of ticket i winning is one over a million.')

In Q1 and Q2, 'certain' means: has subjective probability 1. Q3 makes sure that the agent's set of believed propositions is closed under the usual rules of logical consequence; in particular, if φ is believed, then all of its logical consequences are believed, and if φ and ψ are believed, then so is their conjunction $\varphi \wedge \psi$. For simplicity, let us assume again that we are dealing with a perfectly rational agent who always applies deduction competently and who is always perfectly aware of all the conclusions that can be drawn logically from her beliefs. In all this, I take the usual axioms of probability to be implicit in the term 'probability', which is why I won't state them separately.

From premises Q1–Q5 one can derive: I believe that $(C_1 \vee \ldots \vee C_{1000000}) \wedge \neg(C_1 \vee \ldots \vee C_{1000000})$. So the belief system in question ends up being inconsistent, given the premises.

If compared to Kyburg's famous paradox, our new paradox involves truth (of counterfactuals) where his is about belief, and chance where his deals with credence. And it is crucial to our paradox that I am concerned with conditional notions, not absolute or categorical ones as in Kyburg's paradox. This showed up quite clearly in the last section when I applied a logical rule such as Rational Monotonicity that does not have an 'unconditional' counterpart.

This said, it would of course be possible to reinstate our new paradox in terms of *conditional* belief: belief in a proposition under the supposition of another proposition, as discussed in Chapter 4. And the formal resources of a theory such as the (non-probabilistic) theory of belief revision (cf. Alchourrón et al. 1985, Gärdenfors 1988) would indeed allow us to carry out the derivation from the last section in these conditional doxastic terms. A rule of inference such as Rational Monotonicity that proceeded from the *absence* of a conditional belief state in Chapter 4 is understood above (see P3) to be applied to counterfactuals for which this kind of 'absence' simply corresponds to the falsity of the counterfactual in question. But I will not go into this in any more detail here.

However, the two main differences between Kyburg's and our new paradox lie somewhere else: first of all, where Q4 is nothing but the *right-to-left* direction of the so-called Lockean thesis (cf. Foley 1993) for a threshold of 0.9, that is, the right-to-left direction of

ψ is believed by me iff my subjective probability of ψ is greater than 0.9,

our new paradox relies on premise P4 which is the analogue of the *left-to-right* direction of the Lockean thesis for a threshold of $\frac{1}{2}$. While adding the left-to-right direction of the Lockean thesis to Q1–Q5 from before allows one to strengthen the 'internally' believed inconsistency to a straightforwardly contradictory statement, if taken just by itself the left-to-right direction of the Lockean thesis is perfectly consistent with Q1–Q3 and Q5. In contrast, our new paradox involves the true-counterfactual-to-high-conditional-chance version of the left-to-right direction of the Lockean thesis for $\frac{1}{2}$ as its only bridge principle for counterfactuals and chance, and yet a *logical contradiction* follows from it in conjunction with other plausible assumptions.

Secondly, and even more importantly, Kyburg's classical lottery paradox relies on the presence of the closure of rational belief under conjunction. Indeed, famously, according to Kyburg's own diagnosis of his paradox, it is closure under conjunction that ought to be given up (see e.g. Kyburg 1970). However, in our new Lottery Paradox, the corresponding rule of Agglomeration for conditionals has not been assumed. What I do use instead is Rational Monotonicity, which, as mentioned before, is a rule for conditionals that cannot even be formulated as a closure condition on unconditional belief.[392]

There are other quasi-paradoxical arguments around in the literature on knowledge and chance which do presuppose corresponding knowledge-to-high-chance analogues of the *left-to-right* direction of the Lockean thesis, e.g. in Hawthorne and Lasonen-Aarnio (2009): but in these cases it is typically assumed that some propositions $D_1, \ldots, D_n$ are known, each D_i has a high chance, their conjunction $D_1 \wedge \ldots \wedge D_n$ is also known, but at the same time $D_1 \wedge \ldots \wedge D_n$ is of low chance. The only obvious counterparts of $D_1, \ldots, D_n$ in our paradox are $\neg C_1, \ldots, \neg C_{1000000}$, which in the case of a fair lottery would indeed have high chances. But the counterfactual analogue of knowing each of them—that is, the truth of each counterfactual $A \,\Box\!\!\rightarrow \neg C_i$—is not validated: in fact, the contrary is the case, since I actually derived C1i: $\neg(A \,\Box\!\!\rightarrow \neg C_i)$ for a particular i in the last section.

Finally, one can find related paradoxical arguments in the relevant literature that are concerned immediately with conditional chance and counterfactuals, exactly as is the case in the argument from section D.1. Paradigmatically, consider the argument at the beginning of Hawthorne (2005),[393] which can be stated as such:

R1 If A had been the case, then B would have been the case.
('If I had dropped the plate, it would have fallen to the floor.')

R2 A is possible.
('It could have happened that I dropped the plate.')

R3 The following is not the case: if A had been the case then B would have been the case, and if A had been the case then C might have been the case.
('The following is not the case: if I had dropped the plate it would have fallen to the floor, and if I had dropped the plate it might have flown off sideways.')

392 I should add that the gist of Kyburg's argument is not actually closure under conjunction per se but really any closure condition on rational belief that is at least of the same logical strength as closure under conjunction (modulo some weak background conditions on rational belief that may be defended independently). For instance, closure under conjunction in the Lottery Paradox could be replaced by closure of rational belief under Modus Ponens (cf. Pagin 1990, Sharon and Spectre 2013): indeed, closure under Modus Ponens entails closure under conjunction given the assumption that every tautology is believed; and closure under conjunction entails closure under Modus Ponens given that belief is closed under one-premise logical consequence, that is, valid inference from one premise. However, Rational Monotonicity is a type of rule that differs from all such closure conditions on unconditional belief. I would like to thank an anonymous referee of Leitgeb (2014c) for urging me to comment on this point.

393 Similar arguments can be found in Hájek (n.d.) and Hawthorne and Lasonen-Aarnio (2009).

R4 If the conditional chance of ψ given φ is greater than 0 (and φ is possible), then ⌜if φ, then it might be that ψ⌝ is true.

R5 The conditional chance of C given A is greater than 0.
('The chance of the plate flying off sideways given it had been dropped is greater than 0.')

This set of premises entails a contradiction, and the reasoning is straightforward again.

Once again, I am not interested in evaluating or criticizing this argument. I only want to make clear how it differs from the argument in section D.1. Where Hawthorne's R5 is based on quantum-theoretical considerations—for common sense might simply not have regarded R5 to be true—I did not need any particularly scientific assumptions for my own argument.[394] Instead, I did exploit the logic of counterfactuals to a much greater extent than is the case in Hawthorne's argument. And I did not need to enter any debates on the logical properties of 'might'-counterfactuals, which is clearly an issue in Hawthorne's argument. Indeed, the premises of our argument were spelled out solely in terms of 'would'-counterfactuals and negated 'would'-counterfactuals (as well as statements about possibility, necessity, and chance). I should add that according to David Lewis's (1973) analysis of 'might'-counterfactuals, these are in fact logically equivalent to certain negated 'would'-counterfactuals: but my argument does not rely on this in any way, and one might just as well reject Lewis's analysis of 'might'-counterfactuals.

Most importantly, the reasoning patterns in the two arguments differ substantially, which can be seen clearly if both are reformulated in (roughly) Lewisian semantic terms: whereas Hawthorne derives a contradiction by locating exceptional $A \wedge \neg B$ circumstances in the closest A-worlds from reasoning from conditional chance statements to 'might'-counterfactuals, I derive a contradiction by partitioning the set of closest A-worlds in terms of $C_1, \ldots, C_{1000000}$: I assume the closest A-worlds to be B-worlds (P1), B to be necessarily equivalent to $C_1 \vee \ldots \vee C_{1000000}$ (first conjunct of P2), and there to be some i, such that some $A \wedge C_i$-worlds are amongst the closest A-worlds (second conjunct of P2). Furthermore, there exist A-worlds (as follows from the first conjunct of P5 as well as from the second conjunct of P2), so P1 is non-vacuously true. By the totality or linearity property of Lewisian sphere systems (which is precisely what is expressed by the validity of the rule of Rational Monotonicity in line with P3), the closest $A \wedge C_i$-worlds must then be closer to the actual world than any of the closest $A \wedge \neg B$-worlds. Therefore, the closest $(A \wedge C_i) \vee (A \wedge \neg B)$-worlds must be $A \wedge C_i$-worlds. Comparing the conditional chance of $A \wedge C_i$ given $(A \wedge C_i) \vee (A \wedge \neg B)$ with that of $A \wedge \neg B$ given $(A \wedge C_i) \vee (A \wedge \neg B)$ (using P4 and P5) finally does the trick. The situation is visualized in Figure D.1. The formal derivation in the last section captures this pattern of semantic reasoning without relying on any of the rules of Lewis's logic other than the ones mentioned by P3. In a nutshell: for Hawthorne's argument to proceed it suffices

[394] On the other hand, one would probably find an alternative way of formulating Hawthorne's paradox that would not rely on quantum theory in any way.

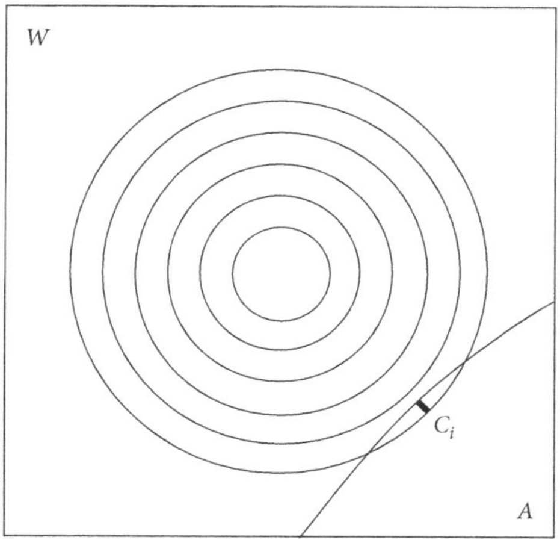

Figure D.1. Comparing the closest $A \wedge C_i$-worlds with $A \wedge \neg B$ (where the closest A-worlds are B-worlds)

to look at the closest A-worlds; but it is crucial to my argument that additionally the closest $(A \wedge C_i) \vee (A \wedge \neg B)$ are being considered.

Note that, given Lewis's (1973) original definition of 'might'-counterfactuals in terms of $\neg(\varphi \mathrel{\Box\!\!\rightarrow} \neg\psi)$, and using standard laws of conditional probability, Hawthorne's R4 from above can be reformulated according to the following equivalences (I suppress the possibility statement for φ):

$$Ch(\psi \mid \varphi) > 0 \rightarrow (\varphi \mathrel{\Diamond\!\!\rightarrow} \psi)$$

$$\leftrightarrow \quad Ch(\psi \mid \varphi) > 0 \rightarrow \neg(\varphi \mathrel{\Box\!\!\rightarrow} \neg\psi)$$

$$\leftrightarrow \quad (\varphi \mathrel{\Box\!\!\rightarrow} \neg\psi) \rightarrow Ch(\psi \mid \varphi) = 0$$

$$\leftrightarrow \quad (\varphi \mathrel{\Box\!\!\rightarrow} \neg\psi) \rightarrow Ch(\neg\psi \mid \varphi) = 1$$

In other words, up to logical equivalence and the analysis of 'might'-counterfactuals, Hawthorne's R4 is the principle

$$(\varphi \mathrel{\Box\!\!\rightarrow} \psi) \rightarrow Ch(\psi \mid \varphi) = 1$$

which is but the extreme version of my bridge principle P4. In fact, it would be possible to run Hawthorne's argument based on any *small* threshold ϵ, where '$Ch(\psi \mid \varphi) \geq \epsilon$' would thus replace the initial '$Ch(\psi \mid \varphi) > 0$' statement, and hence one would end up with any large threshold $1 - \epsilon$, where '$Ch(\psi \mid \varphi) > 1 - \epsilon$' would then replace the '$Ch(\psi \mid \varphi) = 1$' that was stated before. With R4 thus revised, that is, up to logical equivalence and the construal of 'might' again:

$$(\varphi \mathrel{\Box\!\!\rightarrow} \psi) \rightarrow Ch(\psi \mid \varphi) > 1 - \epsilon,$$

and if in R5 'greater than 0' is replaced by '$\geq \epsilon$', accordingly, the original pattern of Hawthorne's argumentation would be preserved. For $\epsilon = \frac{1}{2}$, this revised version of R4 would be exactly our premise P4; however, this choice of a threshold would then no longer be small enough for Hawthorne's original purposes, since the variant of R5 in which 'greater than 0' got replaced by '$\geq \frac{1}{2}$' would no longer be supported by quantum-theoretical considerations for $\epsilon = \frac{1}{2}$.

I conclude that, in spite of some overlap, Hawthorne's argument remains different from the one formulated in section D.1 even if an analysis of might-counterfactuals by means of corresponding $\neg(\varphi \Box\!\!\rightarrow \neg\psi)$ statements is presupposed, and in fact even if all might-counterfactuals in Hawthorne's argument had been replaced by corresponding conditionals of the form $\neg(\varphi \Box\!\!\rightarrow \neg\psi)$ from the start.

D.4 Diagnosis

So what is the problem in the new Lottery Paradox? At least prima facie, there should not be much doubt about the first conjunct of P2 nor about P5, which we can all take to be true quasi-empirical premises about the particular lottery and host in our toy story; and neither of them involves counterfactuals. This leaves us with the obvious options of dropping: P1; or the second conjunct of P2; or P3; or P4. I will first state these options briefly and then criticize them:

- Denying P1: this reaction might come in various different brands. One might object to even *formulating* any claim whatsoever that involves counterfactuals, P1 being just of them; the recommendation might be to restrict oneself just to statements on conditional chance when one is dealing with counterfactual possibilities.[395] Or one does not object to counterfactuals per se—e.g. counterfactuals with probabilistic consequents might be fine—but one regards all *ordinary* counterfactuals as false, which is what Hájek (n.d.) argues for. Accordingly, by being ordinary, P1 would be false. Or one regards counterfactuals not to be true or false at all, as the Suppositional Theory of Conditionals has it (as held e.g. by Ernest Adams and Dorothy Edgington, as touched upon in section 6.2); so P1 would not even be truth-apt, let alone true, though one could still accept P1 in some way other than believing it to be true. Or one takes P1 to be false in view of the additional assumption of P5: for in conjunction with the usual laws of probability (and given that the chance of $(A \wedge C_i) \vee (A \wedge \neg B)$ is greater than 0), P5 entails that the conditional chance of $\neg B$ given A is positive. And maybe a

395 This would be the translation of Richard Jeffrey's rejection of the notion of (qualitative) belief into the present context, as discussed in section 1.3.2. Of course, Jeffrey himself would have liked, in addition, to replace statements on objective chance by statements on subjective probability.

corresponding conditional chance of *not less than* 1 is required for the truth of $A \Box\!\!\rightarrow B$.[396]

- Denying the second conjunct of P2: all of the general worries concerning P1 apply here, too; in the case of the Suppositional Theory of Conditionals, the worries might in fact be greater: for if counterfactuals do not express propositions, it is not clear any more what it even means to negate them. And the second conjunct of P2 would certainly be unacceptable given Robert Stalnaker's famous axiom of conditional excluded middle[397]—that is: $(\varphi \Box\!\!\rightarrow \psi) \vee (\varphi \Box\!\!\rightarrow \neg\psi)$—since it would then entail the counterfactual $A \Box\!\!\rightarrow C_i$ which is clearly false according to our story of the unlucky host.
- Denying P3: here the only salient option would be to drop Rational Monotonicity, as all the other rules are logically very weak and contained in more or less every system of conditional logic in the literature. E.g. Ernest Adams's logic of conditionals, which has been defended by Dorothy Edgington amongst others, does not include Rational Monotonicity as valid; and Lin and Kelly (2012b) have proposed a probabilistic explication of conditional belief that does not validate the rule.[398]
- Denying P4: finally, one might defend the existence of counterfactuals $\varphi \Box\!\!\rightarrow \psi$ that are true but where the conditional chance of ψ given φ is less than or equal to $\frac{1}{2}$. That is, with the usual laws of probability: where the conditional chance of $\neg\psi$ given φ is at least as high as the conditional chance of ψ given φ. Although not stated explicitly, such a view is hinted at by Williamson (2009) who argues for the possibility of divergence between, on the one side, a notion of safety that involves counterfactual possibilities (as one feature of knowledge), and sufficiently high objective chance on the other. From a contextualist understanding of counterfactuals as strict implications that are restricted to contextually determined sets of relevant worlds, one might argue against P4 by pointing out that even high-chance sets of worlds might count as irrelevant in certain contexts. From the Lewisian point of view, one might attack P4 for the reason that it runs counter to Lewis's Strong Centering Axiom scheme (which is not included in our premise P3): $\varphi \wedge \psi \rightarrow (\varphi \Box\!\!\rightarrow \psi)$. For let $\top$ be tautology, and let ψ be a low-chance truth (assuming that there are such truths). By Strong Centering, $\top \Box\!\!\rightarrow \psi$ is true. But the conditional chance of ψ given $\top$ is just the unconditional chance

[396] Leitgeb (2012a) formulates a semantics in which this is the case, even when he argues that his semantics also allows for an interpretation according to which the truth of $A \Box\!\!\rightarrow B$ only requires the conditional chance of B given A to be *close to* 1, as long as 'close to' is understood as a vague term.

[397] See Lewis (1973, pp. 79–81) for a discussion.

[398] One should add that in neither of these theories is any systematic sense being made of nested conditionals or of the application of propositional connectives to conditionals. Even just handling negated conditionals, as in the formulation of Rational Monotonicity, is somewhat problematic in all of these approaches. If $\neg(\varphi \Box\!\!\rightarrow \psi)$ is simply understood as $\varphi \Box\!\!\rightarrow \neg\psi$, as is sometimes the case in suppositional treatments of conditionals, then Rational Monotonicity turns out to be valid again even in Adams's logic of conditionals.

of ψ, which is low. Thus, P4 would fail in these circumstances.[399] Finally, from a Stalnakerian point of view, if the conditional chance of some ψ given some φ is precisely $\frac{1}{2}$, P4 would seem to contradict Stalnaker's additional axiom scheme of conditional excluded middle again: $(\varphi \mathrel{\Box\!\!\rightarrow} \psi) \vee (\varphi \mathrel{\Box\!\!\rightarrow} \neg\psi)$.[400]

This is not the place to deal with any of these options in full detail. Instead I will merely point out briefly why I think that each of them is problematic, after which I will move on and propose a new way out of the dilemma that is raised by the new Lottery Paradox.

About denying P1: talking and reasoning in terms of counterfactuals is so deeply entrenched in common sense, philosophy, and maybe even in the applied corners of science that rejecting the whole scale level of counterfactuals should come with too high a price; similarly, an error theory that regards all ordinary counterfactuals to be false would be so revisionary that it should not amount to more than just an ultimate fallback position. And counterfactuals are so close to e.g. disposition ascriptions, which we do like to think are true or false, that their truth-aptness ought not to be sacrificed easily either. For the same reason, it should also be fine to apply propositional connectives to counterfactuals. Finally, the truth of $A \mathrel{\Box\!\!\rightarrow} B$ should be consistent with the chance of B given A to be less than 1 by some small real-valued margin, for reasons analogous to those for which my belief in A should be compatible with my subjective probability for A to be less than 1 by some small real-valued margin: for otherwise neither the truth of counterfactual assertions nor that of beliefs would be robust enough to survive the presence of even minor uncertainties which almost inevitably occur in real-world cases.

About denying the second conjunct of P2: more or less the same defence applies as in the case of P1. In addition, Stalnaker's conditional excluded middle is problematic in itself: it is not clear why a negated counterfactual of the form $\neg(\varphi \mathrel{\Box\!\!\rightarrow} \psi)$ ought to be logically equivalent to the unnegated counterfactual $\varphi \mathrel{\Box\!\!\rightarrow} \neg\psi$, and famously this has been disputed by David Lewis. What the second conjunct of P2 says is just that a certain counterfactual is not true: if the host had made it to the studio, then ticket i would not have won. A particular instance of counterfactual dependency is being denied. But we are not required to interpret this as telling us that any particular ticket would have won.

About denying P3: Rational Monotonicity is logically valid in David Lewis's and in Robert Stalnaker's semantics for counterfactuals, and it would turn out valid, too, if counterfactuals were analysed as strict conditionals.

399 By means of formal models such as the ones that I will introduce in section D.5, one can show that Lewis's *Weak* Centering axiom scheme—$(\varphi \mathrel{\Box\!\!\rightarrow} \psi) \rightarrow (\varphi \rightarrow \psi)$—is much less problematic in the context of P4.

400 I am grateful to Timothy Williamson for highlighting these points in a discussion and for urging me to comment on them.

As mentioned before, in semantic terms, Rational Monotonicity corresponds to Lewis's similarity or closeness (pre-)orderings[401] $\leq$ being total: for all worlds w, w', it holds that $w \leq w'$ or $w' \leq w$. If totality is dropped, so that overall similarity or closeness is merely assumed to be some partial (pre-)order, then Rational Monotonicity no longer comes out logically valid. Now, let us for the moment disregard the general attractions of total pre-orderings, which are well-known from all the areas in which totality is normally taken as a given, such as decision theory, social choice theory, and belief revision theory; and, say, we also ignore the question of what alternative logical rules for negated counterfactuals ought to replace Rational Monotonicity—for, presumably, there should be *some* rules of inference that are specific for negated counterfactuals. Even then it is still unclear if dropping Rational Monotonicity as a logical rule helps: for even if Rational Monotonicity is not logically, and hence universally, valid, it might still be *locally* truth-preserving. In particular: it might simply be a feature of the story about our unlucky lottery host that the one application of Rational Monotonicity that was required for the formal derivation in section D.2 happens to be truth-preserving. After all, even when a closeness order is not demanded to be total overall, it might still happen to instantiate a pattern of totality *somewhere* in the ordering, if only the (empirical) circumstances are the 'right' ones. Simply tell our toy story such that the original transition from lines 4 and 10 to line 11 by means of Rational Monotonicity is accomplished instead by applying Modus Ponens to a new premise of the material conditional form 'line 4 $\wedge$ line 10 $\rightarrow$ line 11': then the same conclusions as before can be drawn without assuming Rational Monotonicity to be logically valid, and it is difficult to see how the (quasi-empirical) truth of that new premise could be ruled out, once the story has been told in the right manner. Indeed, maybe, one might argue for the premise in terms of Lewis-style similarity reasoning again that would apply just to that special case, even when there would be no guarantee that the same type of reasoning could have been applied universally. And if someone argued that this kind of similarity reasoning in favour of 'line 4 $\wedge$ line 10 $\rightarrow$ line 11' would be trumped by reasoning about chances, and that reasoning about chances would speak *against* the truth of this material conditional, then we will see in section D.5 that this is not necessarily so: my own solution will preserve at the same time reasoning from some kind of similarity relation *and* from chances without there being any contradiction between them, even though it has to be admitted that the similarity relations that I will employ are unlikely to obey the original Lewisian heuristics (cf. Lewis 1979) of what overall similarity or closeness between worlds supposedly consists in.

About denying P4: here is how one might want to defend P4 against the attacks mentioned above. On the contextualist point, one should maybe 'contextualize' the notion of conditional chance accordingly, by which counterfactuals and conditional

[401] Formally, Lewisian sphere systems or similarity ordering are *pre*-orders, since anti-symmetry is not presupposed: two numerically distinct worlds may be of equal rank in such an ordering.

chance would be on par again. As far as Strong Centering is concerned, one response would be to say that it is always possible to choose the assessment point of time for chances differently (and so for counterfactuals). If one chose it to be, say, some time after both the antecedent and the consequent times, then if $\varphi \wedge \psi$ is true, the chances of both φ and ψ will be 1 then, and thus the conditional chance of ψ given φ will be 1, too; hence Strong Centering will not cause problems any more in the presence of P4. In other words: one can have Strong Centering and P4 taken together at least relative to *some* assessment time. Still this would not suffice for Strong Centering to come out as logically valid: but maybe it is not so anyway. Considerations as in Nozick's tracking analysis of knowledge, or on indeterminism (cf. Bennett 2003, section 92), seem to speak against the logical validity of Strong Centering. Also for some true and contingent φ and ψ, one might want $\varphi \mathrel{\Box\!\!\rightarrow} \psi$ to express a counterfactual dependency of ψ on φ, and then to *deny* $\varphi \mathrel{\Box\!\!\rightarrow} \psi$ on these grounds, since φ and ψ might 'merely' describe some causally and conceptually independent and accidental facts. But that natural move would be ruled out from the start by the logicality of Strong Centering. And if the semantics of $\varphi \mathrel{\Box\!\!\rightarrow} \psi$ is to involve some sort of additional ceteris-paribus or normality clause that is to allow for exceptional $\varphi \wedge \neg\psi$-worlds close by the actual world, then one should expect the innermost sphere around the actual world to include worlds other than the actual world, and again Strong Centering would fail. There is one other point that ought to be made about arguments against P4 that are based on considerations on Centering: it is questionable whether they get to the heart of the matter of the paradoxical argument of section D.1. After all, the toy story there concerned counterfactual circumstances: circumstances which did not prevail in the actual world. Assume P4 to be adapted only very slightly in the way that an '... and φ is false' clause is added to its antecedent: hence only *proper counter*factuals would be assumed to entail the conditional chance claim that is the consequent of P4. Lewis's Centering Axioms would be completely unaffected by P4 thus amended, but the same paradoxical argument could still be run. Finally, concerning the last point of criticism that concerned conditional excluded middle: other than rejecting its logical validity, one might simply change the 'greater than' condition in P4 into a 'greater-than-equals' condition, and replace the 'is less than, or equal to' condition of P5 by 'is less than' in compensation. Then once again our argument could proceed as before, the strengthened premise P5 would still be plausible in view of our toy story, and the thus weakened P4 premise would no longer be in conflict with the Stalnakerian principle. Independently, one might hope that some supervaluationist moves would save even the original premise P4 in a Stalnakerian setting.[402]

Over and above defending P4 against these attacks, one might point to some independent reasons for believing it to be true: say, one regards conditional chance to be nothing but the graded version of counterfactual truth, or counterfactual truth to be nothing but the all-or-nothing version of conditional chance, which is certainly not

402 On some of these points, see s. 1 of Leitgeb (2012b) for further details.

an absurd view: then claiming a counterfactual $\varphi \mathrel{\Box\!\!\rightarrow} \psi$ to be true and the conditional chance of $\neg\psi$ given φ to be greater than or equal to the conditional chance of ψ given φ should be necessarily false. That is: P5 should be necessarily true. In fact, one should even expect an analogue of the full Lockean thesis to be necessarily satisfied in this case: the truth of a counterfactual should be necessarily equivalent to the corresponding conditional chance being high enough. Alternatively, if that equivalence is not necessarily the case, the main open question is: what kind of ontic structure is it that the truth condition for counterfactuals is supposed to track? Surely, there must be some answer to the question of what it is 'out there' in the physical world that counterfactuals are describing—something which can be expressed in principle in terms resembling those of the scientists—and if it is not high conditional chance, then finding a good alternative answer constitutes at least an open challenge and a serious worry. Finally, let us focus just on our analogue of the left-to-right direction of the Lockean thesis, that is, P4, and let us assume P4 not to be the case: then how are we to explain that reasoning in terms of counterfactuals seems to be probabilistically reliable? If not a universal claim as in P4, then at least some second-order probabilistic statement ought to hold of the form 'The probability for a counterfactual $\varphi \mathrel{\Box\!\!\rightarrow} \psi$ to be such that the conditional chance of ψ given φ is high given the counterfactual $\varphi \mathrel{\Box\!\!\rightarrow} \psi$ is true, is high'.[403] For if reasoning with counterfactuals is not even probabilistically reliable in such a weaker sense, we simply *should not* engage in it at all, because, if only counterfactually, it will lead to falsity in too many cases.

I conclude that none of the four options so far looks particularly attractive. Therefore, the paradox from section D.1 should constitute a noteworthy challenge to pretty much everyone who is interested in counterfactuals and chance at all.

D.5 A New Way Out

Which leads me to a new proposal for how to cope with this paradox: *contextualism about what counts as a proposition*. This proposal will have the advantage of saving, in a sense to be explained and qualified later, *each premise* of the argument in section D.1 *in at least some context*. However, there won't be a single context that saves all premises simultaneously, even though P3 (a fragment of the logic of counterfactuals)[404] and P4 (the bridge principle for counterfactuals and conditional chance) will be satisfied in every context. And our proposal will not fall prey to the paradoxical reasoning that led us to inconsistency before.

[403] Of course, the interpretation of such second-order probabilities would be in need of serious clarification. Schurz (2001) employs similar second-order probabilistic statements in his explication of the reliability of so-called normic laws in the life sciences, but that is in the context of statistical probability and evolution theory, and even there it is unclear what the appropriate interpretation of the second-order probability measure is meant to be.

[404] In fact I will be able to save much more than just the rules mentioned by P3: we can have all of what Lewis (1973) called the system V of conditional logic if we like.

Of course, contextualist ways out of lottery paradoxes for knowledge and belief have been around for quite some time. But my approach will differ from all of these more standard contextualist solution strategies, and it will do so by relativizing the very notion of proposition to a context.[405] Alternatively, one might say: it won't be important for me to exploit contextualism in the sense that a counterfactual might express different propositions in different contexts—in analogy with some instance of a knowledge or belief ascription that might be taken to have different truth conditions in different contexts[406]—it will only be important whether a counterfactual expresses a proposition in a context at all.

As already mentioned before, there are also contextualist approaches to the semantics for counterfactuals: e.g. recently, Ichikawa (2011) suggested a contextualism about counterfactuals, but that is modelled again after contextualism about knowledge ascriptions; counterfactuals $A \mathrel{\Box\!\!\rightarrow} B$ are strict implications that express that all cases satisfy the material conditional $A \rightarrow B$ (or $A \supset B$), where the intended range of 'all' is determined by the context. However, once again, Ichikawa's account is not about relativizing the space of propositions to the context, and his argument is also independent of considerations on chance.[407]

My own proposal is motivated, in the first place, by probabilistic considerations. In probability theory, in any context in which one intends to consider or apply a probability measure, it is common practice to start from some algebra[408] $\mathfrak{A}$ of events or propositions to which probabilities are then assigned. For any given underlying space W (the sample space), every event or proposition in $\mathfrak{A}$ is required to be a subset of W, but not each and every subset of W is necessarily also a member of $\mathfrak{A}$. As measure theorists say: there may be non-measurable sets (that is, non-measurable subsets of W). In fact, in certain circumstances, it *must* be so that non-measurable sets exist, or otherwise some intrinsically plausible postulates on the measure function in question would not be satisfied.

For instance,[409] any proper 'geometrical' measure of subsets of the real number line that is supposed to extend the intuitive notion of length of intervals to even complicated sets of real numbers ought to have the following properties: (i) for all bounded intervals $[a, b]$ of real numbers, the measure of such an interval ought to coincide with the length $b - a$ of that interval; (ii) the measure function ought to be invariant under geometrical translations; and (iii) the measure function should satisfy all 'quasi-logical axioms' that hold for measures in general, such as monotonicity and

405 The only approach in that area of which I know to come close to what I am going to propose is a part of Levi's (1967) theory of acceptance, in which acceptance is question-dependent (or partition-dependent). But Levi's account is itself a non-standard contextualist one.

406 I discussed this briefly in section 3.3.

407 See Brogaard and Salerno (2008) for another recent contextualist account of counterfactuals.

408 In fact, usually one starts from a so-called σ-algebra of events which is also closed under taking arbitrary countable unions of events. See the beginning of section 4.2.

409 See any typical textbook on measure theory for the details.

countable additivity. One can then *prove* that there is no measure function that satisfies all of these assumptions and which at the same time assigns a measure to *every* subset of the real number line. So we find that in at least some contexts in which a measure space with an infinite sample space W is to be employed, it makes good sense not to require every subset of that sample space to be a member of the algebra $\mathfrak{A}$ of measurable events or propositions. And note that if the intended constraints on the measure had been chosen differently, the class of measurable sets of real numbers might well have been different, too; e.g. if countable additivity is weakened to finite additivity, then there are indeed 'geometrical' measures in the sense above which do assign a measure to every set of real numbers. Now, if one thinks of such constraints on measures in the way that one set of constraints might be salient or required in one context but not in another, then the corresponding classes of measurable sets end up being context-dependent as well. This finding may be expected to extend even to cases in which the members of W are not real numbers but where they should rather be interpreted as possible worlds. Of course, the interpretation of measures in measure theory differs substantially from the intended interpretation of the measure to which the premises in our paradox refer—the former are purely mathematical constructions, the latter is supposed to be a function with an 'empirical' meaning—but the insight may still carry over in terms of its formal pattern: sometimes it may be necessary not to count every subset of the sample space as belonging to the algebra of events or propositions on which a measure is defined, and it may depend on the context whether a set is counted as event/proposition or whether it is not.

Whilst in the case of probability spaces with a *finite* space W of possible worlds, there is no corresponding mathematical need to omit any of the subsets of W from the algebra in question, it is quite obvious that in almost all, if not *all*, concrete applications of any such probability space, the possible worlds in question are far from 'maximally specific' ways the world might be: if anything, they will correspond to more or less coarse-grained partition cells of the class W_{max} of all 'maximally specific' ways the world might be.[410] As far as the intended context of application is concerned, it might simply be sufficient to make a distinction between the different partition cells, while it might not be necessary to draw a wedge between any two different members of one and the same partition cell. Or perhaps, for whatever practical limitations, we might not even be able to make more fine-grained distinctions. In any case, once again, from the viewpoint of the 'real' class W_{max} of maximally fine-grained possible worlds, an algebra that is based on any such set W of worlds that correspond to partition cells of W_{max} will not include each and every proposition, that is, every subclass of the 'real' space W_{max} of possible worlds. And again we might take the context to determine

410 I put the ontological question of whether *there are* such maximally specific ways the world might be *at all* to one side here; let us simply assume, for the sake of the argument, possible worlds in this sense do exist. Accordingly, I will disregard the question of whether the class of all metaphysically possible worlds whatsoever (or the class of all physically possible worlds whatsoever) is a proper class or a set.

the appropriate fineness of grain: in one 'coarse-grained' context, various sets of fine-grained worlds will go missing, while in another 'fine-grained' context, they may all be present.[411]

Overall, I take this context-dependence of the class of events or propositions to be a stable pattern, and an important insight, from measure theory and probability theory.[412] There are good reasons for thinking so even prior to any considerations concerning the new Lottery Paradox.

My next step will be to translate this insight into the domain of counterfactuals.[413] While restricting the algebra of propositions for the chance function will not be important in what follows, restricting the algebra of propositions that can be expressed by counterfactuals will be. (Similarly, while subjective probability was context-*in*sensitive in previous chapters, all-or-nothing belief was not.) In order to show how this might work, I will build a little toy model in which I will be able to evaluate each of the premises of our new Lottery Paradox relative to contexts. Although I will employ the formal structure of a standard Lewis–Stalnaker type semantics for counterfactuals, I do not claim that the usual intended interpretation of this semantics carries over without changes. In particular, the similarity relations between worlds that will be employed will, presumably, not allow for an interpretation in terms of anything like the Lewisian heuristics for weighted similarity or closeness.[414] But I take it that this kind of interpretation of similarity between worlds is problematic anyway (without being able to argue for this here; but see section 2.6 of Leitgeb 2012b). For me it will be more important to save premises such as P4, which relate counterfactuals and chance, and which seem plausible independently of—or maybe even in spite of—Lewis's considerations on similarity. At the same time, sticking to the formal structure of Lewis's models will make sure that the so-called logic V of conditionals comes out valid in each and every context, by which premise P3 will be satisfied as well.

Let us, first of all, assume that every context c in which counterfactuals are to be asserted determines an algebra $\mathfrak{A}_c$ of events or propositions *in c*. If the 'sample space' for $\mathfrak{A}_c$ is the class $W_c = W_{max}$ of all possible worlds whatsoever, then not every subclass of W_{max} will be required to be a member of $\mathfrak{A}_c$; and if the sample space is but a set W_c of worlds that correspond to more or less coarse-grained partition cells of

411 I am grateful to an anonymous referee of Leitgeb (2014c) for emphasizing this point.

412 I should add that some of the problems to do with non-measurable sets can be mitigated by using non-standard probability measures that allow for the assignment of non-standard reals; but there are serious constraints on any such approach which I won't be able to deal with here.

413 Restricting the set of propositions to a proper subalgebra of the full power set algebra of a given set W of possible worlds is *not* a typical move in the possible worlds semantics of modalities. But there are exceptions; see e.g. Segerberg (2001), who bases his semantics of dynamic doxastic logic on some given topological space of propositions. And a relativization to partitions of the underlying set of worlds is to be found in theories such as Levi's (1967) theory of acceptance and Skyrms's (1984) subjectivist theory of objective chance.

414 That is, when determining the kind of similarity required by Lewis: it is of primary importance to minimize violations of laws of nature; it is of secondary importance to . . . ; and so forth.

W_{max}, then $\mathfrak{A}_c$ will not include each and every proposition—each and every subclass of W_{max}—either. What counts as a proposition may vary with the context.

Let also a Lewisian sphere system $\mathfrak{S}_c$ be determined by each context c which fixes for each world $w \in W_c$ a total similarity or closeness (pre-)ordering $\preceq_c^w$ relative to w.[415] I assume that $\mathfrak{A}_c$ and $\mathfrak{S}_c$ are compatible with each other: $\mathfrak{A}_c$ is not just closed under taking complements, unions, intersections (that is, the propositional counterparts of $\neg$, $\vee$, $\wedge$), but also under the propositional counterpart of $\Box\!\!\rightarrow$ as being determined by $\mathfrak{S}_c$ in the usual Lewisian manner. Roughly:[416] for all X, Y in $\mathfrak{A}_c$, there is another proposition, $X \Box\!\!\rightarrow_c Y$,[417] in $\mathfrak{A}_c$, such that for all $w \in W_c$: w is a member of $X \Box\!\!\rightarrow_c Y$ if and only if the set of closest X-worlds relative to w, as being given by $\preceq_c^w$, is a subset of Y. For every proposition Z in $\mathfrak{A}_c$, say that Z is true in w if and only if $w \in Z$.

Now suppose a notion of *expressing a proposition in c* to be given in a compositional manner: in particular, a counterfactual $\varphi \Box\!\!\rightarrow \psi$ expresses a proposition Z in c if and only if φ expresses a proposition X in c, ψ expresses a proposition Y in c, and $Z = X \Box\!\!\rightarrow_c Y$ (which is a member of $\mathfrak{A}_c$ again). If a sentence does not express a proposition in c, call it (and what it expresses) *non-entertainable in c*. It is *not* important for my approach that a sentence might express one proposition in one context and a *different* proposition in another context. For me it will only be relevant *whether a sentence expresses a proposition in a context* at all. Indeed, for my purposes, we may well presuppose that if a sentence expresses a proposition Z in a context c, then, if the same sentence also expresses a proposition in another context c', the proposition that it expresses in c' is just Z again.

Define a sentence to be *true in w, c* if and only if the sentence expresses a proposition Z in c, and Z is true in w. If c were a context in which $\mathfrak{A}_c$ happened to be the algebra of all propositions whatsoever, then truth in c would collapse into truth *simpliciter* again (where Lewisian sphere systems would still be determined by c). More importantly, if $A_1, \ldots, A_n$ are sentences or formulas in the language of conditional logic (quantifiers being omitted), such that all of them express propositions in c, then, by compositionality, also all of their subformulas express propositions in c. And as long as the logical rules of the system V of conditional logic are applied only to sentences that express propositions in c, all of these rules will preserve truth in w, c for all worlds $w \in W_c$ (even in suppositional contexts), since counterfactuals are still having Lewis-style truth conditions in terms of similarity orderings. Let us express this property of these logical rules by means of: *valid in c*.

Finally, we are ready to reconsider the argument from section D.1. I will do so in terms of a little formal toy model that will match the toy story from that section: let

415 This part of our proposal is in line with David Lewis's theory which does acknowledge the sensitivity of similarity orderings to conversational contexts.

416 As in all of our previous informal remarks on Lewis's semantics, I will presuppose the so-called limit assumption in order to simplify the Lewisian truth condition for counterfactuals. But nothing will hang on this.

417 In this context, '$\Box\!\!\rightarrow_c$' does not denote a logical symbol but a logical operation on propositions.

us pretend that the 'real' set W_{max} of 'maximally fine-grained' possible worlds is the set $\{@, w_1, \ldots, w_{1000000}, w^*\}$. I will consider two contexts c and c': let the algebra $\mathfrak{A}_c$ consist of the sets

$$\{@\}, \{w_1, \ldots, w_{1000000}\}, \{w^*\}$$

as well as of all sets that result from taking complements, unions, and intersections of these in arbitrary and maybe iterated manner; hence, $\mathfrak{A}_c$ is a set of $8 = 2^3$ propositions. In contrast, let $\mathfrak{A}_{c'}$ be the power set algebra of W_{max}: so $\mathfrak{A}_{c'}$ includes all $2^{1000002}$ subsets of W_{max}. (It will follow from the considerations below that these algebras may also be regarded closed under $\Box\!\!\rightarrow$ as a propositional operation relativized to the respective context.) Clearly, c will be a context in which only reasonably unspecific propositions are relevant, whereas c' will allow for 'maximally' fine-grained distinctions. Note that, by being the atoms of the algebra $\mathfrak{A}_c$, the sets $\{@\}$,$\{w_1, \ldots, w_{1000000}\}$, $\{w^*\}$ might be said to obtain the role of the more or less coarse-grained possible worlds in the context c. Indeed, we may just as well view $\mathfrak{A}_c$ to be given relative to a set $W_c = \{@, u, w^*\}$ of only *three* worlds, and we may regard every set in $\mathfrak{A}_c$ to be a corresponding subset of W_c, where the singleton $\{u\}$ takes over the role of the set $\{w_1, \ldots, w_{1000000}\}$. In what follows, I will switch back and forth between these two ways of viewing $\mathfrak{A}_c$. On the other hand, $W_{c'}$ will always remain identified with W_{max}.

Now I define a chance measure Ch on the full algebra $\mathfrak{A}_{c'}$ of all subsets of W_{max}. Intuitively, Ch is the chance function of the actual world @, and chances as being given by Ch are meant to be taken at some time shortly before the time of the event described by A, that is, of the host making it to the studio—which, say, is also the time immediately before the accident is to take place. Let $Ch(\{@\}) = \frac{4}{7}$, so the accident, which does take place in the actual world @, is already very likely to happen. Let $Ch(\{w_1\}) = \ldots = Ch(\{w_{1000000}\}) = \frac{2/7}{1000000}$ be the chances of the different tickets to be drawn in the lottery, so that each of the 1,000,000 tickets has the same tiny chance of winning. And let $Ch(\{w^*\}) = \frac{1}{7}$, which will be the small, though not tiny, chance of the host making it to the studio and the lottery still not taking place.[418] If $\mathfrak{A}_c$ is considered to be based on $W_c = \{@, u, w^*\}$, then Ch can be regarded to be defined just as well on the propositions in $\mathfrak{A}_c$ by means of the obvious assignment of $Ch(\{u\})$ to be nothing but $Ch(\{w_1, \ldots, w_{1000000}\}) = \frac{2}{7}$.

Next I determine sphere systems $\mathfrak{S}_c$ and $\mathfrak{S}_{c'}$ for the two contexts; for our purposes, it will be sufficient to determine the similarity or closeness orderings $\leq_c^@$ and $\leq_{c'}^@$ only for the actual world @. In the case of c, let[419]

$$@ <_c^@ u <_c^@ w^*$$

[418] Actually, $\frac{1}{7}$ might be a bit too much given our toy story, but never mind (for simplicity).

[419] Commas separate names for worlds of the same $\leq_c^@$-rank. $<_c^@$ is the strict pre-order that is determined from $\leq_c^@$ by: $w <_c^@ w'$ iff $w \leq_c^@ w'$, but not $w' \leq_c^@ w$. Analogously for c'.

and for c', let

$$@ <^{@}_{c'} w_1, \ldots, w_{1000000}, w^*.$$

This gives us two sphere systems centered on @; the smaller the rank of a world is in the ordering, the closer that world is to the actual world @.

So sphere systems are context-sensitive now, too, just as absolute and conditional belief turned out to be context-sensitive in previous chapters. That kind of context-sensitivity of counterfactuals is a widely appreciated phenomenon: it is quite common to 'call on context ... to resolve part of the vagueness of comparative similarity in a way favorable to the truth of one counterfactual or the other' (Lewis 1973, p. 67)—e.g. by attaching great importance to certain kinds of similarities between worlds in one context and to different kinds of similarities in another. Thus, according to Lewis and others, the context in which a counterfactual is asserted contributes to the truth value of the counterfactual by contributing to which similarity or closeness ordering of worlds is going to be salient for the truth conditions of the counterfactual. I make the same assumption, but in my case the context also contributes *the underlying space of propositions*, which is not part of Lewis's proposal. On the other hand, as mentioned before, I do not assume that chances are context-sensitive: the chance of a proposition at a world will be invariant under change of context (though it may be the case that a proposition is a member of the algebra of propositions of one context but not a member of the algebra of another context).

Both orderings $\leq^{@}_{c}$ and $\leq^{@}_{c'}$ from above satisfy the following salient Compatibility property of total pre-orderings $\leq$ on worlds (or of the strict pre-orderings $<$ of worlds that they determine) with respect to chances:

COMP For all worlds w: the chance of $\{w\}$ is greater than the sum of chances of sets $\{w'\}$ for which $w < w'$.[420]

In the case of c, $\frac{4}{7} > \frac{2}{7} + \frac{1}{7}$, and $\frac{2}{7} > \frac{1}{7}$; and for c': $\frac{4}{7} > 1{,}000{,}000 \cdot \frac{\frac{2}{7}}{1000000} + \frac{1}{7}$. In order for COMP to be satisfied, e.g. we could not have set $w^* <^{@}_{c} u$; nor could we have set $w_i <^{@}_{c'} w^*$ for any world w_i; nor $w^* <^{@}_{c'} w_1, \ldots, w_{1000000}$; nor $w_i <^{@}_{c'} w_j$ for any two worlds w_i, w_j.

Here is a general fact: one can prove that in the case of countably many possible worlds, whenever COMP is satisfied by an ordering $\leq^{w}$ and by the chance function at w, then if a proposition $X \mathrel{\Box\!\!\rightarrow}_c Y$ is true at w, the conditional chance of Y given X at w is greater than $\frac{1}{2}$.[421] In our context, since both $\leq^{@}_{c}$ and $\leq^{@}_{c'}$ satisfy COMP, it follows:

420 This corresponds to the Sum Condition from sections 4.2.3 and 5.2.3. As mentioned in Chapters 4 and 5, a similar compatibility condition on probability measures and strict total *partial orders* (not *pre-orders*) has been formulated in the computer science literature (cf. Snow 1998, Benferhat et al. 1997); so these authors do not allow for ties between worlds. This has the consequence that in their approach only very special probability measures can satisfy COMP, whereas in my approach it is easy to prove that for every probability measure on a countable space there is a total pre-order $\leq$, such that the order satisfies COMP with respect to the given probability measure.

421 This corresponds to Observation 24 in section 4.2.3.

For all $X, Y \in \mathfrak{A}_c$, for all $w \in W_c$: if $X \Box\!\!\rightarrow_c Y$ is true at w, then $Ch(Y|X) > \frac{1}{2}$.
For all $X, Y \in \mathfrak{A}_{c'}$, for all $w \in W_{c'}$: if $X \Box\!\!\rightarrow_{c'} Y$ is true at w, then $Ch(Y|X) > \frac{1}{2}$.

Therefore, the propositional counterpart of premise P4 from section D.1 is satisfied in both contexts.

In fact one can show more. Combining COMP or P4 with the axioms of probability for chance and with a Lewis-style sphere semantics for counterfactuals yields: a counterfactual is true if and only if the corresponding conditional chance is (stably) high. This follows by applying the theorems from Chapter 4 with the required revised interpretation. I will not go into further details on this here—the results carry over more or less term-by-term. Let me instead complete the semantics for our example language and explain how it avoids the paradox from section D.1.

In order to take the ultimate step to sentences or formulas, assume that in both of our contexts c and c', the sentence A expresses $\{w_1, \ldots, w_{1000000}, w^*\}$, and B expresses $\{w_1, \ldots, w_{1000000}\}$. As far as c is concerned, we might say alternatively: A expresses $\{u, w^*\}$, and B expresses $\{u\}$. In both contexts, it follows that $\neg A$ expresses $\{@\}$, $\neg B$ expresses $\{@, w^*\}$, $A \wedge \neg B$ expresses $\{w^*\}$, and so on. On the other hand, assume only in c' that the sentences of the form C_i express propositions $\{w_i\}$, respectively, whereas each such sentence C_i does not express a proposition in c at all. This manner of determining the expressing relation can be completed such that if a sentence expresses a proposition in c and it also expresses a proposition in c', then the propositions in the two cases are identical.

It follows that, e.g. $C_1 \vee \ldots \vee C_{1000000}$ expresses $\{w_1, \ldots, w_{1000000}\}$ in c', $(B \leftrightarrow C_1 \vee \ldots \vee C_{1000000})$ expresses W_{max} in c', and—with an accessibility relation on worlds explained appropriately—$\Box(B \leftrightarrow C_1 \vee \ldots \vee C_{1000000})$ expresses W_{max} in c', too, and precisely the same holds for $\Diamond((A \wedge C_i) \vee (A \wedge \neg B))$. But none of these formulas expresses a proposition in c, by the compositionality of the expressing relation as mentioned before. Thus, c is a context in which only reasonably unspecific sentences, such as B, express propositions, but not specific ones, such as the sentences C_i, which are non-entertainable in c. Perhaps one is interested in c in asserting that $A \Box\!\!\rightarrow B$—if the host had made it to the studio, there would have been the TV lottery that day—but the different possible outcomes of this counterfactual lottery are not being entertained, and indeed not entertain*able*, at all. Accordingly, while $A \Box\!\!\rightarrow B$ follows to be true in $@, c$—as the unique closest A-world in $\leq_c^@$, u, is a B-world—the sentence $A \Box\!\!\rightarrow C_1 \vee \ldots C_{1000000}$ does not express a proposition in c.

What can we say about the truth values of our premises P1–P5 in these contexts c and c' relative to the actual world @ (in which we may suppose our toy story to have taken place)?

Ad c: P1 is true in $@, c$, as just explained. $\Box(B \leftrightarrow C_1 \vee \ldots \vee C_{1000000})$ in P2 is not true in $@, c$, as pointed out, since it does not express a proposition in c; neither does any formula of the form $\neg(A \Box\!\!\rightarrow \neg C_i)$, again for compositionality reasons. P3 is the case if 'valid' is replaced by 'valid in c', also as explained. P4 follows from $\leq_c^@$ satisfying COMP above, once 'true' is replaced by 'true in $@, c$'. The second conjunct of P5 holds by our definition of Ch, while $\Diamond((A \wedge C_i) \vee (A \wedge \neg B))$ does not express a proposition again in c. Hence, all of our five premises except for P2 and the first conjunct of P5 are satisfied.

Ad c': It turns out that P1 is *not* true in $@, c'$: that is because w^* is amongst the closest A-worlds as being given by $\leq_{c'}^{@}$, and w^* is a $\neg B$-world. What happens here is that by splitting up u, or $\{w_1, \ldots, w_{1000000}\}$, into little pieces of the form $\{w_i\}$, these worlds w_i cannot count as more similar to the actual world than w^* any more, or otherwise COMP above would be invalidated, which would thus entail P4 to be invalidated, too. In order for P1 still to hold, it would be necessary for the chance of each proposition $\{w_i\}$ to be greater than $\{w^*\}$. In other words: the chance of B given A would have to be much closer to 1 than it is actually—even though it would still not need to be 1 exactly.

All premises other than P1 are true in $@, c'$. In particular, this applies to the first conjunct of P2 which does express a proposition in c', as pointed out before; and the proposition it expresses in c is true in @. The same holds for the second conjunct of P2 and first conjunct of P5.

So every premise of the argument of section D.1 is satisfied in some context, although not all of them are satisfied in one and the same context. In fact, one can say more: premises P3 and P4, which are the only general statements amongst the premises, are satisfied in *every* context; and in the case of the other three premises, or of statements like them, there might well be *many* contexts in which they are true. In particular, typical counterfactuals such as P1 may be expected to hold in the (coarse-grained, everyday) contexts in which they are typically asserted. No paradoxical conclusion follows from this, as promised.

More generally, the proposal is then: when we assert ordinary counterfactuals, we normally do so in contexts that determine coarse-grained spaces of propositions, since we are normally not interested in making fine distinctions or in considering very special circumstances. This allows us to reason jointly from general principles such as P3 (conditional logic) and P4 (counterfactual-chance bridge principle) and from ordinary counterfactuals, such as P1 (which turns out to be true in the coarse-grained context c). When the context changes with the interests of the subject(s) involved, so that the corresponding space of propositions becomes much more fine-grained—contemplating special outcomes of events, such as the fate of one particular ticket in a lottery, for example—then the general principles remain true. But some ordinary counterfactuals may turn out to be false in such a fine-grained context, unless the conditional chances of their consequents given their antecedents are *very close* to 1. That is exactly what happens to P1 in the fine-grained context c'. In a nutshell: very fine-grained spaces of propositions may demand true counterfactuals to correspond to conditional chances that are very near to 1.

In a sense, something like these results could also have been achieved by merely varying, with the context, which propositions get expressed by sentences at all, but where at the same time the underlying algebra of propositions would always be taken to be the power set algebra of W_{max}. E.g. as long as one makes sure that only sufficiently coarse-grained propositions are expressed in c by sentences of the given language $\mathcal{L}$ (or no proposition is expressed at all), our previous coarse-grained algebra $\mathfrak{A}_c$ from above could simply be identified with the class of propositions *that are expressed by members of* $\mathcal{L}$ *in* c, and if all of the original premises P1–P5 were restricted to members of $\mathcal{L}$, then all of our previous conclusions would go through as before. In other words: *instead of restricting the space of propositions, one might instead restrict the space of propositions that can be expressed by a sentence.* Of course, none of the propositions that could not be expressed by members of $\mathcal{L}$ would have any interesting semantic role to play. And the *propositional* counterpart of P4 would not be guaranteed to hold any more then: for instance, if instead of

$$@ <_c^@ u <_c^@ w^*$$

we would have had

$$@ <_c^@ w_1, \ldots, w_{1000000} <_c^@ w^*$$

in c, with each set $\{w_i\}$ counting as a proposition in c, this would not have mattered as far as A is concerned, since all of $w_1, \ldots, w_{1000000}$ satisfy A and hence a sentence such as A could not drive a wedge between any of the worlds w_i. But there would still be counterfactual *propositions* $X \mathrel{\Box\!\!\rightarrow}_c Y$ (that is, subsets of W_{max}) for which

$$\text{if X } \mathrel{\Box\!\!\rightarrow}_c Y \text{ is true at @, then } Ch(Y|X) > \frac{1}{2}$$

would be false, such as e.g. $\{w_1, w^*\} \mathrel{\Box\!\!\rightarrow}_c \{w_1\}$. Ultimately, the choice between a contextualism about what counts as a proposition and one concerning what propositions are expressible by sentences might be a matter of taste. But if one wants to avoid that a paradoxical argument such as the one from section D.1 could still be run on a propositional level, then one needs to restrict the space of propositions in a context, not just which propositions are expressible by sentences in a context. And restricting the algebra of propositions is certainly much more in line with the practice of most of probability theory, for the simple reason that standard mathematical probability theory does not deal with syntactic items at all. So I stick to my contextualism about propositions.

In contrast, it is not clear at all how something like the results above could have been achieved by merely varying, with the context, which sphere system one is dealing with, without any additional constraints on the space of (expressed) propositions.

D.6 Evaluation and Prospects

I have presented a new kind of Lottery Paradox for counterfactuals. The premises of this paradox are, at least at first glance, plausible assumptions on counterfactuals and conditional chance. Nevertheless a contradiction can be derived from them. The paradox differs in various ways from existing paradoxes in the same ballpark—in particular, no rule of closure under conjunction is being employed—and denying any of its premises seems to be problematic.

In the last section I presented a new proposal for how to deal with the paradox: relativize the notion of proposition to the context of assertion of counterfactuals. We have seen that there are independent reasons for thinking that whenever probabilities are involved, one should not expect every set of possible worlds to count as a proposition. If this probabilistic insight is translated into the semantics of counterfactuals with sufficient care—in particular, so that a nice logic of conditionals comes out valid (and hence also P3), and our initial bridge principle on the truth of counterfactuals and conditional chance (P4) is satisfied—then for each of our premises there is a context in which the premise expresses a true proposition in that context. In order to show how this can be done I employed the formal framework of a Lewis–Stalnaker type semantics, however, I refrained from interpreting the similarity or closeness relations between worlds in this semantics in terms of anything like the usual Lewisian heuristics.

If one compares this to the solutions that were sketched back in section D.3, we find that the new proposal is doing pretty well. In the coarse-grained context c of the last section, all premises are true with the exception of P2 and P5, in particular, with the exception of the first conjunct of P2,

$$\text{Necessarily: } B \text{ if and only if } C_1 \vee \ldots \vee C_{1000000},$$

and the first conjunct of P5,

$$\text{Possibly: } (A \wedge C_i) \vee (A \wedge \neg B).$$

But in that context, presumably, one is not even interested in expressing either of them, since one is not interested in the fine-grained possibilities C_i. Instead, one is interested in asserting 'If A had been the case, B would have been the case' ('If the host had made it to the studio, there would have been the TV lottery that day'), which is perfectly harmless, as that counterfactual is true in that context (relative to the actual world).

Now what happens if one shifts one's attention to the more fine-grained possibilities, that is, to the possible outcomes of the counterfactual TV lottery? This means that one changes the context to something like c' from the last section: in c', one is maybe interested in expressing the first conjunct of P2 and the first conjunct of P5, and indeed they turn out to be true in c'. But at the same time, while all other premises remain intact, P1 happens to be false:

If A had been the case, B would have been the case.

In section D.4, amongst others, I argued that it was implausible to deny P1 on the grounds that the truth of $A \mathrel{\Box\!\!\rightarrow} B$ would require the chance of B given A to be exactly 1, and hence that P1 would have to be false by the simultaneous presupposition of the quasi-empirical premise P5. The truth of $A \Box\!\!\rightarrow B$ should in fact be consistent with the chance of B given A to be less than 1 by some small real-valued margin, since otherwise counterfactuals would be far too sensitive to the possible occurrence of exceptional circumstances. This rebuttal of the initial attack against P1 does not lead to worries concerning what we found to be the case in c': in order for $A \Box\!\!\rightarrow B$ to be true in @, c', and with COMP from the last section still to be satisfied (or P4 would fail), it would be sufficient for the conditional chance of $\neg B$ given A to be positive but super-tiny, that is, less than the chance of each lottery outcome singleton $\{w_i\}$ given A. It is because *that* is not the case that $A \Box\!\!\rightarrow B$ happens to be false in @, c'. In the previous context c this was not an issue because $\{w_i\}$ did not count as a proposition then. This said, of course, if the conditional chance of $\neg B$ given A were in fact less than the chance of each $\{w_i\}$ given A, then something else would have to go in c' (or a contradiction would follow by our original paradoxical argument): in this case, clearly, the second conjunct of P5 would be false then.

But isn't it *still* plausible that P1 should come out as true, even in a context such as c'? That is where the contextualist strategy kicks in: according to the new proposal, the plausibility of P1 arises out of contexts such as c, in which P1 is indeed true. And the plausibility of the first conjuncts of P2 and P5 is due to contexts such as c', in which these conjuncts hold. But one ought to be careful enough not to mix the two contexts and not to mistake the plausibility of one premise in one context for the plausibility of the same premise in a different context.

For some theses, however, it does not matter in which contexts they are being considered; in our case, P3 and P4 are true in whatever context, which is appropriate in view of the universal, and maybe even necessary, character of the two premises.[422] All other premises are specific for the TV lottery and host situation, and hence the context-dependency of their truth values should be less worrying.

That is not to say that contextualization does not come with a price. In particular, there are obvious barriers of inference from some contexts to others, or the original paradox could be reinstated. Indeed, as we have seen, it is possible that one counterfactual is true in one context but not true in another, one set of worlds counts as a proposition in one context but not in another, and one counterfactual expresses a proposition in one context but not in another. Hence, if any of these verdicts were to be transferred from the one context to the other, one would end up with an inconsistent setting again. Which does not mean either that no inferences whatsoever are being permitted to lead from one context to the next: for instance, since I take P3 and P4 for granted universally, if all counterfactuals in a set Φ are true in two contexts (and the actual world), then the results of applying some logically valid rules to members of Φ in the one context can be transferred directly to the other context, as schematically the same rules are valid in both of them (by P3); accordingly, the same conclusions

422 So far as the necessity of P4 is concerned, this will depend on how closely—conceptually or ontologically—counterfactuals and conditional chance are tied to each other.

can be drawn from them on conditional chance (by P4). Additionally, as mentioned in the last section, my approach is perfectly consistent with no sentence expressing distinct propositions in any two contexts in which it expresses propositions at all. So it is still fine to assume that sentences mean the same in any two contexts in which they mean anything at all. However, the meaning of the schematic principles P3 and P4 still varies from context to context: for both principles need to be restricted to sentences that express propositions in the very context in question.

Bibliography

Adams, Ernest W. 1965. 'On the Logic of Conditionals', *Inquiry* 8(1–4), 166–97.

Adams, Ernest W. 1966. 'Probability and the Logic of Conditionals', in Jaakko Hintikka and Patrick Suppes (eds), *Aspects of Inductive Logic*, Studies in Logic and the Foundations of Mathematics, 43. Amsterdam: North-Holland, 265–316.

Adams, Ernest W. 1975. *The Logic of Conditionals: An Application of Probability to Deductive Logic*, Synthese Library, 86. Dordrecht: Reidel.

Alchourrón, Carlos E., Peter Gärdenfors, and David Makinson. 1985. 'On the Logic of Theory Change: Partial Meet Contraction and Revision Functions', *Journal of Symbolic Logic* 50(2), 510–30.

Anscombe, Gertrude. 1957. *Intention*. Oxford: Basil Blackwell.

Arló-Costa, Horacio. 2001. 'Bayesian Epistemology and Subjective Conditionals: On the Status of the Export-Import Laws', *Journal of Philosophy* 98(11), 555–98.

Arló-Costa, Horacio, and Rohit Parikh. 2005. 'Conditional Probability and Defeasible Inference', *Journal of Philosophical Logic* 34(1), 97–119.

Armstrong, David. 1968. *A Materialistic Theory of the Mind*. London: Routledge.

Baltag, Alexandru and Sonja Smets. 2008. 'Probabilistic Dynamic Belief Revision', *Synthese* 165(2), 179–202.

Barber, David. 2012. *Bayesian Reasoning and Machine Learning*. Cambridge: Cambridge University Press.

Battigalli, Pierpaolo. 2002. 'Strong Belief and Forward Induction Reasoning', *Journal of Economic Theory* 106(2), 356–91.

Benferhat, Salem, Didier Dubois, and Henri Prade. 1999. 'Possibilistic and Standard Probabilistic Semantics of Conditional Knowledge', *Journal of Logic and Computation* 9(6), 873–95.

Bennett, Jonathan. 2003. *A Philosophical Guide to Conditionals*. Oxford: Clarendon Press.

Bovens, Luc, and Stephan Hartmann. 2003. *Bayesian Epistemology*. Oxford: Clarendon Press.

Bradley, Richard. 2006. 'Adams Conditionals and Non-Monotonic Probabilities', *Journal of Logic, Language and Information* 15(1–2), 65–81.

Bratman, Michael E. 1987. *Intention, Plans, and Practical Reason*. Cambridge, Mass.: Harvard University Press.

Bratman, Michael E. 1999. *Faces of Intention*. Cambridge: Cambridge University Press.

Brogaard, Berit, and Joe Salerno. 2008. 'Counterfactuals and Context', *Analysis* 68(1), 39–46.

Broome, John. 1991. *Weighing Goods*. Cambridge, Mass.: Blackwells.

Buchak, Lara. 2014. 'Belief, Credence, and Norms', *Philosophical Studies* 169(2), 285–311.

Cantwell, John. 1999. 'Static Justification in the Dynamics of Belief', *Erkenntnis* 50(2), 481–503.

Cariani, Fabrizio. 2016. 'Local Supermajorities', *Erkenntnis* 81(2), 391–406.

Carnap, Rudolf. 1950a. *Logical Foundations of Probability*. Chicago, Ill.: University of Chicago Press.

Carnap, Rudolf. 1950b. 'Empiricism, Semantics, and Ontology', *Revue Internationale de Philosophie* 4, 20–40.

Carnap, Rudolf. 1961. 'On the Use of Hilbert's ϵ-operator in Scientific Theories', in Y. Bar-Hillel, E. I. J. Poznanski, M. O. Rabin, and A. Robinson (eds), *Essays on the Foundations of Mathematics*. Jerusalem: Magnes Press, 156–64.

Carnap, Rudolf. 1963. 'Replies and Expositions', in Paul A. Schilpp (ed.), *The Philosophy of Rudolf Carnap*. La Salle, Ill.: Open Court, 859–1013.

Carr, Jennifer. 2015. 'Subjective Ought', *Ergo* 2(27), 678–710.

Chan, Timothy, ed. 2013. *The Aim of Belief*. Oxford: Oxford University Press.

Chellas, Brian F. 1980. *Modal Logic: An Introduction*. Cambridge: Cambridge University Press.

Chisholm, Roderick. 1989. *Theory of Knowledge*. 3rd edn. Englewood Cliffs, NJ: Prentice-Hall.

Christensen, David. 2004. *Putting Logic in its Place. Formal Constraints on Rational Belief*. Oxford: Clarendon Press.

Churchland, Paul. 1981. 'Eliminative Materialism and the Propositional Attitudes', *The Journal of Philosophy* 78(2), 67–90.

Clarke, Roger. 2013. 'Belief is Credence One (in Context)', *Philosopher's Imprint* 13(11), 1–18.

Coffman, Eldon J. 2012. 'Contextualism and Interest-Relative Invariantism', in Andrew Cullison (ed.), *The Continuum Companion to Epistemology*. London: Continuum, 199–226.

Coffman, Eldon J. 2014. 'Lenient Accounts of Warranted Assertability', in Clayton Littlejohn and John Turri (eds), *Epistemic Norms: New Essays on Action, Belief and Assertion*. Oxford: Oxford University Press, 33–58.

Cohen, Laurence J. 1989. 'Belief and Acceptance', *Mind* 98(391), 367–89.

Cohen, Laurence J. 1992. *An Essay on Belief and Acceptance*. Oxford: Oxford University Press.

Collins, John. 1988. 'Belief, Desire, and Revision', *Mind* 97(387), 333–42.

Cresswell, Max J. 1975. 'Hyperintensional Logic', *Studia Logica* 34(1), 25–38.

Cresto, Eleonora. 2010. 'Belief and Contextual Acceptance', *Synthese* 177(1), 41–66.

Darwiche, Adnan, and Judea Pearl. 1997. 'On the Logic of Iterated Belief Revision', *Artificial Intelligence* 89, 1–29.

Davidson, Donald. 1963. 'Actions, Reasons, and Causes', *Journal of Philosophy* 60(23), 685–700. Reprinted in Donald Davidson, *Essays on Actions and Events*. Oxford: Clarendon Press, 1980, 3–19.

Davidson, Donald. 1970. 'Mental Events', in Lawrence Foster and Joe William Swanson (eds), *Experience and Theory*. New York: Humanities Press, 79–101. Reprinted in Donald Davidson, *Essays on Actions and Events*. Oxford: Clarendon Press, 1980, 207–27.

Davidson, Donald. 1975. 'Thought and Talk', in Samuel D. Guttenplan (ed.), *Mind and Language*. Oxford: Clarendon Press, 7–23. Reprinted in Donald Davidson, *Inquiries into Truth and Interpretation*. Oxford: Clarendon Press, 1984, 154–70.

De Vries, Marc J., and Anthonie W. M. Meijers. 2013. 'Beliefs, Acceptances and Technological Knowledge', in Marc J. de Vries et al. (eds), *Norms in Technology*, Philosophy of Engineering and Technology, 9, Dordrecht: Springer, 55–65.

Dorling, Jon. 1979. 'Bayesian Personalism, the Methodology of Scientific Research Programmes, and Duhem's Problem', *Studies in the History and Philosophy of Science Part A* 10(3), 177–87.

Dorst, Kevin. N.d. 'Lockeans Maximize Expected Accuracy', unpublished draft.

Douven, Igor. 2006. 'Assertion, Knowledge, and Rational Credibility', *Philosophical Review* 115(4), 449–85.

Douven, Igor. 2010. 'The Pragmatics of Belief', *Journal of Pragmatics* 42(1), 35–47.

Douven, Igor, and Timothy Williamson. 2006. 'Generalizing the Lottery Paradox', *British Journal for the Philosophy of Science* 57(4), 755–79.

Dreier, James. 1996. 'Rational Preference: Decision Theory as a Theory of Practical Rationality', *Theory and Decision* 40, 249–76.

Dubois, Didier, Hélène Fargier, Henri Prade, and Régis Sabbadin. 2009. 'A Survey of Qualitative Decision Rules under Uncertainty', ch. 11 in Denis Bouyssou et al. (eds), *Decision-Making Process: Concepts and Methods*. Hoboken, NJ: John Wiley & Sons, 435–73.

Earman, John. 1992. *Bayes or Bust: A Critical Examination of Bayesian Confirmation Theory*. Cambridge, Mass.: MIT Press.

Easwaran, Kenny. 2016. 'Dr. Truthlove or: How I Learned to Stop Worrying and Love Bayesian Probabilities', *Nous* 50(4), 816–53.

Easwaran, Kenny, and Branden Fitelson. 2014. 'Accuracy, Coherence, and Evidence', *Oxford Studies in Epistemology* 5, 61–96.

Edgington, Dorothy. 1995. 'On Conditionals', *Mind* 104(414), 235–329.

Engel, Pascal. 1998. 'Belief, Holding True, and Accepting', *Philosophical Explorations* 1(2), 140–51.

Engel, Pascal, ed. 2000. *Believing and Accepting*. Dordrecht: Kluwer.

Evans, Jonathan St. 2008. 'Dual-Processing Accounts of Reasoning, Judgment, and Social Cognition', *Annual Review of Psychology* 59, 255–78.

Fagin, Ronald, Joseph Y. Halpern, Yoram Moses, and Moshe Y. Vardi. 1995. *Reasoning about Knowledge*, Cambridge, Mass.: MIT Press.

Fantl, Jeremy, and Matthew McGrath. 2009. *Knowledge in an Uncertain World*. Oxford: Oxford University Press.

Festinger, Leon. 1957. *A Theory of Cognitive Dissonance*. Stanford, Calif.: Stanford University Press.

Field, Hartry. 2009. 'What is the Normative Role of Logic?', *Proceedings of the Aristotelian Society* 83, 251–68.

Fitelson, Branden. 2013. 'Contrastive Bayesianism', in Martijn Blaauw (ed.), *Contrastivism in Philosophy*. New York: Routledge, 64–87.

Fitelson, Branden. N.d. 'Coherence', unpublished manuscript.

Fodor, Jerry A. 1975. *The Language of Thought*. New York: Cromwell.

Foley, Richard. 1993. *Working without a Net*. Oxford: Oxford University Press.

Frankish, Keith. 2004. *Mind and Supermind*. Cambridge: Cambridge University Press.

Frankish, Keith. 2009. 'Partial Belief and Flat-Out Belief', in Huber and Schmidt-Petri (2009), 75–93.

Friedman, Michael. 2001. *Dynamics of Reason: The 1999 Kant Lectures at Stanford University*. Stanford, Calif: CSLI Publications.

Gabbay, Dov M. 1984. 'Theoretical Foundations for Non-Monotonic Reasoning in Expert Systems', in Krysztof R. Apt (ed.), *Logics and Models of Concurrent Systems*. Berlin: Springer, 439–58.

Gärdenfors, Peter. 1986a. 'The Dynamics of Belief: Contractions and Revisions of Probability Functions', *Topoi* 5, 29–37.

Gärdenfors, Peter. 1986b. 'Belief Revision and the Ramsey Test for Conditionals', *Philosophical Review* 95(1), 81–93.

Gärdenfors, Peter. 1988. *Knowledge in Flux*. Cambridge, Mass.: MIT Press.

Gärdenfors, Peter, and David C. Makinson. 1994. 'Nonmonotonic Inference Based on Expectations', *Artificial Intelligence* 65(2), 197–245.

Gauker, Christopher. 2010. 'Contexts in Formal Semantics', *Philosophy Compass* 5(7), 568–78.

Gibbard, Allan. 2005. 'Truth and Correct Belief', *Philosophical Issues* 15, 338–50.

Gibbard, Allan. 2007. 'Rational Credence and the Value of Truth', in Tamar Szabo Gendler and John Hawthorne (eds), *Oxford Studies in Epistemology*, vol. 2. Oxford: Oxford University Press, 143–64.

Glymour, Clark. 1980. *Theory and Evidence*. Princeton, NJ: Princeton University Press.

Goldszmidt, Moisés, and Judea Pearl. 1992. 'Rank-Based Systems: A Simple Approach to Belief Revision, Belief Update, and Reasoning about Evidence and Actions', in *Proceedings of the 3rd International Conference on Knowledge Representation and Reasoning*. San Mateo, Calif.: Morgan Kaufmann, 661–72.

Greaves, Hilary, and David Wallace. 2006. 'Justifying Conditionalization: Conditionalization Maximizes Expected Epistemic Utility', *Mind* 115(459): 607–32.

Grice, H. Paul. 1989. *Studies in the Way of Words*, Cambridge, Mass.: Harvard University Press.

Groenendijk, Jeroen, and Martin Stokhof. 1984. *Studies on the Semantics of Questions and the Pragmatics of Answers*, Joint Ph.D. thesis, University of Amsterdam, Department of Philosophy.

Grove, Adam. 1988. 'Two Modellings for Theory Change', *Journal of Philosophical Logic* 17(2), 157–70.

Hájek, Alan. 2003. 'What Conditional Probability Could Not Be', *Synthese* 137(3), 273–323.

Hájek, Alan. 2009. 'Arguments For—or Against—Probabilism?', in Huber and Schmidt-Petri (2009), 229–51.

Hájek, Alan. 2012a. 'The Fall of "Adams Thesis"?', *Journal of Logic, Language and Information* 21(2), 145–61.

Hájek, Alan. 2012b. 'Is Strict Coherence Coherent?', *dialectica* 66(3), 411–24.

Hájek, Alan. N.d. 'Most Counterfactuals are False', unpublished draft.

Halpern, Joseph Y. 2003. *Reasoning about Uncertainty*. Cambridge, Mass.: MIT Press.

Hansson, Sven Ove. 1999. *A Textbook of Belief Dynamics: Theory Change and Database Updating*. Dordrecht: Kluwer.

Hansson, Sven Ove. 2010. 'Objective or Subjective "Ought"?', *Utilitas* 22(1), 33–5.

Harman, Gilbert. 1986. *Change in View*. Cambridge, Mass.: MIT Press.

Hawthorne, James. 2009. 'The Lockean Thesis and the Logic of Belief', in Huber and Schmidt-Petri (2009), 49–74.

Hawthorne, James, and Luc Bovens. 1999. 'The Preface, the Lottery, and the Logic of Belief', *Mind* 108(430), 241–64.

Hawthorne, James, and David Makinson. 2007. 'The Quantitative/Qualitative Watershed for Rules of Uncertain Inference', *Studia Logica* 86(2): 247–97.

Hawthorne, John. 2004. *Knowledge and Lotteries*. Oxford: Oxford University Press.

Hawthorne, John. 2005. 'Chance and Counterfactuals', *Philosophy and Phenomenological Research* 70(2): 396–405.

Hawthorne, John, and Maria Lasonen-Aarnio. 2009. 'Knowledge and Objective Chance', in. Patrick Greenough and Duncan Pritchard (eds), *Williamson on Knowledge*. Oxford: Oxford University Press, 92–108.

Hempel, Carl G. 1962. 'Deductive-Nomological vs Statistical Explanation', in H. Feigl and G. Maxwell (eds), *Minnesota Studies in the Philosophy of Science*, vol. 3. Minneapolis: University of Minnesota Press, 98–169.

Hilpinen, Risto. 1968. *Rules of Acceptance and Inductive Logic, Acta Philosophica Fennica*, 22. Amsterdam: North-Holland.

Hintikka, Jaakko. 1962. *Knowledge and Belief: An Introduction to the Logic of the Two Notions*. Ithaca, NY: Cornell University Press.

Holton, Richard. 2014. 'Intention as a Model of Belief', in Manuel Vargas and Gideon Yaffe (eds), *Rational and Social Agency: Essays on the Philosophy of Michael Bratman*. Oxford: Oxford University Press, 12–37.

Horwich, Paul. 2013. 'Belief-Truth Norms', in Chan (2013), 17–31.

Howson, Colin, and Peter Urbach. 1989. *Scientific Reasoning: The Bayesian Approach*. La Salle, Ill.: Open Court.

Huber, Franz, and Christoph Schmidt-Petri, eds. 2009. *Degrees of Belief*, Synthese Library. Berlin: Springer.

Hume, David. 1978. *A Treatise of Human Nature*. In Lewis Amherst Selby-Bigge (ed.), 2nd edn, with text revised and variant readings by Peter H. Nidditch. Oxford: Oxford University Press.

Ichikawa, Jonathan. 2011. 'Quantifiers, Knowledge, and Counterfactuals', *Philosophy and Phenomenological Research* 82(2), 287–313.

Jackson, Frank. 1979. 'On Assertion and Indicative Conditionals', *Philosophical Review* 88(4), 565–89.

Jackson, Frank. 1981. 'Conditionals and Possibilia', *Proceedings of the Aristotelian Society* 81, 125–37.

James, William. 1896. 'The Will to Believe', in *The Will to Believe and Other Essays in Popular Philosophy*. New York: Dover Publications, 1956, 1–31.

Jeffrey, Richard. 1965. *The Logic of Decision*. New York: McGraw-Hill.

Jeffrey, Richard. 1970. 'Dracula Meets Wolfman: Acceptance vs. Partial Belief', in Swain (1970), 157–85.

Jeffrey, Richard. 2004. *Subjective Probability: The Real Thing*. Cambridge: Cambridge University Press.

Johnson-Laird, Philip N. 1983. *Mental Models*. Cambridge, Mass.: Harvard University Press.

Joyce, James M. 1998. 'A Nonpragmatic Vindication of Probabilism', *Philosophy of Science* 65(4), 575–603.

Joyce, James M. 1999. *The Foundations of Causal Decision Theory*. Cambridge: Cambridge University Press.

Joyce, James M. 2009. 'Accuracy and Coherence: Prospects for an Alethic Epistemology of Partial Belief', in Franz Huber and Christoph Schmidt-Petri (eds), *Degrees of Belief*. Berlin: Springer, 263–300.

Joyce, James M. N.d. 'Why Evidentialists Need Not Worry about the Accuracy Argument for Probabilism', unpublished draft.

Kaplan, Mark. 1981. 'Rational Acceptance', *Philosophical Studies* 40(2), 129–45.

Kaplan, Mark. 1996. *Decision Theory as Philosophy*. Cambridge: Cambridge University Press.

Kim, Jaegwon. 1984. 'Concepts of Supervenience', *Philosophy and Phenomenological Research* 45(2), 153–76.

Koellner, Peter. 2013. 'Large Cardinals and Determinacy', entry in the *Stanford Encyclopedia of Philosophy*, <http://plato.stanford.edu/entries/large-cardinals-determinacy/>.

Krantz, David, Duncan Luce, Patrick Suppes, and Amos Tversky. 1971. *Foundations of Measurement*, vol. 1. *Additive and Polynomial Representations*. New York: New York Academy Press.

Kraus, Sarit, Daniel Lehmann, and Menachem Magidor. 1990. 'Nonmonotonic Reasoning, Preferential Models and Cumulative Logics', *Artificial Intelligence* 44, 167–207.

Kripke, Saul. 1979. 'A Puzzle about Belief', in Avishai Margalit (ed.), *Meaning and Use*, Dordrecht: Reidel, 239–83.

Krombholz, Martin. 2013. *A Joint Theory of Belief and Degrees of Belief and its Implementation*, Bachelor thesis in philosophy, LMU Munich.

Kvart, Igal. 1986. *A Theory of Counterfactuals*. Indianapolis: Hackett.

Kyburg, Henry E. Jr. 1961. *Probability and the Logic of Rational Belief*. Middletown: Wesleyan University Press.

Kyburg, Henry E. Jr. 1970a. 'Conjunctivitis', in Swain (1970), 55–82.

Kyburg, Henry E. Jr. 1970b. *Probability and Inductive Logic*. Toronto: Macmillan.

Kyburg, Henry E. Jr., and Cho Man Teng. 2001. *Uncertain Inference*. Cambridge: Cambridge University Press.

Lange, Marc. 2005. 'Laws and their Stability', *Synthese* 144(3), 415–32.

Lawlor, Krista. 2014. 'Exploring the Stability of Belief: Resiliency and Temptation', *Inquiry* 57(1), 1–27.

Lehmann, Daniel, and Menachem Magidor. 1992. 'What Does a Conditional Knowledge Base Entail?', *Artificial Intelligence* 55(1), 1–60.

Lehrer, Keith. 1983. 'Belief, Acceptance and Cognition', in Herman Paaret (ed.), *On Believing: Epistemological and Semiotic Approaches*. Berlin: de Gruyter, 172–83.

Lehrer, Keith. 1990. *Theory of Knowledge*. London: Routledge.

Lehrer, Keith, and Thomas Paxson. 1969. 'Knowledge: Undefeated Justified True Belief', *Journal of Philosophy* 66(8), 225–37.

Leitgeb, Hannes. 2004. *Inference on the Low Level: An Investigation into Deduction, Nonmonotonic Reasoning, and the Philosophy of Cognition*. Dordrecht: Kluwer, Applied Logic Series.

Leitgeb, Hannes. 2007. 'Beliefs in Conditionals vs. Conditional Beliefs', *Topoi* 26(1), 115–32.

Leitgeb, Hannes. 2010. 'On the Ramsey Test without Triviality', *Notre Dame Journal of Formal Logic* 51(1), 21–54.

Leitgeb, Hannes. 2011. 'God – Moore = Ramsey: (A Reply to Chalmers and Hájek)', *Topoi* 30(1), 47–51.

Leitgeb, Hannes. 2012a. 'A Probabilistic Semantics for Counterfactuals: Part A', *Review of Symbolic Logic* 5(1), 16–84.

Leitgeb, Hannes. 2012b. 'A Probabilistic Semantics for Counterfactuals: Part B', *Review of Symbolic Logic* 5(1), 85–121.

Leitgeb, Hannes. 2012c. 'Metacognition and Indicative Conditionals: A Précis', in M. J. Beran, J. Brandl, J. Perner, and J. Proust (eds), *Foundations of Metacognition*. Oxford: Oxford University Press, 252–66.

Leitgeb, Hannes. 2013a. 'Reducing Belief *Simpliciter* to Degrees of Belief', *Annals of Pure and Applied Logic* 164(12), 1338–89.

Leitgeb, Hannes. 2013b. 'Author's Response', in Van Benthem and Liu (2013), 68–73.

Leitgeb, Hannes. 2013c. 'Scientific Philosophy, Mathematical Philosophy, and All That', *Metaphilosophy* 44(3), 267–75.

Leitgeb, Hannes. 2014a. 'The Stability Theory of Belief', *Philosophical Review* 123(2), 131–71.

Leitgeb, Hannes. 2014b. 'A Way out of the Preface Paradox?', *Analysis* 74(1), 11–15.

Leitgeb, Hannes. 2014c. 'A Lottery Paradox for Counterfactuals without Agglomeration', *Philosophy and Phenomenological Research* 89(3), 605–36.

Leitgeb, Hannes. 2014d. 'The Review Paradox: A Note on the Diachronic Costs of Not Closing Rational Belief under Conjunction', *Nous* 48(4), 781–93.

Leitgeb, Hannes. 2014e. 'Belief as a Simplification of Probability, and What This Entails', in A. Baltag and S. Smets (eds), *Johan van Benthem on Logical and Information Dynamics*, Outstanding Contributions to Logic, 5. Berlin: Springer, 405–17.

Leitgeb, Hannes. 2014f. 'Belief as Qualitative Probability', in Colleen E. Crangle, Adolfo García de la Sienra, and Helen E. Longino (eds), *Foundations and Methods From Mathematics to Neuroscience: Essays Inspired by Patrick Suppes*. Stanford, Calif: CSLI Publications, 119–30.

Leitgeb, Hannes. 2015. 'The Humean Thesis on Belief', *Proceedings of the Aristotelian Society* 89(1), 143–85.

Leitgeb, Hannes. 2016. 'Probability in Logic', in Alan Hájek and Chris Hitchcock (eds), *The Oxford Handbook of Probability and Philosophy*. Oxford: Oxford University Press, 681–704.

Leitgeb, Hannes. N.d. 'From Epistemic Utility to the Lockean Thesis', unpublished manuscript.

Leitgeb, Hannes, and Richard Pettigrew. 2010a. 'An Objective Justification of Bayesianism I: Measuring Inaccuracy', *Philosophy of Science* 77(2), 201–35.

Leitgeb, Hannes, and Richard Pettigrew. 2010b. 'An Objective Justification of Bayesianism II: The Consequences of Minimizing Inaccuracy', *Philosophy of Science* 77(2), 236–72.

Leitgeb, Hannes, and Krister Segerberg. 2007. 'Dynamic Doxastic Logic: Why, How, and Where to?', *Synthese KRA* 155(2), 167–90.

Levi, Isaac. 1967. *Gambling with the Truth: An Essay on Induction and the Aims of Science*. Cambridge, Mass.: MIT Press.

Levi, Isaac. 1980. *The Enterprise of Knowledge. An Essay on Knowledge, Credal Probability and Chance*. Cambridge, Mass.: MIT Press.

Levi, Isaac. 1984. *Decisions and Revisions*. Cambridge: Cambridge University Press.

Levi, Isaac. 1988. 'Iteration of Conditionals and the Ramsey Test', *Synthese* 76(1), 49–81.

Levi, Isaac. 1996. *For the Sake of the Argument: Ramsey Test Conditionals, Inductive Inference, and Nonmonotonic Reasoning*. Cambridge: Cambridge University Press.

Levi, Isaac. 2002. 'Commitment and Change of View', in José Luis Bermúdez and Alan Millar (eds), *Reason and Nature: Essays in the Theory of Rationality*. Oxford: Oxford University Press, 209–32.

Lewis, David K. 1966. 'An Argument for the Identity Theory', *Journal of Philosophy* 63(2), 17–25.

Lewis, David K. 1970. 'How to Define Theoretical Terms', *Journal of Philosophy* 67(13), 427–46.

Lewis, David K. 1973. *Counterfactuals*. Oxford: Blackwell.

Lewis, David K. 1976. 'Probabilities of Conditionals and Conditional Probabilities', *Philosophical Review* 85(3), 297–315. Reprinted with postscript in David K. Lewis, *Philosophical Papers*, vol. 2. Oxford: Oxford University Press, 1986, 133–56.

Lewis, David K. 1979a. 'Attitudes De Dicto and De Se', *Philosophical Review* 88(4), 513–43.

Lewis, David K. 1979b. 'Counterfactual Dependence and Time's Arrow', *Noûs* 13, 455–76.

Lewis, David K. 1983. 'New Work for a Theory of Universals', *Australasian Journal of Philosophy* 61, 343–77.

Lewis, David K. 1986. 'Postscript to "Probabilities of Conditionals and Conditional Probabilities". Indicative Conditionals Better Explained', in *Philosophical Papers*, vol. 2. Oxford: Oxford University Press, 152–6.

Lewis, David K. 1988a. 'Statements Partly about Observation', *Philosophical Papers* 17, 1–31.

Lewis, David K. 1988b. 'Desire as Belief', *Mind* 97(387), 323–32.

Lewis, David K. 1996. 'Elusive Knowledge', *Australasian Journal of Philosophy* 74(4), 549–67.

Lin, Hanti. 2013. 'Foundations of Everyday Practical Reasoning', *Journal of Philosophical Logic* 42(6), 831–62.

Lin, Hanti, and Kevin T. Kelly. 2012a. 'A Geo-Logical Solution to the Lottery Paradox', *Synthese* 186(2), 531–75.

Lin, Hanti, and Kevin T. Kelly. 2012b. 'Propositional Reasoning that Tracks Probabilistic Reasoning', *Journal of Philosophical Logic* 41(6), 957–81.

Lin, Hanti, and Kevin T. Kelly. 2013. 'Comments on "The Stability Theory of Belief. A Summary"', in Van Benthem and Liu (2013), 60–7.

Lindström, Sten, and Wlodek Rabinowicz. 1990. 'Epistemic Entrenchment with Incomparabilities and Relational Belief Revision', in André Fuhrmann and Michael Morreau (eds), *The Logic of Theory Change*, Lecture Notes in Artificial Intelligence 465. Berlin: Springer, 93–126.

Loeb, Louis E. 2001. 'Integrating Hume's Account of Belief and Justification', *Philosophy and Phenomenological Research* 63(2), 279–303.

Loeb, Louis E. 2002. *Stability and Justification in Hume's Treatise*. Oxford: Oxford University Press.

Loeb, Louis E. 2010. *Reflection and the Stability of Belief: Essays on Descartes, Hume, and Reid*. Oxford: Oxford University Press.

Loewer, Barry. 2007. 'Counterfactuals and the Second Law', in Huw Price and Richard Corry (eds), *Causation, Physics, and the Constitution of Reality: Russell's Republic Revisited*. Oxford: Oxford University Press, 293–326.

MacFarlane, John. 2004. 'In What Sense (If Any) is Logic Normative for Thought?', unpublished manuscript.

MacFarlane, John. 2009. 'Nonindexical Contextualism', *Synthese* 166, 231–50.

MacFarlane, John. 2011a. 'What is Assertion?', in Jessica Brown and Herman Cappelen (eds), *Assertion: New Philosophical Essays*. Oxford: Oxford University Press, 79–96.

MacFarlane, John. 2011b. 'Relativism and Knowledge Attributions', in Sven Bernecker and Duncan Pritchard (eds), *Routledge Companion to Epistemology*. London: Routledge, 536–44.

McGee, Vann. 1994. 'Learning the Impossible', in Ellery Eells and Brian Skyrms (eds), *Probability and Conditionals: Belief Revision and Rational Decision*. Cambridge: Cambridge University Press, 179–99.

Maher, Patrick. 1981. 'Probability in Hume's Science of Man', *Hume Studies* 7(2), 137–53.

Maher, Patrick. 1990. 'Acceptance without Belief', in *Proceedings of the Biennial Meeting of the Philosophy of Science Association 1990*, Vol. 1. Contributed Papers. Chicago, Ill.: The University of Chicago Press, 381–92.

Maher, Patrick. 1993. *Betting on Theories*. Cambridge: Cambridge University Press.

Makinson, David C. 1965. 'The Paradox of the Preface', *Analysis* 25(6), 205–7.

Makinson, David C. 2011. 'Conditional Probability in the Light of Qualitative Belief Change', *Journal of Philosophical Logic* 40(2), 121–53.

Makinson, David C. N.d. 'The Scarcity of Stable Belief Sets', unpublished manuscript.

Makinson, David C., and Peter Gärdenfors. 1991. 'Relations between the Logic of Theory Change and Nonmonotonic Logic', in André Fuhrmann and Michael Morreau (eds), *The Logic of Theory Change*. Berlin: Springer, 185–205.

Millikan, Ruth G. 1984. *Language, Thought and Other Biological Categories*. Cambridge, Mass.: MIT Press.

Milne, Peter. 2012. 'Belief, Degrees of Belief, and Assertion', *dialectic* 66(3), 331–49.

Nolan, Daniel. 2014. 'Hyperintensional Metaphysics', *Philosophical Studies* 171(1), 149–60.

Nozick, Robert. 1994. *The Nature of Rationality*. Princeton, NJ: Princeton University Press.

Oaksford, Mike, and Nick Chater. 2007. *Bayesian Rationality. The Probabilistic Approach to Human Reasoning*. Oxford: Oxford University Press.

Olsson, Erik J. 1998. 'Competing for Acceptance: Lehrer's Rule and the Paradoxes of Justification', *Theoria* 64(1), 34–54.

Pagin, Peter. 1990. 'Review of Roy Sorenson, *Blindspots*, Clarendon Press, Oxford 1988', *History and Philosophy of Logic* 11, 243–5.

Paglieri, Fabio. 2009. 'Acceptance as Conditional Disposition', in Alexander Hieke and Hannes Leitgeb (eds), *Reduction: Between the Mind and the Brain*. Heusenstamm: Ontos Verlag, 12–29.

Papineau, David. 1996. 'Theory-Dependent Terms', *Philosophy of Science* 63(1), 1–20.

Pedersen, Arthur P., and Horacio Arló-Costa. 2012. 'Belief and Probability: A General Theory of Probability Cores', *International Journal of Approximate Reasoning* 53(3), 293–315.

Pettigrew, Richard. 2015. 'Pluralism about Belief States', *Proceedings of the Aristotelian Society* 89(1), 187–204.

Pettigrew, Richard. 2016. *Accuracy and the Laws of Credence*. Oxford: Oxford University Press.

Pollock, John L. 1974. *Knowledge and Justification*. Princeton, NJ: Princeton University Press.

Pollock, John L. 1986. *Contemporary Theories of Knowledge*. Savage, MD: Rowman & Littlefield.

Pollock, John L. 1994. 'Justification and Defeat', *Artificial Intelligence* 67, 377–407.

Putnam, Hilary. 1963. ' "Degree of Confirmation" and Inductive Logic', in Paul A. Schilpp (ed.), *The Philosophy of Rudolf Carnap*. La Salle, Ill.: Open Court, 761–83.

Quine, Willard v. O. 1960. *Word and Object*. Cambridge, Mass.: MIT Press.

Ramsey, Frank P. 1931. 'General Propositions and Causality', in Richard B. Braithwaite (ed.), *The Foundations of Mathematics and Other Logical Essays*. London: Kegan Paul.

Roorda, Jonathan. 1995. 'Revenge of Wolfman: A Probabilistic Explication of Full Belief', unpublished <https://www.princeton.edu/~bayesway/pu/Wolfman.pdf>.

Ross, Jacob, and Mark Schroeder. 2014. 'Belief, Credence and Pragmatic Encroachment', *Philosophy and Phenomenological Research* 88(2), 259–88.

Rott, Hans. 2004. 'Stability, Strength and Sensitivity: Converting Belief into Knowledge', *Erkenntnis* 61(2–3), 469–93.

Rott, Hans. 2008. 'A New Psychologism in Logic? Reflections from the Point of View of Belief Revision', *Studia Logica* 88(1), 113–36.

Rott, Hans. 2009. 'Shifting Priorities: Simple Representations for Twenty-Seven Iterated Theory Change Operators', in David Makinson, Jacek Malinowski, and Heinrich Wansing (eds), *Towards Mathematical Philosophy*. Dordrecht: Springer, 269–96.

Russell, Bertrand. 1993. *Theory of Knowledge: The 1913 Manuscript*. New York: Routledge.

Schurz, Gerhard. 1998. 'Probabilistic Semantics for Delgrande's Conditional Logic and a Counterexample to his Default Logic', *Artificial Intelligence* 102(1), 81–95.

Schurz, Gerhard. 2001. 'What is "Normal"? An Evolution-Theoretic Foundation of Normic Laws and Their Relation to Statistical Normality', *Philosophy of Science* 68(4), 476–97.

Schurz, Gerhard, and Hannes Leitgeb. 2008. 'Finitistic and Frequentistic Approximation of Probability Measures with or without σ-Additivity', *Studia Logica* 89(2), 257–83.

Searle, John R. 1969. *Speech Acts*. Cambridge: Cambridge University Press.

Segerberg, Krister. 2001. 'The Basic Dynamic Doxastic Logic of AGM', in Mary-Anne Williams and Hans Rott (eds), *Frontiers in Belief Revision*. Dordrecht: Kluwer, 57–84.

Shapiro, Stewart. 2006. *Vagueness in Context*. Oxford: Clarendon Press.

Sharon, Assaf, and Levi Spectre. 2013. 'Epistemic Closure under Deductive Inference: What is it and Can we Afford it?', *Synthese* 190(14), 2731–48.

Simon, Herbert. 1956. 'Rational Choice and the Structure of the Environment', *Psychological Review* 63(2), 129–38.

Skyrms, Brian. 1977. 'Resiliency, Propensities, and Causal Necessity', *Journal of Philosophy* 74(11), 704–13.

Skyrms, Brian. 1980. *Causal Necessity*. New Haven, CT: Yale University Press.

Skyrms, Brian. 1984. *Pragmatics and Empiricism*. New Haven, CT: Yale University Press.

Smith, Martin. 2010. 'A Generalised Lottery Paradox for Infinite Probability Spaces', *British Journal for the Philosophy of Science* 61(4), 821–31.

Snow, Paul. 1998. 'Is Intelligent Belief Really beyond Logic?', in *Proceedings of the Eleventh International Florida Artificial Intelligence Research Society Conference*. Sanibel Island, FL: American Association for Artificial Intelligence, 430–4.

Snow, Paul. 1999. 'Diverse Confidence Levels in a Probabilistic Semantics for Conditional Logic', *Artificial Intelligence* 113(1–2), 269–79.

Snow, Paul. 2003. 'Belief, Logic, and Partial Truth', *Computational Intelligence* 19(2), 201–13.

Spohn, Wolfgang. 1988. 'Ordinal Conditional Functions: A Dynamic Theory of Subjective States', in William L. Harper and Brian Skyrms (eds), *Causation in Decision, Belief Change, and Statistics*, vol. 2. Dordrecht: Kluwer, 105–13.

Spohn, Wolfgang. 2012. *The Laws of Belief: Ranking Theory and its Philosophical Applications*. Oxford: Oxford University Press.

Staffel, Julia. 2016. 'Beliefs, Buses and Lotteries: Why Rational Belief Can't Be Stably High Credence', *Philosophical Studies* 173(7), 1721–34.

Stalnaker, Robert. 1978. 'Assertion', in Peter Cole (ed.), *Syntax and Semantics*, vol. 9. New York: New York Academic Press, 315–32.

Stalnaker, Robert. 1987. *Inquiry*. Cambridge, Mass.: MIT Press.

Stalnaker, Robert. 2002. 'Common Ground', *Linguistics and Philosophy* 25(5–6), 701–21.

Stalnaker, Robert. 2006. 'On Logics of Knowledge and Belief', *Philosophical Studies* 128(1), 169–99.

Stanley, Jason. 2005. *Knowledge and Practical Interests*. Oxford: Clarendon Press.

Steinberger, Florian. N.d. 'Three Ways in Which Logic Might Be Normative', unpublished manuscript.

Strößner, Corina. 2015. 'Normality and Majority: Towards a Statistical Understanding of Normality Statements', *Erkenntnis* 80(4), 793–809.

Sturgeon, Scott. 2008. 'Reason and the Grain of Belief', *Nous* 42(1), 139–65.
Sturgeon, Scott. 2015. 'The Tale of Bella and Creda', *Philosophers' Imprint* 15(31), 1–9.
Suppes, Patrick. 1994. 'Qualitative Theory of Subjective Probability', in George Wright and Peter Ayton (eds), *Subjective Probability*. Chichester: John Wiley, 17–37.
Swain, Marshall, ed. 1970. *Induction, Acceptance and Rational Belief*. Dordrecht: Reidel, 55–82.
Thomason, Richmond H. 1986. 'The Context-Sensitivity of Belief and Desire', in Michael P. Georgeff and Amy L. Lansky (eds), *Reasoning about Actions and Plans*. Los Altos, Calif.: Morgan Kaufmann, 341–60.
Thomason, Richmond H. 2007. 'Three Interactions between Context and Epistemic Locutions', in Boicho Kokinov et al. (eds), *Modeling and Using Context*, Lecture Notes in Computer Science, 4635. Berlin: Springer, 467–81.
Van Benthem, Johan, Jelle Gerbrandy, and Barteld Kooi. 2009. 'Dynamic Update with Probabilities', *Studia Logica* 93(1), 67–96.
Van Benthem, Johan, and Fenrong Liu, eds. 2013. *Logic across the University: Foundations and Application. Proceedings of the Tsinghua Logic Conference, Beijing*, Studies in Logic, 47. London: College Publications.
Van Ditmarsch, Hans, Wiebe van der Hoek, and Barteld Kooi. 2008. *Dynamic Epistemic Logic*, Synthese Library. Berlin: Springer.
Van Fraassen, Bas C. 1980a. *The Scientific Image*. Oxford: Clarendon Press.
Van Fraassen, Bas C. 1980b. 'Review of Brian Ellis, *Rational Belief Systems*', *Canadian Journal of Philosophy* 10, 497–511.
Van Fraassen, Bas C. 1981. 'A Problem for Relative Information Minimizers in Probability Kinematics', *British Journal for the Philosophy of Science* 32(4), 375–9.
Van Fraassen, Bas C. 1984. 'Belief and the Will', *Journal of Philosophy* 81(5), 235–56.
Van Fraassen, Bas C. 1995. 'Fine-Grained Opinion, Probability, and the Logic of Full Belief', *Journal of Philosophical Logic* 24, 349–77.
Velleman, J. David. 2000. 'On the Aim of Belief', in J. David Velleman, *The Possibility of Practical Reason*. Oxford: Oxford University Press, 244–82.
Weatherson, Brian. 2005. 'Can We Do without Pragmatic Encroachment?', *Philosophical Perspectives* 19, 417–43.
Wedgwood, Ralph. 2002. 'The Aim of Belief', *Philosophical Perspectives* 16, 267–97.
Wedgwood, Ralph. 2012. 'Outright Belief', *dialectica* 66(3), 309–29.
Wedgwood, Ralph. 2013. 'The Right Thing to Believe?', in Chan (2013), 123–39.
Wedgwood, Ralph. 2016: 'Objective and Subjective "Ought"', in Nate Charlow and Matthew Chrisman (eds), *Deontic Modality*. Oxford: Oxford University Press, 143–68.
Weisberg, Jonathan. N.d. 'Belief: Partial and Full', unpublished manuscript.
Wheeler, Gregory. 2007. 'A Review of the Lottery Paradox', in W. Harper and G. Wheeler (eds), *Probability and Inference: Essays in Honor of Henry E. Kyburg, Jr.* London: King's College Publications, 1–31.
Williams, Bernard A. O. 1973. 'Deciding to Believe', in Bernard Williams, *Problems of the Self*. Cambridge: Cambridge University Press, 136–51.
Williamson, Timothy. 1998. 'Conditionalizing on Knowledge', *British Journal for the Philosophy of Science* 49, 89–121.
Williamson, Timothy. 2000. *Knowledge and its Limits*. Oxford: Oxford University Press.

Williamson, Timothy. 2009. 'Reply to John Hawthorne and Maria Lasonen-Aarnio', in Patrick Greenough and Duncan Pritchard (eds), *Williamson on Knowledge*. Oxford: Oxford University Press, 313–29.

Windschitl, Paul D., and Gary L. Wells. 1998. 'The Alternative-Outcomes Effect', *Journal of Personality and Social Psychology* 75(6), 1411–23.

Woodward, Jim. 2006. 'Some Varieties of Robustness', *Journal of Economic Methodology* 13(2), 219–40.

Yablo, Stephen. 2014. *Aboutness*. Princeton, NJ: Princeton University Press.

Yager, Ronald R., and Liping Liu. 2008. *Classic Works of the Dempster–Shafer Theory of Belief Functions*, Studies in Fuzziness and Soft Computing, 219. Berlin: Springer.

Index

The manufacturer's authorised representative in the EU for product safety is Oxford University Press España S.A. of El Parque Empresarial San Fernando de Henares, Avenida de Castilla, 2 - 28830 Madrid (www.oup.es/en or product.safety@oup.com). OUP España S.A. also acts as importer into Spain of products made by the manufacturer.

Printed and bound by CPI Group (UK) Ltd, Croydon, CR0 4YY

12/09/2026

02215929-0001